Lecture Notes in Computer Science 5714

Commenced Publication in 1973
Founding and Former Series Editors:
Gerhard Goos, Juris Hartmanis, and Jan van Leeuwen

Maristella Agosti   José Borbinha
Sarantos Kapidakis   Christos Papatheodorou
Giannis Tsakonas (Eds.)

Research and Advanced Technology for Digital Libraries

13th European Conference, ECDL 2009
Corfu, Greece, September 27 – October 2, 2009
Proceedings

 Springer

Volume Editors

Maristella Agosti
University of Padua, Department of Information Engineering
Via Gradenigo 6/a, 35131 Padova, Italy
E-mail: agosti@dei.unipd.it

José Borbinha
Instituto Superior Técnico
Department of Computer Science and Engineering IST
Av. Rovisco Pais, 1049-001 Lisboa, Portugal
E-mail: jlb@ist.utl.pt

Sarantos Kapidakis
Christos Papatheodorou
Giannis Tsakonas
Ionian University, Department of Archives and Library Sciences
72 Ioannou Theotoki str., 49100 Corfu, Greece
E-mail: {sarantos,papatheodor,gtsak}@ionio.gr

Library of Congress Control Number: 2009934037

CR Subject Classification (1998): H.2.8, H.3.7, H.3, H.5, H.4, J.7

LNCS Sublibrary: SL 3 – Information Systems and Application, incl. Internet/Web and HCI

ISSN   0302-9743
ISBN-10   3-642-04345-3 Springer Berlin Heidelberg New York
ISBN-13   978-3-642-04345-1 Springer Berlin Heidelberg New York

springer.com

© Springer-Verlag Berlin Heidelberg 2009
Printed in Germany

Typesetting: Camera-ready by author, data conversion by Scientific Publishing Services, Chennai, India
Printed on acid-free paper   SPIN: 12753659   06/3180   5 4 3 2 1 0

Preface

We are very pleased to present the proceedings of the 13th European Conference on Research and Advanced Technologies for Digital Libraries (ECDL 2009), which took place on the beautiful island of Corfu, Greece from September 27 to October 2, 2009. Since the first conference in Pisa (1997) ECDL has been held in Heraklion (1998), Paris (1999), Lisbon (2000), Darmstadt (2001), Rome (2002), Trondheim (2003), Bath (2004), Vienna (2005), Alicante (2006), Budapest (2007), and Aarhus (2008). Over the years, ECDL has created a strong interdisciplinary community of researchers and practitioners in the field of digital libraries, and has formed a substantial body of scholarly publications contained in the conference proceedings.

The general theme of the 13th European Conference on Digital Libraries (ECDL 2009) was "Digital Societies" and its organization was assigned to the Laboratory on Digital Libraries and Electronic Publishing of the Department of Archives and Library Sciences, Ionian University, Corfu, Greece.

"Digital societies" encapsulate the involvement of user communities in various aspects of information management stages, such as the creation of new information, enrichment of information artifacts, sharing and distribution of information objects, retrieval and filtering and so on. All these activities are augmented by the exploitation of new approaches and functionalities such as social networking, collaboration and management of annotations and tagging, and they require a thorough examination of metadata issues and services. Under the lens of its general theme, the conference covered a wide range of topics, such as architectures, formal issues and conceptual views of digital libraries, users' studies, Semantic Web and Web 2.0, knowledge organization, metadata, information retrieval, digital curation and preservation. While some of these topics were represented in the general program of the conference, others were extensively addressed in workshops and tutorials.

This year the new concept of a "Special Track" was introduced for plenary sessions, to help high-level works receive the proper exposition to a wide audience. The conference issued four separate Special Tracks calls for papers on Infrastructures, Services, Content and Foundations. While the calls were separate, the assessment of the submissions was unified with the general call, in order to sustain the high levels of integral reviewing. As a consequence of this, only two Special Tracks survived with a minimal number of relevant papers accepted: Services and Infrastructures.

In overall, the total number of submissions was 181, which, according to our memory, is a record in the ECDL series of conferences. In the end of the reviewing process, 28 full papers were accepted (in 129 submissions, thus comprising 21.7% of the submissions). In addition 6 short papers, 20 posters and 16 demos were accepted (these include 17 original submissions to full papers that were moved to other categories). All the areas of the world are represented by the different submissions, including Europe, USA, Canada, South America, South Africa, Asia and Oceania.

Considering the parallel and satellite events, ten workshops (another record for the ECDL) and four tutorials were organized. The workshops were: (a) Digital Curation in the Human Sciences, (b) Digital Libraries: Interoperability, Best Practices and Modelling Foundations; (c) Harvesting Metadata: Practices and Challenges; (d) The 8th European Networked Knowledge Organization Systems (NKOS) Workshop; (e) The 2nd Workshop on Very Large Digital Libraries (VLDL 2009);(f) BooksOnline 2009: 2nd Workshop on Research Advances in Large Digital Book Collections; (g) Workshop on Exploring Musical Information Spaces (WEMIS); (h) 9th International Web Archiving Workshop (IWAW 2009); (i) Conducting a Needs Assessment for a Joint EU-US Digital Library Curriculum; and (j) Cross-Languange Evaluation Forum (CLEF 2009). The four tutorials were: (a) Aggregation and reuse of digital objects' metadata from distributed digital libraries, by Cezary Mazurek and Marcin Werla; (b) Knowledge Organization Systems (KOS) in digital libraries, by Dagobert Soergel; (c) Designing user interfaces for interactive information retrieval systems and digital libraries, by Harald Reiterer and Hans-Christian Jetter; and (d) Digital preservation: Logical and bit-stream preservation using Plato, EPrints and the Cloud, by Adam Field, Hannes Kulovits, Andreas Rauber and David Tarrant. The list of workshops and tutorials represent the continuation of already established scientific entities, as well as an extension to deliberate new themes and topics.

Finally, the conference was also graced by the keynote addresses of Gary Marchionini, University of North Carolina at Chapel Hill, USA and Peter Buneman, University of Edinburgh and Digital Curation Centre, UK. Gary Marchionini's presentation was titled "Digital Libraries as Phenotypes for Digital Societies," while Peter Buneman gave a speech on "Curated Databases."

The success of ECDL 2009 in quantity and quality would not be feasible without the invaluable contributions of all the members of the Program Committee, Organizing Committee, students and volunteers that supported the conference in its various stages. A special mention has to be made regarding the local support of Michalis Sfakakis, without whom a lot of issues would be executed in a deficient mode. Finally, we would like to thank the sponsoring organizations for their significant support, especially in the constrained conditions of our times.

Maristella Agosti
Jose Borbinha
Sarantos Kapidakis
Christos Papatheodorou
Giannis Tsakonas

Organization

ECDL 2009 was organized by the Laboratory on Digital Libraries and Electronic Publishing, Department of Archives and Library Sciences, Ionian University, Corfu, Greece.

Committees

General Chairs

Sarantos Kapidakis   Ionian University, Corfu, Greece
Christos Papatheodorou   Ionian University, Corfu, Greece

Program Chairs

Maristella Agosti   University of Padua, Italy
José Borbinha   IST/INESC-ID

Organizing Chair

Giannis Tsakonas   Ionian University, Corfu, Greece

Special Tracks Chairs

Content
Erik Duval   Katholieke Universiteit Leuven, Belgium
Jane Hunter   University of Queensland, Australia

Foundations
Ee-Peng Lim   Singapore Management University, Singapore
Elaine Toms   Dalhousie University, Halifax, NS, Canada

Infrastructures
Donatella Castelli   ISTI-CNR, Italy
Reagan Moore   UNC, USA

Services
Carl Lagoze   Cornell University, USA
Timos Sellis   Institute for the Management of
                               Information Systems, Greece

Workshop Chairs

Manolis Gergatsoulis	Ionian University, Corfu Greece
Ingeborg Solvberg	Norwegian University of Technology and Science, Norway

Poster and Demo Chairs

Preben Hansen	SICS, Sweden
Laurent Romary	INRIA, France

Tutorial Chairs

Norbert Fuhr	University of Duisburg-Essen, Germany
Laszlo Kovacs	MTA SZTAKI, Hungary

Panel Chairs

Tiziana Catarci	University of Rome "La Sapienza", Italy
Fabio Simeoni	University of Strathclyde, UK

Doctoral Consortium Chairs

Panos Constantopoulos	Athens University of Economics and Business, Greece
Nicolas Spyratos	Université de Paris-Sud, France

Publicity Chairs

George Buchanan	Swansea University, UK
Schubert Foo Shou Boon	Nanyang Technological University, Singapore
Ronald Larsen	University of Pittsburgh, USA
Ee-Peng Lim	Singapore Management University, Singapore
Liddy Nevile	La Trobe University, Australia
Shigeo Sugimoto	University of Tsukuba, Japan

Program Committee

Trond Aalberg	Norwegian University of Technology and Science, Norway
Ricardo Baeza-Yates	Yahoo! Research, Spain
Thomas Baker	Competence Centre for Interoperable Metadata (KIM), Germany
Nicholas Belkin	Rutgers University, USA
Mária Bieliková	Slovak University of Technology in Bratislava, Slovakia
Ann Blandford	University College London, UK

Ray R. Larson	University of California, Berkeley, USA
Ee-Peng Lim	Singapore Management University, Singapore
Clifford Lynch	Coalition for Networked Information, USA
Fillia Makedon	University of Texas at Arlington, USA
Bruno Martins	IST/INESC-ID, Portugal
Carlo Meghini	ISTI-CNR, Italy
András Micsik	MTA SZTAKI, Hungary
Reagan Moore	UNC, USA
John Mylopoulos	University of Toronto, Canada
Michael L. Nelson	Old Dominion University, USA
Liddy Nevile	La Trobe University, Australia
Christos Nicolaou	University of Crete, Greece
Carol Peters	ISTI-CNR, Italy
Dimitris Plexousakis	FORTH-ICS/University of Crete, Greece
Edie Rasmussen	University of British Columbia, Canada
Andreas Rauber	Vienna University of Technology, Austria
Laurent Romary	INRIA, France
Heiko Schuldt	University of Basel, Switzerland
Timos Sellis	Institute for the Management of Information Systems, Greece
Mário J. Silva	University of Lisbon, Portugal
Fabio Simeoni	University of Strathclyde, UK
Dagobert Soergel	University of Maryland, USA
Ingeborg Solvberg	Norwegian University of Technology and Science, Norway
Nicolas Spyratos	Université de Paris-Sud, France
Ulrike Steffens	OFFIS, Germany
Shigeo Sugimoto	University of Tsukuba, Japan
Hussein Suleman	University of Cape Town, South Africa
Alfredo Sánchez	Universidad de las Américas Puebla, Mexico
Atsuhiro Takasu	National Institute of Informatics, Japan
Costantino Thanos	ISTI-CNR, Italy
Helen Tibbo	University of North Carolina at Chapel Hill, USA
Elaine Toms	Dalhousie University, Canada
Herbert Van de Sompel	Los Alamos National Laboratory, USA
Felisa Verdejo	Universidad Nacional de Educación a Distancia, Spain
Ian Witten	University of Waikato, New Zealand

Organizing Committee

Lina Bountouri	Ionian University, Greece
Hara Brindezi	Ionian University, Greece
Nikos Fousteris	Ionian University, Greece
Dimitris Gavrilis	Institute for the Management of Information Systems, Greece

Panorea Gaitanou	Ionian University, Greece
Constantia Kakali	Ionian University, Greece
Spyros Kantarelis	Ionian University, Greece
Eirini Lourdi	Ionian University, Greece
Anna Mastora	Ionian University, Greece
Aggelos Mitrelis	University of Patras, Greece
Maria Monopoli	Bank of Greece Library, Greece
Sofia Stamou	Ionian University, Greece
Michalis Sfakakis	Ionian University, Greece
Ifigeneia Vardakosta	Ionian University, Greece
Spyros Veronikis	Ionian University, Greece

Additional Referees

Aris Anagnostopoulos	Nicola Fanizzi	Thomas Lidy
Roman Arora	Giorgos Flouris	Yong Lin
David Bainbridge	Manuel João Fonseca	Manuel Llavador
José Barateiro	Jakob Frank	Rudolf Mayer
Shariq Bashir	Nuno Freire	Daniel McEnnis
Marenglen Biba	Mark Guttenbrunner	Massimo Mecella
Lina Bountouri	Robert Gwadera	Vangelis Metsis
Mark Carman	Anna Huang	Mayra Ortiz-Williams
Becker Christoph	Matthias Jordan	Benjamin Piwowarski
Eduardo Colaço	Mostafa Keikha	Andrea Schweer
Claudia d'Amato	Haridimos Kondylakis	Fabricio Silva
Michel de Rougemont	Ioannis Konstas	Gianmaria Silvello
Emanuele Di Buccio	Sascha Kriewel	Michael Springmann
Giorgio Maria Di Nunzio	Hannes Kulovits	Paola Velardi
Marco Dussin	Monica Landoni	
Andrea Ernst-Gerlach	Luís Leitão	

Sponsoring Institutions

Alpha Bank
ERCIM - European Research Consortium for Informatics and Mathematics
SWETS Information Services
InterOptics
EBSCO
UniSystems - InfoQuest
Tessella
ARTstor

Table of Contents

Knowledge Organization Systems

Interfaces

Resource Discovery

Architectures

Information Retrieval

Preservation

Evaluation

Panels

Posters

Demos

Digital Libraries as Phenotypes for Digital Societies

Gary Marchionini

University of North Carolina at Chapel Hill, USA
march@ils.unc.edu

Extended Abstract

The research and development community has been actively creating and deploying digital libraries for more than two decades and many digital libraries have become indispensable tools in the daily life of people around the world. Today's digital libraries include interactive multimedia and powerful tools for searching and sharing content and experience. As such, digital libraries are moving beyond personal intellectual prostheses to become much more participative and reflective of social history. Digital libraries not only acquire, preserve, and make available informational objects, but also invite annotation, interaction, and leverage usage patterns to better serve patron needs. These various kinds of usage patterns serve two purposes: first, they serve as context for finding and understanding content, and second, they themselves become content that digital libraries must manage and preserve. Thus, digital library research has expanded beyond technical and informational challenges to consider new opportunities for recommendations, support of affinity groups, social awareness, and cross-cultural understanding, as well as new challenges related to personal and group identity, privacy and trust, and curating and preserving ephemeral interactions. This trend makes digital libraries cultural institutions that reveal and hopefully preserve the phenotypes of societies as they evolve. This talk will illustrate this theoretical perspective with examples from our experience with the Open Video Digital Library over the past decade and with recent extensions (VidArch Project) that harvest YouTube video as a strategy for preserving cultural context for digital collections.

M. Agosti et al. (Eds.): ECDL 2009, LNCS 5714, p. 1, 2009.
© Springer-Verlag Berlin Heidelberg 2009

Curated Databases

Peter Buneman

University of Edinburgh, UK
opb@inf.ed.ac.uk

Extended Abstract

Most of our research and scholarship now depends on *curated databases*. A curated database is any kind of structured repository such as a traditional database, an ontology or an XML file, that is created and updated with a great deal of human effort. For example, most reference works (dictionaries, encyclopaedias, gazetteers, etc.) that we used to find on the reference shelves of libraries are now curated databases; and because it is now so easy to publish databases on the web, there has been an explosion in the number of new curated databases used in scientific research. Curated databases are of particular importance to digital librarians because the central component of a digital library – its catalogue or metadata – is very likely to be a curated database. The value of curated databases lies in the organisation, the annotation and the quality of the data they contain. Like the paper reference works they have replaced, they usually represent the efforts of a dedicated group of people to produce a definitive description of enterprise or some subject area.

Given their importance to our work it is surprising that so little attention has been given to the general problems of curated databases. How do we archive them? How do we cite them? And because much of the data in one curated database is often extracted from other curated databases, how do we understand the provenance of the data we find in the database and how do we assess its accuracy? Curated databases raise challenging problems not only in computer science but also in intellectual property and the economics of publishing. I shall attempt to describe these.

M. Agosti et al. (Eds.): ECDL 2009, LNCS 5714, p. 2, 2009.
© Springer-Verlag Berlin Heidelberg 2009

Leveraging the Legacy of Conventional Libraries for Organizing Digital Libraries

Arash Joorabchi and Abdulhussain E. Mahdi

Department of Electronic and Computer Engineering, University of Limerick, Ireland
{Arash.Joorabchi,Hussain.Mahdi}@ul.ie

Abstract. With the significant growth in the number of available electronic documents on the Internet, intranets, and digital libraries, the need for developing effective methods and systems to index and organize E-documents is felt more than ever. In this paper we introduce a new method for automatic text classification for categorizing E-documents by utilizing classification metadata of books, journals and other library holdings, that already exists in online catalogues of libraries. The method is based on identifying all references cited in a given document and, using the classification metadata of these references as catalogued in a physical library, devising an appropriate class for the document itself according to a standard library classification scheme with the help of a weighting mechanism. We have demonstrated the application of the proposed method and assessed its performance by developing a prototype classification system for classifying electronic syllabus documents archived in the Irish National Syllabus Repository according to the well-known Dewey Decimal Classification (DDC) scheme.

Keywords: Digital library organization, text classification, collective classification, library classification schemes, bibliography.

1 Introduction

Similar to conventional libraries, large-scale digital libraries contain hundreds of thousands of items and therefore require advanced querying and information retrieval techniques to facilitate easy and effective search systems. In order to provide highly precise search results, such search systems need to go beyond the traditional keyword-based search techniques which usually yield a large volume of indiscriminant search results irrespective of their content. Classification of materials in a digital library based on a pre-defined scheme could improve the accuracy of information retrieval significantly and allows users to browse the collection by subject [1]. However, manual classification of documents is a tedious and time-consuming task which requires an expert cataloguer in each knowledge domain represented in the collection, and therefore deemed impractical in many cases. Motivated by the ever-increasing number of E-documents and the high cost of manual classification, Automatic Text Classification/Categorization (ATC) - the automatic assignment of natural language text documents to one or more predefined categories or classes according to their contents - has become one of the key methods to enhance the information retrieval and knowledge management of large digital textual collections.

M. Agosti et al. (Eds.): ECDL 2009, LNCS 5714, pp. 3–14, 2009.

Until the late '80s, the use of rule-based methods was the dominant approach to ATC. Rule-based classifiers are built by knowledge engineers who inspect a corpus of labeled sample documents and define a set of rules which are used for identifying the class of unlabelled documents. Since the early '90s, with the advances in the field of Machine Learning (ML) and the emergence of relatively inexpensive high performance computing platforms, ML-based approaches have become widely associated with modern ATC systems. A comprehensive review of the utilization of ML algorithms in ATC, including the widely used Bayesian Model, k-Nearest Neighbor, and Support Vector Machines, is given in [2]. In general, an ML-based classification algorithm uses a corpus of manually classified documents to train a classification function which is then used to predict the classes of unlabelled documents. Applications of such algorithms include spam filtering, cataloguing news and journal articles, and classification of web pages, just to name a few. However, although a considerable success has been achieved in above listed applications, the prediction accuracy of ML-based ATC systems is influenced by a number of factors. For example, it is commonly observed that as the number of classes in the classification scheme increases, the prediction accuracies of the ML algorithms decreases. It is also well-recognized that using sufficient number of manually classified documents for training influences the prediction performance of ML-based ATC systems considerably. However, in many cases, there is little or no training data available. Hence, over the last few years, research efforts of the machine learning community has been directed towards developing new probability and statistical based ML algorithms that can enhance the performance of the ML-based ATC systems in terms of prediction accuracy and speed, as well as reduce the number of manually labeled documents required to train the classifier.

However, as Golub [3] and Yi [4] discuss, there exits a less investigated approach to ATC that is attributed to the library science community. This approach focuses less on algorithms and more on leveraging comprehensive controlled vocabularies, such as library classification schemes and thesauri that are conventionally used for manual classification of physical library holdings. One of the main applications of this approach to ATC is in the automatic classification of digital library holdings, where using physical library classification schemes is a natural and usually most suitable choice. Another application is in classifying web pages, where due to their subject diversity, proper and accurate labeling requires a comprehensive classification scheme that covers a wide range of disciplines. In such applications using library classification schemes can provide fine-grained classes that cover almost all categories and branches of human knowledge.

A library classification system is a coding system for organizing library materials according to their subjects with the aim of simplifying subject browsing. Library classification systems are used by professional library cataloguers to classify books and other materials (e.g., serials, audiovisual materials, computer files, maps, manuscripts, realia) in libraries. The two most widely used classification systems in libraries around the world today are the Dewey Decimal Classification (DDC) [5] and the Library of Congress Classification (LCC) [6]. Since their introduction in late 18th century, these two systems have undergone numerous revisions and updates.

In general, all ATC systems that have been developed using above library science approach can be categorized into two main categories:

1. String matching-based systems: these systems do not use ML algorithms to perform the classification task. Instead, they use a method which involves string-to-string matching between words in a term list extracted from library thesauri and classification schemes, and words in the text to be classified. Here, the unlabelled incoming document can be thought of as a search query against the library classification schemes and thesauri, and the result of this search includes the class(es) of the unlabelled document. One of the well-known examples of such systems is the Scorpion project [7] by the Online Computer Library Centre (OCLC). Scorpion is an ATC system for classifying E-documents according to the DDC scheme. It uses a clustering method based on term frequency to find the most relevant classes to the document to be classified. A similar experiment was conducted by Larson [8] in early 90's, who built normalized clusters for 8,435 classes in the LCC scheme from manually classified records of 30,471 library holdings and experimented with a variety of term representation and matching methods. For more examples of these systems see [9, 10].

2. Machine learning-based systems: these systems use ML-based algorithms to classify E-documents according to library classification schemes such as DDC and LCC. They represent a relatively new and unexplored trend which aims to combine the power of ML-based classification algorithms with the enormous intellectual effort that has already been put into developing library classification systems over the last century. Chung and Noh [11] built a specialized web directory for the field of economics by classifying web pages into 757 sub-categories of economics category in DDC scheme using k-NN algorithm. Pong et al. [12] developed an ATC system for classifying web pages and digital library holdings based on the LCC scheme. They used both k-NN and Naive Bayes algorithms and compared the results. Frank and Paynter [13] used the linear SVM algorithm to classify over 20,000 scholarly Internet resources based on the LCC scheme. In [14], the authors used both Naïve Bayes and SVM algorithms to classify a dataset of news articles according to the DDC scheme.

In this paper, we propose a new category of ATC systems within the framework of the "library science" approach, which we call Bibliography Based ATC (BB-ATC). The proposed BB-ATC system uses a novel, in-house developed method for automatic classification of E-documents which is solely based on the bibliographic metadata of the references cited in a given document and have been already classified and catalogued in a physical library.

The rest of the paper is organized as follows: Section 2 describes the proposed BB-ATC method. Section 3 describes a prototype ATC system which has been developed based on the proposed method in order to demonstrate the viability of proposed method and evaluate its performance. Section 4 describes details of above evaluation and its experimental results. This is followed by Section 5 which analyses presented results and discusses some of the main factors affecting our classifier's performance. Section 6 provides a conclusion and discusses future work.

2 Outline of Proposed BB-ATC Method

A considerable amount of E-documents have some form of linkage to other documents. For example, it is a common practice in scientific articles to cite other articles and books. It is also common practice for printed books to reference other books, documented law cases to refer to other cases, patents to refer to other patents, and webpages to have links to other webpages. Exploring the potential of leveraging these networks of references/links for ATC opens a new route for investigating the development of ATC systems, which can be linked to the field of collective classification [15]. Our proposed BB-ATC method falls into this route, and aims to develop a new trend of effective and practical ATC systems based on leveraging:

- The intellectual work that has been put into developing and maintaining resources and systems that are used for classifying and organizing the vast amount of materials in physical libraries; and
- The intellectual effort of expert cataloguers who have used above resources and systems to manually classify millions of books and other holdings in libraries around the world over the last century.

With the assumption that materials, such as books and journals, cited/referenced in a document belong to the same or closely relevant classification category(ies) as that of a citing document, we can classify the citing document based on the class(es) of its references/links as identified in one or more existing physical library catalogues. The proposed BB-ATC method is based on automating this process using three steps:

1. Identifying and extracting references in a given document;
2. Searching one or more catalogues of exiting physical libraries for the extracted references in order to retrieve their classification metadata;
3. Allocating a class(es) to the document based on retrieved classification category(ies) of the references with the help of a weighting mechanism.

It should be noted here that this method is applicable to any E-document that cites/references one or more items which have been catalogued in the library catalogues searched by the system. Examples of such E-documents include books, journal and conference papers, learning and teaching materials (such as syllabi and lecture notes), theses and dissertations to name a few.

In order to make a viable ATC system, the proposed method needs to adopt a specific standard library classification scheme. For the purpose of this work, both the Dewey Decimal Classification (DDC) and Library of Congress Classification (LCC) schemes were considered as candidate classification schemes, due to their wide use and subject coverage. We adopted the DDC in our BB-ATC method for two main reasons:

- The DDC scheme is used for classifying library holdings in about 80% of libraries around the world and, therefore, the number of existing items that are classified according to the DDC is much greater than those classified according to the LCC. This makes the DDC scheme a better choice for our method which is based on utilizing the classification metadata of items that have been manually classified according to a library classification scheme.

- The DDC has a fully hierarchical structure while the LCC is not fully hierarchical and usually leans toward alphabetic or geographic sub-arrangements. The hierarchical relationships among topics in the DDC are clearly reflected in the classification numbers which can be continuously subdivided. The hierarchical feature of the DDC allows the development of effective GUI interfaces that enable users to easily browse and navigate the scheme to find the categories that they are interested in without requiring prior knowledge of the classification scheme or its notational representation.

3   System Implementation and Functionality

In order to demonstrate the viability and performance of the proposed BB-ATC method, we have developed a prototype ATC system for classifying electronic syllabus documents for the Irish National Syllabus Repository [16] according to the DDC scheme. Figure 1 shows an overview of the system. As illustrated, the system is effectively a metadata generator comprising a Pre-processing unit, an Information Extractor, a Catalogue-search unit, and a Classifier.

The system has been designed such that it can handle syllabus documents of different formats, such as PDF, MS-Word and HTML. To facilitate this, a pre-processing unit with the help of Open Office Suite [17] and Xpdf [18] extracts the textual content of the arriving documents and passes the results to the rule-based Information Extractor (IE). The IE uses the JAPE transducer from GATE toolkit [19] to identify and extract the ISBN number of each book that has been referenced in the document. After extracting the unique identifiers of all referenced items, the Catalogue-Search unit of the system uses the Z39.50 protocol [20] to create a search query for each reference based on its unique identifier. The search query is then submitted to the OPAC (Online Public Access Catalogue) search engines of the Library of Congress (LOC) and the British Library (BL) to search for matching records. The returned search results contain the records of matching catalogued items in MARC21 format [21]. Each record holds bibliographic and classification metadata about a book including its DDC classification number and the Library of Congress Subject Headings (LCSHs) assigned to it. The Catalogue-Search unit iterates through the retrieved records and extracts the classification numbers and LCSHs of all the referenced materials and passes them to the Classifier.

The task of the Classifier is to use the retrieved classification metadata of references in the document to classify it into one or more classes from the DDC using a simple weighting mechanism. The weighting mechanism works as follows: if the document to be classified contains only one reference belonging to a single class, then the document is assigned to that single class. The same applied if the document contains more than one reference, but all the references belong to the same class. However, if the references used in the document belong to more than one class, then the weight of each class is equal to the number of references in the document which belong to that class. In this case the document is labelled with all the relevant classes and their corresponding weights. This weighting mechanism enriches the classification results by providing not only a list of classes relevant to the subject(s) of the document but also a relevance measure for these relations. In addition to labelling the document with one or more classes from the DDC, the Classifier also assigns a weighted list of the Library of Congress Subject Headings (LCSHs) to the document. This list contains a set of all the LCSHs assigned to the references in the document

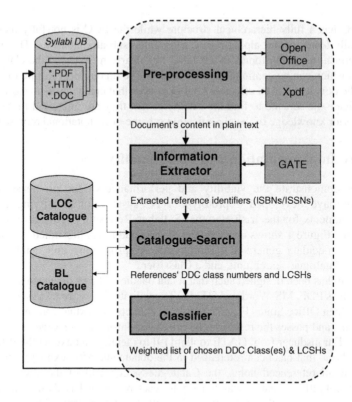

Fig. 1. Overview of the prototype ATC system

and their corresponding weights. The weight of each subject heading is equal to the number of times it has been assigned to the references. The LCSHs attempt to evaluate the subject content of items, whereas the DDC and the LCC rather broadly categorise the item in a subject hierarchy.

4   System Evaluation and Experimental Results

We used one hundred computer science related syllabus documents to evaluate the performance of our prototype ATC system. The documents have already been categorized by the expert cataloguers of the Irish National Syllabus Repository. We used the standard measures of Precision, Recall and their harmonic mean, F1, to evaluate the prediction performance of the system. Precision is the number of times that a class label has been correctly assigned to the test documents divided by the total number of times that class has been correctly or incorrectly assigned to the documents. Recall is the number of times a class label has been correctly assigned to the test documents divided by the number of times that class should have been correctly assigned to the test documents. Accordingly:

$$\text{Recall} = \frac{\#\ \text{Correctly assigned class labels}}{\#\ \text{Total possible correct}} = \frac{\text{TP}}{\text{TP} + \text{FN}} \tag{1}$$

$$\text{Precision} = \frac{\text{\# correctly assigned class labels}}{\text{\# Total assigned}} = \frac{TP}{TP + FP} \tag{2}$$

$$F1 = \frac{2 \times \text{Precision} \times \text{Recall}}{\text{Precision} + \text{Recall}} \tag{3}$$

where the Recall, Precision and F1 are computed in terms of the labels TP (True Positive), FP (False Positive), and FN (False Negative) to evaluate the validity of each class label assigned to a document, such that:

- TPi: refers to the case when both the classifier and human cataloguer agree on assigning class label i to document j;
- FPi: refers to the case when the classifier has mistakenly (as judged by a human cataloguer) assigned class label i to document j;
- FNi: refers to the case when the classifier has failed (as judged by a human cataloguer) to assign a correct class label i to document j.

In order to obtain a true objective evaluation of the classification performance, we applied the micro-average to above target performance measures (i.e., precision, recall, and F1) over all categories. Table 1 summarizes the achieved results.

Table 1. Micro-averaged classification results

TP	FP	FN	Precision	Recall	F1
210	19	26	0.917	0.889	0.902

Table 2 shows the classification results of a sample test syllabus document. The full and individually analyzed classification results of the one hundred test syllabus documents may be viewed online on our webpage[1]. We encourage readers to refer to this document as it provides detailed insight into the performance of the proposed BB-ATC method.

As a common practice in developing a new system, it is always desired to compare its performance to other existing similar systems. However, we would like to note here that it is not possible to conduct a true objective comparison between the performance of our method and other reported ATC systems that use either the string-matching or ML based approaches. This is due to the following:

- Unlike our system which classifies E-documents based on the full DDC scheme, other reported ATC systems, due to their limitations, either adopt only one of the main classes of DDC or LCC as their classification scheme or use an abridged version of DDC or LCC by limiting the depth of hierarchy to second or third level.
- In quite a few reported cases, the performance of the system was evaluated using different measures other than the standard performance measures of precision, recall, and F1 used in our case.

---

[1] http://www.csn.ul.ie/~arash/PDFs/1.pdf

Table 2. A sample syllabus classification results

Syllabus Title: Computer Graphics Programming			
DDC No. **[*Weight*]**	006.696 [1]	006.6 006.69 **006.696**	Computer graphics Special topics in computer graphics **Digital video**
	006.66 [2]	006.6 **006.66**	Computer graphics **Programming**
LCSH **[*Weight*]**	• Computer graphics [3] • OpenGL [2]	• Computer Animation [1] • Three-dimensional display systems [1]	

Despite above, it is possible to provide a relative comparison between the performance of our system and those of similar reported systems. For example, Pong and co-workers [12] used both the Naive Bayes and k-NN ML algorithms to classify 254 test documents based on a refined version of LCC scheme which consisted of only 67 categories. They reported the values of 0.802, 0.825, and 0.781 as the best figures for micro-averaged F1, recall, and precision, respectively, achieved by their system. Also, Chung and Noh [11] reported the development of a specialized economics web directory by classifying a collection of webpages, belonging to the field of economics, into 575 subclasses of the DDC main class of economics. Their unsupervised string-matching based classifier achieved an average precision of 0.77 and their supervised ML based classifier achieved an average precision and recall of 0.963 and 0.901, respectively. In [16] the authors used the naïve Bayes classification algorithm to automatically classify 200 syllabus documents according to the International Standard Classification of Education scheme [22]. The performance of the classifier was measured using one hundred undergraduate syllabus documents and the same number of postgraduate syllabus documents taken from the Irish National Syllabus Repository, achieving micro-average values of 0.75 and 0.60 for precision for each of the above document groups, respectively.

5   Discussion of Results

We examined each individual syllabus document in the test collection in conjunction to its predicted classification categories in more depth in order to obtain an insight into the behavior of the method and underlying factors that affect its performance. We first read the full content of each document and then examined the books that are referenced in the document and the DDC classes that these books have been assigned to by expert cataloguers in the Library of Congress and the British Library. In this section we summarize the findings of this examination.

• The majority of the DDC classes and LCSHs assigned to the documents are quite relevant to the contents of the documents and provide detailed and semantically rich information about the core subjects of the syllabi. For example, in case of the sample syllabus document in Table 2, titled "computer graphics programming", the two classes assigned to the document, *digital video* (descended from *Special topics in computer graphics*) and *programming* (descended from *computer graphics*), objectively cover the two main subjects of the syllabus with the weight assigned to each class logically justifying this classification.

- The number of books referenced in test documents ranges from 1 to 10. In total, the one hundred test documents reference 365 books. 305 of these referenced books were catalogued and classified based on the DDC in either the LOC or the BL catalogue. We did not encounter a case in which none of the referenced books were catalogued. In 38 cases one or more of the referenced books were not catalogued in either the LOC or the BL. The results show that the existence of one or more (up to the majority) un-catalogued references in a document does not always lead to a misclassification of that document. However, as the number of catalogued references in the document increases recall of our system improves and the weights assigned to the classes and LCSHs become more accurate.

- The Dewey classes and the LCSHs are independent from each other. Therefore, if the class label of a referenced book does not match any of the main subjects of the document, i.e. an FP case, this does not mean that the LCSHs assigned to that reference are irrelevant too. In fact, in a substantial number of examined cases, many of the assigned LCSHs were quite relevant despite the fact that the classes of the referenced books were irrelevant to the subject of the documents. This is due to the fact that LCSHs are not bound to DDC classes and they provide subject labels that are not represented in the classification scheme exclusively or are not the main subject of the item being classified. For example, in case of syllabus document No. 48 given in our on-line results, titled "High Performance Computing", one of the references is classified into the class of *microcomputers-architecture* which is considered a misclassification (i.e. an FP). However 2 out of 4 LCSHs assigned to this reference, which are *parallel processing (electronic computers)* and *supercomputers*, are quite relevant to the core subject of this document.

- In a majority of cases, classes assigned to a document can be grouped according to the higher level classes that they descend from. Also LCSHs are usually composed of a combination of different terms (e.g. *Computer networks -- Security measures*) and in many cases some of the LCSHs assigned to a document share the same first term. For example, in case of document No. 8, the subject headings *computer networks - security measures* and *computer networks* each appear once. Since these LCSHs share the same starting term, we can replace both of them with the heading *computer networks* and assign it the accumulated weights of the two original headings. This feature of the LCSHs and the hierarchical structure of the DDC allow us to apply an appropriate tradeoff between the recall and precision of our system according to users' requirements.

6 Conclusions and Future Work

In this paper, we looked at the problem of Automatic Text Classification from the perspective of researchers in the library science community. We specifically highlighted the potential application of controlled vocabularies, which have been initially developed for indexing and organizing physical library holdings, in the development of ATC systems for organizing E-documents in digital libraries and archives. To do so, we first highlighted some of the up-to-date research work in this field and categorized associated ATC systems into two categories; ML-based systems and string matching-based systems, according to the approaches they have taken in leveraging library classification resources. We then proposed a third category of ATC systems based on a new route for leveraging library classification systems and resources, which we referred to as the Bibliography Based ATC (BB-ATC) approach. The proposed approach solely relies on the available classification metadata of publications referenced/cited in an E-document to classify it according to the DDC scheme. Unlike ML-based ATC systems, the new method does not require any training data or bootstrapping mechanism to be deployed. In order to demonstrate and evaluate the performance of proposed method, we developed a prototype ATC system for automatic classification of syllabus documents taken from the Irish National Syllabus Repository. The developed ATC system was evaluated with one hundred syllabus documents and the classification results were presented and analyzed with the aim of quantifying the prediction performance of the system and influencing factors. We reported micro-average values of 0.917, 0.889, and 0.902 for the precision, recall, and F1 performance measures of our system, respectively, and provided a relative comparison between the performance of our system and those of similar reported systems.

Based on above, we believe that we have developed a new robust method for ATC, which offers prediction performance that compares favorably to most other similar systems and outperforms many of them. As for future plans, we have identified a number of issues to be addressed, particularly with regards to enhancing the prediction performance of our method/system:

- Increasing the number of library catalogues that are queried by the catalogue-search unit. This would increase the number of references in the test documents with available classification metadata and, consequently, improve the system's performance in terms of recall. In order to do this, we are currently examining the use of a union catalogue, OCLC's WorldCat [23], which allows users to query the catalogues of 70,000 libraries around the world simultaneously.

- The Information Extractor component of our prototype classification system assumes that each reference entry includes an ISBN/ISSN number. Although this assumption holds for the documents that we tested the system with, in most cases citations in documents only include such information as title, author(s), and publisher's name. In order for our classifier to handle a reference with no ISBN/ISSN number, we need to extend the ability of its Information Extractor component such that it can locate and extract different segments of each reference entry and use them in the search queries submitted to the library catalogues instead of the ISBN/ISSN numbers. To achieve this, we are integrating an open source package called ParsCit [24] into the new version of BB-ATC system which can extract and segment a wide range of bibliographic entries.

- Koch et al. [25] studied users' navigation behaviors in a large web service, Renardus, by means of log analysis. Their study shows Directory-style of browsing in the DDC-based browsing structure to be clearly the dominant activity (60%) in Renardus. Conducting a similar study on users' navigation behaviors in Irish National Syllabus Repository could further justify the application of proposed method in improving information retrieval effectiveness of digital libraries.

- As mentioned in section two, libraries catalogue a wide range of publications including conference proceedings and journals. Therefore, one of the applications of the BB-ATC method is in organizing scientific literature digital libraries and repositories. We are currently working on an enhanced version of BB-ATC to classify about 700,000 scientific papers archived by CiteSeer [26] project.

References

[1] Avancini, H., Rauber, A., Sebastiani, F.: Organizing Digital Libraries by Automated Text Categorization. In: International Conference on Digital Libraries, ICDL 2004, New Delhi, India (2004)

[2] Sebastiani, F.: Machine learning in automated text categorization. ACM Computing Surveys (CSUR) 34(1), 1–47 (2002)

[3] Golub, K.: Automated subject classification of textual Web pages, based on a controlled vocabulary: Challenges and recommendations. New Review of Hypermedia and Multimedia 12(1), 11–27 (2006)

[4] Yi, K.: Automated Text Classification Using Library Classification Schemes: Trends, Issues, and Challenges. In: International Cataloguing and Bibliographic Control (ICBC), vol. 36(4) (2007)

[5] Dewey, M.: Dewey Decimal Classification (DDC) OCLC Online Computer Library Center (1876), http://www.oclc.org/us/en/dewey (cited January 2008)

[6] Putnam, H.: Library of Congress Classification (LCC) Library of Congress, Cataloging Policy and Support Office (1897),
http://www.loc.gov/catdir/cpso/lcc.html (cited January 2008)

[7] Scorpion, OCLC Online Computer Library Center, Inc. (2002),
http://www.oclc.org/research/software/scorpion/default.htm

[8] Larson, R.R.: Experiments in automatic Library of Congress Classification. Journal of the American Society for Information Science 43(2), 130–148 (1992)

[9] Jenkins, C., Jackson, M., Burden, P., Wallis, J.: Automatic classification of Web resources using Java and Dewey Decimal Classification. Computer Networks and ISDN Systems 30(1-7), 646–648 (1998)

[10] Dolin, R., Agrawal, D., Abbadi, E.E.: Scalable collection summarization and selection. In: Proceedings of the fourth ACM conference on Digital libraries, Berkeley, California, United States (1999)

[11] Chung, Y.M., Noh, Y.-H.: Developing a specialized directory system by automatically classifying Web documents. Journal of Information Science 29(2), 117–126 (2003)

[12] Pong, J.Y.-H., Kwok, R.C.-W., Lau, R.Y.-K., Hao, J.-X., Wong, P.C.-C.: A comparative study of two automatic document classification methods in a library setting. Journal of Information Science 34(2), 213–230 (2008)

[13] Frank, E., Paynter, G.W.: Predicting Library of Congress classifications from Library of Congress subject headings. Journal of the American Society for Information Science and Technology 55(3), 214–227 (2004)

[14] Joorabchi, A., Mahdi, A.E.: A New Method for Bootstrapping an Automatic Text Classification System Utilizing Public Library Resources. In: Proceedings of the 19th Irish Conference on Artificial Intelligence and Cognitive Science, Cork, Ireland (August 2008)

[15] Sen, P., Namata, G.M., Bilgic, M., Getoor, L., Gallagher, B., Eliassi-Rad, T.: Collective Classification in Network Data. Technical Report CS-TR-4905, University of Maryland, College Park (2008), http://hdl.handle.net/1903/7546

[16] Joorabchi, A., Mahdi, A.E.: Development of a national syllabus repository for higher education in ireland. In: Christensen-Dalsgaard, B., Castelli, D., Ammitzbøll Jurik, B., Lippincott, J. (eds.) ECDL 2008. LNCS, vol. 5173, pp. 197–208. Springer, Heidelberg (2008)

[17] OpenOffice.org 2.0, sponsored by Sun Microsystems Inc., released under the open source LGPL licence (2007), http://www.openoffice.org/

[18] Xpdf 3.02, Glyph & Cog, LLC., Released under the open source GPL licence (2007), http://www.foolabs.com/xpdf/

[19] Cunningham, H., Maynard, D., Bontcheva, K., Tablan, V.: GATE: A Framework and Graphical Development Environment for Robust NLP Tools and Applications. In: Proceedings of the 40th Anniversary Meeting of the Association for Computational Linguistics (ACL 2002), Philadelphia, US (July 2002)

[20] Z39.50, International Standard Maintenance Agency - Library of Congress Network Development and MARC Standards Office, 2.0 (1992), http://www.loc.gov/z3950/agency/

[21] MARC standards. Library of Congress Network Development and MARC Standards Office (1999), http://www.loc.gov/marc/

[22] ISCED. International Standard Classification of Education -1997 version (ISCED 1997) (UNESCO (1997), http://www.uis.unesco.org (cited July 2008)

[23] WorldCat (Online Computer Library Center (OCLC) (2001)(2008), http://www.oclc.org/worldcat/default.htm (cited January 2008)

[24] Councill, I.G., Giles, C.L., Kan, M.-Y.: ParsCit: An open-source CRF reference string parsing package. In: Proceedings of the Language Resources and Evaluation Conference (LREC 2008), Marrakesh, Morrocco (May 2008)

[25] Traugott, K., Anders, A., Koraljka, G.: Browsing and searching behavior in the renardus web service a study based on log analysis. In: Proceedings of the Proceedings of the 4th ACM/IEEE-CS joint conference on Digital libraries, Tuscon, AZ, USA. ACM Press, New York (2004)

[26] Giles, C.L., Kurt, D.B., Steve, L.: CiteSeer: an automatic citation indexing system. In: Proceedings of the third ACM conference on Digital libraries, Pittsburgh, USA (1998)

Annotation Search: The FAST Way

Nicola Ferro

University of Padua, Italy
ferro@dei.unipd.it

Abstract. This paper discusses how annotations can be exploited to develop information access and retrieval algorithms that take them into account. The paper proposes a general framework for developing such algorithms that specifically deals with the problem of accessing and retrieving topical information from annotations and annotated documents.

1 Introduction

Almost everybody is familiar with annotations and has his own intuitive idea about what they are, drawn from personal experience and the habit of dealing with some kind of annotation in every day life, which ranges from jottings for the shopping to taking notes during a lecture or even adding a commentary to a text. This intuitiveness makes annotations especially appealing for both researchers and users: the former propose annotations as an easy understandable way of performing user tasks, while the latter feel annotations to be a familiar tool for carrying out their own tasks. To cite a few examples, if we deal with the Semantic Web, annotations are considered as metadata [20,22]; in collaborative applications annotations are seen as a discourse [14] and might be considered even like e-mails [15]; in the field of digital libraries annotations are treated as an additional content [1,4]; when we talk about scientific databases, annotations represent both provenance information about the managed data and a way of curating the database itself [10]; in the case of data mining and data visualization, annotations are seen as a means for recording the history of user explorations in visualization environments [18]; finally, in social networks and collaborative tagging, annotations are tags or keywords on different kinds of digital content, e.g. photos or bookmarks [16]. It can be noted as these different notions of annotation partially overlap or share commonalities. For a thorough review of the various viewpoints about annotations, please refer to [3,7].

The *Flexible Annotation Service Tool (FAST)* covers many of the uses and applications of annotations discussed above, since it is able to represent and manage annotations which range from metadata to full content; its flexible and modular architecture makes it suitable for annotating general Web resources as well as digital objects managed by different digital library systems; the annotations themselves can be complex multimedia compound objects, with a varying degree of visibility which ranges from private to shared and public annotations and different access rights. The FAST annotation service has proven its flexibility and adaptability to different applicative contexts in many different ways. It has

M. Agosti et al. (Eds.): ECDL 2009, LNCS 5714, pp. 15–26, 2009.

been integrated into the DelosDLMS [2], the prototype of next generation digital library system developed by DELOS[1], the European network of excellence on digital libraries. It has been used as architectural framework for the DiLAS project [1]. Finally, a completely new and re-engineerd version of it has been recently integrated into The European Library (TEL)[2] development portal.

In this paper, we will focus our attention on this improved version of the FAST annotation service and its completely new query processing and retrieval engine, which allows for searching and retrieving annotations and documents according to both structured and unstructured queries.

The paper is organised as follows: Section 2 presents an intuitive introduction to the annotation model supported by the FAST annotation service; Section 3 discusses our search and retrieval framework; Section 4 describes the query language supported by FAST; Section 4.1 provides an example of use of the proposed search and retrieval framework; lastly, Section 5 draws some conclusions.

2   Annotation Model

FAST adopts and implements the formal model for annotations proposed in [7] which has also been utilised in the reference model for digital libraries[3] developed by DELOS [12] and now carried on by DL.org[4], the European coordination action on digital library interoperability, best practices and modelling foundations. In the following, we discuss the proposed model in an intuitive way and by means of examples; the reader interested in its formal foundations can refer to [7].

Annotations are compound multimedia objects constituted by different *signs of annotation* which materialize the annotation itself. For example, we can have *textual signs*, which contain the textual content of the annotation, *image signs*, if the annotation is made up of images, and so on. In turn, each sign is characterized by one or more *meanings of annotation* which specify the semantics of the sign. For example, we can have a sign whose meaning corresponds to the title field in the *Dublin Core (DC)*[5] metadata schema, in the case of a metadata annotation, or we can have a sign carrying a question of the author about a document whose meaning may be "question" or similar. Moreover, an annotation is uniquely identified by a handle, which usually takes the form of a pair (namespace, identifier), where the namespace provides logical grouping of the identifiers, has a scope which defines its visibility, and can be shared with different groups of users.

Annotations can be linked to digital objects with two main types of links. (1) *annotate link*: an annotation annotates a digital object, which can be either a document or another annotation. The "annotate link" is intended to allow an annotation only to annotate one or more parts of a given digital object. Therefore, this kind of link lets the annotation express *intra-digital object relationships*,

---

[1] http://www.delos.info/

[2] http://www.theeuropeanlibrary.org/

[3] http://www.delos.info/ReferenceModel/

[4] http://www.dlorg.eu/

[5] http://www.dublincore.org/

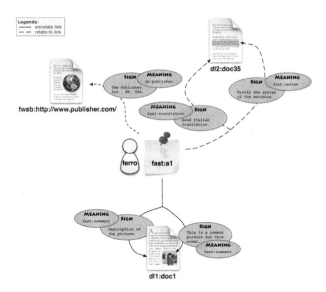

Fig. 1. Example of annotation according to the FAST annotation model

meaning that the annotation creates a relationship between the different parts of the annotated digital object. (2) *relate-to link*: an annotation relates to a digital object, which can be either a document or another annotation. The "relate-to link" is intended to allow an annotation to relate only to one or more parts of other digital objects, but not the annotated one. Therefore, this kind of link lets the annotation express *inter-digital object* relationships, meaning that the annotation creates a relationship between the annotated digital object and the other digital objects related to it.

Figure 1 shows an example of annotation which summarizes the discussion so far. The annotation, whose identifier is `a1` and which belongs to namespace `fast`, is authored by the user `ferro`[6]. It annotates a document containing a novel, whose identifier is `doc1` and which belongs to the namespace `dl1` of a digital library which manages it. The annotation relates to another document containing a translation of the novel, whose identifier is `doc35` and which belongs to the namespace `dl2` of a digital library different from the one which manages `doc1`; in addition, it also relates to the Web page of the publisher of the novel, whose identifier is `http://www.publisher.com/` and which belongs to the namespace `fweb`, used for indicating Web resources.

In particular, `a1` annotates two distinct parts of `doc1`: it annotates an image by using a textual sign whose content is "`This is a common picture for this novel`" and whose meaning is to be a `comment` in the `fast` namespace; it also annotates a sentence by using another textual sign whose content is "`Description of the picture`" and whose meaning is to be a `comment` in the `fast` namespace. In this

---

[6] Also the users of the system are identified by using a pair (identifier, namespace) but for space reasons the namespace for the user `ferro` is not reported in the figure and the following discussion.

way, the annotate link is capable of expressing intra-digital object relationships, as discussed above, since it relates different parts of the same annotated digital object.

In addition, `a1` relates the document `doc1` to its Italian translation by linking to the whole document `doc35` with a textual sign whose content is "`Good Italian translation`" and whose meaning is to be a `translation` in the `fast` namespace; it also relates to a specific sentence of the translation with a textual sign which asks to "`Verify the syntax of the sentence`" and whose meaning is to be a `review` in the `fast` namespace. Finally, `a1` also relates the document to the Web page of the publisher of the novel with a textual sign whose content is "`The Publisher, Inc., NY, USA`" and whose meaning is to be the `publisher` field of the DC metadata schema. In this way, the relate-to link is able to express inter-digital object relationships since, in this example, it relates to a novel with both its Italian translation and its publisher Web page.

This example illustrates how annotations can range from metadata to content and they can also mix metadata and content, as in the case of annotation `a1`. The example also shows another important feature of the annotations: they can take a part of a hypertext, as observed by [9,19,23], since they enable the creation of new relationships among existing digital objects, by means of links that connect annotations together with existing digital objects. The hypertext between annotations and annotated objects can be exploited for providing alternative navigation and browsing capabilities. In addition, it can span and cross the boundaries of the single digital library, if users need to interact with the information resources managed by diverse information management systems. In the example above, the annotation `a1` links the document `doc1` which belongs to the digital library `dl1` to both the document `doc35` which belongs to the digital library `dl2` and the Web page `http://www.publisher.com/`. This latter possibility is quite innovative, because it offers the means for interconnecting various digital libraries in a personalized and meaningful way for the end-user and this is a major challenge for the next generation digital libraries [21] and their interoperability [8].

3   The FAST Search Framework

The problem of information access and retrieval by exploiting annotations comprises two different issues: the first concerns the search and retrieval of the annotations themselves; the second regards the search and retrieval of annotated documents. The first case requires the design and development of algorithms able to express complex queries which take into account both the different features of the annotations and the context to which annotations belong. The second case calls for algorithms able to estimate the relevance of annotated documents with respect to a user information need on the basis of the annotations on them.

Previous research has addressed this topic from different viewpoints. [13] deals with the problem of full-text retrieval by providing proximity search operators able to match queries "where some of the words might be in the main text and others in an annotation". [17] intends annotations as a kind of automatic query expansion by adding text from highlighted passages in a document. Both these examples consider annotations as instrumental to document retrieval. [11] proposes a "social validation" approach where the degree of global confirmation or refutation of a debate carried out by annotations is evaluated; this example mainly focuses on the information access for annotations themselves. Lastly, [15] proposes an object-oriented, probabilistic logical model which takes into account annotation polarity, as in the previous case, in order to search and retrieve both documents and annotations; this latter example takes into consideration both aspects discussed above.

We propose a framework which allows users to express both simple free-text queries and complex semi-structured queries in order to retrieve either annotations or documents. The objective of our work is to design a framework able to present relevant results (documents or annotations) based on their topicality without considering, at the same time, their polarity, as done in [11] and [15]. Both positive and negative arguments can be considered as relevant to the topic entailed by a user information need and we would like to avoid "hard wiring" the reasoning about polarity in the retrieval process. For example, some retrieval tasks, such as ad-hoc search or bibliographic search, may need to be carried out without taking into consideration polarity of possible annotations on relevant documents or catalogue records. Moreover, for many retrieval tasks, users have acquired an intuitive mental model of what is going on and this helps them to easily interpret the retrieved results and reformulate the query. The introduction of reasoning about polarity requires a different mental model for the users and this might impair the search and retrieval process if the users do not easily understand how documents and annotations are ranked. We therefore prefer to treat polarity as an advanced option, that the user may apply on demand, and leave it as a possible extension to the present search and retrieval framework.

When we design search functionalities that allow users to retrieve annotations and documents according to their information needs, we have to take into consideration two key points:

- *annotations are constituted by both structured and unstructured content*: information such as the author or the media type of the annotation is part of the structured content of the annotation, while the comments and remarks contained in the body of the annotation are unstructured content;
- *documents and annotations constitute a hypertext*: annotations enable the creation of new relationships among existing documents, by means of links that connect annotations together and with existing documents.

The presence of both structured and unstructured content within an annotation calls for different types of search functionalities, since structured content can be dealt with exact match searches while unstructured content can be dealt with best match searches. These two different types of searches may need to be merged together in a query if, for example, the user wants to retrieve annotations by a given author about a given topic; this could be expressed by a boolean AND query which specifies both the author and the content of the annotations to be searched. Nevertheless, boolean clauses are best suited for dealing with exact match searches and they need to be somewhat extended to also deal with best match searches. Therefore, we need to envision a search strategy able to express complex conditions that involve both exact and best match searches. This is discussed in section 3.1 where the proposed approach stems from our previous work [6] which points out the need for both structured and unstructured search with annotations.

The hypertext that connects documents to annotations calls for a search strategy that takes it into consideration and allows us to modify the score of annotations and/or documents according to the paths in the hypertext. For example, we could consider that an annotation, retrieved in response to a user query, is more relevant if it is part of a thread where other annotations have also been retrieved in response to the same query rather than if it is part of a thread where it is the only annotation that matches the query. This is discussed in section 3.2 and revises and improves our previous work [5], where we proposed an initial algorithm for taking into account the hypertextual nature of annotations.

3.1   Annotation Extended Boolean Retrieval

The P-norm extended boolean model proposed by [24] is capable of dealing with and mixing both exact and best match queries, since it is an intermediate between the traditional boolean way of processing queries and the vector space processing model. Indeed, on the one hand, the P-norm model preserves the query structure inherent in the traditional boolean model by distinguishing among different boolean operators (and, or, not); on the other hand, it allows us to retrieve items that would not be retrieved by the traditional boolean model due to its strictness, and to rank them in decreasing order of query-document similarity. Moreover, the P-norm model is able to express queries that range from pure boolean queries to pure vector-space queries, thus offering great flexibility to the user.

Consider a set of terms $t_1, t_2, \ldots, t_n$ and let $\text{sim}(a, t_i) \in [0, 1]$ be the similarity score of term t_i with respect to annotation a; $\text{sim}(a, t_i) = 0$ if the term t_i is not present in the annotation a.

Let $p \geq 1$ be a real number indicating the degree of strictness of the boolean operator. A generalized **or**-query is expressed as $q_{\text{or}(p)} = [t_1 \, \textbf{or}^p \, t_2 \, \textbf{or}^p \cdots \textbf{or}^p \, t_n]$; a generalized **and**-query is expressed as $q_{\text{and}(p)} = [t_1 \, \textbf{and}^p \, t_2 \, \textbf{and}^p \cdots \textbf{and}^p \, t_n]$. The **extended boolean similarity scores** between an annotation and a query are defined as:

$$\text{sim}_p^{\textbf{or}}(a, q) = \left[\frac{\text{sim}(a, t_1)^p + \text{sim}(a, t_2)^p + \cdots + \text{sim}(a, t_n)^p}{n} \right]^{\frac{1}{p}}$$

$$\text{sim}_p^{\textbf{and}}(a, q) = 1 - \left[\frac{(1 - \text{sim}(a, t_1))^p + (1 - \text{sim}(a, t_2))^p + \cdots + (1 - \text{sim}(a, t_n))^p}{n} \right]^{\frac{1}{p}}$$

where t_i indicates a generic term of the query q. Note that for **not**-queries you have to substitute $1 - \text{sim}(a, t_i)$ to $\text{sim}(a, t_i)$ as term weight.

By varying the value of p between 1 and ∞, it is possible to obtain a query processing intermediate between a pure vector-processing model ($p = 1$) and a traditional boolean processing ($p = \infty$), as discussed in [24] to which the reader can refer for further details.

3.2   Annotation Hypertext-Driven Retrieval

Consider the document-annotation hypertext $H_{da} = (DO, E)$ where DO is a set of digital objects (either documents or annotations) and E is a set of edges indicating that an annotation is annotating a digital object, as introduced in [5,7]. The **hypertext similarity score** between an annotation and a query is defined as:

$$\mathrm{sim}_\alpha^{ht}(a, q) = \frac{1}{\alpha}\mathrm{sim}(a, q) + \frac{\alpha - 1}{\alpha} \cdot \frac{1}{|\mathrm{succ}(a)|} \sum_{a_k \in \mathrm{succ}(a)} \frac{\mathrm{sim}(a_k, q) + \mathrm{sim}_\alpha^{ht}(a_k, q)}{2}$$

where $\mathrm{sim}(a, q) \in [0, 1]$ is a generic similarity function between an annotation and a query, $\mathrm{succ}(a)$ is a function that returns the set of successors of an annotation a_j and α is a real number called the *annotation thread damping* factor. We consider that $\mathrm{sim}(a_j, q) = 0$ for those annotations that do not match the query. $\mathrm{sim}_\alpha^{ht}(a, q)$ computes the weighted average between $\mathrm{sim}(a, q)$, the similarity score of an annotation with respect to a query, and the similarity scores which come from the thread to which the annotation belongs. In particular, the thread similarity scores are given by the average between the similarity scores of the successors of a and the hypertext similarity scores of the successors of a; in other words, the hypertext similarity score recursively averages the similarity scores of the annotations that belong to the same thread of the given annotation a. Furthermore, $\mathrm{sim}_\alpha^{ht}(a, q)$ penalizes similarity scores which come from lengthy paths, because for a path $P = a_0 \ldots a_k$ the similarity score $\mathrm{sim}(a_k, q)$ of a_k is weighted $\frac{1}{2^k}$.

By varying the value of α between 0 and ∞, it is possible to obtain a query processing intermediate between a traditional information retrieval model ($\alpha = 1$), when $\mathrm{sim}_1^{ht}(a, q) = \mathrm{sim}(a, q)$ and only the similarity between the annotation and the query is taken into account, and a pure hypertext driven retrieval model ($\alpha = \infty$), when $\mathrm{sim}_\infty^{ht}(a, q) = \frac{1}{|\mathrm{succ}(a)|} \sum_{a_k \in \mathrm{succ}(a)} \frac{\mathrm{sim}(a_k,q) + \mathrm{sim}_\infty^{ht}(a_k,q)}{2}$ and only the thread to which the annotation belongs is taken into account.

Note that the above definition of the hypertext similarity score provides us with an additional degree of freedom in the choice of the actual function to be used for computing the similarity between a term and the annotation, which allows us to plug in different retrieval models for annotations besides the one based on extended boolean retrieval proposed in Section 3.1.

Finally, the hypertext-driven retrieval model allows us to compute a similarity score also for the documents that have been annotated, so that it is possible to search and retrieve documents in response to a user query by means of their annotations. The **similarity score by annotation** between the document and a query is defined as:

$$\mathrm{sim}_\alpha^a(d, q) = \frac{1}{|\mathrm{succ}(d)|} \sum_{a \in \mathrm{succ}(d)} \mathrm{sim}_\alpha^{ht}(a, q)$$

Basically, the similarity score by annotation of a document averages the hypertext similarity scores of the annotations that are annotating the document.

4   The FAST CQL Context Set

The search framework discussed in the previous section has been fully implemented in the running prototype of FAST by developing a query language based on the *Contextual Query Language (CQL)* syntax[7], version 1.2, which provides conformance to CQL up to level 2. The FAST Context Set supports all the search capabilities discussed in the previous section and provides users with a uniform query syntax.

Indexes. The FAST context set defines about 40 indexes and corresponds to the annotation and hypertext model discussed in Section 2, according to which annotations and documents can be searched. For example, `fast.annotation.text` matches annotations on the basis of their textual content and provides full text search capabilities in 14 different languages; `fast.document.byAnnotation` enables the search for documents on the basis of a complete CQL query on their annotations; `fast.annotation.language` indicates a search for annotation based on their language; `fast.annotation.author.email` indicates a search for annotation based on the email address of their author, and so on.

Relations and Relations Modifiers. The FAST context set does not introduce new relations but relies on those defined in the grammar of CQL. Nevertheless, it introduces the following relation modifier, called `thread`, which expressed the kind of hypertext similarity strategy to be applied when evaluating the relations, as discussed in Section 3.2.

The `thread` relation modifier can assume the following values: `noThread` ($\alpha = 1$), i.e. no hypertext structure has to be taken into account into the scoring but only the content of the digital object; `halfThread` ($\alpha = 2$), i.e. the hypertext structure influences the scoring as much as the content of the digital object; `almostThread` ($\alpha = 5$), i.e. the hypertext structure influences the scoring much more than the content of the digital object; `onlyThread` ($\alpha = \infty$), i.e. only the hypertext structure has to be taken into account for the scoring and not the content of the digital object.

Finally, the * and ? wildcards are supported to mask zero or more than one characters, the former, and a single character, the latter.

---

[7] `http://www.loc.gov/standards/sru/specs/cql.html`

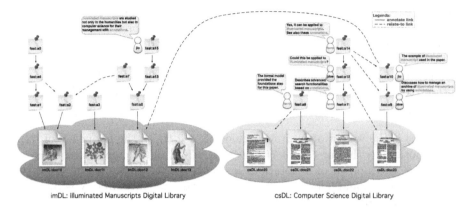

Fig. 2. Example of annotation search

Boolean Operators and Boolean Modifiers. The FAST context set does not introduce new boolean operators, as these can only be defined by the CQL grammar. Nevertheless, it introduces the following boolean modifier, called `match`, which specifies the kind of matching to be applied when computing the boolean expression, according to the different extended boolean retrieval strategies discussed in Section 3.1.

The `match` relation modifier can assume the following values: `bestMatch` $(p = 1)$, i.e. a best matching has to be performed; `looseMatch` $(p = 2)$, i.e. a very approximate matching has to be performed; `fuzzyMatch` $(p = 5)$, i.e. a fuzzy matching has to be performed; `exactMatch` $(p = \infty)$, i.e. a strict boolean matching has to be performed.

4.1 Annotation Search Example

Consider the following example, shown in figure 2: the user is interested in finding annotations that deal with digital archives for illuminated manuscripts. In particular, the user is interested in the use of annotations for linking illuminated manuscripts and remembers that the author corresponding to the user `ferro` worked in this field. This information need could be expressed with the following query: $q =$ `illuminated manuscript and`^{p} (`ferro or`^{p} `annotation`) which looks for annotations whose content is about $t_1 =$ `illuminated manuscripts` and either are authored by user $t_2 =$ `ferro` or also talk about $t_3 =$ `annotations`.

We can therefore compute the similarity between an annotation and the query as follows:

$$\text{sim}_\alpha^p(a, q) = 1 - \left[\frac{\left(1 - \text{sim}_\alpha^{ht}(a, t_1)\right)^p + \left(1 - \left(\frac{\text{sim}_\alpha^{ht}(a,t_2)^p + \text{sim}_\alpha^{ht}(a,t_3)^p}{2}\right)^{\frac{1}{p}}\right)^p}{2} \right]^{\frac{1}{p}}$$

where we used the hypertext similarity scores of section 3.2 to compute the similarity between an annotation and the different terms of the query and we used the extended boolean operators of section 3.1 to combine the results of the previous computations. Note that the above equation expresses a whole family of similarity scores, since it is parametric in both p and α.

Let us use the following weights: for the exact match clause, a weight 1 indicates that `ferro` is the author of the annotation, 0 that somebody else is the author; in the best match clause, the normalized term frequency (tf) / inverse document frequency (idf) weights are used: $w_{ij} = \dfrac{\text{tf}_{ij} \cdot \log_2 \frac{N}{\text{df}_i}}{\max_{h,k}(\text{tf}_{hk}) \cdot \max_h\left(\log_2 \frac{N}{\text{df}_h}\right)}$, where w_{ij} is the weight of term i in annotation j; tf_{ij} is the frequency of term i in annotation j (in our case it is always 1); df_i is the document frequency of term i, i.e. the number of annotations in which the term i is present (4 for the term `annotation` and 5 for the term `illuminated manuscript`); N is the total number of annotations, i.e. 15. We use $\text{sim}(a_j, t_i) = w_{ij}$.

For example, if we set $p = 2$ (`looseMatch` boolean modifier) and $\alpha = 2$ (`halfThread` relation modifier), which correspond to the following query according to the FAST context set and the CQL syntax:

```
(fast.annotation.text =/thread=halfThread "illuminated manuscript")
  and/match=looseMatch
(
  (fast.annotation.author.identifier =/thread=halfThread ferro)
  or/match=looseMatch
  (fast.annotation.text =/thread=halfThread annotation)
)
```

we obtain the following ranking: `fast:a12` with score 0.52; `fast:a9` with score 0.50; `fast:a11` with score 0.48; `fast:a14` with score 0.46; `fast:a15` with score 0.38; `fast:a10` with score 0.18; `fast:a18` with score 0.16;

As an additional example, if we set $p = 2$ (`looseMatch` boolean modifier) and $\alpha = \infty$ (`onlyThread` relation modifier), we obtain the following ranking: `fast:a12` with score 0.46; `fast:a11` with score 0.41; `fast:a9` with score 0.18; `fast:a8`, `fast:a10`, `fast:a14`, and `fast:a15` with score 0.

As you can note the `onlyThread` relation modifier can produce some unexpected results since, for example, the annotation `fast:a14`, which is the only one that matches all the query clauses, get a similarity score of 0 being a leaf in the tree. On the other hand, the `halfThread` relation modifier produces a ranking where all the annotations are present but relevance is still moved from annotations that are leaves of the tree to inner annotations. This is still the case of `fast:a14` which is not the top ranked annotation while `fast:a12` gets ranked before it, being part of a thread. Moreover, lengthy paths are penalized: `fast:a11` is in the same thread of `fast:a14` and `fast:a12` but it does not benefit from the score obtained from `fast:a14` as much as `fast:a12`, due the lengthy path, and gets ranked after `fast:a12`.

The examples above show how flexible and powerful are the querying capabilities available for the user in order to search and retrieve annotations and how much they actually impact and modify the ranking and result list returned to the user, providing him with different views on the managed annotations.

5   Conclusions

We have discussed the issues related to the design and development of retrieval algorithms capable of taking into account annotations and their peculiar features. In particular, we have introduced a general framework which allows us to develop such retrieval algorithms based on a combination of extended boolean retrieval operators and hypertext driven information retrieval.

The proposed framework is quite general since it can express queries which range from pure boolean queries to unstructured queries as well as formulate queries which range from considering only the content of the annotations to considering only the structure of the hypertext between documents and annotations. Moreover, the proposed framework allows us to seamlessly plug different retrieval models into it.

Acknowledgements

The work reported has been partially supported by the TELplus Targeted Project for digital libraries, as part of the *eContentplus* Program of the European Commission (Contract ECP-2006-DILI-510003).

References

1. Agosti, M., Albrechtsen, H., Ferro, N., Frommholz, I., Hansen, P., Orio, N., Panizzi, E., Pejtersen, A.M., Thiel, U.: DiLAS: a Digital Library Annotation Service. In: Proc. Int. Workshop on Annotation for Collaboration – Methods, Tools, and Practices (IWAC 2005), pp. 91–101. CNRS - Programme société de l'information (2005)
2. Agosti, M., Berretti, S., Brettlecker, G., Del Bimbo, A., Ferro, N., Fuhr, N., Keim, D.A., Klas, C.-P., Lidy, T., Milano, D., Norrie, M.C., Ranaldi, P., Rauber, A., Schek, H.-J., Schreck, T., Schuldt, H., Signer, B., Springmann, M.: Delos-DLMS - the integrated DELOS digital library management system. In: Thanos, C., Borri, F., Candela, L. (eds.) Digital Libraries: Research and Development. LNCS, vol. 4877, pp. 36–45. Springer, Heidelberg (2007)
3. Agosti, M., Bonfiglio-Dosio, G., Ferro, N.: A Historical and Contemporary Study on Annotations to Derive Key Features for Systems Design. International Journal on Digital Libraries (IJDL) 8(1), 1–19 (2007)
4. Agosti, M., Ferro, N.: Annotations: Enriching a digital library. In: Koch, T., Sølvberg, I.T. (eds.) ECDL 2003. LNCS, vol. 2769, pp. 88–100. Springer, Heidelberg (2003)
5. Agosti, M., Ferro, N.: Annotations as context for searching documents. In: Crestani, F., Ruthven, I. (eds.) CoLIS 2005. LNCS, vol. 3507, pp. 155–170. Springer, Heidelberg (2005)
6. Agosti, M., Ferro, N.: Search strategies for finding annotations and annotated documents: The FAST service. In: Larsen, H.L., Pasi, G., Ortiz-Arroyo, D., Andreasen, T., Christiansen, H. (eds.) FQAS 2006. LNCS (LNAI), vol. 4027, pp. 270–281. Springer, Heidelberg (2006)
7. Agosti, M., Ferro, N.: A Formal Model of Annotations of Digital Content. ACM Transactions on Information Systems (TOIS) 26(1), 3:1–3:57 (2008)

8. Agosti, M., Ferro, N.: Annotations: a way to interoperability in DL. In: Christensen-Dalsgaard, B., Castelli, D., Ammitzbøll Jurik, B., Lippincott, J. (eds.) ECDL 2008. LNCS, vol. 5173, pp. 291–295. Springer, Heidelberg (2008)

9. Agosti, M., Ferro, N., Frommholz, I., Thiel, U.: Annotations in digital libraries and collaboratories – facets, models and usage. In: Heery, R., Lyon, L. (eds.) ECDL 2004. LNCS, vol. 3232, pp. 244–255. Springer, Heidelberg (2004)

10. Buneman, P., Khanna, S., Tan, W.-C.: On Propagation of Deletions and Annotations Through Views. In: Proc. 21st ACM Symposium on Principles of Database Systems (PODS 2002), pp. 150–158. ACM Press, New York (2002)

11. Cabanac, G., Chevalier, M., Chrisment, C., Julien, C.: Collective Annotation: Perspectives for Information Retrieval Improvement. In: Proc. 8th Conference on Information Retrieval and its Applications (RIAO 2007) (2007), http://riao.free.fr/papers/99.pdf

12. Candela, L., et al.: The DELOS Digital Library Reference Model. Foundations for Digital Libraries (2007), http://www.delos.info/files/pdf/ReferenceModel/DELOS_DLReferenceModel_0.98.pdf

13. Fraenkel, A.S., Klein, S.T.: Information Retrieval from Annotated Texts. Journal of the American Society for Information Science (JASIS) 50(10), 845–854 (1999)

14. Frommholz, I., Brocks, H., Thiel, U., Neuhold, E.J., Iannone, L., Semeraro, G., Berardi, M., Ceci, M.: Document-centered collaboration for scholars in the humanities – the COLLATE system. In: Koch, T., Sølvberg, I.T. (eds.) ECDL 2003. LNCS, vol. 2769, pp. 434–445. Springer, Heidelberg (2003)

15. Frommholz, I., Fuhr, N.: Probabilistic, Object-oriented Logics for Annotation-based Retrieval in Digital Libraries. In: Proc. 6th ACM/IEEE-CS Joint Conference on Digital Libraries (JCDL 2006), pp. 55–64. ACM Press, New York (2006)

16. Golder, S., Huberman, B.A.: Usage Patterns of Collaborative Tagging Systems. Journal of Information Science 32(2), 198–208 (2006)

17. Golovchinsky, G., Price, M.N., Schilit, B.N.: From Reading to Retrieval: Freeform Ink Annotations as Queries. In: Proc. 22nd Annual Int. ACM SIGIR Conference on Research and Development in Information Retrieval (SIGIR 1999), pp. 19–25. ACM Press, New York (1999)

18. Groth, D.P., Streefkerk, K.: Provenance and Annotation for Visual Exploration Systems. IEEE Transactions on Visualization And Computer Graphics 12(6), 1500–1510 (2006)

19. Halasz, F.G., Moran, T.P., Trigg, R.H.: Notecards in a Nutshell. In: Proc. Conference on Human Factors in Computing Systems and Graphics Interface (CHI 1987), pp. 45–52. ACM Press, New York (1987)

20. Handschuh, S., Staab, S. (eds.): Annotation for the Semantic Web. IOS Press, Amsterdam (2003)

21. Ioannidis, Y., et al.: Digital library information-technology infrastructures. International Journal on Digital Libraries 5(4), 266–274 (2005)

22. Kahan, J., Koivunen, M.-R.: Annotea: an open RDF infrastructure for shared Web annotations. In: Proc. 10th Int. Conference on World Wide Web (WWW 2001), pp. 623–632. ACM Press, New York (2001)

23. Marshall, C.C.: Toward an Ecology of Hypertext Annotation. In: Proc. 9th ACM Conference on Hypertext and Hypermedia (HT 1998): links, objects, time and space-structure in hypermedia systems, pp. 40–49. ACM Press, New York (1998)

24. Salton, G., Fox, E.A., Wu, H.: Extended Boolean Information Retrieval. Communications of the ACM (CACM) 26(11), 1022–1036 (1983)

wikiSearch – From Access to Use

Elaine G. Toms, Lori McCay-Peet, and R. Tayze Mackenzie

Centre for Management Informatics
Dalhousie University
6100 University Ave
Halifax, Nova Scotia, Canada
`etoms@dal.ca, mccay@dal.ca, mackenziert@dal.ca`

Abstract. A digital library (DL) facilitates a search workflow process. Yet many DLs hide much of the user activity involved in the process from the user. In this research we developed an interface, wikiSearch, to support that process. This interface flattened the typical multi-page implementation into a single layer that provided multiple memory aids. The interface was tested by 96 people who used the system in a laboratory to resolve multiple tasks. Assessment was through use, usability testing and closed and open perception questions. In general participants found that the interface enabled them to stay on track with their task providing a bird's eye view of the events – queries entered, pages viewed, and pertinent pages identified.

Keywords: Digital libraries; user interface; information task; search; BookBag; interactivity tools; search workflow.

1 Introduction

The design of interfaces to digital libraries (DLs) is strongly embedded in information retrieval (IR) traditions. Like the early IR systems, the first generation of DLs provided a simple search box to receive user queries, and had more limited functionality than a 1970s command-driven search system; help was nonexistent, and any special features (e.g., syntax) were hidden from the user. Systems resources were focused on extracting documents from huge repositories.

But DLs support a larger task – a work process – that may have multiple inter-related sub-tasks. For a student, that larger task may be writing a term paper for a course. For the shopper, this may mean understanding the nuances of mp3 players and comparing possible options to make a purchase. For the citizen, it may be separating fact from fiction during elections in order to decide how to vote. Rarely has the full spectrum of tasks been delineated such that the work flow and the information flow(s) within the larger work task are operationalized, and systems constructed to support the entire process (although efforts are emerging, [e.g., 3, 17]). Regardless of the type of work task or domain, multiple information sub-tasks are invoked during that workflow process to extract and manipulate the raw material required to complete that work task. In this research, we focused specifically on one aspect of that workflow, articulating how one type of information task – search – may be integrated into the workflow. We describe our design and its subsequent evaluation.

M. Agosti et al. (Eds.): ECDL 2009, LNCS 5714, pp. 27–38, 2009.

2   Previous Work

Embedding search within applications has been widely promoted [5, 10, 17]. But the research underlying support for users of DLs and the design of that support is founded partially in information seeking and retrieval (IS&R) research, partially in human computer interaction (HCI), and partially in the application domain.

While the stand-alone DL started with a simple search interface, these have evolved to aid the user: techniques for searching/scanning/examining information objects [e.g., 10, 2, 11], enabling social engagement [6] and personalizing the user's experience [14]. But many of these initiatives have not considered the "information journey" [1], and the overall workflow that underpins that process.

One of the earliest implementations to integrate search into the work process was perhaps the first mathematics digital library [12] which from an interface perspective is not unlike those that followed, providing targeted and integrated support tools such as automatic query reformulation. An interesting contrast in design is a comparison of [17] and [13]'s prototypes to aid student writing. Twidale *et al* [17] deployed multiple intelligent agents to suggest keywords and related documents. Reimer *et al*'s [13] prototype worked with already found documents. Twidale *et al* automated much of the search process leaving the user no longer in control while Reimer *et al* dealt only with "information assimilation." In both cases, the job is incomplete – knowing the *what* does not necessarily indicate the *how*.

For the past 20 years, HCI has been developing best practices and principles for interface design [15]. Notably core principles include maximizing visibility and minimizing search time while not overloading the user's working memory, providing logically structured displays while at the same time providing only relevant information, and enabling the reversal of actions. The user needs to feel "in charge of the interface" which challenges our basic instinct to simplify the task of the user by automating as much as possible. Yet most search systems and indeed many digital libraries fail on these basic principles. Consider the classic IS&R system, Google, which hides much of the search workflow process from the user. As a result the user has to keep track of that process as well as keep track of the original work task progression. Typically, a labyrinth of pages must be navigated while the user hunts for the needed information requiring the user to remember the core landmarks that had previously been visited, the information acquired, as well as what is completed and still left to do. The usual solution for information found is to bookmark the page, print the page or produce endless lists of post-it notes. DLs do much the same.

Our development considered the basic HCI design principles in concert with IS&R conceptual models (e.g., Kuhlthau's [7, 8] modified by Vakkari [18,19], and Marchionini's [9]) to represent search as a workflow process.

3   Design of wikiSearch

The design process involved brainstorming sessions to identify possible approaches while being mindful of design principles. These sessions included eight lab personnel: undergraduate and graduate students and researchers. A series of prototypes emerged and the result of this iterative process was the interface illustrated in Figure 1.

Our implementation, which we call wikiSearch, accesses the Wikipedia at present, although the design is generic and applicable to other types of information resources/repositories. The interface was driven by the principle to "structure and sequence the display" so that the grouping and ordering was logical to the user's task. As a result we ordered the interface into three core components. The first column reflects task-based activities, the second relates to the search itself, and the third is at a high level of granularity – detailed viewing of documents.

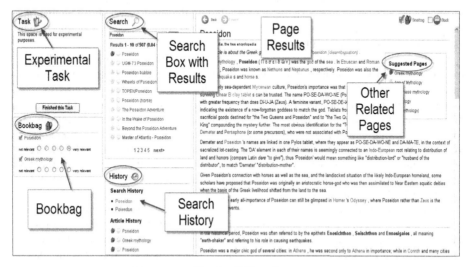

Fig. 1. wikiSearch

The **Task** (extreme left) section contains the contents of the experimental task that participants are assigned. However, we envision this as workspace for task completion. Below the task description is a BookBag, akin to the shopping cart in online shopping environments, which is used to collect information objects deemed useful to the task. Pages can be removed from the BookBag, and to assist our experimental process, pages can be rated for likely relevance to the task. Pages can be added to the BookBag from several places in the interface. While browser bookmarks enable the identification of useful websites, they do not support the principle of maximized visibility provided by the BookBag.

The second column, the **Search** section, contains three sub sections: one devoted to entering a query, one to displaying results and a third to displaying history of search activities. The search box remains unchanged from other systems at this point. Remaining attentive to our principles meant a compromise in displaying search results (both the current and historic). While most systems provide a single independent page that is part of a multi-page labyrinth, we wanted to avoid overloading the user's memory so that results and their use are always visible. To conserve space, the results section contains a list of ten titles with the option to display others by selecting a button forward (or backward). Given user prevalence for looking only at the first page of results, displaying ten simultaneously seems sufficient. "Mousing" over each title invokes a pop-up window that displays a search word-in-context summary of each information object. The third, the history sub section, keeps track of all queries issued and all information objects viewed. Both can easily be used by a mouse click. As a result we have made visible to users all prior actions – a useful memory aid.

The third column displays the **information object**, a scrollable wiki page. Each page contains ordinary hypertext links and links to search within the wiki. A page may be loaded from a link in the Search results or History section, or from links stored in the BookBag. In addition to the page display, a text box provides a further list of Suggested Pages. This set of page links is created by entering the first paragraph of that page as a search string, and displaying the top five results. This list had the potential to provide more specific pages about the topic, or be distracting (much like the Suggestions provided by Toms [16]).

In addition, pages may also be blocked so that they never need to be viewed again for this particular task. This is much like a Boolean negative, ensuring that the workspace is not cluttered with unneeded, irrelevant information objects.

These elements and the collapsed/flattened design removes and limits the number of mouse clicks required to do a task, eliminates the 'labyrinth' effect, introduces space for tracking useful items, makes visible as many actions as possible, enables a low level Boolean negative, and introduces the concept of serendipity while searching/scanning. Reversal of actions was deemed irrelevant as it is impossible for any action to interfere with the process; all actions are visible. Enabling page blocking meant that irrelevant and/or unneeded pages never need be viewed a second time.

Our first design challenge – managing the search process – was rendered by column 1 and 2, and the second challenge by our History and BookBag functions. As a result, the simplified interface condensed what normally would have appeared in four interface windows or browser web pages into a single-layer display that reflected the search process while adhering to standard design principles. At the same time, it introduced a natural workflow: task is contained to the right as well as the work associated with that task, and the actions associated with active searching are contained in the centre with the view of items to the extreme right. Previous activities – queries entered, pages viewed, and pertinent items to retain – are always visible to the user, requiring less time to traverse a labyrinth of pages. Other than the magical black box – the retrieval engine – all actions are also controlled by the user.

4   Evaluation of wikiSearch

To test our design, initially we ran basic usability studies with 22 participants and then used the resulting design for the conduct of other research studies. In these studies the interface accessed a locally-stored version of Wikipedia using a retrieval system based on Lucene 2.2, an open source search engine. In one of these studies (N=96) we additionally assessed the wikiSearch interface by adding post session questions which enabled participants to reflect on their experience with the system and to assess their perception of its features. In this section, we describe the experiment used to collect the data.

4.1   Participants

The 96 participants (M=49, F=47) were primarily (90%) students from the university community, and from mixed disciplines. 25% held undergraduate degrees and

12% graduate or other degrees. 84.4% were under 27. They were an experienced search group with 86.5% searching for something one or more times a day, and also relatively frequent users of the Wikipedia (54% use it at least weekly).

4.2 Tasks

Prior to assessing the system, participants completed three tasks from a set of 12 tasks that were developed according to a set of principles: 1) no task could be completed using a single page; 2) the task required searchers to actively make a decision about what information was truly relevant in order to complete a task. The tasks were used in the INEX 2006 Interactive Track. Each task required that pages used to respond to the task be added to the BookBag and rated. Task topics varied from environmental issues surrounding logging and mining to the hazards of red ants and the impressionism movement. One example is:

> "As a tourist in Paris, you have time to make a single day-trip outside the city to see one of the attractions in the region. Your friend would prefer to stay in Paris, but you are trying to decide between visiting the cathedral in Chartres or the palace in Versailles, since you have heard that both are spectacular. What information will you use to make an informed decision and convince your friend to join you? You should consider the history and architecture, the distance and different options for travelling there."

Three tasks were assigned to each participant, such that all tasks were performed by 24 people. Order of task was counterbalanced to control for learning effects.

4.3 Metrics

The assessment used the System Usability Scale (SUS) [4] to assess user perception of usability. User events were logged to assess whether and how the features were used, and finally users responded to a series of closed and open-ended questions concerning their perception of the system.

4.4 Procedure

Data collection took place in a laboratory setting where 5 to 7 people were processed simultaneously. A research assistant was always present. Participants were simply told that we were assessing how people search. They were presented with the following steps in a series of self-directed webpages: 1) Introduction which introduced the study, 2) Consent Form that outlined the details of participation, 3) Demographics and Use Questionnaire to identify prior knowledge and experience, 4) Tutorial and practice time using the wikiSearch system, 5) Pre-Task Questionnaire, 6) Assigned task to be completed using wikiSearch integrated into the interface as illustrated in Figure 1, 7) Post-Task Questionnaire, 8) Steps 5 to 7 were repeated for the other two tasks, 9) Post-Session Questionnaire to identify user perception of the system, 10) SUS Questionnaire, and 11) Thank-you for participating page.

5 Results

After completing the three tasks and working with the system for 21 minutes, on average (and not including the tutorial or completion of experimental questionnaires), the 96 participants responded to sets of questions that dealt specifically with the interface and selected tools, as well as the 10 item SUS questionnaire. All questionnaires used the same seven point scale with the left side labeled "Strongly Disagree" and the right side, "Strongly Agree."

SUS Questionnaire: This questionnaire [4] is composed of five positively worded statements and five negatively worded statements. Participants indicated the degree to which each agreed with the statements about the wikiSearch system.

As illustrated in Table 1, the level of agreement with the positively worded statements varied from 5.5 to 6.23 on a seven point scale. In general participants found the system easy to use and easy to learn, and expressed an interest in using it more frequently and felt confident in using the system.

Table 1. Positively expressed statement on the SUS questionnaire

#	*Positive Statements*	Mean	SD
1	I think that I would like to use wikiSearch frequently	5.50	1.47
3	I thought wikiSearch was easy to use	6.11	1.09
5	I found the various functions in wikiSearch were well integrated	5.66	1.00
7	I think most people would learn to use wikiSearch very quickly	6.23	0.86
9	I felt very confident using wikiSearch	5.70	1.27

Similarly, the level of disagreement with negative statements varied from 1.44 to 2.93 on the same scale indicating general disagreement with the statements. Participants tended not to find the system complex, or to require technical support, or difficult to learn or cumbersome to use (see Table 2).

Table 2. Negatively expressed statement on the SUS questionnaire

#	*Negative Statements*	Mean	SD
2	I found wikiSearch unnecessarily complex	2.22	1.04
4	I think that I would need the support of a technical person to be able to use wikiSearch	1.44	0.86
6	I thought there was too much inconsistency in wikiSearch	2.93	1.43
8	I found wikiSearch very cumbersome to use	2.50	1.34
10	I needed to learn a lot of things before I could get going with wikiSearch	1.83	1.19

Overall, results indicate positive response on the System Usability Scale indicating that the system meets at least a level of usability.

Use of the System. Customized logging software logged all user activity. Among the possible actions recorded per task were issuing queries (mean = 6.7), viewing pages (mean = 11), reviewing pages from history (mean = 1.5), viewing results pages (mean = 7.7) and reviewing results pages two or higher (mean = 7.8), accessing pages via an internal page link (mean=2.1), adding pages to the BookBag and rating their relevance (mean=4.5), and blocking pages (mean=0.5). Overall, participants executed 38 actions per task (which does not including scrolling forward and backward within a single page view or using any other scrolling action). As a result, we concluded that their level of activity was sufficient for them to assess a novel interface.

Perception of the BookBag. The BookBag, much like the shopping cart in e-commerce, is particularly pertinent to digital libraries. As illustrated in Table 3, participants found the BookBag useful in gathering pertinent pages and keeping track of those pages. In addition, participants speculated on their potential future use, indicating 6.38 on average agreement in using it again if it were available.

Table 3. Responses concerning the BookBag

Statements	Mean	SD
I found the BookBag useful in helping me collect the pages that I needed.	6.47	0.75
I found the BookBag helped me keep track of the pages that I found useful.	6.50	0.71
If presented with the opportunity, I would use the BookBag feature again.	6.38	0.92

After the closed questions, participants responded to two open-ended questions asking when they found the BookBag the most useful and when they found it not useful.

Table 4. When the BookBag was most useful and when particpants would not use it

Most useful for:	N=86	Would not use:	N=71
Organization of search	45	For certain task or search types	59
Task completion	32	When not planning to return/irrelevant or insufficient content	12
Navigation of web pages	32	When other tools are available to perform the same function	2
Substitution of other tools	31	When privacy is a concern	1

Participants found the BookBag usefully served several functions (Table 4):

1. Organization of search: Participants indicated that the BookBag was useful for helping keep track of, save and organize their search results. The BookBag allowed participants to search and collect web pages and then move on to task completion.
2. Task completion: The BookBag's value with regard to task completion included reviewing, analysing, clarifying, cross-referencing, synthesizing, determining relevancy, and comparing web content. "When I had read everything through, I could look at my results, and weigh them with the information provided, and change my mind as I went along" (P596).
3. Navigation of web pages: The organizational and task completion functionality of the BookBag were closely related to navigation. The BookBag was useful when "comparing and contrasting information. It was easy to have the web pages in front of

me instead of click back and forth from the favorites page" (P619). The BookBag also prevented participants from getting "lost" while exploring other pages.

4. Substitution of other tools or techniques: Participants made direct comparisons between the BookBag and other tools and techniques they currently use. These included the use of the BookBag as a "memory aid" (P605). The BookBag eliminated the need for pen and paper, an email to self, copying and pasting, and printing results. The BookBag also replaced browser functions such as history, favourites, bookmarks, back-and-forth navigation, as well as right clicking, the use of drop-down menus, the need to keep multiple windows open, and repeating searches.

Participants would *not* use the BookBag for the following reasons (Table 4):

1. For certain task and search types: Participants indicated that the usefulness of the BookBag was dependent on the length and complexity of the task or search. The BookBag's usefulness would be low if conducting short or simple tasks while "this feature is most useful when completing searches or research on a complex, multi-part task" (P564). Others indicated it would not be useful for general interest searching or browsing, and particularly for non-school/research related searching.

2. Content: Some responses related to the functionality of the BookBag feature, with several suggesting it would not be useful if the web page content was insufficient or "if I did not think the information was useful for me in the future" (P689).

3. Competing tools: Two participants expressed their preference for other browser features they currently use including tabs and search history.

4. Privacy: One participant indicated that the BookBag function would not be useful "when privacy is an issue" (P576).

Perception of the Interface. A core design element of this interface was the integration of search box, results and page display as well as history and BookBag on a single display. Participants responded to three questions regarding this element. In general, participants found the side-by-side display useful, saved time and kept them on topic as illustrated in Table 5.

Table 5. Response about the interface display

#	Statement	Mean	SD
1	I found the presentation of search results side-by-side with the display of a single page useful	6.13	1.00
2	I found the presentation of search results side-by-side with the display of a single page saved time	6.09	1.21
3	I found the presentation of search results side-by-side with the display of a single page kept me on topic	5.77	1.29

Perception of the Mouse-over Summaries. The compromise in design was the implementation of search result summaries, or snippets, as 'mouse-over' elements. This is a significant change from typical practice where a page of result summaries is the norm. The two questions asked slightly different versions of the same question (see Table 6). Participants found the 'mouse-over' easy to use, but at the same time were not as strong in their agreement about the mouse-over feature.

Table 6. Response to the Mouse-over Summaries

#	Statement	Mean	SD
1	I found the 'mouse over' summaries easy to use	5.82	1.50
2	I would prefer to see each summary displayed at the same time on a single page	3.46	1.89

Interface preference: As a final question participants chose between the wikiSearch style and Google-style of search interface. As illustrated in Table 7, 74% preferred the single, collapsed interface to the multiple page solution.

Table 7. Preference for interface style

Items	N (96)	%
wikiSearch interface: search box, search results and the display of a single web page on a single screen or page; the search summaries are not all visible at the same time	71	74
Google-style interface: search box and search results are on the same page; you need to click backward and forward to select and display a new webpage; search results are all visible on a single page	20	21
Neither	5	5

In addition, participants identified their reason(s) for their preferences which are identified in Table 8. The 96 responses were coded using one or more of the codes.

Table 8. Number of participants who identified reasons for preference for a wikiSearch or Google-style interface

Code	Wikisearch (N)	Google-like (N)
Ease, speed, efficiency	43	2
Results display	39	5
Navigation	26	2
Task focus/ organization	29	0
More accustomed to Google	n/a	7

While the overall preference was for the wikiSearch interface, participants had difficulty separating content from interface. Eleven participants perceived a reduction in the quantity of results and the quality and reliability of results content. Comments included "Wiki info is not credible" (P721) and "the amount of useful information I found was not the same as google" (P579). The remainder of this analysis is limited to the usefulness of the interface rather than content.

1. Ease, speed, efficiency: Almost half (43) participants made positive comments specifically relating to the ease, speed, and efficiency of the wikiSearch, preferring the collapsed interface "because it was so easy to stay on task and there was little distraction. It was much more time efficient" (P548). Two participants preferred the Google-like interface for its "simpler interface" (P598) and because "it's easy to use" (P630).

2. Results display: Forty-three participants who preferred the wikiSearch interface commented that they liked the search results display. Most of these comments related to the single page display of the wikiSearch making it "easier and quicker to always see search results on the same page instead of having to go back and forth" (P547).

3. Navigation: Participants indicate that they preferred the navigational functions of wikiSearch "because everything I needed was on one page, I didn't need to go back and forward between different search results" (P592). Two participants, however, indicated that they preferred the Google-like search interface because it is easier to locate specific pages (for which they have forgotten the URL) through Google and due to a preference for opening links in new tabs.

4. Task focus/organization: twenty-nine participants made comments regarding the usefulness of the Wikisearch to help them organize their search and maintain their focus on the task at hand. WikiSearch showed participants "alot of what you want not what you don't want" (P709), helped them to avoid getting "lost in many google pages" (P684) and "definitely kept me focused on my task" (P661).

5. More accustomed to Google: Seven participants indicated their preference for the Google-style interface by qualifying it with a statement that they were more accustomed to Google. "The format was new so it didn't seem as instinctive as Google" (P571).

6   Analysis and Discussion

The goal of our work is to develop an interface that emulates the flow of activities within the search process, while speculating on the further integration of that search process within the larger workflow. Our collapsed interface supports both the user's right to be in control and to have visible and available all the tools of the trade. Much like the desktop metaphor used to reference the Apple and Windows environments, this interface creates a "desktop" to support the search process. The evaluation completed by 96 potential users of the technology gave the design high usability scores. In addition, they made significant use of all of the features so as to gain some exposure to the system, and they did this for three tasks. Some of the features were used more than others as would be expected, but that use provided them with sufficient experience to assess its effectiveness.

As part of their assessment, we asked participants to contrast this interface with a Google-like interface. Some responses related both to the content as well as to the interface. Three-quarters preferred wikiSearch and this was for the navigational and organizational functions that it provided. Google-like interfaces tended to be preferred for the detailed search results and quality of the content; students seem trained to perceive Wikipedia content as sub-standard at our University.

Overall wikiSearch enabled people to stay on course with their task by providing a bird's eye view of the events – queries entered, pages viewed, and pertinent pages identified. This relatively simple, two-dimensional interface simplified navigation, and prevented people from getting lost in the labyrinth of pages. As a result, it demonstrated the value of providing visibility to the activities required in the search workflow process, and this visibility additionally support the core cognitive abilities, e.g., Recall and Summarize, noted by Kuhlthau [8] in the seeking and use of information.

The BookBag and Query History – mere lists – challenge the current superiority of tabbed-based browsers. Which one is the most effective remains to be seen. A limitation of our design was the inability to examine two pages in parallel, although participants noted the capability to quickly re-load pages from the BookBag and history. The BookBag was perceived the least useful for short, simple searches or when participants simply wanted to scan; it was perceived the most useful for more complex tasks that would involve multiple queries – the types of tasks performed by students. The BookBag replaced less reliable or cumbersome techniques they normally employed including memory, favourites, search history, copy and paste, clicking back and forward, and keeping multiple windows open.

The design decision to enable mouse-over summaries rather than full presentation met with mixed responses. While the presentation did not actually interfere with use, the user perception was not as positive. Whether this format is novel and thus has a learning curve, or whether it is truly a barrier to assessing results needs further research. Presumably, as search engines improve precision, results can be more reliably assumed to be relevant.

Although the interface was successful, it was used with a limited resource – the Wikipedia. However we believe the technology to be scalable to larger, e.g., the Web, or multiple, e.g., scholarly journal publishers,' repositories.

While this is a first attempt at delineating and supporting the search workflow, we see the potential for augmentation and improvements. For example, our history display separates queries and page views; would an integration of the two be more valuable, allowing the user to discern which queries were the most useful? The Bookbag, as noted by participants, is particularly useful for school projects and research. But how might it be enhanced? For example, adding note-taking capability (not unlike Twidale et al [17]) would at the same time support those significant cognitive activities noted by Kuhlthau [8]. In addition, what other functions does the student, in particular, need while using a digital library to complete a term paper? How much control is left to the user, and how much can be assumed by the system before the user feels out of control? Perhaps we need a stronger requirements specification that is both seated in what is ostensibly the user's work task process as well as in our information models and frameworks.

7 Conclusion

Our research is working toward an improved search interface with appropriate support for search workflow and its integration with the larger work task. We found strong support for enabling better visibility of basic activities within the search workflow while leaving the user in control of the process which are fundamental design guidelines. While much of DL interface development has followed implementations in information retrieval systems, it is time to consider how search is connected to the larger work process.

Acknowledgments. Chris Jordan implemented the search engine. Alexandra MacNutt, Emilie Dawe, Heather O'Brien, and Sandra Toze participated in the design and implementation of the study in which the data was collected. Research was supported by the Canada Foundation for Innovation, Natural Science and Engineering Research Council Canada (NSERC), and the Canada Research Chairs Program.

References

1. Adams, A., Blandford, A.: Digital libraries support for the user's information journey. In: Proceedings of the 5th ACM/IEEE-CS JCDL, JCDL 2005, Denver, CO, USA, June 07 - 11, 2005, pp. 160–169 (2005)
2. Baldonado, M., Wang, Q.: A user-centered interface for information exploration in a heterogeneous digital library. JASIS&T 51(3), 297–310 (2000)
3. Bartlett, J.C., Neugebauer, T.: A task-based information retrieval interface to support bio-informatics analysis. In: Proceedings of the Second IIiX Symposium, vol. 348, pp. 97–101 (2008)
4. Brook, J.: SUS - A quick and dirty usability scale,
   http://www.usabilitynet.org/trump/documents/Suschapt.doc
5. Hendry, D.G.: Workspaces for search. JASIS&T 57(6), 800–802 (2006)
6. Krafft, D.B., Birkland, A., Cramer, E.J.: NCore: Architecture and implementation of a flexible, collaborative digital library. In: Proceedings of the ACM International Conference on Digital Libraries, pp. 313–322 (2008)
7. Kuhlthau, C.: Inside the search process: Information seeking from the user's perspective. JASIS 42(5), 361–371 (1991)
8. Kuhlthau, C.: Seeking Meaning: A Process Approach to Library and Information Services. Libraries Unlimited, Westport (2004)
9. Marchionini, G.: Information Seeking in Electronic Environments. Cambridge, NY (1995)
10. Marchionini, G., White, R.: Find what you need, understand what you find. Int. J. Hum-Compt. Int. 23(3), 205–237 (2007)
11. Komlodi, A., Soergel, D., Marchionini, G.: Search Histories for User Support in User Interfaces. JASIST 57(6), 803–807 (2006)
12. McAlpine, G., Ingwersen, P.: Integrated information retrieval in a knowledge worker support system. In: Proceedings of the 12th Annual international ACM SIGIR Conference on Research and Development in information Retrieval, Cambridge, Massachusetts, United States, June 25 - 28, 1989, pp. 48–57 (1989)
13. Reimer, Y.J., Brimhall, E., Sherve, L.: A study of student notetaking and software design implications. In: Proceedings of the 5th IASTED International Conference on Web-Based Education, Puerto Vallarta, Mexico, January 23 - 25, 2006, pp. 189–195 (2006)
14. Renda, M.E., Straccia, U.: A personalized collaborative Digital Library environment: A model and an application. Inform ProcessManag, An Asian Digital Libraries Perspective 41(1), 5–21 (2005)
15. Shneiderman, B., Plaisant, C.: Designing the User Interface: Strategies for Effective Human-Computer Interaction, 4th edn. Pearson, Boston (2004)
16. Toms, E.G.: Understanding and facilitating the browsing of electronic text. IJHCS 52(3), 423–452 (2000)
17. Twidale, M.B., Gruzd, A.A., Nichols, D.M.: Writing in the library: Exploring tighter integration of digital library use with the writing process. Inform Process Manag. 44(2), 558–580 (2008)
18. Vakkari, P.: A theory of the task-based information retrieval process. JDOC 51(1), 44–60 (2001)
19. Vakkari, P.: Changes in search tactics and relevance judgements when preparing a research proposal: A summary of the findings of a longitudinal Study. Information Retrieval 4(3-4), 295–310 (2001)

Adding Quality-Awareness to Evaluate Migration Web-Services and Remote Emulation for Digital Preservation

Christoph Becker[1], Hannes Kulovits[1], Michael Kraxner[1], Riccardo Gottardi[1],
Andreas Rauber[1], and Randolph Welte[2]

[1] Vienna University of Technology, Vienna, Austria
http://www.ifs.tuwien.ac.at/dp
[2] University of Freiburg, Germany
http://www.uni-freiburg.de/

Abstract. Digital libraries are increasingly relying on distributed services to support increasingly complex tasks such as retrieval or preservation. While there is a growing body of services for migrating digital objects into safer formats to ensure their long-term accessability, the quality of these services is often unknown. Moreover, emulation as the major alternative preservation strategy is often neglected due to the complex setup procedures that are necessary for testing emulation. However, thorough evaluation of the complete set of potential strategies in a quantified and repeatable way is considered of vital importance for trustworthy decision making in digital preservation planning.

This paper presents a preservation action monitoring infrastructure that combines provider-side service instrumentation and quality measurement of migration web services with remote access to emulation. Tools are monitored during execution, and both their runtime characteristics and the quality of their results are measured transparently. We present the architecture of the presented framework and discuss results from experiments on migration and emulation services.

1 Introduction

Digital library systems today rely on web services and employ distributed infrastructures for accomplishing increasingly complex tasks and providing shared access to distributed content. The amount and the heterogeneity of content that is contained in these digital libraries is increasing rapidly; the new wave of web archives being built around the world will further accelerate this development in the near future.

However, this content is inherently ephemeral, and without concrete measures it will not be accessible in the very near future, when the original hardware and software environments used to create and render the content are not available any more. Thus, digital preservation tools have become a major focus of research. The two principle approaches to preservation are migration and emulation. Migration transforms the representation of a digital object by converting it to a

M. Agosti et al. (Eds.): ECDL 2009, LNCS 5714, pp. 39–50, 2009.

different format that is regarded as safer, more stable and thus better suited for long-term access. Emulation simulates the original technical environment in which the object is known to function. The optimal treatment for a specific object depends on numerous factors ranging from technical aspects such as scalability, performance and platform independence, user needs to access objects, with respect to utilizing the object and familiarity with certain software environments to the concrete qualities of migration tools and the importance of authenticity and trust in these qualities. The selection process for one particular course of action is a key question of *preservation planning*.

To enable accountable decision making in preservation planning, we need to rely not on subjective judgement of the quality and suitability of a potential preservation action, but instead on repeatable and transparent measurements that can be considered a solid evidence base. These measurements can be on the one hand obtained through emerging community efforts such as the Planets Testbed[1], which is building up an infrastructure for repeatable tests and accumulating institution-independent evidence on preservation action services. This approach can provide a fundamental body of knowledge for digital preservation research and practice. However, it will usually not take into account the specific peculiarities of a planning scenario in an organisation. Thus, measurements specific to the situation at hand are used during the preservation planning procedure by applying potential actions to the specific digital objects at hand and measuring the outcomes. Service-oriented computing as means of arranging autonomous application components into loosely coupled networked services has become one of the primary computing paradigms of our decade. Web services as the leading technology in this field are widely used in increasingly distributed systems. Their flexibility and agility enable the integration of hetereogeneous systems across platforms through interoperable standards. However, the thus-created networks of dependencies also exhibit challenging problems of interdependency management. Some of the issues arising are service discovery and selection, or the question of service quality and trustworthiness of service providers. Of core interest is the problem of measuring quality-of-service (QoS) attributes and using them as means for guiding the selection of the optimal service for consumption at a given time and situation.

Migration as the conversion of content to a different representation is a straightforward in/out operation. It lends itself naturally for being made accessible as a web service at least for evaluation purpose when setting up all potential migration tools may be too costly or complex. (We focus primarily on evaluation phase. For production level deployment tools will usually be installed locally for data protection and performance reasons.) Emulation as the second major preservation strategy is not as readily convertible. While considerable work has been presented on services for migrating content, emulation is still mostly used in non-distributed environments. In many cases, the effort needed just to *evaluate* an emulation software is viewed as prohibitive due to the often complex setup procedures that are required for rendering a single file.

Even considering migration services, there is still a severe lack regarding the modelling and measurement of quality-of-service (QoS). When deciding which migration path to take and which tool to use in a repository, these quality aspects are of vital importance. Tool characteristics such as the processing time and memory needed to convert digital content to different formats are an important criterion guiding the evaluation of a migration tool; even more important is often the *quality* of transformation, i.e. a quantified answer to the question 'How much information was *lost* during this transformation?'.

To obtain answers to these questions of quality is a necessary precursor to the deployment of any tool on a digital repository system. The preservation planning process described in [12,4] relies on controlled experimentation to obtain measurements of the actual qualities of different preservation action tools. While it is possible to carry out these experiments on dedicated infrastructures, it is also very time-consuming and effort-intensive. Distributed services greatly speed up this process by providing action services that can be directly invoked inside the preservation planning application *Plato*[1].

However, measuring quality attributes of web services is inherently difficult due to the very virtues of service-oriented architectures: The late binding and flexible integration ideals ask for very loose coupling, which often implies that little is known about the actual quality of services and even less about the confidence that can be put into published service metadata, particularly QoS information. Ongoing monitoring of these quality attributes is a key enabler of reliable preservation services and a prerequisite for building confidence and trust in services to be consumed.

This paper presents a generic architecture and reference implementation for non-invasively measuring quality and performance of migration tools and instrumenting the corresponding web services on the provider side. We demonstrate how a service wrapper that makes a tool installed on a specific machine accessible as a web service can monitor the execution of this tool and provide additional benchmark information as metadata together with the actual transformation result. We investigate different profiling approaches for this purpose and analyse benefits and challenges. We further describe a web service engine consisting of a registry which facilitates service discovery, metadata schemas and interfaces, and services for migration and emulation, and demonstrate how these components can be integrated through a distributed service oriented preservation planning architecture. Remote emulation is integrated into this framework by providing transparent access to virtualised emulators on remote host machines through the standard web browser interface.

The rest of this paper is structured as follows. The next section outlines related work in the areas of web service QoS modelling, performance measurement, and distributed digital preservation services. Section 3 describes the overall architectural design, the remote emulation framework and the monitoring engines, while Section 4 analyses the results of practical applications of the implemented framework. Section 5 discusses implications and sets further directions.

---

[1] http://www.ifs.tuwien.ac.at/dp/plato

2   Related Work

The initially rather slow takeup of web service technology has been repeatedly attributed to the difficulties in evaluating the quality of services and the corresponding lack of confidence in the fulfillment of non-functional requirements. The lack of QoS attributes and their values is still one of the fundamental drawbacks of web service technology [11,10].

Web service selection and composition heavily relies on QoS computation [9,6]. A considerable amount of work has been dedicated to modelling QoS attributes and web service performance, and to ranking and selection algorithms. A second group of work is covering infrastructures for achieving trustworthiness, usually by extending existing description models for web services and introducing certification roles to the web service discovery models. Most of these approaches assume that QoS information is known and can be verified by the third-party certification instance. While this works well for static quality attributes, variable and dynamically changing attributes are hard to compute and subject to change. Platzer et al. [10] discuss four principle strategies for the continuous monitoring of web service quality: provider-side instrumentation, SOAP intermediaries, probing, and sniffing. They further separate performance into eight components such as network latency, processing and wrapping time on the server, and round-trip time. While they state the need for measuring all of these components, they focus on round-trip time and present a provider-independent bootstrapping framework for measuring performance-related QoS on the client-side.

Wickramage et al. analyse the factors that contribute to the total round trip time (RTT) of a web service request and arrive at 15 components that should ideally be measured separately to optimize bottlenecks. They focus on web service frameworks and propose a benchmark for this layer [13]. Woods presents a study on migration tools used for providing access to legacy data, analysing the question of migration-on-request versus a-priori migration from a performance perspective [14]. Clausen reports on experiments for semi-automated quality assurance of document conversions by comparing attributes such as page count and differences in rendered images of the contained pages [5].

To support the processes involved in digital preservation, current initiatives are increasingly relying on distributed service oriented architectures to handle the core tasks in a preservation system [8,7,3]. The core groups of services in such a digital preservation system are

- the identification of object types and formats,
- analysis and characterisation of digital objects,
- the transformation of content to different representations, and
- the validation of the authenticity of objects in different formats [2].

In the Planets preservation planning environment, planning decisions are taken following a systematic workflow supported by the web-based planning application *Plato* which serves as the frontend to a distributed architecture of preservation services [4]. The architecture enables flexible integration of migration services from different sources and service providers [3]. However, emulation so far had to be carried out manually in an external environment.

The work presented here builds on this architecture and takes two specific steps further. We present a migration and emulation engine for provider-side monitoring and QoS-aware provisioning of migration tools wrapped as web services. In this architecture, processing time and resource usage of migration tools are monitored transparently and non-invasively on a variety of different platforms. These runtime characteristics are combined with an assessment of migration quality using characterisation functionality of tools such as ImageMagick and emerging approaches such as the eXtensible Characterisation Languages (XCL) for comparing original and transformed objects. The quality information obtained thereby is provided to the client together with the migration result. In parallel, the engine enables users to render digital objects from their own environment transparently in their browser using the remote emulation framework GRATE.

3   A Preservation Action Monitoring Infrastructure

3.1   Measuring QoS in Web Services

Figure 1 shows a simplified abstraction of the core elements of the monitoring design and their relations. The key elements are **Services**, **Engines**, and **Evaluators**, which are all contained in a **Registry**. Each **Engine** specifies which aspects of a service it is able to measure in its **MeasurableProperties**. The property definition includes the scale and applicable metrics for a property, which are used for creating the corresponding **Measurements**.

Each **Engine** is deployed on a specific hardware **Environment** that shows a certain performance. This performance is captured by the score of a **Benchmark** which is a specific configuration of services and **Data**, aggregating measurements over these data to produce a representative *score* for an environment. The benchmark scores of the engines' environments are provided to the clients as part of the service execution metadata and can be used to normalise performance data of migration tools running on different hardware platforms.

The **Services** contained in a registry are not invoked directly, but run inside a monitoring engine to enable performance measurements. This monitoring accumulates **Experience** for each service, which is collected in each successive call to a service and used to aggregate information over time. It thus enables continuous monitoring of performance and migration quality.

CompositeEngines are a flexible form of aggregating measurements obtained in different monitoring environments. This type of engine dispatches the service execution dynamically to several engines to collect information. This is especially useful in cases where measuring code in real-time actually changes the behaviour of that code. For example, measuring the memory load of Java code in a profiler usually results in a much slower performance, so that simultaneous measurement of memory load *and* execution speed leads to skewed results. Fortunately, in this case there is a way around this uncertainty relation – forking and distributing the execution leads to correct results.

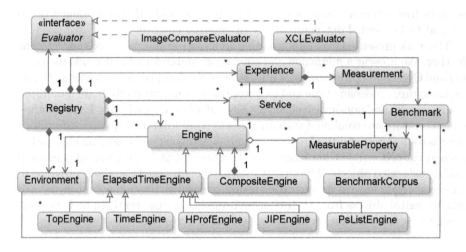

Fig. 1. Core elements of the monitoring framework

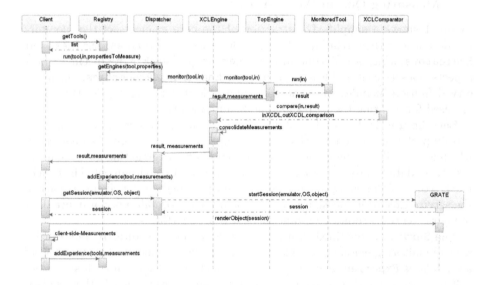

Fig. 2. Exemplary interaction in an evaluation scenario

The bottom of the diagram illustrates some of the currently deployed monitoring engines. Additional engines and composite engine configurations can be added dynamically at any time. The ElapsedTimeEngine is a simple default implementation measuring wall-clock time. Several engines have been implemented to support performance measurement of migration tools on different operating systems and platforms. The TopEngine is based on the Unix tool *top* and used for measuring memory load and execution speed of wrapped applications installed on the server. The TimeEngine uses the Unix call *time* to measure the

CPU time used by a process. On Windows servers, the `PsListEngine` relies on *PsList*[2]. Monitoring the performance of Java tools is accomplished in a platform-indepent manner by a combination of the `HProfEngine` and `JIPEngine`, which use the *HPROF*[3] and *JIP*[4] profiling libraries for measuring memory usage and timing characteristics, respectively.

In contrast to performance-oriented monitoring through engines, `Evaluators` are used for comparing input and output of migration tools to compute similarity measures and judge migration quality. The `ImageCompareEvaluator` relies on the *compare* utility of ImageMagick[5] to compute distance metrics for pairs of images. The more generic `XCLEvaluator` uses the eXtensible Characterisation Languages (XCL) which provide an abstract information model for digital content which is independent of the underlying file format [2], and compares different XCL documents for degrees of equality.

Figure 2 illustrates an exemplary simplified flow of interactions between service requesters, the registry, the engines, and the monitored tools, in the case of an engine measuring the execution of a migration tool through the Unix tool *top*. The dispatcher is responsible for selecting the appropriate engine for a given tool and set of properties to measure; it calls all configured evaluators after the successful migration. Both the dispatcher and the client can contribute to the accumulated experience of the registry. This allows the client to add round-trip information, which can be used to deduct network latencies, or quality measurements computed on the result of the consumed service. A single-use key prevents spamming of the experience base. The bottom of the sequence shows in a simplified form how an `EmulationService` connects the planning client to the remote emulation services provided by *GRATE*. The connector uploads the content to be rendered and provides the client with the obtained session key, i.e. a URL for gaining access to an emulator running on a remote machine via a web browser. Section 3.2 will explain this mechanism in detail.

3.2   GRATE: Global Remote Access to Emulation

GRATE is a webservice written in Java/PHP/Perl and JavaScript (Ajax) and allows for location-independent remote access to designated emulation-services over the Internet. Figure 3 shows a high-level overview of the main components of the distributed emulation service infrastructure. Not shown in this diagram are the planning tool and emulation connector services described above.

The GRATE client consists of two components, the GRATE Java applet and a Java Tight VNC client applet, embedded in PHP/JavaScript/Ajax code. Tight VNC is used for remote desktop access to the emulator[6]. Since Java applets are platform independent, every Java-enabled web-browser is suitable for running

---

[2] http://technet.microsoft.com/en-us/sysinternals/bb896682.aspx
[3] http://java.sun.com/developer/technicalArticles/Programming/HPROF.html
[4] http://jiprof.sourceforge.net/
[5] http://www.imagemagick.org/Usage/compare/
[6] http://www.tightvnc.com/

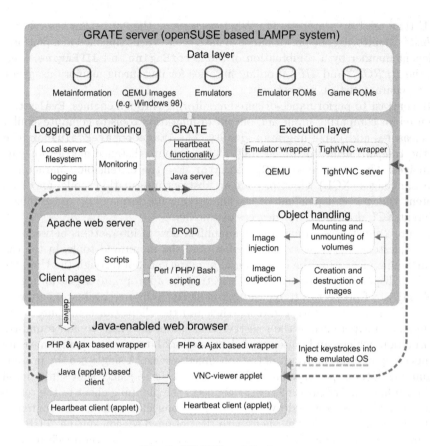

Fig. 3. GRATE architecture

the GRATE client. This client communicates with the GRATE server component, which is responsible for session management (establishing and terminating VNC sessions, executing emulators, delivering meta-information, etc.) as well as transporting uploaded digital objects into the emulated environments. GRATE is Java-based and therefore portable. It is currently running on a Linux-Apache-MySQL-Perl-PHP (LAMPP) system.

Client-server communication takes place via TCP; it is possible to input key commands into the browser, which are remotely injected into the running emulator. Injecting digital objects to be rendered is accomplished by mounting virtual drives containing the objects to be rendered. The actual emulated image, e.g. of a Windows 95 installation, then contains a listener which automatically opens the encountered object. Table 1 gives an overview of some of the formats currently supported by the Windows 98 images running on QEMU[7].

---

[7] Note that for some formats such as RAW, only specific camera profiles are supported, while EXE and DLL means that the contained applications can be displayed. For video formats, a number of codecs are currently installed, but not listed here.

Table 1. Formats supported by the Windows images currently deployed in GRATE

Video/Audio	Images	Documents
AIF, AU, SND, MED, MID, MP3, OGG, RA, WAV, ASF, AVI, MOV, MP4, MPG, MPEG, WMA, WMV	ANI, CUR, AWD, B3D, BMP, DIB, CAM, CLP, CPT, CRW/CR2, DCM/ACR/IMA, DCX, DDS, DJVU, IW44, DXF, DWG, HPGL, CGM, SVG, ECW, EMF, EPS, PS, PDF, EXR, FITS, FPX, FSH, G3, GIF, HDR, HDP, WDP, ICL, EXE, DLL, ICO, ICS, IFF, LBM, IMG, JP2, JPC, J2K, JPG, JPEG, JPM, KDC, LDF, LWF, Mac PICT, QTIF, MP4, MNG, JNG, MRC, MrSID, SID, DNG, EEF, NEF, MRW, ORF, RAF, DCR, SRF/ARW, PEF, X3F, NLM, NOL, NGG, PBM, PCD, PCX, PDF, PGM, PIC, PNG, PPM, PSD, PSP, PVR, RAS, SUN, RAW, YUV, RLE, SFF, SFW, SGI, RGB, SIF, SWF, FLV, TGA, TIF, TIFF, TTF, TXT, VTF, WAD, WAL, WBMP, WMF, WSQ, XBM, XPM	PDF, ODT, OTT, SXW, STW, DOC, DOCX, DOT, TXT, HTML, HTM, LWP, WPD, RTF, FODT, ODS, OTS, SXC, STC, XLS, XLW, XLT, CSV, ODP, OTP, SXI, STI, PPT, PPS, POT, SXD, ODG, OTG, SXD, STD, SGV

Table 2. Supported in/out formats of currently integrated migration tools

FLAC	LAME	OpenOffice.org	GIMP	ImageMagick	JavaIO
WAV, FLAC	WAV, MP3	PPT, SXI, SDX, ODP, ODG, ODT, SXW, DOC, RTF, TXT, ODS, SXC, XLS, SLK, CSV, ODG, SXD	BMP, JPG, PNG, TIFF	BMP, GIF, JPG, PNG, ICO, MPEG, PS, PDF, SVG, TIFF, TXT	BMP, GIF, JPG, PNG, TIFF

This combination of virtual machine allocation on a pre-configured server with remote access to the emulators reduces the total amount of time needed for evaluating a specific emulation strategy from many hours to a single click.

4   Experiment Results

We run a series of experiments to evaluate different aspects of both the tools and the engines themselves: We compare different performance measurement techniques; we demonstrate the accumulation of average experience on tool behaviour; and we compare performance and quality of image conversion tools.

Table 2 gives an overview of currently deployed migration tools and supported formats[8].

4.1   Measurement Techniques

The first set of experiments compares the exactness and appropriateness of performance measurements obtained using different techniques and compares these values to check for consistency of measurements. We monitor a Java conversion tool using all available engines on a Linux machine. The processing time measured by top, time, and the JIP profiler are generally very consistent, with an empirical correlation coefficient of 0.997 and 0.979, respectively. Running HProf on the same files consistently produces much longer execution times due to the processing overhead incurred by profiling the memory usage.

Figure 4(a) shows measured memory values for a random subset of the total files to visually illustrate the variations between the engines. The virtual memory assigned to a Java tool depends mostly on the settings used to execute the JVM and thus is not very meaningful. While the resident memory measured by Top includes the VM and denotes the amount of physical memory actually used

---

[8] To save space we list only the format's file suffix. For a detailed description of the supported format versions, please refer to the web page of the respective tool.

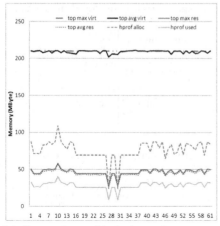

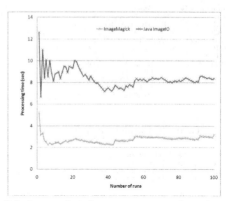

(a) Comparison of techniques for measuring memory

(b) Accumulated average processing time per MB

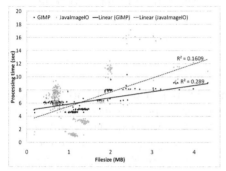

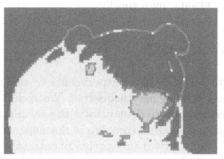

(c) Runtime behaviour of two conversion services

(d) Visualisation of an examplary conversion error

Fig. 4. Experiment results

during execution, HProf provides figures for memory used and allocated within the VM. Which of these measurements are of interest in a specific preservation planning scenario depends on the integration pattern. For Java systems, the actual memory within the machine will be relevant, whereas in other cases, the virtual machine overhead has to be taken into account as well.

When a tool is deployed as a service, a standard benchmark score is calculated for the server with the included sample data; furthermore, the monitoring engines report the average system load during service execution. This enables normalisation and comparison of a tool across server instances.

4.2   Accumulated Experience

An important aspect of any QoS management system is the accumulation and dissemination of experience on service quality. The described framework automatically tracks and accumulates all numeric measurements and provides aggregated averages with every service response. Figure 4(b) shows how processing time per MB quickly converges to a stable value during the initial bootstrapping sequence of service calls on benchmark content.

4.3   Tool Performance and Quality

Figure 4(c) shows the processing time of two conversion tools offered by the same service provider on 312 JPEG 1.01 image files. Simple linear regression shows the general trend of the performance relation, revealing that the Java tool is faster on smaller images but outperformed on larger images. It has to be noted that the *conversion quality* offered by GIMP is certainly higher. Figure 4(d) shows a visualisation of the conversion error introduced by the simple Java program when converting an image with transparent background from GIF to JPG. While changed pixels are shown in grey in the figure, the darker parts indicate the transparent layer that has been lost during migration. ImageMagick *compare* reports that 59.27% of the pixels are different in the migrated file, with an RMSE of 24034. In most cases, the information loss introduced by a faster tool will be considered much more important than its speed.

A comparison of the different evaluation results shows that the ImageMagick `compare` tool reports several errors when converting jpg to tiff with GIMP. The reason is that GIMP in the deployed configuration incorporates the EXIF orientation setting into the migrated picture, while ImageMagick fails to do so. While it is often possible to tweak the behaviour of tools through appropriate configuration, cases like this illustrate why there is a need for *independent* quality assurance in digital preservation: A tool must not judge its own migration quality, but instead be evaluated by an independent method. Current efforts such as the XCL languages strive to deliver this QA in the future.

5   Discussion and Conclusion

We described a preservation action monitoring infrastructure combining monitored migration web services with remote emulation. Quality measurements of migration services are transparently obtained through a flexible architecture of non-invasive monitoring engines, and the services are instrumented by the provider to deliver this quality information to the client. These quality measures include performance figures from engines as well as similarity measures obtained by evaluators by comparing input and result of migration tools. We demonstrated the performance monitoring of different categories of applications wrapped as web services and discussed different techniques and the results they yield.

The resulting measurements are of great value for transparent and reliable decision making in preservation planning. Together with the ease of access that is provided through the remote emulation system GRATE, the effort needed for planning the preservation of digital collections can be greatly reduced. Part of our current work is the introduction of flexible benchmark configurations that support the selection of specifically tailored benchmarks, e.g. to calculate scores for objects with certain characteristics; and the extension of measurements to arrive at a fully trackable set of measurements covering the entire range of factors that influence decision making in digital preservation.

Acknowledgements

Part of this work was supported by the European Union in the 6th Framework Program, IST, through the PLANETS project, contract 033789.

References

1. Aitken, B., Helwig, P., Jackson, A., Lindley, A., Nicchiarelli, E., Ross, S.: The Planets Testbed: Science for Digital Preservation. Code4Lib 1(5) (June 2008), http://journal.code4lib.org/articles/83
2. Becker, C., Rauber, A., Heydegger, V., Schnasse, J., Thaller, M.: Systematic characterisation of objects in digital preservation: The extensible characterisation languages. Journal of Universal Computer Science 14(18), 2936–2952 (2008)
3. Becker, C., Ferreira, M., Kraxner, M., Rauber, A., Baptista, A.A., Ramalho, J.C.: Distributed preservation services: Integrating planning and actions. In: Christensen-Dalsgaard, B., Castelli, D., Ammitzbøll Jurik, B., Lippincott, J. (eds.) ECDL 2008. LNCS, vol. 5173, pp. 25–36. Springer, Heidelberg (2008)
4. Becker, C., Kulovits, H., Rauber, A., Hofman, H.: Plato: A service oriented decision support system for preservation planning. In: Proc. JCDL 2008, Pittsburg, USA. ACM Press, New York (2008)
5. Clausen, L.R.: Opening schrödingers library: Semi-automatic QA reduces uncertainty in object transformation. In: Kovács, L., Fuhr, N., Meghini, C. (eds.) ECDL 2007. LNCS, vol. 4675, pp. 186–197. Springer, Heidelberg (2007)
6. Dustdar, S., Schreiner, W.: A survey on web services composition. International Journal of Web and Grid Services 1, 1–30 (2005)
7. Ferreira, M., Baptista, A.A., Ramalho, J.C.: An intelligent decision support system for digital preservation. International Journal on Digital Libraries 6(4), 295–304 (2007)
8. Hunter, J., Choudhury, S.: PANIC - an integrated approach to the preservation of complex digital objects using semantic web services. International Journal on Digital Libraries 6(2), 174–183 (2006)
9. Menascé, D.A.: QoS issues in web services. IEEE Internet Computing 6(6), 72–75 (2002)
10. Platzer, C., Rosenberg, F., Dustdar, S.: Enhancing Web Service Discovery and Monitoring with Quality of Service Information. In: Securing Web Services: Practical Usage of Standards and Specifications. Idea Publishing Inc. (2007)
11. Ran, S.: A model for web services discovery with QoS. SIGecom Exch. 4(1), 1–10 (2003)
12. Strodl, S., Becker, C., Neumayer, R., Rauber, A.: How to choose a digital preservation strategy: Evaluating a preservation planning procedure. In: Proc. JCDL 2007, pp. 29–38. ACM Press, New York (2007)
13. Wickramage, N., Weerawarana, S.: A benchmark for web service frameworks. In: 2005 IEEE International Conf. on Services Computing, vol. 1, pp. 233–240 (2005)
14. Woods, K., Brown, G.: Migration performance for legacy data access. International Journal of Digital Curation 3(2) (2008)

Functional Adaptivity for Digital Library Services in e-Infrastructures: The gCube Approach

Fabio Simeoni[1], Leonardo Candela[2], David Lievens[3],
Pasquale Pagano[2], and Manuele Simi[2]

[1] Department of Computer and Information Sciences, University of Strathclyde, Glasgow, UK
fabio.simeoni@cis.strath.ac.uk
[2] Istituto di Scienza e Tecnologie dell'Informazione "Alessandro Faedo", CNR, Pisa, Italy
{leonardo.candela,pasquale.pagano,manuele.simi}@isti.cnr.it
[3] Department of Computer Science, Trinity College, Dublin 2, Ireland
david.lievens@cs.tcd.ie

Abstract. We consider the problem of e-Infrastructures that wish to reconcile the generality of their services with the bespoke requirements of diverse user communities. We motivate the requirement of *functional adaptivity* in the context of *gCube*, a service-based system that integrates Grid and Digital Library technologies to deploy, operate, and monitor *Virtual Research Environments* defined over infrastructural resources.

We argue that adaptivity requires mapping service interfaces onto multiple implementations, truly alternative interpretations of the same functionality. We then analyse two design solutions in which the alternative implementations are, respectively, full-fledged services and local components of a single service. We associate the latter with lower development costs and increased binding flexibility, and outline a strategy to deploy them dynamically as the payload of *service plugins*. The result is an infrastructure in which services exhibit multiple behaviours, know how to select the most appropriate behaviour, and can seamlessly learn new behaviours.

1 Introduction

gCube[1] is a distributed system for the operation of large-scale scientific infrastructures. It has been designed from the ground up to support the full lifecycle of modern scientific enquiry, with particular emphasis on application-level requirements of information and knowledge management. To this end, it interfaces pan-European Grid middleware for shared access to high-end computational and storage resources [1], but complements it with a rich array of services that collate, describe, annotate, merge, transform, index, search, and present information for a variety of multidisciplinary and international communities. Services, information, and machines are infrastructural resources that communities select, share, and consume in the scope of collaborative *Virtual Research Environments* (VREs).

To gCube, VREs are service-based applications to dynamically deploy and monitor within the infrastructure. To the users, they are self-managing, distributed Digital

---

[1] http://www.gcube-system.org

M. Agosti et al. (Eds.): ECDL 2009, LNCS 5714, pp. 51–62, 2009.

Libraries that can be declaratively defined and configured, for arbitrary purposes and arbitrary lifetimes. In particular, gCube is perceived as a *Digital Library Management System* (DLMS) [5], albeit one that is defined over a pool of infrastructural resources, that operates under the supervision of personnel dedicated to the infrastructure, and that is built according to Grid principles of controlled resource sharing [8]. The ancestry of gCube is a pioneering service-based DLMS [7] and its evolution towards Grid technologies took place in the testbed infrastructure of the Diligent project [4]. Five years after its inception, gCube is the control system of D4Science, a production-level infrastructure for scientific communities affiliated with the broad disciplines of Environmental Monitoring and Fishery and Aquaculture Resources Management[2].

The infrastructural approach to DLMS design is novel and raises serious challenges, both organisational and technical. Among the latter, we notice core requirements of *dynamic service management* and *extensive development support*. The first is the very premise of the approach; a system that does not transparently manage its own services it is not a service-based DLMS. The second requirement constrains how one plans to accommodate the first; a DLMS that achieves transparencies for users but denies them to its developers is prone to error, is hard to maintain, and has little scope for evolution.

In this paper, we comment briefly on service management and development complexity, overviewing relevant parts of the gCube architecture in the process (cf. Figure 1). Our goal is not to present the solutions adopted in gCube, for which we refer to existing literature. Rather, we wish to build enough context to discuss in more detail a third requirement: the ability of the system to reconcile the generality of its services with the specific demands of its communities of adoption.

Functional adaptivity is key to the usefulness of the system; a DLMS that does not serve a wide range of user communities, and does not serve each of them well, is simply not adopted. In particular, we argue that functional adaptivity requires services that: (*i*) can simultaneously manage multiple implementations of their own interface; (*ii*) can autonomically match requests against available implementations; and most noticeably: (*iii*) can acquire new implementations at runtime. Effectively, we advocate the need for services that exhibit multiple behaviours, know how select the most appropriate behaviour, and can *learn* new behaviours.

The rest of the paper is organised as follows. We set the context for the discussion on functional adaptivity in Section 2 and show why a general solution requires multiple implementations of service interfaces in Section 3. We motivate our choice to accommodate this multiplicity *within* individual services in Section 4, and then illustrate our deployment strategy in Section 5. Finally, we draw some conclusions in Section 6.

2 Context

In a DLMS that commits to service-orientation, managing resources is tantamount to managing services that virtualise or manipulate resources. In such system, services do not run in pre-defined locations and are not managed by local administrators. Rather, they are resources of the infrastructure and they are dynamically deployed and redeployed by the system that controls it, where and when it proves most convenient. As an

---

[2] http://www.d4science.eu

implication of dynamic deployment, the configuration, staging, scoping, monitoring, orchestration, and secure operation of services become also dynamic and a responsibility of the system. Essentially, the baseline requirement is for a state-of-the-art *autonomic* system.

In gCube, the challenge of autonomicity is met primarily in a layer of *Core Services* that are largely independent from the DL domain and include:

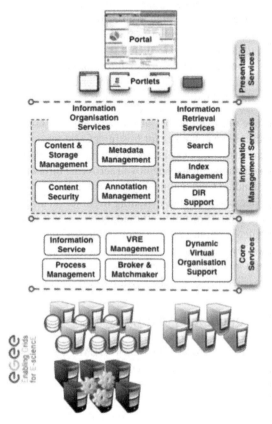

Fig. 1. The gCube Architecture

- *Process Management Services* execute declaratively specified workflows of service invocations, distributing the optimisation, monitoring, and execution of individual steps across the infrastructure [6];
- *Security Services* enforce authentication and authorisation policies, building over lower-level technologies to renew and delegate credentials of both users and services;
- *VRE Management Services* host service implementations and translate interactive VRE definitions into declarative specifications for their deployment and runtime maintenance [2];
- *Brokering and Matchmaking Services* inform deployment strategies based on information about the available resources that a network of *Information Services* gathers and publishes within the infrastructure.

The need for autonomicity bears on a second problem, the complexity of service development and the impact of the associated costs on the scope and evolution of the system. A system that aspires to manage its own services needs to raise non-trivial requirements against their runtime behaviour, and thus on how this behaviour is implemented. This adds to the complexity already associated with service development, whether generically related to distributed programming (e.g. concurrency, performance-awareness, tolerance to partial failure) or specifically introduced by open technologies (e.g. reliance upon multiple standards, limited integration between development tools, inadequate documentation). At highest risk here are the services of a second layer of the gCube architecture, the *Information Management Services* that implement DL functionality, including:

- a stack of *Information Organisation Services* rooted in a unifying information model of binary relationships laid upon storage replication and distribution services. Services higher up in the stack specialise the semantics of relationships to model compound information objects with multiple metadata and annotations, and to group such objects into typed collections.
- a runtime framework of *Information Retrieval Services* that execute and optimise structured and unstructured queries over a federation of forward, geo-spatial, or inverted indices of dynamically selected collections [13].

gCube offers a number of tools to tame the complexity of service development, most noticeably a container for hosting gCube services and an application framework to implement them. These are the key components of *gCore*[3], a minimal distribution of gCube which is ubiquitously deployed across the infrastructure [9]. Built as an ad-hoc extension of standard Grid technology[4], gCore hides or greatly simplifies the systemic aspects of service development, including lifetime, security, scope, and state management; at the same time, it promotes the adoption of best practices in multiprogramming and distributed programming. This guarantees a qualitative baseline for gCube services and allows developers to concentrate on domain semantics. Equally, it allows changes related to the maintenance, enhancement, and evolution of the system to sweep transparently across its services at the rhythm of release cycles.

3   Functional Adaptivity

The infrastructural approach to DLMS design emphasises the generality of services, i.e. the need to serve user communities that operate in different domains and raise different modelling and processing requirements. Communities, on the other hand, expect the infrastructure to match the specificity of those requirements and are dissatisfied with common denominator solutions. Accordingly, the DLMS needs to offer generic DL services that can adapt to bespoke requirements. In gCube, we speak of the functional adaptivity of services and see it as an important implication of the autonomicity that is expected from the system.

Abstraction, parameterisation, and composition are the standard design principles in this context. One chooses generic data models to represent arbitrary relationships and generic formats to exchange and store arbitrary models. One then defines parametric operations to customise processes over the exchanged models and provides generic mechanisms to compose customised processes into bespoke workflows. In gCube, content models, format standards, service interfaces, and process management services follow precisely these design principles (cf. Section 1).

Often, however, we need novel algorithmic behaviour. In some cases we can specify it declaratively, and then ask some service to execute it. Data transformations and distributed processes, for example, are handled in this manner in gCube, through services

---

[3] http://wiki.gcore.research-infrastructures.eu
[4] http://www.globus.org

that virtualise XSLT and WS-BPEL engines, respectively. In other cases, parameterisation and composition do not take us far enough.

Consider for example the case of the *DIR Master*, a gCube service used to optimise the evaluation of content-based queries across multiple collections [12]. The service identifies collections that appear to be the most promising candidates for the evaluation of a given query, typically based on the likelihood that their content will prove relevant to the underlying information need (*collection selection*). It also integrates the partial results obtained by evaluating the queries against selected collections, thus reconciling relevance judgements based on content statistics that are inherently local to each collection (*result fusion*). For both purposes, the service may derive and maintain summary descriptions of the target collections (*collection description*). Collectively, the tasks of collection description, collection selection, and result fusion identify the research field of (content-based) *Distributed Information Retrieval* (DIR) [3].

Over the last fifteen years, the DIR field has produced a rich body of techniques to improve the effectiveness and efficiency of distributed retrieval, often under different assumptions on the context of application. Approaches diverge most noticeably in the degree of cooperation that can be expected between the parties that manage the collections and those that distribute searches over them. With cooperation, one can guarantee efficient, timely, and accurate gathering of collection descriptions; similarly, selection and merging algorithms can be made as effective as they would be if the distributed content was centralised. Without cooperation, collection descriptions are approximate and require expensive content sampling and size estimation techniques; as a result, collection selection and fusion are based on heuristics and their performance may fluctuate across queries and sets of collections.

In gCube, we would like to cater for both scenarios, and ideally be able to accommodate any strategy that may be of interest within each scenario. However, DIR strategies may diverge substantially in terms of inputs and in the processes triggered by those inputs. As an example, cooperative fusion strategies expect results to carry content statistics whereas uncooperative strategies may expect only a locally computed score (if they expect anything at all). Similarly, some collection description strategies may consume only local information and resources (e.g. results of past queries); others may require interactions with other services of the infrastructure, and then further diverge as to the services they interact with (e.g. extract content statistics from indices or perform query-based sampling of external search engines).

It is unclear how these strategies could be declaratively specified and submitted to a generic engine for execution. In this and similar contexts within the infrastructure, functional adaptivity seems to call for the deployment of multiple implementations, a set of truly alternative interpretations of DIR functionality. As we would like to preserve uniform discovery and use of the DIR functionality, we also require that the alternative implementations expose the same DIR Master interface.

4   The Design Space

Once we admit multiple implementations of the same interface, we need to decide on the nature of the implementations and on the strategy for their deployment. There are at least two desiderata in this respect. Firstly, we would like to minimise the cost of developing multiple implementations, ideally to the point that it would not be unreasonable to leverage expertise available within the communities of adoption. Secondly, we would like to add new implementations dynamically, without interrupting service provision in a production infrastructure. The convergence of these two goals would yield an infrastructure that is open to third party enhancements and is thus more sustainable.

4.1 Adaptivity with Multiple Services

The obvious strategy in a service-oriented architecture is to identify multiple implementations with multiple services, e.g. allow many 'concrete' services to implement an 'abstract' DIR Master interface. gCube would standardise the interface and provide one or more concrete services for it; the services would publish a description of their distinguishing features within the infrastructure; clients would dynamically discover and bind to services by specifying the interface and the features they require. The approach is appealing, for gCube already supports the dynamic deployment of services and gCore cuts down the cost of developing new ones.

There are complications, however. Firstly, full-blown service development is overkill when implementations diverge mildly and in part. Many DIR strategies, for example, share the same approach to collection description but diverge in how they use descriptions to select among collections and to merge query results. Server-side libraries can reduce the problem by promoting reuse across service implementations [11]. Yet, service development remains dominated by configuration and building costs that may be unduly replicated.

Secondly, multiple services must be independently staged. This is inefficient when we wish to apply their strategies to the same state and becomes downright problematic if the services can change the state and make its replicas pair-wise inconsistent.

Thirdly, we would like the system to discover and select implementations on behalf of clients. This is because they make no explicit choice, or else because their choice is cast in sufficiently abstract terms that some intelligence within the system can hope to resolve it automatically. In the spirit of autonomicity, the requirement is then for some form of *matchmaking* of implementations within the infrastructure.

In its most general form, matchmaking concerns the dynamic resolution of processing requirements against pools of computational resources. Grid-based infrastructures use it primarily for hardware resources, i.e. to allocate clusters of storage and processors for high-performance and high-throughput applications; in gCube, we employ it routinely to find the best machines for the deployment of given services (cf. Section 2).

Matchmaking of software resources is also possible. There is an active area of research in *Semantic Web Services* that relies upon it to flexibly bind clients to services, most often in the context of service orchestration. The assumption here is that (*i*) clients will associate non-functional, structural, and behavioural requirements with their requests, and (*ii*) distinguished services within the system will resolve their requirements against service descriptions available within the infrastructure [10,15].

In practice, however, the scope of application for service-based matchmaking remains unclear. We would like to use it for arbitrary interactions between services, not only in the context of process execution. We would also like to use it as late as possible, based on evidence available at call time rather than mostly-static service descriptions. Its impact on the operation of the infrastructure and the design of the system that controls raise also some concerns. The ubiquitous mediation of a remote matchmaker at runtime is likely to increase latencies; similarly, the necessity to distribute its load would complicate considerably its design and model of use. Finally, just-in-time descriptions seem costly to produce and publish, and the expressiveness of these descriptions is likely to introduce novel standardisation costs. Service-based matchmaking remains an interesting option for gCube, but the invasiveness of the approach and the current state of the art does not encourage us to deploy it in a production infrastructure.

4.2   Adaptivity with Multiple Processors

We have explored another strategy for functional adaptivity which promises a smoother integration with the current system. The underlying assumption is different: the diverse implementations of a given functionality are mapped onto multiple components of a single service, rather than onto multiple services. These components are thus local *processors* that compete for the resolution of clients requests on the basis of functional and non-functional requirements.

From a service perspective, the strategy may be summarised as follows (cf. Figure 2). A distinguished *gatekeeper* component receives requests and passes them to a local *matchmaker* as implicit evidence of client requirements. The matchmaker cross-references this evidence with information about the available processors, as well as with any other local information that might bear on the identification of suitable processors. As there might be many such processors for any given

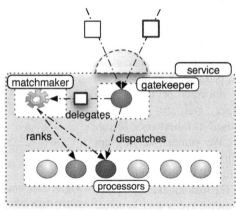

Fig. 2. Matchmaking requests and processors

request, the matchmaker ranks them all from the most suitable to the least suitable, based on some quantification of its judgement. It then returns the ranking to the gatekeeper. The gatekeeper can then apply any selection and dispatch strategy that is compatible with the semantics of the service: it may dispatch the request to the processor with the highest rank, perhaps only if scored above a certain threshold; it may broadcast it to all processors, it may dispatch it to the first processor that does not fail, and so on.

Placing multiple implementations within the runtime of a single service is unconventional but yields a number of advantages. The definition of processors is framed within a model of local development based on the instantiation of service-specific application frameworks. The model is standard and incremental with respect to overall service behaviour; compared with full-blown service development, it promises lower costs on average, and thus lends itself more easily to third-party adoption. Problems of staging and state synchronisation across multiple services are also alleviated, as the application of multiple strategies over the same state can occur in the context of a single service. Finally, matchmaking has now a local impact within the infrastructure and can easily exploit evidence available at call-time, starting from the request itself. In particular, it can be introduced on a per-service basis, it does no longer require the definition, movement, and standardisation of service descriptions, and it does not raise latencies or load distribution issues. Overall, the requirement of functional adaptivity is accommodated across individual services, where and when needed; as mentioned in Section 2, the infrastructure becomes more autonomic because its individual services do. Indeed, the strategy goes beyond the purposes of functional adaptivity; the possibility of multi-faceted behaviours enables services to adapt functionally (to requests) but also non-functionally (e.g. to processor's failures, by dispatching to the next processor in the ranking).

As to the matchmaking logic, the design space is virtually unbound. In previous work, we have presented a matchmaker that chooses processors based on the *specificity* with which they can process a given request [14]. In particular, we have considered the common case in which processors specialise the input domains of the service interface, and are themselves organised in inheritance hierarchies. We summarise this approach to matchmaking here and refer to [14] for the technical details. The matchmaker compares the runtime types that annotate the graph structure of request in-

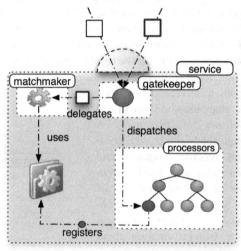

Fig. 3. Matchmaking for hierarchies of processors of increased specificity

puts and those that annotate the graph structures of *prototypical* examples of the inputs expected by the available processors. A processor matches a request if the analysis concludes that, at each node, the runtime type of the input is a subtype of the runtime type of the prototype (cf. Figure 3). The analysis is quantified in terms of the distance between matching types along the subtype hierarchy, though the combination of distances is susceptible of various interpretations and so can yield different metrics; in [14], we show a number of metrics, all of which can be injected into the matchmaking logic as configuration of the matchmaker.

5   Service Plugins

Services that are designed to handle multiple processors can be seamlessly extended by injecting new processors *in their runtime*. This challenges another conventional expectation, that the functionality of the services ought to remain constant after their deployment; not only can services exhibit multiple behaviours, their behaviours can also grow over time. This amplifies the capabilities of both the service — which can serve in multiple application scenarios — and the infrastructure — which maximises resources exploitation.

To achieve this in an optimal manner, we need to look beyond the design boundary of the single service, and consider the role of the infrastructure in supporting the development, publication, discovery, and deployment of *plugins* of service-specific processors. In this Section, we overview the high-level steps of a strategy whereby service plugins become a new kind of *resource* that the infrastructure makes available for the definition of Virtual Research Environments (VREs).

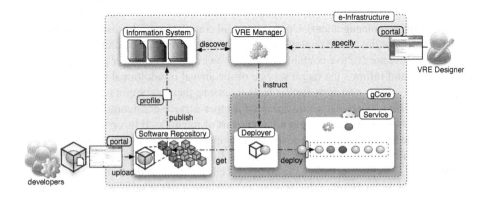

Fig. 4. Plugins publication, discovery and deployment

With reference to Figure 4, the lifecycle of a plugin begins with development and packaging phases. Development is supported by programming abstractions included in gCore and is framed by service-specific libraries (e.g. define service-specific plugin and processor interfaces). Packaging is dictated by infrastructure requirements; plugin code and dependencies are collected in a *software archive* that includes all the metadata needed by the Core Services to manage later phases of the plugin's lifetime. Like for services, in particular, the software archive includes a *profile* that identifies the plugin and its target service, and that describes its logical and physical dependencies.

The plugin enters the infrastructure when it is uploaded to the *Software Repository*, a distinguished Core Service that hosts all the software resources available in the infrastructure. The Software Repository verifies the completeness and correctness of the software archive and then publishes the plugin profile with the *Information Services*, so as to inform the rest of the infrastructure of the availability of the new resource.

The deployment of the plugin may then occurs asynchronously, in the context of the definition or update of a VRE. The process originates in the actions of a distinguished user, the *VRE Designer*, that interacts with a portal to select the content and functionality required for the VRE. gCube translates the human choice into declarative VRE specifications and feds them to dedicated Core Services for the synthesis of a deployment plan. The plan includes the identification of the software, data, and hardware resources that, collectively, guarantee the delivery of the expected functionality at the expected quality of service [2]. In this context, the plugin is additional functionality to directly present to the VRE Designer or to rely upon when translating the choice of a higher-level functionality into a deployment plan (e.g. the functionality can be implemented by deploying the service augmented with the plugin).

The actuation of the deployment plan is responsibility of another Core Service, the *VRE Manager*. The VRE Manager may inform services that already operate in the scope of existing VREs that their functionality ought to be made available in a new scope. Equally, it may instruct the deployment of new software archives on selected nodes. The instructions are carried out by the *Deployer* service, a Core Service that is distributed with gCore to deploy and undeploy software locally to any node of the infrastructure. The Deployer obtains the archives from the Software Repository and subjects them to a number of tests to ascertain their compatibility with the local environment (i.e. dependency and version management). Upon successful completion of the tests, the Deployer unpackages the archives and deploys their content in the running container.

When a software archive contains a plugin, however, the Deployer performs an additional task and informs the target service of the arrival of additional functionality. In response, the service performs further tests to ensure that the plugin has the expected structure and then stores its profile along with other service configuration, so as to re-establish the existence of the plugin after a restart of the node. If this process of activation is successful, the processors inside the plugin are registered with the matchmaker and become immediately available for the resolution of future requests. The service profile itself is modified to reflect the additional capabilities for the benefit of clients.

Crucially to the viability of the strategy, the behaviour of the service in the exchange protocol with the Deployer is largely pre-defined in gCore. The developer needs to act only upon plugin activation, so as to register the processor the plugin with its matchmaker of choice. Further, the type-based matchmaker summarised in Section 4.2 is also pre-defined in gCore, though service-specific matchmakers defined anew or by customisation of pre-defined ones can be easily injected within the protocol for plugin activation.

6   Conclusions

e-Infrastructures that provide application services for a range of user communities cannot ignore the diversity and specificity of their functional requirements. Yet, the ability to functionally adapt to those requirements, as dynamically and cost-effectively as possible, has received little attention so far.

We have encountered the problem in gCube, a system that enables Virtual Research Environments over infrastructural resources with the functionality and transparencies of conventional Digital Library Management Systems. In this context, adaptivity requires the coexistence of multiple implementations of key DL functionality, the possibility to add new implementations on-demand to a production infrastructure, and enough intelligence to automate the selection of implementations on a per-request basis.

In this paper, we have argued that such requirements can be conveniently met by embedding multiple implementations within individual services, i.e. by making them polymorphic in their run-time behaviour. In particular, we have shown how mechanisms and tools already available in gCube can be employed to, respectively, reduce the cost of developing such services and enable their extensibility at runtime.

Acknowledgments. This work is partially funded by the European Commission in the context of the D4Science project, under the 1st call of FP7 IST priority, and by a grant from Science Foundation Ireland (SFI).

References

1. Appleton, O., Jones, B., Kranzmuller, D., Laure, E.: The EGEE-II project: Evolution towards a permanent european grid inititative. In: Grandinetti, L. (ed.) High Performance Computing (HPC) and Grids in Action. Advances in Parallel Computing, vol. 16, pp. 424–435. IOS Press, Amsterdam (2008)
2. Assante, M., Candela, L., Castelli, D., Frosini, L., Lelii, L., Manghi, P., Manzi, A., Pagano, P., Simi, M.: An extensible virtual digital libraries generator. In: Christensen-Dalsgaard, B., Castelli, D., Ammitzbøll Jurik, B., Lippincott, J. (eds.) ECDL 2008. LNCS, vol. 5173, pp. 122–134. Springer, Heidelberg (2008)
3. Callan, J.: Distributed Information Retrieval. In: Advances in Information Retrieval, pp. 127–150. Kluwer Academic Publishers, Dordrecht (2000)
4. Candela, L., Akal, F., Avancini, H., Castelli, D., Fusco, L., Guidetti, V., Langguth, C., Manzi, A., Pagano, P., Schuldt, H., Simi, M., Springmann, M., Voicu, L.: Diligent: integrating digital library and grid technologies for a new earth observation research infrastructure. Int. J. Digit. Libr. 7(1), 59–80 (2007)
5. Candela, L., Castelli, D., Ferro, N., Ioannidis, Y., Koutrika, G., Meghini, C., Pagano, P., Ross, S., Soergel, D., Agosti, M., Dobreva, M., Katifori, V., Schuldt, H.: The DELOS Digital Library Reference Model - Foundations for Digital Libraries. In: DELOS: a Network of Excellence on Digital Libraries (February 2008) ISSN 1818-8044 ISBN 2-912335-37-X
6. Candela, L., Castelli, D., Langguth, C., Pagano, P., Schuldt, H., Simi, M., Voicu, L.: On-Demand Service Deployment and Process Support in e-Science DLs: the DILIGENT Experience. In: ECDL Workshop DLSci06: Digital Library Goes e-Science, pp. 37–51 (2006)
7. Castelli, D., Pagano, P.: OpenDLib: A digital library service system. In: Agosti, M., Thanos, C. (eds.) ECDL 2002. LNCS, vol. 2458, pp. 292–308. Springer, Heidelberg (2002)
8. Foster, I., Kesselman, C.: The Grid 2: Blueprint for a new computing infrastructure, 2nd edn. Morgan Kaufmann Publishers, San Francisco (2004)
9. Pagano, P., Simeoni, F., Simi, M., Candela, L.: Taming development complexity in service-oriented e-infrastructures: the gcore application framework and distribution for gcube. Zero-In e-Infrastructure News Magazine 1 (to appear, 2009)
10. Paolucci, M., Kawamura, T., Payne, T.R., Sycara, K.P.: Semantic matching of web services capabilities. In: Horrocks, I., Hendler, J. (eds.) ISWC 2002. LNCS, vol. 2342, pp. 333–347. Springer, Heidelberg (2002)
11. Simeoni, F., Azzopardi, L., Crestani, F.: An application framework for distributed information retrieval. In: Sugimoto, S., Hunter, J., Rauber, A., Morishima, A. (eds.) ICADL 2006. LNCS, vol. 4312, pp. 192–201. Springer, Heidelberg (2006)
12. Simeoni, F., Bierig, R., Crestani, F.: The DILIGENT Framework for Distributed Information Retrieval. In: Kraaij, W., de Vries, A.P., Clarke, C.L.A., Fuhr, N., Kando, N. (eds.) SIGIR, pp. 781–782. ACM Press, New York (2007)

13. Simeoni, F., Candela, L., Kakaletris, G., Sibeko, M., Pagano, P., Papanikos, G., Polydoras, P., Ioannidis, Y.E., Aarvaag, D., Crestani, F.: A Grid-based Infrastructure for Distributed Retrieval. In: Kovács, L., Fuhr, N., Meghini, C. (eds.) ECDL 2007. LNCS, vol. 4675, pp. 161–173. Springer, Heidelberg (2007)
14. Simeoni, F., Lievens, D.: Matchmaking for Covariant Hierarchies. In: ACP4IS 2009: Proceedings of the 8th workshop on Aspects, components, and patterns for infrastructure software, pp. 13–18. ACM Press, New York (2009)
15. Sycara, K.P.: Dynamic Discovery, Invocation and Composition of Semantic Web Services. In: Vouros, G.A., Panayiotopoulos, T. (eds.) SETN 2004. LNCS (LNAI), vol. 3025, pp. 3–12. Springer, Heidelberg (2004)

Managing the Knowledge Creation Process of Large-Scale Evaluation Campaigns

Marco Dussin and Nicola Ferro

University of Padua, Italy
{dussinma,ferro}@dei.unipd.it

Abstract. This paper discusses the evolution of large-scale evaluation campaigns and the corresponding evaluation infrastructures needed to carry them out. We present the next challenges for these initiatives and show how digital library systems can play a relevant role in supporting the research conducted in these fora by acting as virtual research environments.

1 Introduction

Large-scale evaluation initiatives provide a significant contribution to the building of strong research communities, advancement in research and state-of-the-art, and industrial innovation in a given domain. Relevant and long-lived examples from the Information Retrieval (IR) field are the Text REtrieval Conference (TREC)[1] in the United States, the Cross-Language Evaluation Forum (CLEF)[2] in Europe, and the NII-NACSIS Test Collection for IR Systems (NTCIR)[3] in Japan and Asia. Moreover, new initiatives are growing to support emerging communities and address specific issues, such as the Forum for Information Retrieval and Evaluation (FIRE)[4] in India.

These initiatives impact not only the IR field itself but also related fields which adopt and apply results from it, such as the Digital Library (DL) one. Indeed, the information access and extraction components of a DL system, which index, search and retrieve documents in response to a user's query, rely on methods and techniques taken from the IR field. In this context, large-scale evaluation campaigns provide qualitative and quantitative evidence over the years as to which methods give the best results in certain key areas, such as indexing techniques, relevance feedback, multilingual querying, and results merging, and contribute to the overall problem of evaluating a DL system [14].

This paper presents a perspective on the evolution of large-scale evaluation campaigns and their infrastructures and the challenges that they will have to face in the future. The discussion will provide the basis to show how these emerging challenges call for an appropriate consideration and management of the

---

[1] http://trec.nist.gov/
[2] http://www.clef-campaign.org/
[3] http://research.nii.ac.jp/ntcir/
[4] http://www.isical.ac.in/~clia/

M. Agosti et al. (Eds.): ECDL 2009, LNCS 5714, pp. 63–74, 2009.

knowledge creation process involved by these initiatives, and how DL systems can play an important role in the evolution of large-scale evaluation campaigns and their infrastructures by acting as virtual research environments.

The paper is organized as follows: Section 2 summarizes the evolution of large-scale evaluation campaigns, the challenges for their future and our vision of the extension to the current evaluation methodology to address these challenges; Sections 3 and 4 discuss the DIKW hierarchy as a means of modeling the knowledge creation process of an evaluation campaign; Section 5 presents the DIRECT digital library system to show how the previously introduced concepts can be applied; finally, Section 6 draws some conlusions.

2   Evolution of Large-Scale Evaluation Campaigns and Infrastructures

Large-scale evaluation campaigns have been a driver of research and innovation in IR since the early 90s, when TREC was launched [15]. They have been relying mainly on the traditional Cranfield methodology [6], which focuses on creating comparable experiments and evaluating their performance. During their life span, large-scale evaluation campaigns have produced a great amount of research not only on specific IR issues – such has indexing schemes, weighting functions, retrieval models, and so on – but also on improving the evaluation methodology itself, for example, with respect to the effective and efficient creation of reliable and re-usable test collections, the proposal and study of appropriate metrics for assessing a task, or the application of suitable statistical techniques to validate and compare the results.

As part of recent efforts to shape the future of large-scale evaluation campaigns [3,12], more attention has been paid to evaluation infrastructures, meant as the information management systems that have to take care of the different steps and outcomes of an evaluation campaign. The need for appropriate evaluation infrastructures which allows for better management and exploitation of the experimental results has been highlighted also by different organizations, such the European Commission in the i2010 Digital Library Initiative [10], the US National Scientific Board [16], and the Australian Working Group on Data for Science [19].

In this context, we have proposed an extension to the traditional evaluation methodology in order to explicitly take into consideration and model the valuable scientific data produced during an evaluation campaign [2,5], the creation of which is often expensive and not easily reproducible. Indeed, researchers not only benefit from having comparable experiments and a reliable assessment of their performances, but they also take advantage of the possibility of having an integrated vision of the scientific data produced, together with their analyses and interpretations, as well as benefiting from the possibility of keeping, re-using, preserving, and curating them. Moreover, the way in which experimental results are managed, made accessible, exchanged, visualized, interpreted, enriched and referenced is therefore an integral part of the process of knowledge transfer and sharing towards relevant application communities, such as the DL one, which needs to properly understand these experimental results in order to create and assess their own systems.

Therefore, we have undertaken the design of an evaluation infrastructure for large-scale evaluation campaigns and we have chosen to rely on DL systems in order to develop it, since they offer content management, access, curation, and enrichment functionalities. The outcome is a DL system, called Distributed Information Retrieval Evaluation Campaign Tool (DIRECT)[5], which manages the scientific data produced during a large-scale evaluation campaign, as well as supports the archiving, access, citation, dissemination, and sharing of the experimental results [7,8,9]. DIRECT has been used, developed and tested in the course of the annual CLEF campaign since 2005.

2.1   Upcoming Challenges for Large-Scale Evaluation Campaigns

Since large-scale evaluation campaign began, the associated technologies, services and users of information access systems have been in continual evolution, with many new factors and trends influencing the field. For example, the growth of the Internet has been exponential with respect to the number of users and languages used regularly for global information dissemination. With the advance of broadband access and the evolution of both wired and wireless connection modes, users are now not only information consumers, but also information producers: creating their own content and augmenting existing material through annotations (e.g. adding tags and comments) and cross-referencing (e.g. adding links) within a dynamic and collaborative information space. The expectations and habits of users are constantly changing, together with the ways in which they interact with content and services, often creating new and original ways of exploiting them. Moreover, users need to be able to co-operate and communicate in a way that crosses language boundaries and goes beyond simple translation from one language to another. Indeed, language barriers are no more perceived simply as an "obstacle" to retrieval of relevant information resources, they also represent a challenge for the whole communication process (i.e. information access and exchange). This constantly evolving scenario poses new challenges to the research community which must react to these new trends and emerging needs.

From a glance at Figure 1, it can be noted that large-scale evaluation campaigns initially assumed a user model reflecting a simple information seeking behavior: the retrieval of a list of relevant items in response to a single query that could then be used for further consultation in various languages and media types. This simple scenario of user interaction has allowed researchers to focus their attention on studying core technical issues for information access systems and associated components. If we are to continue advancing the state-of-the-art in information access technologies, we need to understand a new breed of users, performing different kinds of tasks within varying domains, often acting within communities to find and produce information not only for themselves, but also

---

[5] http://direct.dei.unipd.it/

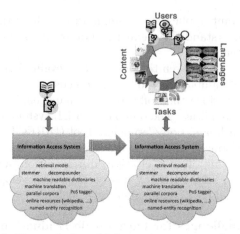

Fig. 1. The evolution of information access technologies

to share with other users. To this end, we must study the interaction among four main entities: users, their tasks, languages, and content to help understand how these factors impact on the design and development of information access systems.

2.2   Upcoming Challenges for Large-Scale Evaluation Infrastructures

The future challenges for the evaluation campaigns will require an increased attention for the knowledge process entailed by an evaluation campaign. The complexity of the tasks and the interactions to be studied and evaluated will produce, as usual, valuable scientific data, which will provide the basis for the analyses and need to be properly managed, curated, enriched, and accessed. Nevertheless, to effectively investigate these new domains, not only the scientific data but also the information and knowledge derived from them will need to be appropriately treated and managed, as well as the cooperation, communication, discussion, and exchange of ideas among researchers in the field. As a consequence, we have to further advance the evaluation methodologies in order to support the whole knowledge creation process entailed by a large-scale evaluation campaign and to deal with the increasing complexity of the tasks to be evaluated. This requires the design and development of evaluation infrastructures which offer better support for and facilitate the research activities related to an evaluation campaign.

A first step in this direction, which is also the contribution of the paper, is to approach and study the information space entailed by an evaluation campaign in the light of the Data, Information, Knowledge, Wisdom (DIKW) hierarchy [1,20], used as a model to organize the information resources produced during it. The study contributes to creating awareness about the different levels and increasing complexity of the information resources produced during an evaluation campaign and indicates the relationships among the different actors involved in it, their tasks, and the information resources produced. The outcomes of this study are then applied in the design and development of the DIRECT system in order to validate their usefulness and effectiveness in the context of CLEF, which represents a relevant example of large-scale evaluation campaign with about 100 participating research groups per year.

In the perspective of the upcoming challenges, our final goal is to turn the DIRECT system from a DL for scientific data into a kind of virtual research environment, where the whole process which leads to the creation, maintenance, dissemination, and sharing of the knowledge produced during an evaluation campaign is taken into consideration and fostered. The boundaries between *content producers* – evaluation campaign organizers who provide experimental collections, participants who submit experiments and perform analyses, and so on – and *content consumers* – students, researchers, industries and practicioners who use the experimental data to conduct their own research or business, and to develop their own systems – are lowered by the current technologies: considering that we aim at making DIRECT an active communication vehicle for the communities interested in the experimental evaluation. This can be achieved by extending the DL for scientific data with advanced annotation and collaboration functionalities in order to become not only the place where storing and accessing the experimental results take place, but also an active communication tool for studying, discussing, comparing the evaluation results, where people can enrich the information managed through it with their own annotations, tags, ... and share them in a sort of social evaluation community. Indeed, the annotation of digital content [4,11] which ranges from metadata, tags, bookmarks, to comments and discussion threads, is the ideal means for fostering the active involvement of user communities and is one of the advanced services which the next generation digital libraries aim at offering.

3 The DIKW Hierarchy

The Data, Information, Knowledge, Wisdom (DIKW) hierarchy is a widely recognized model in the information and knowledge literature [1,18,20]. The academic and professional literature supports diversified meanings for each of the four concepts, discussing the number of elements, their relations, and their position in the structure of hierarchy. In particular, [18] summarizes the original articulation of the hierarchy and offers a detailed and close examination of the similarities and differences between the subsequent interpretations, and [13] identifies the good and the bad assumptions made about the components of the hierarchy. The four layers can summarized as follows:

- at the *data layer* there are raw, discrete, objective, basic elements, partial and atomized, which have little meaning by themselves and no significance beyond their existence. Data are defined as symbols that represents properties of objects, events and their environment, are created with facts, can be measured, and can be viewed as the building blocks of the other layers;
- the *information layer* is the result of computations and processing of the data. Information is inferred from data, answers to questions that begin

with *who, what, when* and *how many.* Information comes from the form taken by the data when they are grouped and organized in different ways to create relational connections. Information is data formatted, organized and processed for a purpose, and it is data interpretable and understandable by the recipient;

- the *knowledge layer* is related to the generation of appropriate actions, by using the appropriate collection of information gathered at the previous level of the hierarchy. Knowledge is *know what* and *know that,* articulable into a language, more or less formal, such as words, numbers, expressions and so on, and transmittible to others (also called *explicit knowledge* [17]), or *know how,* not necessarily codifiable or articulable, embedded in individual experience, like beliefs or intuitions, and learned only by experience and communicated only directly (*tacit knowledge* [17]).

- the *wisdom layer* provides interpretation, explanation, and formalization of the content of the previous levels. Wisdom is the faculty to understand how to apply concepts from one domain to new situations or problems, the ability to increase effectiveness, and it adds value by requiring the mental function we call judgement. Wisdom is not one thing: it is the highest level of understanding, and a uniquely human state. The previous levels are related to the past, whereas with wisdom people can strive for the future.

Those four layers can be graphically represented as a continuum linear chain or as the *knowledge pyramid,* where the wisdom is identified as the pinnacle of the hierarchy, and it is possible to see some transitions between each level in both directions [18]. There is a consensus that data, information, and knowledge are to be defined in terms of one another, but less agreement as to the conversion of one into another one. According to [18], moreover, wisdom is a very elusive concept in the literature about DIKW hierarchy, because the a limited discussion of its nature, "and even less discussion of the organizational processes that contribute to the cultivation of wisdom", despite its position at the pinnacle of the hierarchy.

4 Applying the DIKW Hierarchy to Large-Scale Evaluation Campaigns

Our aim is to define a relationship between the elements of the DIKW hierarchy and the knowledge process carried out by the actors involved in an evaluation campaign. Indeed, each step of a campaign and its outcomes can be coupled with specific actors and with one or more elements of the hierarchy. The result is a chain linking each step with a particular information resource, such as *experiments, performance measurements, papers,* etc., and the actors involved. Note that wisdom "has more to do with human intuition, understanding, interpretation and actions, than with systems" [18], but passing through the chain, each campaign become a spiral staircase connected to the other campaigns, allowing the user to create their own path to move towards wisdom supported by a system able to support and make explicit each step.

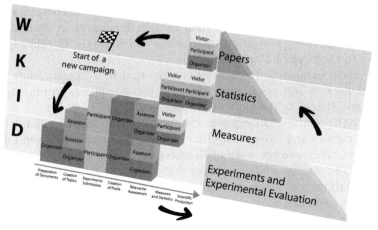

Fig. 2. DIKW knowledge pyramid applied to large-scale evaluation campaigns

Figure 2 frames the different types of information resources, actors, and main steps involved in an evaluation campaign into the pyramid of the DIKW hierarchy.

The left facet of the pyramid is created by a table that summarizes the relationships between the main steps of an evaluation campaign, shown in chronological order on the horizontal axis, the elements of the DIKW hierarchy, shown on the vertical axis, and the main actors involved in an evaluation campaign. For practical reasons, the D, I, K, and W layers are represented as separated, but each step can produce resources that belong to more than one layer.

The right facet summarizes the information resources given at the end of the campaign at each level of the hierarchy: in this way, we can talk about the *experimental collections* and the *experiments* as *data*, since they are raw elements: in fact, an experiment is useless without a relationship with the experimental collection with respect to which the experiment has been conducted. The *performance measurements*, by associating meaning to the data through some kind of relational connection, and being the result of computations and processing on the data, are *information*; the *descriptive statistics* and the *hypothesis tests* are *knowledge* since they are carried by the performance measurements and could be used to make decisions and take further actions about the scientific work. Finally, *wisdom* is provided by *theories, models, algorithms, techniques*, and *observations*, communicated by means of papers, talks, and seminars to formalize and explain the content of the previous levels.

The arrows in Figure 2 explain how each campaign is a step of a cycle where information resources generated in the past are used to allow the user to move towards wisdom as on a spiral staircase. The role of different actors is central to this process since their interactions make it possible to pass from one layer to another.

5   The DIRECT Digital Library System

DIRECT has successfully adopted in the CLEF campaigns since 2005 and has allowed us to:

- CLEF 2005: manage 530 experiments submitted by 30 participants spread over 15 nations and assess more than 160,000 documents in seven different languages, including Bulgarian and Russian which use the Cyrillic alphabet, thanks to the work of 15 assessors;
- CLEF 2006: manage 570 experiments submitted by 75 participants spread over 25 nations and assess more than 200,000 documents in nine different languages, thanks to the work of 40 assessors;
- CLEF 2007: manage 430 experiments submitted by 45 participants spread over 18 nations and assess more than 215,000 documents in seven different languages, thanks to the work of 75 assessors;
- CLEF 2008: manage 490 experiments submitted by 40 participants spread over 20 nations and assess more than 250,000 documents in seven different languages, including Farsi which is written from right to left, thanks to the work of 65 assessors.

In the following, we present the architecture and one example of the functionalities of the DIRECT system.

5.1 Architecture

DIRECT has been designed to be cross-platform and easily deployable to end users; to be as modular as possible, clearly separating the application logic from the interface logic; to be intuitive and capable of providing support for the various user tasks described in the previous section, such as experiment submission, consultation of metrics and plots about experiment performances, relevance assessment, and so on; to support different types of users, i.e. participants, assessors, organizers, and visitors, who need to have access to different kinds of features and capabilities; to support internationalization and localization: the application needs to be able to adapt to the language of the user and their country or culturally dependent data, such as dates and currencies.

Figure 3 shows the architecture of the system. It consists of three layers:

- *data logic*: this deals with the persistence of the different information objects coming from the upper layers. There is a set of "storing managers" dedicated to storing the submitted experiments, the relevance assessments and so on. The Data Access Object (DAO) pattern implements the access mechanism required to work with the underlying data source, acting as an adapter between the upper layers and the data source. Finally, on top of the various DAOs there is the "DIRECT Datastore" which hides the details about the storage management to the upper layers. In this way, the addition of a new DAO is totally transparent for the upper layers.
- *application logic*: this layer deals with the flow of operations within DIRECT. It provides a set of tools capable of managing high-level tasks, such as experiment submission, pool assessment, and statistical analysis of an experiment. For example, the "Performance Measures and Statistical Analyses" tool offers the functionalities needed to conduct a statistical analysis on a set of experiments. In order to ensure comparability and reliability, the tool makes

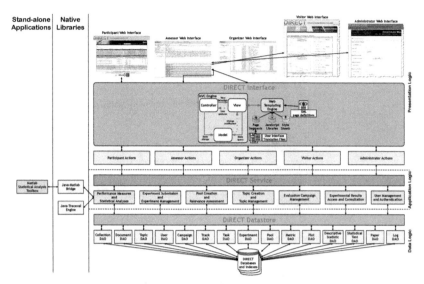

Fig. 3. Architecture of the DIRECT system

uses of well-known and widely used tools to implement the statistical tests, so that everyone can replicate the same test, even if they have no access to the service. In the architecture, the MATLAB Statistics Toolbox has been adopted, since MATLAB is a leader application in the field of numerical analysis which employs state-of-the-art algorithms, but other software could have been used as well. Finally, the "DIRECT Service" provides the interface logic layer with uniform and integrated access to the various tools. As in the case of the "DIRECT Datastore", thanks to the "DIRECT Service" the addition of new tools is transparent for the interface logic layer.

- *interface logic*: this is a Web-based application based on the Model-View-Controller (MVC) approach in order to provide modularity and a clear separation of concerns. Moreover, being Web-based, the user interface is cross-platform, easily deployable, and accessible without the need of installing any software on the end-user machines.

5.2   Topic Creation: An Example of DIKW for DIRECT

Figure 4 presents the main page for the management of the topic creation process which allows the assessors to create the topics for the test collection.

The interface manages information resources which belong to different levels of the DIKW hierarchy and relates them in a meaningful way. Assessor and organizers can access the *data* stored and indexed in DIRECT in the form of collections of documents, and shown in relevance order after a search, and the *data* produced by assessors themselves, i.e. the informations about the topics, such as the title, description, and narrative, and the history of the changes made on those values. The latter, in particular, is shown as a branch of a tree where each node is related at the timestamp of the change made. DIRECT

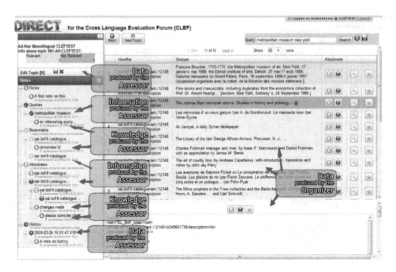

Fig. 4. DIRECT: creation of topics

automatically updates the tree each time a change is made, nesting the nodes related to the same topic and putting the newest near the root of the tree. This is an example of how the system can support and make explicit the creation of information resources at the *data* layer without forcing the user to taking care of the details.

You can also see how *information* and *knowledge* are produced by assessors who can save the queries used to create the topic, bookmark specific documents relevant to the topic, and save an aboutness judgement about a document in relation to the current topic. All these information resources are *information*, creating relational connections between documents and topics. Notes, comments, and discussion made by assessors are instead *knowledge*, which is created over the previous *information* and articulates into a language, and can also be attached to queries, bookmarks, and aboutness judgments.

In addition to easing the topic creation task, all these information resources are then available for conducting experiments and gaining qualitative and quantitative evidence about the pros and cons of different strategies for creating experimental collections and, thus, contribute to the advancement of the research in the field.

Finally, the possibility of interleaving and nesting different items in the hierarchy together with the ability of capturing and supporting the discussions among assessors represent, in concrete terms, a first step in the direction of making DIRECT a communication vehicle which acts as a kind of virtual research environment where the research about experimental evaluation can be carried out.

6 Conclusions

We have presented the next challenges for large-scale evaluation campaigns and their infrastructures and we have pointed out how they call for appropriate management of the knowledge process that they entail. In particular, we have discussed how digital library systems can play a key role in this scenarios and we have applied the DIKW hierarchy in the design and development of the DIRECT digital library system for scientific data.

Future work will concern the extension of the DIRECT system by adding advanced annotation functionalities in order to better support the cooperation and interaction among researchers, students, industrial partners and practicioners.

Acknowledgments

The authors would like to warmly thank Carol Peters, coordinator of CLEF and Project Coordinator of TrebleCLEF, for her continuous support and advice. The authors would like to thank Maristella Agosti and Giorgio Maria Di Nunzio for the useful discussions on the topics addressed in this chapter.

The work reported has been partially supported by the TrebleCLEF Coordination Action, as part of the Seventh Framework Programme of the European Commission, Theme ICT-1-4-1 Digital libraries and technology-enhanced learning (Contract 215231).

References

1. Ackoff, R.L.: From Data to Wisdom. Journal of Applied Systems Analysis 16, 3–9 (1989)
2. Agosti, M., Di Nunzio, G.M., Ferro, N.: A Proposal to Extend and Enrich the Scientific Data Curation of Evaluation Campaigns. In: Proc. 1st International Workshop on Evaluating Information Access (EVIA 2007), pp. 62–73. National Institute of Informatics, Tokyo (2007)
3. Agosti, M., Di Nunzio, G.M., Ferro, N., Harman, D., Peters, C.: The Future of Large-scale Evaluation Campaigns for Information Retrieval in Europe. In: Kovács, L., Fuhr, N., Meghini, C. (eds.) ECDL 2007. LNCS, vol. 4675, pp. 509–512. Springer, Heidelberg (2007)
4. Agosti, M., Ferro, N.: A Formal Model of Annotations of Digital Content. ACM Transactions on Information Systems (TOIS) 26(1), 3:1–3:57 (2008)
5. Agosti, M., Ferro, N.: Towards an Evaluation Infrastructure for DL Performance Evaluation. In: Evaluation of Digital Libraries: An Insight to Useful Applications and Methods. Chandos Publishing (2009)
6. Cleverdon, C.W.: The Cranfield Tests on Index Languages Devices. In: Readings in Information Retrieval, pp. 47–60. Morgan Kaufmann Publisher, San Francisco (1997)
7. Di Nunzio, G.M., Ferro, N.: DIRECT: a System for Evaluating Information Access Components of Digital Libraries. In: Rauber, A., Christodoulakis, S., Tjoa, A.M. (eds.) ECDL 2005. LNCS, vol. 3652, pp. 483–484. Springer, Heidelberg (2005)
8. Dussin, M., Ferro, N.: Design of a Digital Library System for Large-Scale Evaluation Campaigns. In: Christensen-Dalsgaard, B., Castelli, D., Ammitzbøll Jurik, B., Lippincott, J. (eds.) ECDL 2008. LNCS, vol. 5173, pp. 400–401. Springer, Heidelberg (2008)

9. Dussin, M., Ferro, N.: The Role of the DIKW Hierarchy in the Design of a Digital Library System for the Scientific Data of Large-Scale Evaluation Campaigns. In: Proc. 8th ACM/IEEE-CS Joint Conference on Digital Libraries (JCDL 2008), p. 450. ACM Press, New York (2008)

10. European Commission Information Society and Media. i2010: Digital Libraries (October 2006), http://europa.eu.int/information_society/activities/digital_libraries/doc/brochures/dl_brochure_2006.pdf

11. Ferro, N.: Digital Annotations: a Formal Model and its Applications. In: Information Access through Search Engines and Digital Libraries, pp. 113–146. Springer, Heidelberg (2008)

12. Ferro, N., Peters, C.: From CLEF to TrebleCLEF: the Evolution of the Cross-Language Evaluation Forum. In: Proc. 7th NTCIR Workshop Meeting on Evaluation of Information Access Technologies: Information Retrieval, Question Answering and Cross-Lingual Information Access, pp. 577–593. National Institute of Informatics, Tokyo (2008)

13. Fricke, M.: The Knowledge Pyramid: a Critique of the DIKW Hierarchy. Journal of Information Science 35(2), 131–142 (2009)

14. Fuhr, N., et al.: Evaluation of Digital Libraries. International Journal on Digital Libraries, 8(1):21–38 (2007)

15. Harman, D.K., Voorhess, E.M. (eds.): TREC. Experiment and Evaluation in Information Retrieval. MIT Press, Cambridge (2005)

16. National Science Board. Long-Lived Digital Data Collections: Enabling Research and Education in the 21st Century (NSB-05-40). National Science Foundation (NSF) (September 2005), http://www.nsf.gov/pubs/2005/nsb0540/

17. Nonaka, I., Takeuchi, H.: The knowledge-creating company: How Japanese companies create the dynamics of innovation. Oxford University Press, USA (1995)

18. Rowley, J.: The Wisdom Hierarchy: Representations of the DIKW Hierarchy. Journal of Information Science 33(2), 163–180 (2007)

19. Working Group on Data for Science. FROM DATA TO WISDOM: Pathways to Successful Data Management for Australian Science. Report to the Primw Minister's Science, Engineering and Innovation Council (PMSEIC) (September 2006), http://www.dest.gov.au/sectors/science_innovation/publications_resources/profiles/Presentation_Data_for_Science.htm

20. Zeleny, M.: Management Support Systems: Towards Integrated Knowledge Management. Human Systems Management 7(1), 59–70 (1987)

Hear It Is: Enhancing Rapid Document Browsing with Sound Cues

Parisa Eslambochilar, George Buchanan, and Fernando Loizides

Future Interaction Laboratory, Swansea University, UK
Centre for HCI Design, City University, UK
{p.eslambochilar,f.loizides}@swansea.ac.uk,
george.buchanan.1@city.ac.uk

Abstract. Document navigation has become increasingly commonplace as the use of electronic documents has grown. Speed–Dependent Automatic Zooming (SDAZ) is one popular method for providing rapid movement within a digital text. However, there is evidence that details of the document are overlooked as the pace of navigation rises. We produced a document reader software where sound is used to complement the visual cues that a user searches for visually. This software was then evaluated in a user study that provides strong supportive evidence that non-visual cues can improve user performance in visual seeking tasks.

1   Introduction

Document navigation has received increasing coverage from researchers in recent years. This paper reports a project that develops the Speed–Dependent Automatic Zooming method of navigation [14]. The SDAZ method applies to all digital navigation (e.g. of maps, images, etc.) but has been used with particular success in the task of navigation with digital texts [1,14].

This paper reports an initial investigation into how users' document navigation could be enhanced through the deployment of simple audio cues. We begin by exploring the navigation of documents where vision is most hindered; namely when navigating on a small screen device. There are many cases where this technique would be inappropriate (e.g. where the sound was made within a library's reading room), but when a user is on the move, or sitting at their own computer wearing headphones, additional support can be given to what is known to be a surprisingly challenging task. Such situations are increasingly commonplace as the use of digital library materials moves from being solely on the desktop.

The results show that the audio cue was sufficient for the users to distinguish where the target is. Results show improved timings for the completion of tasks and increased reliability in the successful location of sections. We also regularly witness a lag in reaction times when a user overshoots the target location. The paper commences with a related works section, followed by a description of our sound-enhanced document reader software. We then report an initial user study, conducted to evaluate the impact of this technique on simple navigational tasks. The paper progresses to a detailed discussion of our findings and work that

M. Agosti et al. (Eds.): ECDL 2009, LNCS 5714, pp. 75–86, 2009.

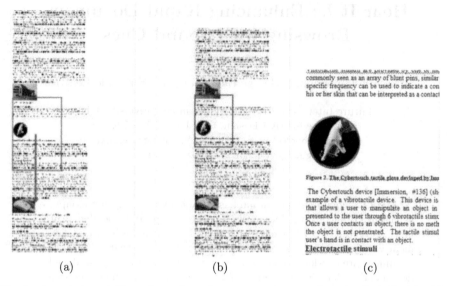

(a)   (b)   (c)

Fig. 1. A user is scrolling through a document using touch-controlled SDAZ, (a) Scrolling through the text document at maximum speed, (b) the user has released the stylus. Zero scroll speed in document. Vertically falling to 100% magnification, and (c) finished zooming. Document is at 100% magnification.

closely relates to it. We conclude with a review of future work in this area, and the opportunities that exist to make further improvements to user's interaction with digital texts.

2   Related Work

SDAZ is a simple technique that automatically varies the current level of zooming used in a view with the speed of the user's movement. Rapid movement results in the document view zooming outwards to view a larger area in less detail, while when the user's motion slows, the view zooms in to see a small area in fine relief (see Fig. 1). This intuitive model for a navigator first was proposed and implemented by Igarashi and Hinckley [14] for browsing large documents and maps on Desktop PCs. Cockburn et al. [1] suggested and successfully evaluated the refinements of this approach for navigating digital documents such as PDF files on Desktop PCs. Eslambolchilar and Murray-Smith [6,7] suggested a control-centric approach to control both zooming and panning by tilting alone or stylus alone on mobile platforms.

Even within the mobile domain, there is currently minimal research on how to use the very same cues for other navigation tasks. [9] presents the use of multimodal feedback in a document browsing task based on tilt-controlled SDAZ on a mobile platform. They assigned audio and haptic (tactile) feedback to important structures in the document such as headers, figures, tables etc.. They showed that audio or tactile feedback can support intermittent interaction, i.e. allowing movement-based interaction techniques to continue while the user is simultaneously involved with real life tasks.

Researchers in mobile devices have been exploring many other methods of enhancing navigation in general. One popular area of research is exploring what utility non-visual cues provide for navigation in the physical world. One example is Warren et al's use of music to indicate a user's position compared to their chosen destination [17]. In that case, if the target is to the user's left, the music is panned in that direction; if the user turns to face in its direction, the music gradually moves to the user's centre of attention.

Surprisingly, the progress made in using non–visual cues for navigation in the physical world has made little impact on document navigation. We therefore explore the utilisation of sonification to aid the document navigation process. Sonification is defined as the use of non-speech audio to convey information [10]. More specifically, sonification is the transformation of data relations into perceived relations in an acoustic signal for the purposes of facilitating communication or interpretation [13].

It is known that during visual search of documents, users often initially overshoot a target, and subsequently need to back-track to return to their chosen location. This knowledge has even been used to successfully backtrack automatically [16]. Whereas optical search in a dense visual target is problematic, we anticipated that an audio cue could assist the user adjust their navigation strategy in the presence of a potential target: e.g. slowing down their scrolling rate to identify the headings on display.

3   Sound–Enhanced Document Navigation

Rather than embark on providing a large number of different techniques that would be difficult to evaluate systematically at the same time, we focused on one particular element of user navigation that is relatively well documented. Researchers have indicated [2,5,15], that users utilise document headings for multiple tasks during digital navigation. Headings provide a cue for locating other material (e.g. from previous reading, a user may know that a particular detail is found in Section 4.3), and are also used to support skimming a document (determining if a section is likely to contain text of interest). Recent research has demonstrated that enhancing the visibility of headings makes a number of improvements to document navigation [3].

We followed this well–established direction of research by providing simple cues for the presence of headings. As an intuitive model of the sonification process, we can imagine the text on the screen to be embossed on the surface. This embossed type excites some object (an elastic band or sharp striking of speed ramps, for example) as it is dragged over the text. This physically motivated model is similar in nature to the model-based sonifications described by Hermann and Ritter [12]. As each single heading passes into view, a simple "click" sound is heard by the user. If a user navigates rapidly up–and–down a document, they will experience a sequence of clicks as different headings become visible.

Fig. 2. One of participants who plays with the sonic-enhanced document navigator on a PDA

For the implementation of our sound–enhanced document navigator we pursued the refinements of SDAZ approach suggested by [6,7] and Cockburn et al. [1]. We developed a stylus-controlled SDAZ–based document reader on a Personal Digital Assistant (PDA) device. This document reader utilises sonic cues (Sonification, see Sec. 2) to assist the reader's navigation. The current reader is implemented in C++, and uses the fmod API[1] and GAPI Draw library[2] to load and display each document. A document is then parsed to identify the location of each heading. These locations are interpreted into scroll positions, and as the user scrolls up and down the document, a click is produced when a heading passes into view.

4   User Study

To evaluate the impact of our sonic cues on user navigation we undertook a laboratory-based study. Our participants consisted of 24 university undergraduate and postgraduate students in a computer science department. We carried out a within-subjects design, allowing for 12 participants in each group to use one of two available variations of the SDAZ software on a PDA device with a touch screen facility, using a stylus to navigate. The PDA used throughout was a Dell AX70 (Fig. 2).

The first group used SDAZ without sonic cues whereas the second utilised the available sonic cue additions to the SDAZ software. Users were given a pre-study questionnaire to establish familiarity with small screen devices and to list the difficulties experienced in using these devices for reading, navigating and searching within documents on a small screen device.

Following the pre-study questionnaire all participants were given the same 50 page document containing plain text and 5 headings only. The document dimensions were 20000 height by 250 width(measured in pixels). The first heading contained the word "Bob", the second the word "Sam", third "Tom", fourth "Tim" and the final heading contained the word "Kim". These headings, which

---

[1] Music and sound effect system www.fmod.org
[2] Graphic api draw www.gapi.org

Heading 1	Heading 2	Heading 3	Heading 4	Heading 5

Fig. 3. Heading locations respective to document

were located in different pages of the document (see Fig. 3), were distinguishable clearly from a close zoom but semi-obscured when scrolling fast from a distant zoom. At maximum zoom, the title was almost, yet not impossible, to distinguish from the text. Participants were introduced to the system and allowed time to familiarise themselves with its navigation capabilities.

The participants were then given a set of 8 tasks to undertake, one at a time, using the software assigned to them. They were encouraged to complete each task in as little time as possible. The study was aimed at exploring behaviour for realistic within-document information seeking tasks. The first task was identifying the number of headings in the document. This simulated an information seeker trying to get an overview of document content during document triage before more in depth search and reading. The subsequent four tasks were to find specific headings, just as seekers would try to locate relevant sections they had previously found. The final three tasks were to find the first, second and final headings. These tasks emulated an information seeker locating the introduction, the subsequent section and conclusions of document. It should be noted here that after completing the first task participants were also asked if they could recall the heading titles and order. It was found that 22 participants could remember the first and final titles but were uncertain about the titles and order of the remaining three. Participants' navigational patterns were recorded in real time by the software. This data included position within the document, zoom values and speed of scrolling. Notes were taken during the tasks and participants were asked to comment on their behaviour after every task was completed. After all tasks were completed, we conducted a semi-structured interview asking participants to express the positive aspects they found, as well as any difficulties that hindered their tasks from being achieved efficiently. Participants from both groups were then presented with the alternate software and asked to choose between the two in terms of usefulness.

5   Results

We evaluated the navigation results for three following sets of tasks: finding number of headings in the document, locating a heading by content), and locating the heading by order (e.g. second, last). Before comparing the figures we take a look at a common behaviour seen in the navigational behaviour of the participants using the audio cues.

5.1   Overshooting and Backtracking

A common occupance between the participants with audio assistance is that of overshooting titles while scrolling and needing to backtrack to land on the

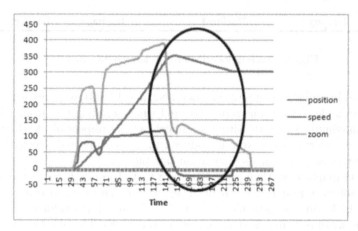

Fig. 4. Overshooting and backtracking in sonic-enhanced document navigator

title. This is due to the reaction time between the participant hearing the audio cue and the time it took to release the stylus from the screen. Participants when backtracking were also moving at a decreased speed so as not to miss the title again and consequently at a much lower zoom percentage. We notice this behaviour clearly from all the participants using the audio cues and all have commented on it as a disadvantage. One participant noted that it "would be nice to have an indication of a heading coming up rather than going too far". It should also be noted that this also affects the timings for audio cues in a negative way, since users, if provided with a predictive model to land successfully on the titles are likely to improve performance. The common overshooting and backtracking behaviour can be seen in Fig. 4.

5.2   Finding Number of Headings

The first task that users were required to undertake was to identify the total number of headings in the example document. This was an overview–style task that also familiarised the participant with the document.

The 12 participants using the audio cue all completed the task correctly. Their navigational behaviour was linear, progressing through the document in an orderly manner (see Fig. 5(a)). There were no instances of backtracking: all motion was downward through out the document. The mean zoom level was 307% across this task for all participants (sd=48.03%). The participants assisted by the audio cues immediately commented saying that the audio cues are "very handy for skim reading titles". They were sure that they did " not miss titles and therefore useful sections on the way".

In contrast, the 12 participants using the silent interface all demonstrated some backtracking behaviour: six showing localised backtracking of less than 100 pixels, the others showing large-scale movement (see Fig. 5(b)). For example, silent participant 4 moved forward to the end of the document, reversed the

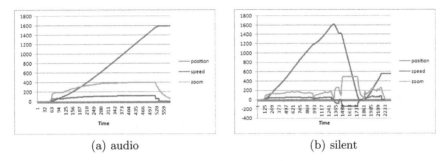

(a) audio   (b) silent

Fig. 5. Between group navigation comparison for the first task

entire length of the document, and then again moved forward to the end. The mean zoom level was a mere 129% (sd=37.09%).

We analysed the differences for this first task by applying Student's t–test for equal variance with degrees of freedom (df) set at 22. For the average zoom level per participant, this yielded p=0.0079 (t=2.9242). The same test for the maximum zoom level gave p=0.58 (t=0.5614), indicating that there was no significant difference in the maximum zoom level. As the zoom level is related to the current speed of movement, similar results could be expected for the comparison of speed. Ignoring the direction of movement, the average movement rate for the audio group was 85.48 pixels/second (sd=15.51), whilst the average for the silent group was 36.57 (sd=10.29); p=0.0154 (t=2.2677).

We also analysed the time taken to find all headings in both scenarios. Data show that participants in the audio group (sd=4.67,average=19.2 seconds) performed significantly faster than participants in the silent group(sd=47, average=86 seconds); p=0.009 (t=-4.91). Similarly, the total distance traveled by the audio group (sd=841.4, average=47800 pixels) was shorter than the silent group (sd=85291, average=101000); p=0.0408(t=2.1728).

5.3   Locating by Heading Content

There were four tasks that required the participants to find a given heading, based on the content. The data is presented in Table 1. Data in columns "Audio" and "Silent" present the average and standard deviation. Data in column "Significant" present the p-value and t-value using Student's t-test (df=22). Zoom levels are in percentages and the speed given in pixels per second.

The initial goal of the four was to find the heading that contains "Bob". There was no significant difference in the traveled distance between the audio and silent group.

The second heading navigation task was to find the heading "Sam". There was no significant difference in the traveled distance between the audio and silent group. Backtracking behaviour was more common, and observed in 70% of participants in the audio group. However, 40% of participant in the silent group also showed this behaviour.

Table 1. Data analysis for locating headings by content. "Audio" column: average (sdev), "Silent" column: average (sdev), and "Significant" column: Student's t–test; p-value (t-value).

Bob	Audio	Silent	Significant
Time (secs)	4.85 (0.86)	13.7 (24.34)	0.17 (t=1.4)
Zoom (Average)	86.62 (44.57)	57.41 (58.15)	0.091 (t=1.38)
Zoom (Max)	229.04 (94.33)	189.63 (139.89)	0.21 (t=0.81)
Speed (Average)	7.38 (1.75)	6.11 (3.17)	0.12 (t=1.21)
Speed (Max)	66.19 (21.64)	55.73 (31.32)	0.17 (t=0.95)
Sam	Audio	Silent	Significant
Time (secs)	18.4 (18.41)	11.6 (4.21)	0.23 (t=1.24)
Zoom (Average)	166.23 (60.30)	98.42 (28.85)	0.0009 (t=3.51)
Zoom (Max)	341.65 (107.30)	235.05 (52.39)	0.002 (t=3.09)
Speed (Average)	24.13 (10.68)	22.82(5.75)	0.35 (t=0.37)
Speed (Max)	94.75 (24.38)	69.79 (15.74)	0.003 (t=2.98)
Tom	Audio	Silent	Significant
Time (secs)	16.7 (8.85)	16.4 (5.51)	0.93 (t=0.084)
Zoom (Average)	199.67 (35.21)	147.98 (35.01)	0.00078 (t=3.61)
Zoom (Max)	359.29 (90.76)	250.04 (75.04)	0.002 (t=3.21)
Speed (Average)	43.94 (12.52)	39.68(11.00)	0.19 (t=0.88)
Speed (Max)	106.21 (25.52)	75.71 (22.04)	0.0024 (t=3.13)
Kim	Audio	Silent	Significant
Time (secs)	25.71 (3.7)	31.2 (12.7)	0.158 (t=1.46)
Zoom (Average)	247.82 (43.00)	178.46 (67.18)	0.0032 (t=3.01)
Zoom (Max)	413.19 (77.80)	286.54 (113.35)	0.0021 (t=3.19)
Speed (Average)	73.26 (62.13)	48.30(15.11)	0.095 (t=1.35)
Speed (Max)	139.95 (76.65)	85.68 (34.97)	0.018 (t=2.23)

The third heading to find was "Tom". There was no significant difference in the traveled distance between the audio and silent group. Similar backtracking behaviour to "Sam" heading was observed, but even stronger in this case: 85% of participants in the audio case showed this behaviour.

In the last heading navigation task, participants were asked to locate a heading that contained "Kim". Due to backtracking in the audio mode there was a difference in the traveled distance: the average traveled distance in the audio group was 49100 pixels (sd=3506) vs 41600 pixels (sd=4315) in the silent group. The t–test result was p=0.042 (t=2.157).

5.4 Locating by Heading Order

There were three tasks that required the participant to find a given heading, based on the order of its appearance. Basic results are seen in Table 2. The data layout in this table is similar to Table 1. Zoom levels are measured in percentage and the speed is measured in pixels per second.

The initial goal of the three was to find the first heading in the document. The average traveled distance in the audio group was 2080 pixels (sd=693) vs 3140

Table 2. Data analysis for locating headings by content. "Audio" column: average (sdev), "Silent" column: average (sdev), and "Significant" column: Student's t–test; p-value (t-value).

1st Heading	Audio	Silent	Significant
Time	4.33 (0.745)	4.95 (5.089)	0.6769 (t=0.422)
Zoom (Average)	109.61 (14.09)	79.29 (32.11)	0.0052 (t=3.08)
Zoom (Max)	210.73 (50.25)	140 (64.72)	0.0061 (t=2.99)
Speed (Average)	20.75 (3.68)	18.52 (8.07)	0.197 (t=0.897)
Speed (Max)	67.62 (16.32)	42.09 (15.08)	0.0006 (t=4.24)
2nd Heading	**Audio**	**Silent**	**Significant**
Time	2.44 (0.374)	2.73 (0.666)	0.20 (t=1.32)
Zoom (Average)	178.18(29.55)	123(34.23)	0.000017(t=4.21)
Zoom (Max)	30.46 (62.25)	203.73 (61.69)	0.0011 (t=3.99)
Speed (Average)	43.92(5.99)	33.66(8.83)	0.0015(t=3.33)
Speed (Max)	93.46 (19.02)	63.03 (19.40)	0.0017 (t=3.70)
Last Heading	**Audio**	**Silent**	**Significant**
Time	5.71 (1.22)	7.04 (2.11)	0.037 (t=1.87)
Zoom (Average)	331.37 (46.44)	261.16 (51.83)	0.001 (t=3.50)
Zoom (Max)	469.39 (49.21)	437.06 (48.86)	0.045 (t=1.85)
Speed (Average)	90.11(11.78)	73.78(15.95)	0.0047(t=2.843)
Speed (Max)	140.33 (14.09)	128.55 (13.28)	0.031 (t=2.07)

pixels in the silent group (sd=642). Student's t–test analysis produced p=0.58 (t=0.57) meaning no significant difference was observed in the average traveled distances between two groups. No significant difference was observed in our other log data.

The next heading navigation task was to find the second heading in the document. While the timing difference is not conclusive, the navigational patterns do suggest a difference in behaviour. There was no significant difference in the traveled distance between the audio and silent group; average traveled distance in the modes was: audio 10600 pixels (sd=1004); silent 9900 pixels (sd=191). Student's t–test produced p=0.43 (t=1.746).

In the last–heading navigation task, there was no significant difference in the traveled distance between the audio and silent group; average traveled distance in the modes was: audio 50300 pixels (sd=4301); silent 49200 pixels (sd=5869). Student's t–test produced p=0.6 (t=2.07). As with our overview task, the overall performance figures gave the advantage to the audio mode. However, in this task, backtracking was performed by all audio mode participants, whereas eight silent mode participants demonstrated some backtracking. The distribution giving two values of less than five, applying the Chi–squared test is not appropriate.

6 Discussion

These studies show that the audio cue improved participants' performance in locating headings, i.e. participants in the audio group performed significantly faster and the distance they traveled was shorter than participants in the silent group. Also, two participants in the silent group reported 4 titles and two reported 6 titles. All users in the audio group reported correct number of headings. Participants liked the selected audio cue for SDAZ and found it intuitive.

One participant stated that "I like that you don't have to read something to know that it's there." and another participant commented "I need to pay more attention if I use it without sound; I have to go back and verify if they are correct."

In locating by heading content there was no significant difference in the traveled distance between the audio and silent group. Backtracking was a common navigational behaviour where participants were asked to search for headings by contents or order. The further down the heading was in the document (e.g. "Tom", "Kim"), participants in the audio group took more advantage of the zoom feature. The average zoom level was significantly different when the headings were in further distance from the beginning of the document.

The issue of within–document navigation has previously been investigated within digital library research. One such example is the enhanced scrollbar technique suggested by Byrd [4]. This used a visual cue – coloured indicators on the scrollbar – to indicate the presence of words that matched a previous query. This technique could be combined with our method, or our audio cues simply used in place of the visual tiles used by Byrd.

David Harper [11] also uses visual cues to suggest the presence of words matching a query, but Harper's technique is to provide a visual document profile above the document display that complements the main document view and is separate from the usual scrollbar navigation.

Under evaluation, Byrd's technique has not yielded conclusive results, and it may not be an effective solution. Harper, on the other hand, did demonstrate some effect. Clearly secondary visual cues can have a positive effect on user navigation, so this supports our hypothesis that supplementary information can give users tacit task support. However, there is insufficient data at present to compare our non–visual approach to the techniques of Byrd or Harper.

The DL community's research on within–document navigation is, at present, young, and we are not aware of any major work in this domain other than Byrd's and Harper's. In general, DL–centred HCI work has focussed on primary task performance, rather than the effect of secondary cues, and thus there is also a lack of more general work to contrast Harper, Byrd's and our own work against. The techniques used for within–document navigation could well be used to assist users in other tasks. One possible example would be when to support then user when they are triaging document result sets to identify the texts they will actually view in detail.

7   Future Work

Our tasks encouraged participants to do skim reading, paying little attention to the content of the document. Although participants in the audio group reported that with audio they cannot miss targets, they did not like overshooting targets and backtracking. One possible solution is having an indication of a heading coming up or predicating what is coming next. For this we can adopt the predictive model described in [8]. The initial study in that paper suggests that a predictive model could reduce users' overshoots when landing on a target: a common problem in audio/haptic interaction.

Another area for future work is to compare the functionality of different cues. In the previous section, we noted the potential of haptic cues. One more simple comparison is to look at the comparative performance of secondary visual cues. For example, a sign may flash on the screen when a heading comes into view. The likely problem of that technique would be that users' visual attention would have to constantly readapt and refocus between different points – most likely reproducing the existing problems of visual search. However, a systematic understanding of the impact of using secondary visual cues versus the non–visual techniques we have discussed in this paper would be of immense value both within document navigation specifically, and for navigation research generally.

In our laboratory, we have briefly investigated other means of providing cues for navigation. One alternative method that seems particularly promising is haptic, or tactile, feedback. This has the advantage that in a public space, sensory cues can be given that have minimal impact on those sharing the same space. One simple example we developed is used where a user is navigating with a "Shake" input device [18]. A "Shake" device is matchbox–sized, and can control navigation using tilt sensors (e.g. tilting forward may move down a document, and tilting forward move up the text). Shakes also include a vibration device that can be triggered by software. Hence, as the human reader uses the shake to navigate, a vibration is given when a heading moves into view, as an alternative to the sonic cue we used here. In principle, the same method could be used with an adapted mouse, for example.

8 Conclusion

Our straightforward addition of sonic cues to document navigation yielded clear benefits for our users. When sound was used, user behaviour measurably changed, and navigation times fell. Users were more able to use the zooming features of SDAZ, and moved rapidly at a high zoom level between points where the audio cue marked the presence of a heading. When the "click" was heard that a heading was in view, they rapidly closed in to view the document in detail. In contrast, without this indicator, movement strategies were more controlled and conservative. Users maintained a low level of zoom, and often resorted to backtracking to ensure that they had not missed anything.

References

1. Andy Cockburn, J.S., Wallace, A.: Tuning and testing scrolling interfaces that automatically zoom. In: CHI 2005: Proceedings of the SIGCHI conference on Human factors in computing systems, pp. 71–80. ACM Press, New York (2005)

2. Buchanan, G., Loizides, F.: Investigating document triage on paper and electronic media. In: Kovács, L., Fuhr, N., Meghini, C. (eds.) ECDL 2007. LNCS, vol. 4675, pp. 416–427. Springer, Heidelberg (2007)
3. Buchanan, G., Owen, T.: Improving skim reading for document triage. In: Proceedings of the Symposium on Information Interaction in Context (IIiX). British Computer Society (2008)
4. Byrd, D.: A scrollbar-based visualization for document navigation. In: DL 1999: Proceedings of the fourth ACM conference on Digital libraries, pp. 122–129. ACM Press, New York (1999)
5. Cool, C., Belkin, N.J., Kantor, P.B.: Characteristics of texts affecting relevance judgments. In: 14th National Online Meeting, pp. 77–84 (1993)
6. Eslambolchilar, P., Murray-Smith, R.: Tilt-based Automatic Zooming and Scaling in mobile devices-a state-space implementation. In: Brewster, S., Dunlop, M.D. (eds.) Mobile HCI 2004. LNCS, vol. 3160, pp. 120–131. Springer, Heidelberg (2004)
7. Eslambolchilar, P., Murray-Smith, R.: Control centric approach in designing scrolling and zooming user interfaces. In: Brewster, S., Oulasvirta, A. (eds.) International Journal of Human-Computer Studies (IJHCS), Special issue on Mobility. Elsevier, Amsterdam (2008)
8. Eslambolchilar, P., Murray-Smith, R.: Model-based Target Sonification in Small Screen Devices: Perception and Action. In: Handbook of Research on User Interface Design and Evaluation for Mobile Technology, February 2008, pp. 478–506. Idea Group Reference (2008)
9. Eslambolchilar, P., Murray-Smith, R.: Interact, excite, and feel. In: Schmidt, A., Gellersen, H., van den Hoven, E., Mazalek, A., Holleis, P., Villar, N. (eds.) Second International Conference on Tangible and Embedded Interaction, TEI 2008, Bonn, Germany, February 2008, pp. 131–138. ACM, New York (2008)
10. Gaver, W.W.: Auditory Interfaces. In: Handbook of Human-Computer Interaction, 2nd edn. (1997)
11. Harper, D.J., Koychev, I., Sun, Y., Pirie, I.: Within-document retrieval: A user-centred evaluation of relevance profiling. Information Retrieval 7(3-4), 265–290 (2004)
12. Hermann, T., Hansen, M., Ritter, H.: Principal curve sonification. In: Proceedings of International Conference on Auditory Displays, ICAD 2000, USA, April 2000, pp. 81–86 (2000)
13. Hermann, T., Hunt, A.: The discipline of interactive sonification. In: Hermann, T., Hunt, A. (eds.) Proceedings of the International workshop on interactive sonification, Bielefeld, Germany (January 2004)
14. Igarashi, T., Hinckley, K.: Speed-dependent automatic zooming for browsing large documents. In: UIST 2000: Proceedings of the 13th annual ACM symposium on User interface software and technology, pp. 139–148. ACM Press, New York (2000)
15. Liu, Z.: Reading behavior in the digital environment. Journal of Documentation 61(6), 700–712 (2005)
16. Sun, L., Guimbretière, F.: Flipper: a new method of digital document navigation. In: CHI 2005: extended abstracts on Human factors in computing systems, pp. 2001–2004. ACM Press, New York (2005)
17. Warren, N., Jones, M., Jones, S., Bainbridge, D.: Navigation via continuously adapted music. In: CHI 2005: CHI 2005 extended abstracts on Human factors in computing systems, pp. 1849–1852. ACM Press, New York (2005)
18. Williamson, J., Murray-Smith, R., Hughes, S.: Shoogle: excitatory multimodal interaction on mobile devices. In: CHI 2007: Proceedings of the SIGCHI conference on Human factors in computing systems, pp. 121–124. ACM Press, New York (2007)

Creating Visualisations for Digital Document Indexing

Jennifer Pearson, George Buchanan, and Harold Thimbleby

FIT Lab, Swansea University
{j.pearson,g.r.buchanan,h.w.thimbleby}@swan.ac.uk

Abstract. Indexes are a well established method of locating information in printed literature just as FIND is a popular technique when searching in digital documents. However, document reader software has seldom adopted the concept of an index in a systematic manner. This paper describes an implemented system that not only facilitates user created digital indexes but also uses colour and size as key factors in their visual presentation. We report a pilot study that was conducted to test the validity of each visualisation and analyses the results of both the quantitative analysis and subjective user reviews.

Keywords: Document Triage, Indexing, Information Visualisation.

1 Introduction

Searching for relevant information within a document is a tiresome but necessary procedure in the information retrieval process. In physical books the process of locating relevant information is supported by the index; a classic structure that references key terms within a document for easy navigation. The problems associated with this method are that they are author created which restricts them to static terms that exist when the book is created and that they take time to physically navigate to.

One dynamic user-prompted method of information seeking on electronic documents is text search (Ctrl+f). By facilitating user-defined search terms, Ctrl+f has established itself as being a considerable advantage in electronic documents. However, in many document readers and web browsers this function does not return a list. Instead, it simply steps linearly through the document, highlighting each occurrence one at a time. This sequential interaction is slow and cumbersome if the main aim of the search is to get an overview of the location of text matches in a document. This concept of keyword overview can be a useful feature when performing document triage as it gives the user a solid indication of the areas of the document to observe first.

The limitations of linear search are underlined by a known discrepancy between actual behaviour and self reported use [3]. In fact, these shortcomings often force users to make use of other tools such as Internet search engines to overcome these problems. More sophisticated document reader software such as Preview for the Mac provide search facilities with a different presentation, giving

M. Agosti et al. (Eds.): ECDL 2009, LNCS 5714, pp. 87–93, 2009.

a list of results as well as a small snippet of text surrounding them. Several occurrences of a keyword on the same page then yeild multiple result lines. However, even this improved interaction is not a perfect solution. For example, this type of visualisation relies on the users' ability to mentally 'group' these same-page occurrences to decide where there are relevant clusters of information.

This paper introduces a system that improves information seeking by incorporating both user-prompted searching and indexing in a custom index-builder. The Visual Index System (VIS) both creates user defined indexes and visualises them in different ways to illustrate the relevance of each page/cluster.

2   Related Work

Although there has been much research into general document retrieval, there seems to be a relatively limited amount of exploration into within-document search techniques. Harper et al. [2] have investigated the differences between two such systems: FindSkim which has been designed to mimic standard search (Ctrl+f) tools found in most text processing and web browsing applications and SmartSkim, a visual interpretation based on 'relevance profiling'. Their results concluded that SmartSkim; a bar chart visualisation that constructs 'bars' based on word (not phrase) occurrences in text chunks was, in general a more effective method of precision and recall than FindSkim.

Byrd [1] discusses visualisations for whole- and within- document retrieval, making use of multi-coloured 'dots' that appear in the scroll-bar, and that correspond to the locations of keywords within the document. The user can use this display to identify clusters of a keyword by close proximity of dots of one colour.

3   Design and Implementation

The main aim of VIS is to allow users to create their own index which in turn gives a visual overview of the most relevant parts of the document based on the keywords entered. It groups the results that appear on the same page (and also occurrences that appear in clustered pages) and visualises their occurrences by means of colour and size. Creating a graphical representation of the data increases cognitive activity allowing users to better understand the underlying

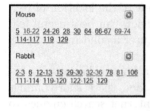 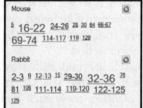 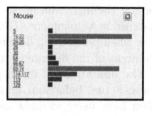

Fig. 1. The Three Types of Visualisation: Colour Tag, Tag Cloud and Graph

information [4] giving a clear overview of relevant sections by illustrating where the highest occurrences of each keyword appear.

VIS provides three separate index visualisations (see Fig 1) and allows users to toggle between them easily by means of a radio button set on the taskbar:

Colour Tag: The Colour Tag system is the same as a traditional index list layout in terms of size, but we have also coloured each link depending upon the number of occurrences of the word on that page/cluster.

Tag Cloud: The Tag Clouds system is an alphabetically sorted, size-weighted list of links that allow users to easily see each page/clusters relevance by means of their size and/or colour.

Graph: Harper et al produced the SmartSkim interface (see Section 2) which produced a vertical interactive bar graph representing the document and each section's relative retrieval status values. Working with this idea in mind, we decided to incorporate a simple graph type in the visual indexing solution. The simple horizontal bar chart implemented into the system represents the page/clusters versus the number off occurrences of the keyword/phrase.

The basic approach of the VIS system includes three functional features which are used in all of the visualisations we are exploring:

Hyperlinks: Each of the search results in the system will be in the form of a hyperlink which when clicked will take the user to the appropriate page and highlight all occurrences of the keyword/phrase on that page.

Page Clusters: One feature that the program possesses is its ability to 'cluster' page hits. For example, in a catalogue when you are looking for all references to sofas the index may look something like this: Sofa: 345, 467, **1067-1098**.

Tool Tips: Each link has a tool tip containing the occurrence information of the particular keyword/phrase,which pops up when the mouse hovers over it.

These visualisation methods use two distinct visual cues:

Colour: To minimise the time taken for a user to process the information presented by VIS, we used a system of colour coding to indicate areas of the document with most occurrences of the particular keyword or phrase. The colours used for this visualisation have been selected to mimic a temperature gauge; i.e. links with a low number of occurrences will be blue in colour whereas links with a high number of occurrences will be red.

Size: In our three new visualisations, we use different size-related cues to indicate the number of matches in a set of pages. In the Tag Cloud mode, the size of the text indicates the number of matches; in Graph mode, the length of the bars; the colour tag mode does not use size.

4   Pilot Study

To investigate the issues described in section 3, a pilot study was performed which focused on qualitative data in the form of subjective user questionnaires, as well as quantitative data obtained from task timings. Our hypothesis was that a custom indexing solution would in itself be an improvement over traditional linear indexing and furthermore, that a visual representation (colour and size) would prove to be a valuable asset in triage efficiency. In order to test the effectiveness of the implemented visualisations, two additional searching methods were written as a basis to test against. These two techniques were based on long-established methods of document searching:

Linear Search: The linear search method implemented in the program is based on the standard Windows within-document find feature that allows users to sequentially progress through a document one keyword at a time.

Traditional Indexing: The traditional indexing method has been designed to look like the classic index structure i.e. all entries being the same size and colour. It does however have the same 'build' feature as the visual index methods and also includes hyperlinks and tool tips.

The participants chosen for the study were selected from a set of postgraduate researchers due to their increased knowledge of online and document searching. In total 14 participants were selected; 11 male and 3 female, all between the ages of 22 and 37 and all with normal colour perception. All recorded data was kept anonymous, the study itself lasted on average around 30 minutes and the participants were given a £5 gift voucher in return for their time.

The structure of the study comprised of a pre-study questionnaire designed to gain general information about each participants searching habits, followed by a set of tasks and finally a post-study questionnaire devised to obtain subjective views with regards to the tested systems. The tasks provided were designed to time how long it takes to find the most relevant parts of a document based on a particular keyword(s). Users were given 3 PDFs and asked to perform 5 separate searches on each; one for each of the search methods. They were then asked to discover the part of a specific document that best matched a given query.

Due to the nature of the study, there were several factors that could affect the time taken to complete the tasks. These include the length of the document and the number of occurrences of the keyword/phrase in the document. To minimise the effects of these factors, each search term was assigned on a latin-square design to balance orderings and the logged time data across the 3 separate PDFs was averaged. To reduce any bias resulting from the different PDFs and any possible learning effects, the orderings of the documents and tasks were varied.

Table 1. The Averaged Timed Data (in seconds)

	Linear Search	Traditional Index	Colour Tag	Tag Cloud	Graph
AVERAGE PDF 1	99.9	37.7	21.2	14.4	14.5
AVERAGE PDF 2	92.1	33.9	13.2	10.9	9.9
AVERAGE PDF 3	134.1	32.8	15.6	11.6	11.2
AVERAGE ALL	108.7	34.8	16.7	12.3	11.9
SD ALL	59.33	22.78	11.18	9.54	10.20

4.1 Results

The results of the timed tasks (see Table 1) performed on each search method concluded that the traditional index is approximately 3 times faster than the standard linear search for locating relevant information in a document. Furthermore, it also confirmed that the use of colour and size is a further improvement with the average time for completion being at least half when using the visual systems (Colour Tag, Tag Cloud and Graph) over the traditional index. These statistics are also backed up by the subjective user ratings as shown in Fig 2b.

To assess the statistical significance of the timed test results a single-factor ANOVA test was performed. This test produced results of $p < 0.0001$, $df = \{4, 205\}$, $F = 81.915$ and $Fcrit = 2.416$, concluding that the resulting data was statistically significant. Due to the non-similar variances of the linear search versus the index builders, a Welch's t-test with bonferroni corrections was conducted upon the data to pin-point the areas of major significance.

The results of the bonferroni t-test (Fig 2a) confirm the significance of the difference between linear search compared against the 4 indexing methods, consistently yielding $p < 0.01$ and t values ranging from 11.470 to 15.030. Furthermore, comparing the traditional indexing method to each of the visual indexing methods (Colour Tag, Tag Cloud and Graph) has also been proven to have statistical significance with $p < 0.05$ and t values ranging from 2.816 to 3.560.

The tests performed clearly defined the major differences between the five implemented systems. Unsurprisingly then, the systems that performed similarly in the time tests (i.e. the visual systems) have yielded non-significant results from this data. A more precise study of these methods will be required to make a more concrete analysis of the differences between the visual index systems themselves.

As well as determining the different speeds in which each system can locate relevant information in a document, the study also determined the precision. In the linear tasks users were able to locate the most relevant section(s) of a document only 40.47% of the time, whereas the traditional index method yielded a result of 73.81%. We applied the Chi-squared test to these two modes, giving a significant result of p=0.002 (chi=9.528). Furthermore, The Colour Tag system allowed users to find the most relevant section 95.23% of the time and the Tag Cloud and Graph methods both resulted in 100% accuracy. It is clear from these results that not only is a custom index builder a large improvement in relevance accuracy than traditional linear searching, but also substantiates the theory that

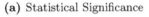

	Traditional Index	Colour Tag	Tag Cloud	Graph
Linear Search	p < 0.01	p < 0.01	p < 0.01	p < 0.01
Traditional Index		p < 0.05	p < 0.05	p < 0.05
Colour Tag			✘	✘
Tag Cloud				✘

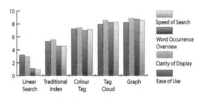

(a) Statistical Significance   **(b)** User Ratings

Fig. 2. Results: a) Performed using student t-test with bonferroni corrections. **x** indicates a non-significant result b) Graph of User Ratings (out of 10)

colour and size add further precision to the search. A global Chi-squared test (all 5 interfaces) produced chi=98.96; p < 0.00001.

The averaged user scores for each of the five types of search systems are highlighted in Fig 2b. The subjective ratings given for the visual systems were consistently higher than those for Linear Search and Traditional Index for every characteristic. Furthermore, 93% of the participants concurred that the visual systems are either a major (12 out of 14) or minor (1 out of 14) improvement over traditional indexing. Comments such as "The relevance of pages can be shown easily in a graphical manner" and "These (The Visual Systems) are much faster and more dynamic – I don't have to plough through pages of indexes to find the one I want" further substantiate this improvement.

This data confirms our underlying hypothesis; namely, that colour and size both play a productive roll in the effective visualisation of keyword occurrences. Using a 7 point Likert scale, we asked the participants to rate the temperature colour system and the different sizes (tag clouds and graphs) for how well they illustrate the number of occurrences of a keyword on a particular page. The results from this yielded average results of 5 out of 7 and 6 out of 7 for colour and size respectively. The popularity of the visual systems are also backed up by the positive comments made by some of the participants. One for example, describes the graph system as "Very clear, big, bold and picture perfect" whereas another said "It's obvious which choices are the best and you can see them very quickly" with regards to the Tag Cloud. One participant even went as far as saying that the colour system was "common sense" and the size system "gives emphasis to the pages where there are more occurrences".

The analysis on the task timings confirmed that there is no statistical significance to suggest any differences in performance between the 3 visual systems. However, the post-study questionnaires have produced some interesting subjective results indicating a combination of both size and colour is the optimum method of visualisation. When asked which system they favour, all 14 of the participants selected either Tag Cloud (5/14) or Graph (9/14) and justified it with answers such as "It was easy to use and it appealed to me because I could visualise it" and "the colour and size of the page numbers make it easy to see".

5   Conclusions

This paper has explored the concept of custom index builders for use in digital document readers. It has described three unique visualisations designed to aid users in locating the most relevant sections in a document. These three approaches have utilised colour and size in an attempt to tap into existing cognitive mappings and provide a visual overview of keyword occurrences in a document. A provisional study of these implemented solutions concluded both quantitatively and subjectively that custom indexing is a large improvement over traditional linear searching. In addition to this, it also confirms that colour and size play a constructive role in their visualisation by proving that users favour them over traditional index systems.

Acknowledgements

Jennifer Pearson is sponsored by Microsoft Research and Harold Thimbleby is a Royal Society-Leverhulme Trust Senior Research Fellow. We gratefully acknowledge their support. This research is supported by EPSRC Grant GR/S84798.

References

1. Byrd, D.: A scrollbar-based visualization for document navigation. In: DL 1999: Fourth ACM conference on Digital libraries, pp. 122–129. ACM Press, New York (1999)
2. Harper, D.J., Koychev, I., Sun, Y.: Query-based document skimming: A user-centred evaluation of relevance profiling. In: Sebastiani, F. (ed.) ECIR 2003. LNCS, vol. 2633, pp. 377–392. Springer, Heidelberg (2003)
3. Loizides, F., Buchanan, G.R.: The myth of find: user behaviour and attitudes towards the basic search feature. In: Proc. JCDL 2008, pp. 48–51. ACM Press, New York (2008)
4. Ware, C.: Information Visualization: Perception for Design. Morgan Kaufmann, San Francisco (2004)

Document Word Clouds: Visualising Web Documents as Tag Clouds to Aid Users in Relevance Decisions

Thomas Gottron

Institut für Informatik
Johannes Gutenberg-Universität Mainz
55099 Mainz, Germany
gottron@uni-mainz.de

Abstract. Information Retrieval systems spend a great effort on determining the significant terms in a document. When, instead, a user is looking at a document he cannot benefit from such information. He has to read the text to understand which words are important. In this paper we take a look at the idea of enhancing the perception of web documents with visualisation techniques borrowed from the tag clouds of Web 2.0. Highlighting the important words in a document by using a larger font size allows to get a quick impression of the relevant concepts in a text. As this process does not depend on a user query it can also be used for explorative search. A user study showed, that already simple TF-IDF values used as notion of word importance helped the users to decide quicker, whether or not a document is relevant to a topic.

1 Introduction

Research on Information Retrieval (IR) systems aims at helping users to satisfy their information needs. One important task in the underlying theoretical models is to determine the – generally speaking – important terms in an indexed document. This knowledge is then used to compute the relevance of a document to a given query.

But, once the user has selected a document from the result list of a search engine, he cannot access any longer the IR system's notion of word importance. He has to take a look at the document and judge by himself, whether it really is relevant to his information need. This is even more the case for explorative search, where the user browses along some reference structures and does not formulate a query at all.

Web designers, however, are aware that users usually do not really *read* documents on the web. Initially a user only *scans* a document [1,2]. This means to pass with the eyes over the screen quickly and pick up bits and pieces of layout, images or words here and there. Only if this quick *scanning* provides the user with the impression that it is worthwhile to actually read the entire text, he proceeds.

M. Agosti et al. (Eds.): ECDL 2009, LNCS 5714, pp. 94–105, 2009.

On the other hand, a study on users' information seeking behaviour on the web [3] revealed, that the text contents are the most useful feature in web documents to judge relevance. Unfortunately, when scanning a web document, its plain text parts usually do not attract much attention.

This paper proposes to support the user in the task of visually scanning a document employing techniques similar to the ones used in tag clouds. Based on a TF-IDF model, we calculate the importance of each word in a web document and display it with a respectively larger or smaller font size. These *document word clouds* allow a user to perceive much faster, which words in a document distinguish its content. We implemented a desktop http proxy server to analyse web documents on-the-fly and convert them into word clouds. Hence, users can use this visualisation enhancement transparently while browsing the web. For eventually reading a document, the system allows to convert it back into its normal state. A small experiment showed, that with this kind of document visualisation, users can decide quicker, whether or not a document is relevant to a given topic.

We proceed as follows: after a brief look at related work in 2, we describe our notion of word importance based on TF-IDF in 3. In section 4 we use the word importance to turn web documents into document word clouds. This section also explains the design of our http proxy implementation. In 5 we finally analyse users' experience with this document representation before concluding the paper in 6.

2   Related Work

The study of Tombros, Ruthven and Jose [3] analysed which factors influence a user's relevance decision in a web context. Examining the information seeking behaviour of 24 users, they found out that features in the text and structure of a web document are considered most useful in relevance judgements. Another interesting result: while a document's text in general was mentioned in 44% of the cases to be a useful relevance hint, the sub-categories *titles/headlines* and *query terms* were mentioned only in 4.4% and 3.9% respectively.

Highlighting the query terms in retrieved documents with a coloured background is a common way to support a user in scanning a document for relevance. Google's web search, for instance, is highlighting query terms in documents retrieved from its cache. Ogden et al. [4] analysed how such a highlighting helped users in a relevance judgements in combination with a thumbnail visualisation of the document. Dziadosz and Raman [5] tried to help users to make their relevance decision already a step earlier. They extended a classical result list of a web search engine with thumbnail previews of the documents. The combination of thumbnail and text summaries allowed the users to make more reliable decisions about a document really being relevant to their information need. All these approaches, however, depend on a user formulated query and are not suitable for explorative search.

Tag clouds are a common visualisation method in the Web 2.0 community. They alphabetically list user annotations (tags) of documents, pictures, links or other online contents. The importance of the tags, i.e. how often they have been used to annotate contents, is visualised, too. More important tags are represented with a larger font size or different colours. In this way, tag clouds provide a very quick impression of trends or "hot topics".

Research on tag clouds is relatively scarce. Publications in this context often merely use the tags as a resource for data-mining tasks (e.g. in [6]). The visual effects used in tag clouds were analysed by Bateman, Gutwin and Necenta [7]. They found that font-size, font-weight and intensity had the strongest visual influence on a user's perception of importance. Viégas and Wattenberg [8] discuss tag clouds from the point of view of visualisation techniques. Not originating from a visualisation research background, they say, tag clouds break some "golden rules". However, given their success and wide adoption, one has to recognise their effectiveness.

Our method to determine word importance is based on classical TF-IDF weights. This concept can be found in virtually every standard IR book. For details and further reading we refer to the very good introduction of Manning, Raghavan and Schütze [9]. Though a classical method it is still applied in current research to determine important words in web documents. In [10], document terms are connected to Wikipedia categories to determine the general document topic. The influence of the categories assigned to each term is weighted by TF-IDF values of the terms. So, the authors imply a correlation between the TF-IDF values of the terms and their importance for the topic of the document.

To determine which parts of a document to include in the calculation of word importance, we use content extraction techniques. We adopted the fast TCCB algorithm [11] with optimised parameter settings [12]. For term normalisation we used stemmer implementations provided by the Snowball project (http://snowball.tartarus.org/): the Porter stemmer [13] for English documents and the project's stemmer for German language.

3   Determining Word Importance

Our aim is to highlight those words in a document, which are more important than others, i.e. which distinguish a particular document from other documents.

The concept of word importance in a document can be mapped onto term weighting in an IR system. Effectively, here we are going to use a simple TF-IDF scheme. For each term t we determine its document frequency $df(t)$, i.e. in how many documents of a corpus of size N it appears (we come to the question of which corpus to use in 4.2). For a given document d we then determine the term frequency $tf_d(t)$, i.e. we count how often the term appears in this particular document. The TF-IDF weight for term t in document d is defined as:

$$\text{TF-IDF}_d(t) = tf_d(t) \cdot \log \frac{N}{df(t)}$$

This formula describes a classical weighting scheme for terms in a vector space IR model. If a query term matches an index term with a high TF-IDF value, the corresponding documents obtain a higher relevance score. The intention behind this scoring is, that a term with a high TF-IDF score describes the document very well – especially in comparison to other documents in the corpus. Hence, we adopt TF-IDF values of the terms in a document as a notion of word importance. Note, that we can do this without having the user have formulated a query.

4 Creating Document Word Clouds

The idea of document word clouds is to transfer the visualisation idea of tag clouds to web documents. Instead of visualising tags, we modify the font size of the words contained in the document itself. Instead of using the frequency of tag assignments to determine how large to write a word, we use the above explained notion of word importance. And instead of sorting the terms alphabetically – as done in tag clouds – we leave their order unchanged.

To actually turn a web document into a document word cloud, we implemented an http proxy server for on-the-fly analysis and modification of documents. Embedding the system in a proxy server is a very flexible solution as it is entirely independent of both sides: the browser client and the web server. Figure 1 sketches the system and we proceed with a detailed explanation on how it works.

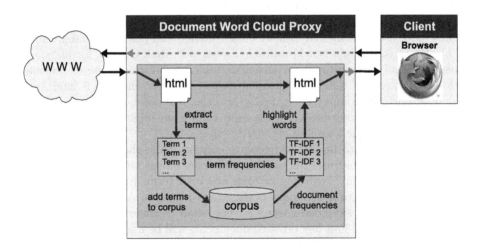

Fig. 1. Our proxy analyses and modifies documents on-the-fly

4.1 Document Preprocessing

After forwarding a client's request to a server and receiving the according response, the proxy does not deliver the document directly back to the client, but first analyses and eventually modifies it.

Therefore, we first need to determine the terms in the document. Exploiting the inherent document structure of HTML, we tokenise the contents into words, using white space characters, sentence and paragraph delimiters and paying attention to some particularities like abbreviations. The resulting tokens are then normalised via case folding. With a simple and fast heuristic we determine the language of the document: we assume the document to be in the language in which it contains the most stopwords. The implemented system so far contains stop word lists for English and German, but can easily be extended to other languages[1]. Once we know which language we are dealing with, we apply a stemmer to finally obtain the terms we use for indexing and TF-IDF calculation.

4.2   Corpus

To calculate TF-IDF for the terms in the document we need a corpus over which to calculate the document frequencies $df(t)$. Actually, as we distinguish between documents in different languages, we need several corpora: one for each language.

One option is to provide a standard corpus for each language. These corpora would have to provide representative frequencies for a large choice of words. This approach has two disadvantages. First, from a practical point of view, it is difficult to provide and handle such a corpus. Probably all available corpora have a bias towards some topics. Further, to store and look up document frequencies in such a large corpus, would result in a higher demand for computational and storage resources. This would make it more difficult to use the system as a desktop proxy and might require a dedicated high-end machine. The second disadvantage is more user specific. Even if we could provide and efficiently handle such an ideal corpus, it might not be suitable for an individual user. If, for instance, a user is mainly interested in a particular topic, the terms in the documents he looks at have very different document frequencies. Say, the user is interested mainly in portable computers. In this case terms like notebook, laptop or netbook are much less informative to the user, than what might be deduced from their document frequencies in a standard corpus.

An alternative is to build the reference corpora while the user browses the web. As the proxy tokenises and analyses each document anyway, we can keep track of the document frequency of each observed term. In this way, we can operate on a corpus that also reflects the user's interests. As in the beginning such a continuously extended corpus is very small considering the number of terms seen so far, the first few browsed documents will be analysed on a basis of pretty distorted document frequencies for the contained words. Accordingly, also the TF-IDF values will be distorted. Hence, a conversion into document word clouds makes sense only after the corpus has reached a certain size. Empirically we found out, that already a corpus of around 3,000 to 4,000 unique terms was sufficient to obtain reasonable results in document visualisation.

Both alternatives have their advantages and disadvantages. For our proxy system we chose the latter option of building the corpus during runtime. However, a third option would be to combine both approaches: to start with a small general corpus and to extend it constantly with newly browsed documents.

---

[1] Provided the concept of stop words exists in these languages.

4.3   Document Rewriting

On the technical side of actually rewriting a document d we first need to calculate the actual TF-IDF values of its terms. The term frequencies $tf_d(t)$ are obtained by counting the term appearances in the document. The document frequencies $df(t)$ are stored in central data structure along with the corpus' size. This is all the data needed to compute the importance of a particular word in the document.

Once we know the importance of each word in the document, we can turn the document into a word cloud. Those words with a relative high importance, i.e. TF-IDF value, will be written larger, while those with low values are written smaller.

To obtain always similar results in font size and in the distribution of large and small words, we normalise the TF-IDF values into k classes. Within each of these importance classes, we will display all words with the same font size. The terms with the lowest importance are always assigned to the lowest class, the highest TF-IDF value in the document corresponds to the highest class. The parameter k can be set by the user. For our user test we found a setting of $k = 10$ to provide enough importance classes.

The assignment of terms into the classes follows a logarithmic scheme. Given the highest TF-IDF value $w_{\max}$ and the lowest value $w_{\min}$, a term with a TF-IDF value of t is assigned to the class:

$$class(t) = \left\lfloor \left(\frac{t - w_{\min}}{w_{\max} - w_{\min}} \right)^{\beta} \cdot k \right\rfloor$$

The parameter β influences the distribution into the importance classes. The higher the value the smaller is the proportion of larger written words. In our tests we used a value of $\beta = 1.2$ which produced well balanced document word clouds.

In order to change the font size, each word in the document body is wrapped into span tags. These span elements all have a class attribute with a value representing the importance class of the contained word. Once the words are marked in this way, the document is extended with CSS directives. The CSS causes the font size of the texts inside the span elements to be increased or decreased according to the importance class and relative to its normal font size. After these transformations the proxy serialises the document into an http message and returns it to the client who requested it in the first place.

Concerning runtime, the whole process is unproblematic. The analysis of the document, its tokenisation into terms, computation of TF-IDF values, the annotation of the words, the document's extension with CSS directives and its serialisation usually takes less than a second. So, the users do not feel big delays when browsing the web via our proxy.

4.4   The User's Side

In the client browser the documents look like the one shown in figure 2. The screenshot demonstrates nicely how important words are written larger, while, for instance, stop words are very small.

CALL ᶠᵒᴿ CONTRIBUTIONS

Aim ₐₙd Scope

Close to the turn of the first decade of the third millennium, digital libraries are facing critical challenges that lead to major transformations. The expansion of social networking applications is an important development, which has lead to the creation of new user communities and the cohesion of already existing ones. Although user communities have been under the research lens of the digital library community, they never had higher interest than nowadays. User communities have abandoned pathetic participation in information environments and have developed an active behavior expressed in a multitude of ways.

In the same time, after a decade of solidification the issue of metadata re-emerges to address the new challenges. Annotations and tagging has been an edge-leading theme for digital libraries, which now can be viewed under a different perspective. The implication of user communities in various aspects of information management stages, such as creation of new information, enrichment of information artifacts, sharing and distribution of information objects, filtering of relevant items and so on, require a thorough examination of the metadata issues and services that augment all these activities.

In this intense environment ECDL 2009, under the general title "Digital Societies", invites submissions for the proliferation of scientific and research osmosis in the following categories: Full Papers, Short Papers, Posters and Demonstrations, Workshops and Tutorials, Panels and Doctoral Consortium. All submissions will be reviewed on the basis of relevance, originality, importance and clarity in a triple peer review process.

Fig. 2. Part of the ECDL 2009 call for contribution converted into a document word cloud

For the purpose of reading the document, the proxy additionally embeds some JavaScript and HTML code. A small button calls a JavaScript function which resets the font sizes of all words to their normal value. This *zooming back* is realised in a smooth, gradual way. Such a smooth transition to the normal document representation turned out to be less confusing for the user.

Alternatively to the embedded button, the proxy can also be configured to let the documents zoom back into their normal state automatically after a preset timeout.

4.5   Optimisation

When analysing our first document word clouds we noticed some unwanted effects. Particularly those terms in headlines showed relatively high document frequencies. Hence, they were displayed extraordinarily small. The reason was, that headlines are often used as anchor texts in the navigation menus or related links lists of other documents. Similarly, the terms in navigation menus also distorted the term frequencies within a document and caused insignificant words to be displayed larger. In other cases, this happened because a term appeared in a legal disclaimer or in advertisements.

To overcome these problems, we restricted the analysis of term frequency and document frequency to those parts of the documents which contain the main text content. Therefore, we used the TCCB content extraction algorithm [11,12]. Though, TCCB does not always outline the main content precisely, its inclusion in the preprocessing phase helped to overcome the problems we observed. Further, we also used it to determine whether a document contains a long text content at all. If the main content is composed of too few words, we do not modify the document at all, as it does not have a textual main content or the text is too short for a reasonable analysis.

5   User Experience

In order to find out, whether the document word cloud approach really supports users in relevance decisions, we conducted a small user test. The users were shown a series of documents in a normal or a word cloud representation. Their task was to decide as fast as possible if a document belonged to a given topic. So, the users had to make a simple yes/no decision. They were not told, that larger words in the document word clouds represented significant words. In order to find out, how fast and reliable the users could judge the relevance of a document for a given topic we measured the time they needed to make their decision and whether their decision was correct.

As documents we used articles taken from a German online news website. The corpus for calculating word importance was a larger collection of 1,000 news articles from the same website. All documents consisted of a headline, a teaser paragraph and the full article body. The topic categories were cars (test reports), cell phones (hardware, service providers), economics (international economic policies, large companies) and tv/movies (actors, directors, films). For all documents it was rather clear whether they actually belonged to the topic or not. However, for some documents already the headline contained good hints for a decision, for others it was necessary to read parts of the article body.

We had 14 participants in the test. They were all familiar with documents in a web environment, several knew the concept of tag clouds. Each user was given five documents for each of the four topics. So, they had to judge a total of 20 documents each. We divided the users in two groups of equal size. The first group was provided with documents in the standard format (large headline, a bold written teaser paragraph and plain text for the article) for the topics tv/movies and economics, while the document for the topics cars and cell phones were shown as document word clouds. Group two had the same documents, but based on the respectively other presentation style.

Table 1. Time needed for relevance decisions (in seconds). Group 1 saw the categories cars and cell phones as document word clouds, group 2 the categories tv/movies and economics.

	documents	U1	U2	U3	U4	U5	U6	U7	U8	U9	U10	U11	U12	U13	U14
				user group 1								user group 2			
tv/movies	D1	3.01	3.30	2.39	2.18	2.96	2.54	3.93	2.48	1.68	1.65	2.99	1.66	2.96	1.32
	D2	2.46	2.25	2.35	2.20	1.87	2.67	2.06	7.77	3.58	1.51	2.30	2.36	5.88	2.29
	D3	2.61	3.48	1.69	3.69	2.11	4.86	3.25	3.97	3.86	1.43	3.19	2.68	2.97	2.62
	D4	2.22	3.94	3.13	6.67	4.26	4.16	4.77	3.26	2.75	1.71	2.13	3.77	3.20	1.81
	D5	3.14	8.17	3.78	2.13	2.41	2.23	3.34	3.39	2.20	1.75	3.28	4.25	2.63	4.43
economics	D6	1.99	2.48	3.00	4.27	1.68	2.54	2.21	2.68	2.96	1.72	5.20	3.49	2.46	2.72
	D7	2.51	2.99	2.47	2.32	1.61	2.63	1.73	2.64	1.78	1.29	2.30	2.06	1.42	1.42
	D8	3.27	3.65	2.60	2.55	1.42	3.86	2.42	3.55	4.01	1.57	2.46	2.36	2.58	3.81
	D9	2.59	3.79	2.57	2.91	2.23	6.79	3.11	2.82	6.44	1.38	2.50	3.62	1.47	4.51
	D10	2.64	2.50	3.85	3.21	1.92	4.20	3.45	4.44	3.66	1.78	3.57	3.15	2.06	3.20
	avg.	*2.64*	*3.65*	*2.78*	*3.21*	*2.25*	*3.65*	*3.03*	*3.70*	*3.29*	*1.58*	*2.99*	*2.94*	*2.76*	*2.81*
cars	D11	4.42	3.11	3.64	2.96	1.21	2.40	2.89	4.36	3.53	2.77	3.88	2.80	3.77	3.76
	D12	3.35	5.04	2.82	2.71	1.55	2.42	3.37	2.64	2.59	2.01	3.79	3.55	3.43	2.68
	D13	2.92	3.90	1.60	2.43	1.47	4.03	3.69	3.95	3.01	2.75	4.24	5.52	3.91	2.68
	D14	1.76	1.84	1.48	1.78	1.53	2.90	2.48	2.97	3.11	2.19	2.06	2.49	2.10	3.52
	D15	2.78	2.18	2.37	1.19	1.35	3.14	2.19	2.36	3.65	1.67	2.44	2.02	2.44	2.13
cell phones	D16	3.47	5.96	2.68	2.47	2.62	3.36	4.94	8.70	6.38	2.22	2.93	11.11	3.10	3.06
	D17	2.78	2.48	1.76	1.42	1.63	1.76	1.88	1.84	2.23	1.25	1.73	2.15	2.07	1.73
	D18	2.19	2.03	1.42	2.02	1.98	2.32	2.34	2.17	4.01	1.64	2.04	3.56	1.85	2.17
	D19	2.17	2.58	1.92	2.53	2.15	2.51	3.34	2.15	2.12	1.36	4.44	2.78	2.67	3.82
	D20	2.50	3.13	2.50	1.85	1.53	4.28	2.64	3.28	2.00	1.51	3.35	3.37	3.67	2.19
	avg.	*2.83*	*3.23*	*2.22*	*2.14*	*1.70*	*2.91*	*2.98*	*3.44*	*3.26*	*1.94*	*3.09*	*3.93*	*2.90*	*2.77*
tendency		-	+	+	+	+	+	0	-	0	+	0	+	+	0

Table 1 lists the time the users needed for their relevance decision. It lists the users as U1 to U14 and the documents D1 to D20 for the purpose of reference in the following discussion of the results. The table also indicates the tendency, whether a user made his decision faster (+), slower (-) or more or less in the same time (0) when presented with document word clouds. Note, that the second user group actually saw the documents D11 to D20 first and in their normal representation, before getting documents D1 to D10 as word clouds.

We can observe, that document word clouds allow a quicker relevance decision. On a global average, the candidates took their relevance decision on document word clouds about 0.32 seconds faster. Given the average decision time of 3.1 seconds on the normal document format, this means a 10% quicker decision. Looking at the average time each individual user needed to make her or his judgement in figure 3, we see that two users (U1, U8) took longer for their decision with the cloud representation, four (U7, U9, U11, U14) took more or less the same time and eight were faster. For five users the improvements are statistically significant.

Also with the focus on the single documents in figure 4 the improvement can be measured. For two documents (D2, D5) the decision took longer if they were

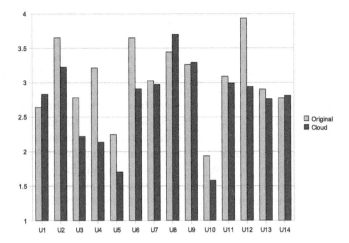

Fig. 3. Average time users needed for relevance judgements on original documents and document word clouds

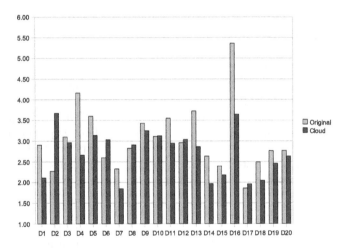

Fig. 4. Average time needed for the documents to be judged depending on their representation in original or cloud format

presented as document word clouds, for four (D8, D10, D12, D17) there was no notable difference, and for the remaining 14 the responses came quicker. Document D2 took significantly longer to assess relevance in its cloud representation. The problem was, that most of the highlighted words (though specific for the actual content of the document) did not allow a clear negative answer, which was expected and eventually given for this document. However, for five documents the time improvements are statistically significant.

In following-up interviews, most users said, the larger font-size attracted their attention and made them focus on theses words. Accordingly, they also mentioned the importance of highlighting those words in a document, which actually aid a decision. Some users (U1, U3, U8, U9) said, they did not "trust" the larger words. So, they tried to read the text in the standard fashion left to right and top down. As we did not provide the option to switch to the normal representation during the test, reading was more difficult and most of those users took longer or were at least not faster in their decision.

The users made relatively few mistakes in their decision. Only two misclassification were made on the original documents, and four on the cloud representation. Three of the latter mistakes occurred in the topic of economy news, because an off-topic document (D9) about operating systems mentioned Apple and Microsoft. In this case, the users took those names as a hint for a business report about those companies. Also other users mentioned, that names of companies or persons are not always useful in a relevance judgement. So, their usually high TF-IDF values do not necessarily correspond to their importance for relevance assessment.

Finally, some users said, they were initially confused by the unusual cloud layout of the document. However, they also said, that given a little more time they would probably get used to it – and then it might become very helpful.

6   Conclusions

The concept of document word clouds helps users to get a quick idea of what a document is about. Inspired by tag clouds we write important words using a larger font-size, while reducing the font-size of less significant words. As a measure for word importance we used TF-IDF values. The whole process is independent of a user query, hence, can also be used for explorative search. We implemented a http proxy to convert web documents on-the-fly into document word clouds. The system was used to create documents for a user test, in which document word clouds allowed the users to make relevance decisions quicker.

The developed system can be extended and improved in several directions. First of all, it would be interesting to realise document word clouds on different, more sophisticated IR models. Techniques to detect named entities can be used to recognise names of persons or companies. They usually have a high TF-IDF values but are not always supportive for relevance decisions. Another problem are longer documents in which the important words appear towards the end. In this case, the users do not see any large written words unless they scroll down. A summary at the top or an embedded thumbnail might solve this problem. A time-limit for keeping documents in the corpus might be an idea for future research, as well. It could improve the systems performance when the interest of a user changes. More practical issues would be to finetune the determination of suitable relevance classes and to include a duplicate or near-duplicate detection to avoid indexing the same document more than once.

References

1. Krug, S.: Don't make me think – Web Usability, 2nd edn. mitp, Heidelberg (2006)
2. Lindgaard, G., Fernandes, G., Dudek, C., Brown, J.: Attention web designers: You have 50 milliseconds to make a good first impression! Behaviour & Information Technology 25(2), 115–126 (2005)
3. Tombros, A., Ruthven, I., Jose, J.M.: How users assess web pages for information seeking. J. Am. Soc. Inf. Sci. Technol. 56(4), 327–344 (2005)
4. Ogden, W.C., Davis, M.W., Rice, S.: Document thumbnail visualization for rapid relevance judgments: When do they pay off? In: TREC, pp. 528–534 (1998)
5. Dziadosz, S., Chandrasekar, R.: Do thumbnail previews help users make better relevance decisions about web search results? In: SIGIR 2002: Proceedings of the 25th annual international ACM SIGIR conference on Research and development in information retrieval, pp. 365–366. ACM, New York (2002)
6. Noll, M.G., Meinel, C.: Exploring social annotations for web document classification. In: SAC 2008: Proceedings of the 2008 ACM symposium on Applied computing, pp. 2315–2320. ACM, New York (2008)
7. Bateman, S., Gutwin, C., Nacenta, M.: Seeing things in the clouds: the effect of visual features on tag cloud selections. In: HT 2008: Proceedings of the nineteenth ACM conference on Hypertext and hypermedia, pp. 193–202. ACM, New York (2008)
8. Viégas, F.B., Wattenberg, M.: Tag clouds and the case for vernacular visualization. Interactions 15(4), 49–52 (2008)
9. Manning, C.D., Raghavan, P., Schütze, H.: Introduction to Information Retrieval. Cambridge University Press, Cambridge (2008)
10. Schönhofen, P.: Identifying document topics using the wikipedia category network. In: WI 2006: Proceedings of the 2006 IEEE/WIC/ACM International Conference on Web Intelligence, pp. 456–462. IEEE Computer Society Press, Los Alamitos (2006)
11. Gottron, T.: Content code blurring: A new approach to content extraction. In: Bhowmick, S.S., Küng, J., Wagner, R. (eds.) DEXA 2008. LNCS, vol. 5181, pp. 29–33. Springer, Heidelberg (2008)
12. Gottron, T.: An evolutionary approach to automatically optimise web content extraction. In: IIS 2009: Proceedings of the 17th International Conference Intelligent Information Systems (in preparation, 2009)
13. Porter, M.F.: An algorithm for suffix stripping. Program 14(3), 130–137 (1980)

Exploratory Web Searching with Dynamic Taxonomies and Results Clustering

Panagiotis Papadakos, Stella Kopidaki,
Nikos Armenatzoglou, and Yannis Tzitzikas

Institute of Computer Science, FORTH-ICS, Greece
Computer Science Department, University of Crete, Greece
{papadako,skopidak,armenan,tzitzik}@ics.forth.gr

Abstract. This paper proposes exploiting both *explicit* and *mined* metadata for enriching Web searching with *exploration* services. On-line results clustering is useful for providing users with overviews of the results and thus allowing them to restrict their focus to the desired parts. On the other hand, the various metadata that are available to a WSE (Web Search Engine), e.g. domain/language/date/filetype, are commonly exploited only through the advanced (form-based) search facilities that some WSEs offer (and users rarely use). We propose an approach that combines both kinds of metadata by adopting the interaction paradigm of dynamic taxonomies and faceted exploration. This combination results to an effective, flexible and efficient exploration experience.

1 Introduction

Web Search Engines (WSEs) typically return a ranked list of documents that are relevant to the query submitted by the user. For each document, its title, URL and *snippet* (fragment of the text that contains keywords of the query) are usually presented. It is observed that most users are impatient and look only at the first results [1]. Consequently, when either the documents with the intended (by the user) meaning of the query words are not in the first pages, or there are a few dotted in various ranks (and probably different result pages), it is difficult for the user to find the information he really wants. The problem becomes harder if the user cannot guess additional words for restricting his query, or the additional words the user chooses are not the right ones for restricting the result set.

One solution to these problems is *results clustering* [25] which provides a quick overview of the search results. It aims at grouping the results into topics, called *clusters*, with predictive names (labels), aiding the user to locate quickly documents that otherwise he wouldn't practically find especially if these documents are low ranked (and thus not in first result pages). Another solution is to exploit the various metadata that are available to WSEs (like domain, dates, language, filetype, etc). Such metadata are usually exploited through the advanced search facilities that some WSEs offer, but users very rarely use these services. A more flexible and promising approach is to exploit such metadata in the context of the interaction paradigm of *faceted and dynamic taxonomies* [18,21], a paradigm

M. Agosti et al. (Eds.): ECDL 2009, LNCS 5714, pp. 106–118, 2009.

that is used more and more nowadays. Its main benefit is that it shows only those terms of the taxonomy that lead to non-empty answer sets, and the user can gradually restrict his focus using several criteria by clicking. In addition this paradigm allows users to switch easily between searching and browsing.

There are works [1,14] in the literature that compare automatic results clustering with guided exploration (through dynamic faceted taxonomies). In this work we propose combining these two approaches. In a nutshell, the contribution of our work lies in: (a) proposing and motivating the need for exploiting both explicit and mined metadata during Web searching, (b) showing how automatic results clustering can be combined with the interaction paradigm of dynamic taxonomies, by clustering on-demand the top elements of the user focus, (c) providing incremental evaluation algorithms, and (d) reporting experimental results that prove the feasibility and the effectiveness of the approach.

To the best of our knowledge, there are no other WSEs that offer the same kind of information/interaction. A somehow related interaction paradigm that involves clustering is Scatter/Gather [4,8]. This paradigm allows the users to select clusters, subsequently the documents of the selected clusters are clustered again, the new clusters are presented, and so on. This process can be repeated until individual documents are reached. However, for very big answer sets, the initial clusters apart from being very expensive to compute on-line, will also be quite ambiguous and thus not very helpful for the user. Our approach alleviates this problem, since the user can restrict his focus through the available metadata, to a size that allows deriving more specific and informative cluster labels.

The rest of this paper is organized as follows. Section 2 discusses requirements, related work and background information. Section 3 describes our approach for dynamic coupling clustering with dynamic taxonomies. Section 4 describes implementation and reports experimental results. Finally, Section 5 concludes and identifies issues for further research.

2   Requirements and Background

Results Clustering. Results clustering algorithms should satisfy several requirements. First of all, the generated clusters should be characterized from high intra-cluster similarity. Moreover, results clustering algorithms should be efficient and scalable since clustering is an online task and the size of the retrieved document set can vary. Usually only the $top - C$ documents are clustered in order to increase performance. In addition, the presentation of each cluster should be concise and accurate to allow users to detect what they need quickly. *Cluster labeling* is the task of deriving readable and meaningful (single-word or multiple-word) names for clusters, in order to help the user to recognize the clusters/topics he is interested in. Such labels must be predictive, descriptive, concise and syntactically correct. Finally, it should be possible to provide high quality clusters based on small document snippets rather than the whole documents.

In general, clustering can be applied either to the original documents (like in [4,8]), or to their (query-dependent) snippets (as in [25,20,6,26,7,22]). Clustering meta-search engines (e.g. clusty.com) use the results of one or more WSEs, in order to increase coverage/relevance. Therefore, meta-search engines have direct access only to the snippets returned by the queried WSEs. Clustering the snippets rather than the whole documents makes clustering algorithms faster. Some clustering algorithms [6,5,23] use internal or external sources of knowledge like Web directories (e.g. DMoz[1]), Web dictionaries (e.g. WordNet) and thesauri, online encyclopedias and other online knowledge bases. These external sources are exploited to identify significant words/phrases, that represent the contents of the retrieved documents or can be enriched, in order to optimize the clustering and improve the quality of cluster labels.

One very efficient and effective approach is the *Suffix Tree Clustering (STC)* [25] where search results (mainly snippets) can be clustered fast (in linear time), incrementally, and each cluster is labeled with a phrase. Overall and for the problem at hand, we consider important the requirements of *relevance, browsable summaries, overlap, snippet-tolerance, speed* and *incrementality* as described in [25]. Several variations of STC have emerged recently (e.g. [3,11,22]).

Exploratory Search and Information Thinning. Most WSEs are appropriate for focalized search, i.e. they make the assumption that users can accurately describe their information need using a small sequence of terms. However, as several user studies have shown, this is not the case. A high percentage of search tasks are exploratory [1], the user does not know accurately his information need, the user provides 2-5 words, and focalized search very commonly leads to inadequate interactions and poor results. Unfortunately, available UIs do not aid the user in query formulation, and do not provide any exploration services. The returned answers are simple ranked lists of results, with no organization.

We believe that modern WSEs should guide users in exploring the information space. *Dynamic taxonomies* [18] (faceted or not) is a general knowledge management model based on a multidimensional classification of heterogeneous data objects and is used to explore and browse complex information bases in a guided, yet unconstrained way through a visual interface. Features of faceted metadata search include (a) display of current results in multiple categorization schemes (facets) (e.g. based on metadata terms, such as size or date), (b) display categories leading to non-empty results, and (c) display of the count of the indexed objects of each category (i.e. the number of results the user will get by selecting this category). An example of the idea assuming only one facet, is shown in Figure 1. Figure 1(a) shows a taxonomy and 8 indexed objects (1-8). Figure 1(b) shows the dynamic taxonomy if we restrict our focus to the objects {4,5,6}. Figure 1(c) shows the browsing structure that could be provided at the GUI layer and Figure 1(d) sketches user interaction.

The user explores or navigates the information space by setting and changing his *focus*. The notion of focus can be *intensional* or *extensional*. Specifically, any set of terms, i.e. any conjunction of terms (or any boolean expression of terms) is a possible *focus*. For example, the initial focus can be the empty, or the top term

---

[1] www.dmoz.org

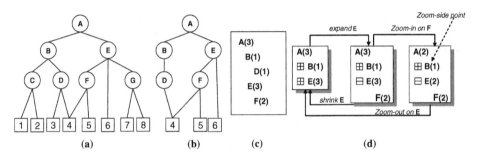

Fig. 1. Dynamic Taxonomies

of a facet. However, the user can also start from an arbitrary set of objects, and this is the common case in the context of a WSE and our primary scenario. In that case we can say that the focus is defined extensionally. Specifically, if A is the result of a free text query q, then the interaction is based on the restriction of the faceted taxonomy on A (Figure 1(b) shows the restriction of a taxonomy on the objects $\{4,5,6\}$). At any point during the interaction, we compute and provide to the user the immediate zoom-in/out/side points along with count information (as shown in Figure 1(d)). When the user selects one of these points then the selected term is added to the focus, and so on. Note that the user can exploit the faceted structure and at each step may decide to select a zoom point from different facet (e.g. filetype, modification date, language, web domain, etc).

Examples of applications of faceted metadata-search include: e-commerce (e.g. ebay), library and bibliographic portals (e.g. DBLP), museum portals (e.g. [10] and Europeana[2]), mobile phone browsers (e.g. [12]), specialized search engines and portals (e.g. [16]), Semantic Web (e.g. [9,15]), general purpose WSEs (e.g. Google Base), and other frameworks (e.g. mSpace[19]).

Related Work. Systems like [24,9,13,15,2] support multiple facets, each associated with a taxonomy which can be predefined. Moreover, the systems described in [24,9,15] support ways to configure the taxonomies that are available during browsing based on the contents of the results. Specifically, [24] enriches the values of the object descriptions with more broad terms by exploiting WordNet, [9] supports rules for deciding which facets should be used according to the type of data objects based on RDF, and [15] supports reclassification of the objects to predefined types. However, none of these systems apply content-based results clustering, re-constructing the cluster tree taxonomy while the user explores the answer set. Instead they construct it once per each submitted query.

3   On-Demand Integration

Dynamic taxonomies can load and handle thousands of objects very fast (as it will be described in Section 4.1). However, the application of results clustering on

[2] http://www.europeana.eu

thousands of snippets would have the following shortcomings: (a) *Inefficiency*, since real-time results clustering is feasible only for hundreds of snippets, and (b) *Low cluster label quality*, since the resulting labels would be too general. To this end we propose a *dynamic (on-demand) integration* approach. The idea is to *apply the result clustering algorithm only on the top-C (usually $C = 100$) snippets of the current focus.* This approach not only can be performed fast, but it is expected to return more informative cluster labels. Let q be the user query and let $Ans(q)$ be the answer of this query. We shall use A_f to denote top-K (usually $K < 10000$) objects of $Ans(q)$ and A_c to denote top-C objects of $Ans(q)$. Clearly, $A_c \subseteq A_f \subseteq Ans(q)$. The steps of the process are the following:

(1) The snippets of the elements of A_c are generated.
(2) Clustering is applied on the elements of A_c, generating a cluster label tree *clt*.
(3) The set of A_f (with their metadata), as well as *clt*, are loaded to `FleXplorer`, a module for creating and managing the faceted dynamic taxonomy. As the facet that corresponds to automatic clustering includes only the elements of A_c, we create an additional artificial cluster label, named "REST" where we place all objects in $A_f \setminus A_c$ (i.e. it will contain $K - C$ objects).
(4) `FleXplorer` computes and delivers to the GUI the (immediate) zoom points.

The user can start browsing by selecting the desired zoom point(s). When he selects a zoom point or submits a new query, steps (1)-(4) are performed again.

3.1   Results Clustering: HSTC

We adopt an extension of STC [25] that we have devised called HSTC. As in STC, this algorithm begins by constructing the suffix tree of the titles and snippets and then it scores each node of that tree. HSTC uses a scoring formula that favors the occurrences in titles, does not merge base clusters and returns an hierarchically organized set of cluster labels. In brief, the advantages of HSTC are: (a) the user never gets unexpected results, as opposed to the existing STC-based algorithms which adopt overlap-based cluster merging, (b) it is more configurable w.r.t. desired cluster label sizes (STC favors specific lengths), (c) it derives hierarchically organized labels, and (d) it favors occurrences in titles. The experimental evaluation showed that this algorithm is more preferred by users[3] and it is around two times faster that the plain STC (see Section 4.1). We do not report here the details, because any result clustering algorithm could be adopted. However, it is worth noticing that the hierarchy of cluster labels by HSTC, can be considered as a subsumption relation since it satisfies $c < c' \implies Ext(c) \subseteq Ext(c')$, where $Ext(c_i)$ denotes the documents that belong to cluster c_i. This property allows exploiting the interaction paradigm of *dynamic taxonomies*. HSTC, like STC, results in overlapping clusters.

---

[3] http://groogle.csd.uoc.gr:8080/mitos/files/clusteringEvaluation/userStudy.html

3.2   Dynamic Taxonomies: FleXplorer

FleXplorer is a main memory API (Application Programmatic Interface) that allows managing (creating, deleting, modifying) terms, taxonomies, facets and object descriptions. It supports both finite and infinite terminologies (e.g. numerically valued attributes) as well as explicitly and intensionally defined taxonomies. The former can be classification schemes and thesauri, the latter can be hierarchically organized intervals (based on the *inclusion* relation), etc. Regarding user interaction, the framework provides methods for setting the focus and getting the applicable zoom points.

3.3   Incremental Evaluation Algorithm

Here we present an incremental approach for exploiting past computations and results. Let A_f be the objects of the current focus. If the user selects a zoom point he moves to a different focus. Let A'_f denote the top-K elements of the new focus, and A'_c the top-C of the new focus. The steps of the algorithm follow.

(1) We set $A_{c,new} = A'_c \setminus A_c$ and $A_{c,old} = A_c \setminus A'_c$, i.e. $A_{c,new}$ is the set of the new objects that have to be clustered, and $A_{c,old}$ is the set of objects that should no longer affect clustering.
(2) The snippets of the objects in $A_{c,new}$ are generated (those of $A_{c,old}$ are available from the previous step). Recall that snippet generation is expensive.
(3) HSTC is applied *incrementally* to $A_{c,new}$.
(4) The new cluster label tree clt' is loaded to FleXplorer.
(5) FleXplorer computes and delivers to the GUI the (immediate) zoom points for the focus with contents A'_f.

Let's now focus on Step (3), i.e. on the incremental application of HSTC. Incremental means that the previous suffix tree sf is preserved. Specifically, we extend sf with the suffixes of the elements in the titles/snippets of the elements in $A_{c,new}$, exploiting the incremental nature of STC. Let sf' denote the extended suffix tree. To derive the top scored labels, we have to score again all nodes of the suffix tree. However we should not take into account objects that belong to $A_{c,old}$. Specifically, scoring should be based on the extension of the labels that contain elements of A'_c only. The preserved suffix tree can be either the initial suffix tree or the pruned suffix tree. Each node of the initial tree corresponds to a single word, while the pruned tree is more compact in the sense that if a node contains only one child node and both nodes contain the same objects, they are collapsed to one single node that has as label the concatenation of the labels of the constituent nodes. Scoring is done over the pruned suffix tree. However to add and delete objects to/from a pruned suffix tree sometimes requires "splitting" nodes (due to the additions) and pruning extra nodes (due to the deletions). On the other hand, if the unpruned suffix tree is preserved, then additions and deletions are performed right away

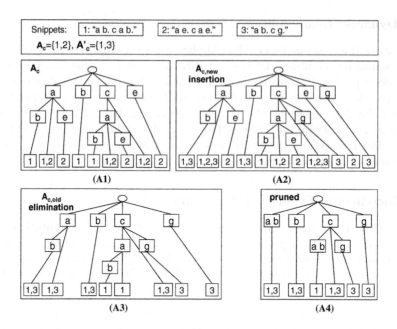

Fig. 2. Incremental Evaluation

and pruning takes place at the end. Independently of the kind of the preserved suffix tree, below we discuss two possible approaches for updating the suffix tree:

- *Scan*-approach

  We scan the nodes of the suffix tree sf' and delete from their extensions all elements that belong to $A_{c,old}$. Figure 2 illustrates an example for the case where $A_c = \{1,2\}$ and $A'_c = \{1,3\}$.
- *Object-to-ClusterLabel Index*-approach

  An alternative approach is to have an additional data structure that for each object o in A_c it keeps pointers to the nodes of the suffix tree to whose extension o belongs. In that case we do not have to scan the entire suffix tree since we can directly go to the nodes whose extension has to be reduced. The extra memory space for this policy is roughly equal to the size of the suffix tree. However the suffix tree construction process will be slower as we have to maintain the additional data structure too.

We have to note that sf can be considered as a cache of snippets and recall that snippet generation is more expensive than clustering. The gained speedup is beneficial both for a standalone WSE as well for a Meta WSE, since fetching and parsing of snippets are reused. The suffix tree sf has to be constructed from scratch whenever the user submits a new query and is incrementally updated while the user browses the information space by selecting zoom points.

4   Implementation and Experimental Evaluation

The implementation was done in the context of Mitos[4] [17], which is a prototype WSE[5]. FleXplorer is used by Mitos for offering general purpose browsing and exploration services. Currently, and on the basis of the top-K answer of each submitted query, the following five facets are created and offered to users:

- the hierarchy or clusters derived by HSTC
- web domain, a hierarchy is defined (e.g. *csd.uoc.gr* < *uoc.gr* < *gr*),
- format type (e.g. pdf, html, doc, etc), no hierarchy is created in this case
- language of a document based on the encoding of a web page and
- (modification) date hierarchy.

When the user interacts with the clustering facet we do not apply the re-clustering process (i.e. steps (1) and (2) of the on-demand algorithm). This behavior is more intuitive, since it preserves the clustering hierarchy while the user interacts with the clustering facet (and does not frustrate the user with unexpected results). In case the user is not satisfied by the available cluster labels for the top-C objects of the answer, he can enforce the execution of the clustering algorithm for the next top-C by pressing the *REST* zoom-in point as it has already been mentioned (which keeps pointers to $K - C$ objects).

4.1   Experimental Results

Clustering Performance. It is worth noting that the most time consuming subtask is not the clustering itself but the extraction of the snippets from the cached copies of textual contents of the pages[6]. To measure the performance of the clustering algorithm and the snippet generation, we selected 16 queries and we counted the average times to generate and cluster the top-{100, 200, 300, 400, 500} snippets. All measurements were performed using a Pentium IV 4 GHz, with 2 GB RAM, running Linux Debian.

Table 1. Top-C Snippet Generation and Clustering Times (in seconds)

Measured Task	100	200	300	400	500
Time to generate snippets	0.793	1.375	1.849	2.268	2.852
Time to apply STC	0.138	0.375	0.833	1.494	2.303
Time to apply HSTC	0.117	0.189	0.311	0.449	0.648

Table 1 shows snippet generation times and the clustering algorithms performance (measured in seconds). Notice that snippet generation is a slow operation and is the bottleneck in order to provide fast on-demand clustering, for a big

---

[4] http://groogle.csd.uoc.gr:8080/mitos/

[5] Under development by the Department of Computer Science of the University of Crete and FORTH-ICS.

[6] The snippets in our experiments contain up to two sentences (11 words maximum each) where the query terms appear most times.

top-C number ($C > 500$). We should mention though, that our testbed includes a rather big number of large sized files (i.e. pdf, ppt), which hurt snippet generation times. Moreover, notice that HSTC is at least two times faster than STC. This is because HSTC does not have to intersect and merge base clusters.

Dynamic Taxonomies Performance. Loading times of `FleXplorer` have been thoroughly measured in [21]. In brief, the computation of zoom-in points with count information is more expensive than without. In 1 sec we can compute the zoom-in points of 240.000 results with count information, while without count information we can compute the zoom-in points of 540.000 results.

Overall Performance. In this experiment we measured the overall cost, i.e. cluster generation times (snippet generation and clustering algorithm execution) and the dynamic taxonomies times (to compute the zoom points and and to load the new clustering labels to the corresponding facet). Moreover, we compare

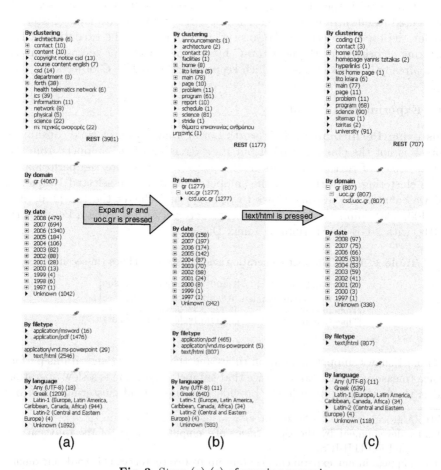

Fig. 3. Steps (a)-(c) of running scenario

the non-incremental with the incremental algorithm, which preserves the initial suffix tree and the elimination of old objects is done using the Scan-approach. The scenario we used includes: (a) the execution of the query *crete* which returns 4067 results, (b) the expansion of the *gr* zoom point of the *By domain* facet and the selection of the *uoc.gr* (1277) zoom-in point from the hierarchy revealed from the expansion, and (c) the selection of the *text/html* (807) zoom-in point of the *By filetype* facet. Let c_a, c_b and c_c be snippets of the $top - C$ elements in the steps (a), (b) and (c) respectively. Figure 3 shows the facet terms after steps (a), (b) and (c), as they are displayed in the left bar of the WSE GUI. We set $K = 10000$ (i.e. the whole answer set is loaded) and repeated the above steps for the following values of C:100, 200 ... 500. We do not measure the cost of the query evaluation time. In all experiments `FleXplorer` computes count information.

Table 2. Top-C Comparison of Incremental/Non-Incremental Algorithms (in seconds)

	‖ Step (a)	Step (b)	Step (c)
top-100	‖$|c_a| = 100$	$|c_a \cap c_b| = 43$, overlap=43%	$|c_b \cap c_c| = 85$, overlap=85%
Non-Incr.	0.914	0.443	0.204
Incr.	0.931	0.431	0.101
top-200	‖$|c_a| = 200$	$|c_a \cap c_b| = 71$, overlap=35.5%	$|c_b \cap c_c| = 113$, overlap=56.5%
Non-Incr.	1.266	1.245	0.789
Incr.	1.245	0.965	0.68
top-300	‖$|c_a| = 300$	$|c_a \cap c_b| = 74$, overlap=24.6%	$|c_b \cap c_c| = 201$, overlap=67.7%
Non-Incr.	1.676	2.534	1.383
Incr.	1.65	2.527	0.761
top-400	‖$|c_a| = 400$	$|c_a \cap c_b| = 85$, overlap=21.5%	$|c_b \cap c_c| = 252$, overlap=63%
Non-Incr.	2.246	3.067	1.944
Incr.	2.118	3.335	0.942
top-500	‖$|c_a| = 500$	$|c_a \cap c_b| = 97$, overlap=19.4%	$|c_b \cap c_c| = 324$, overlap=64.8%
Non-Incr.	2.483	3.495	2.001
Incr.	2.493	3.652	0.751

Table 2 shows the intersection of A_c and A'_c for steps (a), (b) and (c) and the execution times that correspond to the integration of `FleXplorer` and results clustering using the non-incremental and an incremental approach of HSTC, for the $top - C$ elements. It is evident that for top-100 and top-200 values, the results are presented to the user almost instantly (around 1 second), making the proposed on demand clustering method suitable as an online task. Moreover, we can see that there is a linear correlation between time cost and the top-C value. Finally, calculating and loading clusters for the top-500 documents, costs around 3 seconds making even big top-C configurations a feasible configuration.

Comparing the incremental and the non-incremental algorithm, we observe a significant speedup whenever the overlap is more than 50%, for our scenario. At step (a) the suffix tree construction is the same for both algorithms as the suffix tree sf has to be constructed from scratch. For step (b) there are small variations due to the small overlap, so the time saved from the snippets generation/parsing is compensated by the time needed for eliminating old objects. Specifically, the incremental algorithm is faster for the top-200 case and slower for the top-$\{400, 500\}$ cases which have the lowest overlap. For the other cases performance is almost the same. Notice that although the top-100 case has the biggest overlap of all, there are no differences in the execution time of the two algorithms. This is probably due to the fact that the overlapping documents have fast snippet generation times, while the rest are big sized. At step (c) the benefit from the incremental approach is clear, since it is almost twice as fast as the non incremental one. Specifically, the best speedup is in the case of top-500, where overlap reaches 65% and the execution time of the non-incremental is 2.001, while for the incremental is just 0.751.

5   Conclusion

The contribution of our work lies in: (a) proposing and motivating the need for exploiting both explicit and mined metadata during Web searching, (b) showing how automatic results clustering can be combined effectively and efficiently with the interaction paradigm of dynamic taxonomies by applying top-C clustering on-demand, (c) providing incremental evaluation approaches for reusing the results of the more computationally expensive tasks, and (d) reporting experimental results that prove the feasibility and the effectiveness of this approach. In the future we plan to conduct a user study for investigating what top-C value most users prefer. Finally, we plan to continue our work on further speeding up the incremental algorithms presented.

References

1. Special issue on Supporting Exploratory Search. Communications of the ACM 49(4) (April 2006)
2. Ben-Yitzhak, O., Golbandi, N., Har'El, N., Lempel, R., Neumann, A., Ofek-Koifman, S., Sheinwald, D., Shekita, E., Sznajder, B., Yogev, S.: Beyond basic faceted search. In: Procs. of the Intern. Conf. on Web Search and Web Data Mining (WSDM 2008), Palo Alto, California, USA, February 2008, pp. 33–44 (2008)
3. Crabtree, D., Gao, X., Andreae, P.: Improving web clustering by cluster selection. In: Procs. of the IEEE/WIC/ACM Intern. Conf. on Web Intelligence (WI 2005), Compiegne, France, September 2005, pp. 172–178 (2005)
4. Cutting, D.R., Karger, D., Pedersen, J.O., Tukey, J.W.: Scatter/Gather: A cluster-based approach to browsing large document collections. In: Procs. of the 15th Annual Intern. ACM Conf. on Research and Development in Information Retrieval (SIGIR 1992), Copenhagen, Denmark, June 1992, pp. 318–329 (1992)
5. Dakka, W., Ipeirotis, P.G.: Automatic extraction of useful facet hierarchies from text databases. In: Procs. of the 24th Intern. Conf. on Data Engineering (ICDE 2008), Cancún, México, April 2008, pp. 466–475 (2008)
6. Ferragina, P., Gulli, A.: A personalized search engine based on web-snippet hierarchical clustering. In: Procs. of the 14th Intern. Conf. on World Wide Web (WWW 2005), Chiba, Japan, May 2005, vol. 5, pp. 801–810 (2005)
7. Gelgi, F., Davulcu, H., Vadrevu, S.: Term ranking for clustering web search results. In: 10th Intern. Workshop on the Web and Databases (WebDB 2007), Beijing, China (June 2007)

8. Hearst, M.A., Pedersen, J.O.: Reexamining the cluster hypothesis: Scatter/Gather on retrieval results. In: Procs. of the 19th Annual Intern. ACM Conf. on Research and Development in Information Retrieval (SIGIR 1996), Zurich, Switzerland, pp. 76–84 (August 1996)
9. Hildebrand, M., van Ossenbruggen, J., Hardman, L.: /facet: A browser for heterogeneous semantic web repositories. In: Cruz, I., Decker, S., Allemang, D., Preist, C., Schwabe, D., Mika, P., Uschold, M., Aroyo, L.M. (eds.) ISWC 2006. LNCS, vol. 4273, pp. 272–285. Springer, Heidelberg (2006)
10. Hyvönen, E., Mäkelä, E., Salminen, M., Valo, A., Viljanen, K., Saarela, S., Junnila, M., Kettula, S.: MuseumFinland – Finnish museums on the semantic web. Journal of Web Semantics 3(2), 25 (2005)
11. Janruang, J., Kreesuradej, W.: A new web search result clustering based on true common phrase label discovery. In: Procs. of the Intern. Conf. on Computational Intelligence for Modelling Control and Automation and Intern. Conf. on Intelligent Agents Web Technologies and International Commerce (CIMCA/IAWTIC 2006), Washington, DC, USA, November 2006, p. 242 (2006)
12. Karlson, A.K., Robertson, G.G., Robbins, D.C., Czerwinski, M.P., Smith, G.R.: FaThumb: A facet-based interface for mobile search. In: Procs. of the Conf. on Human Factors in Computing Systems (CHI 2006), Montréal, Québec, Canada, April 2006, pp. 711–720 (2006)
13. Kules, B., Kustanowitz, J., Shneiderman, B.: Categorizing web search results into meaningful and stable categories using fast-feature techniques. In: Procs. of the 6th ACM/IEEE-CS Joint Conf. on Digital Libraries (JCDL 2006), pp. 210–219. Chapel Hill, NC (2006)
14. Kules, B., Wilson, M., Schraefel, M., Shneiderman, B.: From keyword search to exploration: How result visualization aids discovery on the web. Human-Computer Interaction Lab Technical Report HCIL-2008-06, University of Maryland, pp. 2008–06 (2008)
15. Mäkelä, E., Hyvönen, E., Saarela, S.: Ontogator - a semantic view-based search engine service for web applications. In: Cruz, I., Decker, S., Allemang, D., Preist, C., Schwabe, D., Mika, P., Uschold, M., Aroyo, L.M. (eds.) ISWC 2006. LNCS, vol. 4273, pp. 847–860. Springer, Heidelberg (2006)
16. Mäkelä, E., Viljanen, K., Lindgren, P., Laukkanen, M., Hyvönen, E.: Semantic yellow page service discovery: The veturi portal. In: Gil, Y., Motta, E., Benjamins, V.R., Musen, M.A. (eds.) ISWC 2005. LNCS, vol. 3729. Springer, Heidelberg (2005)
17. Papadakos, P., Theoharis, Y., Marketakis, Y., Armenatzoglou, N., Tzitzikas, Y.: Mitos: Design and evaluation of a dbms-based web search engine. In: Procs. of the 12th Pan-Hellenic Conf. on Informatics (PCI 2008), Greece (August 2008)
18. Sacco, G.M.: Dynamic taxonomies: A model for large information bases. IEEE Transactions on Knowledge and Data Engineering 12(3), 468–479 (2000)
19. Schraefel, M.C., Karam, M., Zhao, S.: mSpace: Interaction design for user-determined, adaptable domain exploration in hypermedia. In: Procs of Workshop on Adaptive Hypermedia and Adaptive Web Based Systems, Nottingham, UK, August 2003, pp. 217–235 (2003)
20. Stefanowski, J., Weiss, D.: Carrot2 and language properties in web search results clustering. In: Menasalvas, E., Segovia, J., Szczepaniak, P.S. (eds.) AWIC 2003. LNCS (LNAI), vol. 2663, pp. 240–249. Springer, Heidelberg (2003)
21. Tzitzikas, Y., Armenatzoglou, N., Papadakos, P.: FleXplorer: A framework for providing faceted and dynamic taxonomy-based information exploration. In: 19th Intern. Workshop on Database and Expert Systems Applications (FIND 2008 at DEXA 2008), Torino, Italy, pp. 392–396 (2008)

22. Wang, J., Mo, Y., Huang, B., Wen, J., He, L.: Web search results clustering based on a novel suffix tree structure. In: Rong, C., Jaatun, M.G., Sandnes, F.E., Yang, L.T., Ma, J. (eds.) ATC 2008. LNCS, vol. 5060, pp. 540–554. Springer, Heidelberg (2008)
23. Xing, D., Xue, G.R., Yang, Q., Yu, Y.: Deep classifier: Automatically categorizing search results into large-scale hierarchies. In: Procs. of the Intern. Conf. on Web Search and Web Data Mining (WSDM 2008), Palo Alto, California, USA, February 2008, pp. 139–148 (2008)
24. Yee, K., Swearingen, K., Li, K., Hearst, M.: Faceted metadata for image search and browsing. In: Procs. of the Conf. on Human Factors in Computing Systems (CHI 2003), Ft. Lauderdale, Florida, USA, April 2003, pp. 401–408 (2003)
25. Zamir, O., Etzioni, O.: Web document clustering: A feasibility demonstration. In: Procs. of the 21th Annual Intern. ACM Conf. on Research and Development in Information Retrieval (SIGIR 1998), Melbourne, Australia, August 1998, pp. 46–54 (1998)
26. Zeng, H.J., He, Q.C., Chen, Z., Ma, W.Y., Ma, J.: Learning to cluster web search results. In: Procs. of the 27th Annual Intern. Conf. on Research and Development in Information Retrieval (SIGIR 2004), Sheffield, UK, July 2004, pp. 210–217 (2004)

Developing Query Patterns

Panos Constantopoulos[1,2], Vicky Dritsou[1,2,*], and Eugénie Foustoucos[1]

[1] Dept. of Informatics, Athens University of Economics and Business, Athens, Greece
[2] Digital Curation Unit / Athena Research Centre, Athens, Greece
{panosc,vdritsou,eugenie}@aueb.gr

Abstract. Query patterns enable effective information tools and provide guidance to users interested in posing complex questions about objects. Semantically, query patterns represent important questions, while syntactically they impose the correct formulation of queries. In this paper we address the development of query patterns at successive representation layers so as to expose dominant information requirements on one hand, and structures that can support effective user interaction and efficient implementation of query processing on the other. An empirical study for the domain of cultural heritage reveals an initial set of recurrent questions, which are then reduced to a modestly sized set of query patterns. A set of Datalog rules is developed in order to formally define these patterns which are also expressed as SPARQL queries.

1 Introduction

A common feature of the informational function of most digital library systems is that they must serve users who may not be familiar with the subject matter of a domain, or, even if they are, may be ignorant of the organization of the information in a given source. This poses the requirement of supporting users to explore sources and discover information. Moreover, it is desirable for users to have access to a repertoire of model questions, complex as well as simple, which can be trusted to represent important information about objects of interest, facilitate the formulation of particular questions and efficiently implement those.

Streamlining the formulation of questions has been addressed through several approaches ([1]). In [5] authors have developed a catalog of question templates to facilitate query formulation, where users must define a corresponding scenario for each query. Similarly in [8] questions are matched to specific question templates (called patterns) for querying the underlying ontology. In [15] a template-based querying answering model is proposed. In the area of Data Mining, and more precisely in association rule mining, many studies are based on the idea of the discovery of frequent itemsets ([14,2]). The effectiveness of such approaches can be enhanced if templates accommodate frequent query patterns. Algorithms for discovering such patterns have been developed in areas with a large amount of

---

* Vicky Dritsou was supported by E.U.-European Social Fund(75%) and the Greek Ministry of Development-GSRT(25%) through the PENED2003 programme.

M. Agosti et al. (Eds.): ECDL 2009, LNCS 5714, pp. 119–124, 2009.

Table 1. From propositions to tacit questions

Proposition	Tacit question
This statue was found at the same place with the statue No 1642.	Which other objects have been found at the same place with the current one?
This object is a copy of a famous scultpure made by Kalimachos.	Who created the original object whose copy is the current one?
This object was exhibited at the museum of Aegina until 1888.	Where was this object exhibited at during other periods of time?

data, such as biology and medicine ([9,4]) and social networks ([12]), but with no separation between the syntactic and the semantic layers of these patterns.

In this paper we make the case that the "right questions" in a domain can be captured through query patterns that are characteristic of the domain of discourse. More precisely we address the development of query patterns at successive representation layers; this layered abstraction (from semantics to structure) can elegantly support the subsequent development of information search and delivery services. We ground our work on the analysis of a specific domain, namely cultural heritage, through a selected sample of collection and exhibition catalogs, where we consider each statement as an answer to a tacit question. A large, yet not vast, set of recurrent tacit questions is revealed from our study: we then focus on reducing it to a set of query patterns. These patterns determine the structure of the questions and disclose the most interesting information within the specific domain. As research has shown that digital libraries can benefit from semantic technologies and tools ([10,3]), these query patterns are conceived as RDF-like graphs and they are shown to be generated from an even smaller set of graph patterns, here called signatures. Finally, we express our query patterns into two well-known query languages, namely Datalog and SPARQL.

2   Empirical Study

In order to conduct a grounded analysis, we considered determining query patterns in the cultural heritage domain first. Such patterns, together with their relative frequency of occurrence, can be taken to represent the largest part of the information needs of users in the domain. Rather than conducting a user study, we turned to an indirect source, namely descriptive texts written by experts about objects in specific collections. These texts contain statements about objects and concepts, which we interpret as answers to tacit questions: questions that could be answered by the statements at hand. Query patterns, abstracted from this set of tacit questions, essentially capture the domain experts' perceptions of what is important to know in a given domain.

Our study comprised five collection catalogs published by the Greek Ministry of Culture (see references in [6]). In total the descriptions of some 1250 objects were studied and a list of tacit questions underlying the object descriptions was compiled. Examples of this question extraction process are shown in Table 1. We contend that these questions lead to information considered as important by the experts. We further contend that these questions also correspond to possible queries that interested users should be able to execute in order to retrieve important information. Therefore, the execution of such queries should be facilitated by cultural heritage digital library systems.

The initial outcome of this study was the recording of 82 distinct questions. Studying these further, we noticed that most of them recur throughout many different descriptions. This recurrence is, of course, expected of the who, where, when, what queries that stand as universal for all exhibits and all types of resources. The remaining questions show a document frequency of at least 25%.

3 Patterns

In the current work, we first set out to analyze the recurrence of queries that we noticed among our 82 queries. As a first step we partitioned the set of these queries into classes of queries that share a common structure at the schema level: this analysis revealed 8 different structures, each one represented as an RDFS graph called "signature" in what follows (see Fig. 1). As a second step we refined the previous partition into a finer one, thus producing 16 structures, each one also represented as an RDF graph called "graph pattern" in the sequel. These 16 graph patterns are given in [6]. Each of them consists of two subgraphs: the first one is a signature among the 8 ones of Fig. 1 and the second one is an "instance" of that signature. An example of such a graph pattern is given in Fig. 4. Each graph pattern describes the common structure of queries in a finer way since it captures not only their common structure at the schema level but also at the instance level. Signatures as well as graph patterns give structural patterns.

In order to better describe the structure of a query, we have to assign to each element of a graph pattern either the characterization "*known*", if it corresponds to an input element of the query, or otherwise the characterization "*unknown*" or "*searching*". Thus, starting from our 16 graph patterns, we develop all possible queries that can be expressed through them. Considering all possible combinations (on the same graph pattern) of known and unknown elements leads to structures richer than the graph pattern itself: we call these structures query patterns.

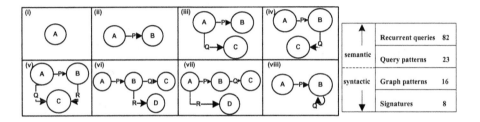

Fig. 1. Signatures of recurrent questions   **Fig. 2.** Pattern types

However, not all such possible combinations represent meaningful queries in the domain of discourse; our goal then is to identify the valid query patterns, where the validity of graph/query patterns is judged by whether they correspond to at least one recurrent question recorded in our empirical study. The above process resulted in the development of 23 query patterns that are valid for the domain of cultural heritage. From now on we call query patterns only the valid ones. A schematic representation of the abstraction hierarchy of all the aforementioned pattern types and their frequencies is shown in Fig. 2.

The developed query patterns have been expressed as Datalog rules on the one hand, following the recent researches on the development of a Semantic Web layer using rules expressed in logic languages with nonmonotonic negation ([7]). On the other hand, the impact of SPARQL together with current studies that suggest translations of SPARQL to Datalog in order to express rules ([11,13]), lead us to express these query patterns as SPARQL queries as well. Due to space limitations, we do not present here the theoretical foundation of our work and we therefore refer interested readers to [6]. Instead, we study in the sequel a complicated example and show how to define its pattern.

4   Defining and Using a Query Pattern: An Example

Consider the following query: "Find all objects that depict the same figure as the current object, but which have been created by different persons" (query I). Our goal here is, first, to express this query in terms of a query pattern and, second, formulate it as a Datalog rule and as a corresponding SPARQL query.

Query expression I comprises four main concepts, namely *Object*, *Depiction*, *Figure*, *Creator*; these four concepts are particular instances of the unspecified concepts A, B, C, D respectively. Moreover, it can be checked that A, B, C, D are interrelated according to signature (vii) of Fig. 1. Thus signature (vii) gives the structural pattern of our query at the schema level; by properly instantiating signature (vii) we obtain the RDFS graph of Fig. 3, which constitutes the semantical pattern of the query at the schema level. A finer structural pattern is given in Fig.4; this graph pattern is obtained by enriching signature (vii) with an instance level.

From our current query we produce the graph pattern of Fig.4 which has two copies at the instance level. In order to understand how we produced this

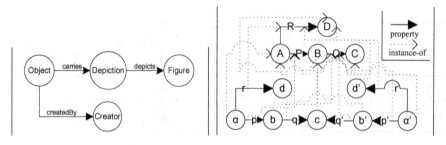

Fig. 3. RDFS graph of sign. (vii)   **Fig. 4.** Graph Pattern I

complex graph pattern from query I, consider the following reformulation of the query: "For the given object a [the current object] in $A : Object$, for the given object b [depiction of the current object] in $B : Depiction$, for the given object c [figure of the depiction of the current object] in $C : Figure$, for the given object d in $D : Creator$, and for the given properties $p : P : carries$, $q : Q : depicts$, $r : R : createdBy$, find all objects a' in $A : Object$, a' different from a, such that a' $createdBy$ d' in $D : Creator$, d' different from d, and a' $carries$ the depiction b' which $depicts$ c." Clearly, the known variables are $A, P, B, Q, C, R, D, a, p, b, q, c, r, d$, the unknown are a', p', b', q', r', d' and the searching variable is a'. That reformulation makes explicit the semantical pattern of the query: it constitutes an "instance" of the graph pattern of Fig. 4. If we remove the semantics from the above reformulation of our query, we obtain its query pattern. Therefore, the Datalog rule that retrieves the desired answer of query I (and of every query having the same query pattern) is:

$$S(x_A) \leftarrow P_{AB}(a, p, b), Q_{BC}(b, q, c), R_{AD}(a, r, d), P_{AB}(x_A, x_P, x_B),$$
$$x_A \neq a, Q_{BC}(x_B, x_Q, c), R_{AD}(x_A, x_R, x_D), x_D \neq d$$

This Datalog rule can also be easily translated in SPARQL as follows:

SELECT $?x_A$
WHERE { $?x_A$ $ns{:}P$ $?x_B$; $rdf{:}type$ $ns{:}A$.
         $?x_B$ $rdf{:}type$ $ns{:}B$; $ns{:}Q$ $ns{:}c$.
         $?x_A$ $ns{:}R$ $?x_D$.
         $?x_D$ $rdf{:}type$ $ns{:}D$.
         FILTER $(?x_A \neq ns : a$ && $?x_D \neq ns : d)$. }

where ns is the namespace prefix of the ontology.

5   Testing and Discussion

In an attempt to test the adequacy of the query patterns to express arbitrary queries submitted by users, we used a set of dialogs between a robot in the role of an online guide of a museum and its online visitors. These dialogues were collected during Wizard of Oz experiments of project INDIGO (see http://www.ics.forth.gr/indigo/) and a total of 23 distinct tours were examined, each one of them containing from 10 to 50 questions. The interesting result is that 93% of the recorded user questions are covered by our developed list of query patterns and the respective Datalog rules and/or SPARQL queries, giving evidence in support of the validity and adequacy of the query pattern development process that we employed.

Query patterns can be exploited to provide user guidance in the difficult task to express a question correctly (a) in terms of the underlying schema and (b) in a given formal query language. Moreover, patterns reveal features of the domain that the user may have not been previously aware of, or filter the available information in such a way, as to highlight the most interesting features. From the developer's perspective, systems can be developed in such a way, so as to provide quick results to these frequent questions. Additionally, the recurrent questions we have established can be considered as predetermined views over the conceptual models of different information sources, calling for the implementation of special mappings and shortcuts. Efficient shortcut definition is a field we are currently exploring. Finally, we believe that query patterns for other domains, sharing certain commonalities, can be developed in analogous ways and we intend to investigate this further.

References

1. Andrenucci, A., Sneiders, E.: Automated Question Answering: Review of the Main Approaches. In: Proc. of the 3rd IEEE International Conference on Information Technology and Applications (ICITA 2005), Sydney, Australia (2005)
2. Bayardo, R.J.: Efficiently mining long patterns from database. In: Proc. of the 1998 ACM SIGMOD Conference, Seattle, USA (1998)
3. Bloehdorn, S., Cimiano, P., Duke, A., Haase, P., Heizmann, J., Thurlow, I., Völker, J.: Ontology-Based Question Answering for Digital Libraries. In: Kovács, L., Fuhr, N., Meghini, C. (eds.) ECDL 2007. LNCS, vol. 4675, pp. 14–25. Springer, Heidelberg (2007)
4. Borgelt, C., Berthold, M.: Mining Molecular Fragments: Finding Relevant Substructures of Molecules. In: Proc. of the 2002 ICDM Conference, Japan (2002)
5. Clark, P., Chaudhri, V., Mishra, S., Thoméré, J., Barker, K., Porter, B.: Enabling domain experts to convey questions to a machine: a modified, template-based approach. In: Proc. of the K-CAP 2003 Conference, Florida, USA (2003)
6. Constantopoulos, P., Dritsou, V., Foustoucos, E.: Query Patterns: Foundation and Analysis. Technical Report AUEB/ISDB/TR/2008/01 (2008),
   `http://195.251.252.218:3027/reports/TR_2008_01.pdf`
7. Eiter, T., Ianni, G., Polleres, A., Schindlauer, R., Tompits, H.: Reasoning with Rules and Ontologies. In: Barahona, P., Bry, F., Franconi, E., Henze, N., Sattler, U. (eds.) Reasoning Web 2006. LNCS, vol. 4126, pp. 93–127. Springer, Heidelberg (2006)
8. Hao, T., Zeng, Q., Wenyin, L.: Semantic Pattern for User-Interactive Question Answering. In: Proc. of the 2nd International Conference on Semantics, Knowledge, and Grid, Guilin, China (2006)
9. Hu, H., Yan, X., Huang, Y., Han, J., Zhou, X.J.: Mining coherent dense subgraphs across massive biological networks for functional discovery. Bioinformatics 21(1), 213–221 (2005)
10. Jones, K.S.: Information Retrieval and Digital Libraries: Lessons of Research. In: Proc. of the 2006 International Workshop on Research Issues in Digital Libraries, Kolkata, India (2006)
11. Polleres, A.: From SPARQL to rules (and back). In: Proc. of the 16th International Conference on World Wide Web (WWW 2007), Alberta, Canada (2007)
12. Sato, H., Pramudiono, I., Iiduka, K., Murayama, T.: Automatic RDF Query Generation from Person Related Heterogeneous Data. In: Proc. of the 15th International World Wide Web Conference (WWW 2006), Scotland (2006)
13. Shenk, S.: A SPARQL Semantics Based on Datalog. In: Proc. of the 30th Annual German Conference on Artificial Intelligence, Osnabrück, Germany (2007)
14. Shrikant, R., Agrawal, R.: Mining quantitative association rules in large relational tables. In: Proc. of the 1996 ACM SIGMOD Conference, Quebec, Canada (1996)
15. Sneiders, E.: Automated Question Answering: Template-Based Approach. PhD thesis, Stockholm University, Sweden (2002)

Matching Multi-lingual Subject Vocabularies

Shenghui Wang[1,2], Antoine Isaac[1,2], Balthasar Schopman[1], Stefan Schlobach[1],
and Lourens van der Meij[1,2]

[1] Vrije Universiteit Amsterdam
[2] Koninklijke Bibliotheek, den Haag
{swang,aisaac,baschopm,schlobac,lourens}@few.vu.nl

Abstract. Most libraries and other cultural heritage institutions use controlled knowledge organisation systems, such as thesauri, to describe their collections. Unfortunately, as most of these institutions use different such systems, unified access to heterogeneous collections is difficult. Things are even worse in an international context when concepts have labels in different languages. In order to overcome the multilingual interoperability problem between European Libraries, extensive work has been done to manually map concepts from different knowledge organisation systems, which is a tedious and expensive process.

Within the TELplus project, we developed and evaluated methods to automatically discover these mappings, using different ontology matching techniques. In experiments on major French, English and German subject heading lists Rameau, LCSH and SWD, we show that we can automatically produce mappings of surprisingly good quality, even when using relatively naive translation and matching methods.

1   Introduction

Controlled knowledge organisation systems, such as thesauri or subject heading lists (SHLs), are often used to describe objects from library collections. These vocabularies, specified at the semantic level using dedicated relations—typically *broader*, *narrower* and *related*—can be of help when accessing collections, *e.g.*, for guiding a user through a hierarchy of subjects, or performing automatic query reformulation to bring more results for a given query.

However, nearly every library uses its own subject indexing system, in its own natural language. It is therefore impossible to exploit the semantically rich information of controlled vocabularies over several collections simultaneously. This greatly hinders access to, and usability of the content of The European Library [1], which is one of important problems to address in the TELplus project [2]. A solution to this issue is the semantic linking (or *matching*) of the concepts present in the vocabularies. This solution has been already investigated in the Cultural Heritage (CH) domain, as in the MACS [3], Renardus [4] and CrissCross [5] projects. MACS, in particular, is building an extensive set of manual links between three SHLs used respectively at the English (and American), French and German national libraries, namely LCSH, Rameau and SWD.

M. Agosti et al. (Eds.): ECDL 2009, LNCS 5714, pp. 125–137, 2009.

These links represent most often equivalence at the semantic level between concepts and can, *e.g.*, be used to *reformulate* queries from one language to the other. For example, an equivalence link between `Sprinting`, `Course de vitesse` and `Kurzstreckenlauf` will allow to transform a query for sprints, that would only give results in the British Library catalogue, into equivalent queries that will have matching results in the French and German catalogues, respectively.

A crucial problem is the cost of building such manual *alignments* of vocabularies. While some reports mention that around 90 terms may be matched per day by a skilled information professional dealing with concepts in a same language [6], the vocabularies to match often contain hundreds of thousands of concepts. In this paper we will show that automatic matching methods can be viable instruments for supporting these efforts in an effective way.

Methodology. We have implemented four straightforward methods for ontology alignment, two based on lexical properties of the concept labels, and two based on the extensions of the concepts, *i.e.* the objects annotated by them. The simplest approach lexically compares the labels of concepts without translating them; a more sophisticated version uses a simple translation service we could deploy out-of-the-box. A simple extensional method makes use of the fact that all three collections have joint instances (shared books), which can be determined by common ISBN numbers. Finally, we extend our previous work of matching based on instance similarity [7] to the multilingual case.

Research questions. In line with TELplus objectives, which include establishing practical solutions as well as guidelines for the participating libraries, the research questions we want to address in this paper are (i) whether an automatic vocabulary matching approach is feasible in a multilingual context, and (ii) what kind of matching approach performs best.

Experiments. To answer the questions above we pairwise applied our four matching methods to the three SHLs Rameau, LCSH and SWD, for which the MACS project has over the years gathered significant amounts of methodological expertise and reference alignments which we can compare newly produced ones with. This comparison is possible because our case comes with large amounts of collection-level data from the libraries using these vocabularies.

Results. The general results show a relatively good precision of all four methods with respect to reproducing manual mappings from the MACS project. Even stronger, the lexical methods produce a high coverage of those mappings, which indicates that the use of such very simple algorithms can already support the creation of such links significantly. Interestingly enough, the extensional mappings produce results that are non-identical to the MACS mappings, which indicates that intensional and extensional semantics (*i.e.*, the meaning of a concept attached by people and its actual use in the libraries) differ significantly.

Structure of the paper. In Section 2 we describe the context of our research, and the problem we started out to solve. Section 3 describes the matching methods in more detail; our experiments are summarised in Section 4, before we conclude.

2   Problem and Context

The TELplus project aims at adding content to The European Library, but also at improving access and usability, notably by investigating full-text indexing and semantic search.

One crucial issue is that collections—and their metadata—come in different languages, which hampers the access to several of them at a same time. A first solution to this issue relies on using *cross-language information retrieval* methods over the different collections at hand, as currently investigated, *e.g.*, in the Cacao project [8]. This approach is promising for cases where full-text content is available, and standard retrieval of documents is sought. However it may require some complement for the cases where only structured metadata records are available, or when one aims at semantically enriched or personalised access to CH objects, as in the Europeana Thought Lab demo [9].

As collection objects are described by controlled knowledge organisation systems, another promising solution is to identify the semantic links between such systems. Currently the TELplus and MACS projects are collaborating so that manually built MACS mappings between the LCSH, Rameau and SWD SHLs can be used to provide automatic query reformulation to bring more results from the three corresponding national library collections.

Meanwhile, automated alignment methods have been investigated in other domains, such as anatomy [10] and food [11]. Particularly, the mapping between AGROVOC and the Chinese Agricultural Thesaurus [12] exemplifies the difficulties of multilingual matching. In the Semantic Web community, efforts related to Ontology Matching [13], like the OAEI campaigns [14], have already dealt with the library and food domains, as well as with multilingual settings. In such context, our work is to investigate the feasibility and potential benefit of applying automated matching techniques for semantic search in TELplus.

3   Ontology Matching Methods Applied

3.1   SKOS Lexical Mapper

Many lexical mappers—that is, mappers exploiting lexical information such as labels of concepts—are only dedicated to English. To palliate this, we have adapted a lexical mapper first developed for Dutch [15] to French, English and German languages. It is mostly based on the CELEX [16] database, which allows to recognise lexicographic variants and morphological components of word forms. This mapper produces *equivalence* matches between concepts, but also *hierarchical* (broader) matches, based on the morphological (*resp.* syntactic) decomposition of their labels.

The different lexical comparison methods used by this mapper give rise to different confidence values: using exact string equivalence is more reliable than using lemma equivalence. Also, the mapper considers the status of the lexical features it compares. It exploits the SKOS model for representing vocabularies [17], where concepts can have *preferred* or *alternative* labels. The latter ones can be approximate synonyms. For two concepts, any comparison based on them is therefore considered less reliable than a comparison based on preferred labels. The combination of these two factors—different comparison techniques and different features compared—results in a grading of the produced mappings, which can be used as a confidence value.

Our mapper only considers one language at a time. To apply it in a multilingual case, we translated the vocabularies beforehand. For each vocabulary pair (*e.g.*, Rameau and LCSH), we translate each vocabulary by adding new labels (preferred or alternative) that result from translating the original labels, using the Google Translate service [18]. We then run the mapper twice, once for each language of the pair. In the Rameau-LCSH case, the translation of Rameau to English is matched (in English) to the original LCSH version, and the translation of LCSH in French is matched (in French) to the original Rameau version. The obtained results are then merged, and we keep only the *equivalence* links.

3.2   Instance-Based Mapping

Instance-based matching techniques determine the similarity between concepts by examining the extensional information of concepts, that is, the instance data they classify. The idea behind such techniques, already used in a number of works like [19], is that the similarity between the extensions of two concepts reflects the semantic similarity of these concepts. This is a very natural approach, as in most ontology formalisms the semantics of the relations between concepts is defined via their instances. This also fits the notion of literary warrant that is relevant for the design of controlled vocabularies in libraries or other institutes.[1]

Using overlap of common instances. A first and straightforward method is to measure the *common extension* of the concepts—the set of objects that are simultaneously classified by both concepts [19,21]. This method has a number of important benefits. Contrary to lexical techniques, it does not depend on the concept labels, which is particularly important when the ontologies or thesauri come in different languages. Moreover, it does not depend on a rich semantic structure; this is important in the case of SHLs, which are often incompletely structured.

The basic idea is simple: the higher the ratio of co-occurring instances for two concepts, the more related they are. In our application context, the instances of a

---

[1] As Svenonius reportedly wrote in [20] "As a name of a subject, the term Butterflies refers not to actual butterflies but rather to the set of all indexed documents about butterflies. [...] In a subject language the extension of a term is the class of all documents about what the term denotes, such as all documents about butterflies."

concept c, noted as $e(c)$, are the set of books related to this concept via a `subject` annotation property. For each pair of concepts, the overlap of their instance sets is measured and considered as the confidence value for an equivalence relation. Our measure, shown below, is an adaption of the standard Jaccard similarity, to avoid very high scores in the case of very few instances: the 0.8 parameter was chosen so that concepts with a single (shared) instance obtain the same score as concepts with, in the limit, infinitely many instances, 20% of which co-occur. This choice is relatively arbitrary, but this measure has shown to perform well on previous experiments in the library domain for the STITCH project [21].

$$overlap_i(c_1, c_2) = \frac{\sqrt{|e(c_1) \cap e(c_2)|} \times (|e(c_1) \cap e(c_2)| - 0.8)}{|e(c_1) \cup e(c_2)|}$$

Note that one concept can be related to multiple concepts with different confidence values. In our experiments, we consider the concept with the highest confidence value as the candidate mapping for evaluation.

Using instance matching. Measuring the common extension of concepts requires the existence of sufficient amounts of shared instances, which is very often *not* the case. However, as instances—in our cases, books—have their own information, such as authors, titles, etc., it is possible to calculate the similarity between them. Our assumption is that similar books are likely to be annotated with similar subject headings, no matter they are from different collections or described in different languages.

The instance matching based method first compares books from both collections. For each book from Collection A, i_a, there is a most similar book from Collection B, i_b. We then consider that i_a shares the same subject headings as i_b does. In other words, i_a is now an instance of all subject headings which i_b is annotated with. This matching procedure is carried out on both directions. In this way, we can again apply measures on common extensions of the subject headings, even if the extensions have been enriched artificially.

There are different ways to match instances. The simplest way is to consider instances as documents with all their metadata as their feature, and apply information retrieval techniques to retrieve similar instances (documents). We use the tf-idf weighting scheme which is often exploited in the vector space model for information retrieval and text mining [22]. Obviously, the quality of the instance matching has an important impact on the quality of concept mappings. One may argue that the whole process is questionable: it is in fact one aim of this paper to investigate it.

To apply such a method in a multilingual context, automated translation is crucial. We take a naive approach, using the Google Translate service to translate book metadata, including subject labels. The translation was done offline on a word-to-word level. We created a list of all unique words in each book collection. Batches of words were sent via an API to the Google Translate service. Every word for which we obtained a translation was stored in a translation table. During the instance matching process, we translate every word of that instance by looking it up in the translation table and replacing it with the translation if available. We then calculate book similarity scores within a same language.

4   Experiments and Evaluation

4.1   Data Preprocessing and Experiment Settings

The three SHLs at hand, namely, LCSH in English, Rameau in French and SWD in German, have been converted into the standard SKOS format [17]—see [23] for an overview of how this was done for LCSH. Collections annotated by these three SHLs, respectively, are gathered from the British Library (BL), the French National Library (BnF) and the German National Library (DNB).

In order to apply instance-based matching methods, the link between each book record and its subjects, *i.e*, concepts from the SHL SKOS conversions, should be properly identified. Instead of using unique identifiers of subjects, librarians often use simple *string* annotations. This introduces many issues, for example, using the lower case version of a concept label, or using alternative labels instead of the preferred ones. This has to be addressed by using a simple string look-up and matching algorithm to identify the book-concept links, using the concept labels found in the SKOS representations.

Furthermore, it is also necessary to tackle the pre-coordination issue. Librarians often combine two thesaurus concepts into a single complex subject to annotate books, *e.g.*, `France--History--13th century`. Some of these combinations are so often used that they are included into the subject vocabulary later, while some are generated only at annotation time. In our data preprocessing step, we applied the following strategy: if the subject combination cannot be recognised as an existing concept, it is then separated into single (existing) concepts, and the book is considered to be annotated by each of these concepts.

We are well aware that this choice is certainly not neutral: hereby, a concept's extension, beyond the instances simply annotated by it, also contains the instances indexed with a compound subject that includes it, if this combination is not an existing concept in the vocabulary. However, it also brings more instances for concepts, which is very important for the statistical validity of the instance-based methods we employ here. Indeed this is made even more important by the low number of annotations we identified from the collections. Not every concept is used in the collections we have, *cf.* Tables 1 and 2. This issue, which is mostly caused by the SHLs being designed and used for several collections, will cause a mapping coverage problem for the instance-based methods, which we will discuss later.

Another related, important decision we made is to restrict ourselves to match only individual concepts, or combinations that are reified in the vocabulary files. This drastically reduces the problem space, while keeping it focused on

Table 1. Size of SHLs and number of concepts used to annotate books in collections

	Total concepts	Concepts used in collection
LCSH	339,612	138,785
Rameau	154,974	87,722
SWD	805,017	209,666

Table 2. Three collections and identified records with valid subject annotations, *i.e.* , there is at least one link between these books and one SHL concept

	Total records	Rec. with valid subject annot.	Individual book-concept links
English	8,430,994	2,448,050	6,250,859
French	8,333,000	1,457,143	4,073,040
German	2,385,912	1,364,287	4,258,106

Table 3. Common books between different collections

Collection pair	Common books
French–English	182,460
German–English	83,786
German–French	63,340

the arguably more important part of the potential book subjects. In fact this is rather in line with what is done in MACS, where very few mappings (up to 3.8% for SWD mappings) involve coordinations of concepts that are not already in the vocabularies.

A last step of preprocessing we need is identifying the common books in two collections. The ISBN number is a unique identifier of one book. By comparing the ISBNs in both collections, we found three dually annotated datasets between the three pairs of SHLs, as shown in Table 3. The amount of common books is extremely small compared to the size of collections. This is not unexpected, but certainly causes a serious problem of concept coverage for the simple instance-based method that completely relies on these common instances.

4.2   Comparing with MACS Manual Mappings

Method. The MACS manual mappings were used as reference mappings. This gold standard is however not complete, as MACS is still work in progress. Table 4 gives the concept coverage of mappings between each pair of vocabularies.

Obviously, there is a serious lack in terms of concept coverage if using MACS as a gold standard. For example, only 12.7% of LCSH concepts and 27.0% of Rameau concepts are both involved in MACS mappings and used to annotate books in the collections we gathered. The situation is much worse for the other two pairs of thesauri, where only 1 to 3% concepts are both considered by MACS and used to annotate books.

To perform a relatively fair evaluation on our matchers' *accuracy*, that is, taking into account the sheer coverage of the MACS gold standard, we separated the generated mappings as "judgeable" and "non-judgeable." A mapping is judgeable if at least one concept of the pair is involved in a MACS manual mapping, otherwise, it is not judgeable—that is, no data in MACS allows us to say whether the mapping is correct of not. We measure *precision* as the proportion of the correct mappings over all generated and judgeable mappings.

Table 4. Simple statistics of MACS manual mappings and concepts involved

	Total MACS mappings	Concepts involved
LCSH – Rameau	57,663	16.4% of LCSH and 36.1% of Rameau
LCSH – SWD	12,031	3.2% of LCSH and 1.4% of SWD
Rameau – SWD	13,420	7.8% of Rameau and 1.6% of SWD

To measure the *completeness* of the found alignments, we would need to compute *recall*, that is, the proportion of the correct mappings over all possible correct mappings. Unfortunately it is very difficult to get all correct mappings in practice. Manual alignment efforts are time consuming and result in a limited amount of mappings if the two vocabularies to align are big. Despite the lack in concept coverage for MACS, we decided that these manual mappings were still useful to exploit. Indeed, measuring how well we can reproduce manual mappings with the help of automatic tools is valuable *per se*. As a proxy for completeness, we thus measure the *coverage* of MACS, that is, the proportion of MACS mappings we find in the automatically produced alignments.

As already hinted, our matchers return candidate mappings with a confidence value—based on linguistic considerations or the extensional overlap of two concepts. This allows us to rank the mappings, and, moving from the top of the ranked list, to measure the above two measurements up to certain ranks.

Results. Fig. 1 gives the performance of four different mapping methods on the task of matching LCSH and Rameau. Here the x-axis is the global rank of those mappings—by "global ranking," we take the non-judgeable mappings into account; however, they are not considered when computing precision. Note that our lexical mapper provides three confidence levels. Mappings with the same value are given the same rank; they are therefore measured together.

The lexical method applied on non-translated LCSH and Rameau gives a very limited amount of mappings: in total, 86% of these mappings are in MACS, but they only represent 13% of the MACS mappings. By naively using Google

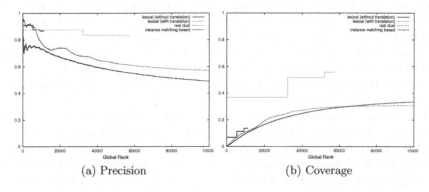

(a) Precision   (b) Coverage

Fig. 1. Performance of different methods on matching LCSH and Rameau

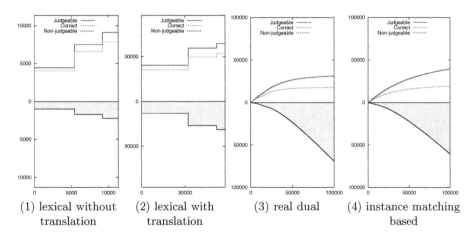

(1) lexical without   (2) lexical with   (3) real dual   (4) instance matching
      translation   translation   based

Fig. 2. The distribution of mappings generated by different methods — LCSH *vs.* Rameau

Translate, the automated lexical method already recovers 56% of MACS mappings, while the precision decreases by 3%. The main reason for this precision decrease is that the translation is not perfect nor stable. For example, in SWD the German name `Aachen` occurs in several subject headings. However, it was sometimes (rightly) translated to the French name `Aix-la-Chapelle` and in other cases it was not translated at all.

From Fig.1, the precision and coverage of the first 7K instance-based mappings generated from the real dually annotated dataset (1% of total book records in two collections) are similar to the lexical method on non-translated thesauri. Gradually, the precision decreases and the coverage increases, and both level after approximately 60K mappings.

The sheer amount of instances inevitably influences the performance of this method. Another possible reason is that instance-based methods focus on the extensional semantics of those concepts, *i.e.*, how they are used in reality. Some mappings are not really intensionally equivalent, but they are used to annotate the same books in two collections. For example, according to MACS, the Rameau concept `Cavitation` is mapped to the LCSH concept `Cavitation`; however, our instance-based method maps it to another LCSH concept `Hydraulic machinery`, because they both annotate the same books. Such mappings could therefore be very useful in the query reformation or search applications, and of course would require further evaluation. This also indicates that the intensional and the extensional semantics (*i.e.*, the meaning of a concept attached by people and its actual use in the libraries) may differ significantly.

The method based on instance matching performed worse here. The loss of nearly 10% in precision could have two reasons: 1) the translation of book information is not good enough; 2) the similarity between books is calculated purely based on weighted common words, where we ignore the semantic distinction between different metadata field, which could potentially help to identify similar

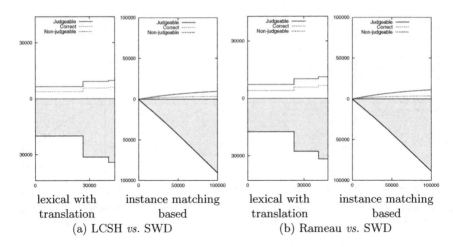

lexical with translation	instance matching based	lexical with translation	instance matching based
(a) LCSH *vs.* SWD		(b) Rameau *vs.* SWD	

Fig. 3. Coverage issue for the LCSH-SWD and Rameau-SWD cases

books. Meanwhile, by matching similar instances on top of using real dually annotated books, we gradually include new concepts, which increases coverage.

As introduced earlier, not every mapping can be evaluated, as neither of their concepts are considered in MACS before. For example, up to rank 50K, 29% of MACS mappings (16,644) between LCSH and Rameau are found, and the precision is 63%. However, only less than 26K mappings were actually judgeable. Fig. 2 compares the distribution of different kinds of mappings of each method, where the shaded area shows the amount of non-judgeable mappings. The coverage issue for the SWD-related cases is more serious, even for lexical mappings, as shown in Fig. 3. Among those non-judgeable mappings, we expect to find valid ones given the precision of the judgeable mappings around them.

We carried out manual evaluation of non-judgeable mapping samples. For instance-based mappings, we first ranked them based on their confidence values, and then chose every 10th mapping among the first 1000 mappings, every 100th mapping from 1000 to 10,000 mappings, and every 1000th mappings from 10,000 to 100,000 mappings. For lexical mappings, we took 50 random mappings within each of the three confidence levels. In all cases, we kept for manual evaluation only the mappings that are not judgeable according to MACS. Depending on the actual sample size, the corresponding error bar was also calculated.

Fig. 4 shows the precision of the manual evaluation proportionally combined with that from comparing with MACS reference mappings. For the LCSH–Rameau case, the precision, which is consistent with Fig 1 (a), indicates that our methods indeed provide a significant amount of valid, and more importantly, complementary mappings to MACS manual mappings. For the LCSH–SWD and Rameau–SWD cases, the global precision is also comparable with the one obtained using MACS alone. It also confirms that all methods perform worse in these two cases, which we can relate to the fact that LCSH and Rameau headings are quite similar in the way they are designed and used, and less similar to

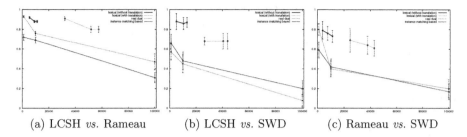

(a) LCSH *vs.* Rameau   (b) LCSH *vs.* SWD   (c) Rameau *vs.* SWD

Fig. 4. Overall precision combining MACS and manual evaluation

SWD. Finally, the performances of the two instance-based methods cannot be distinguished anymore. Yet, due to the small sample size, it is impossible to say whether this is due to statistical uncertainty, or to that fact that the method using instance matching would suffer less from a very small overlap of collections. We will investigate more this aspect in the future.

5   Conclusion

We have explored methods to automatically align multilingual SHLs in a realistic setting, using different techniques known from ontology matching, exploiting labels of concepts or book-level information. We also reported various problems with applying existing techniques and the possible (and simple) solutions to them. In particular, we use the Google Translate service to address the translation problem. We are well aware that other, more sophisticated approaches exist, but deploying them requires thorough expertise in the use of linguistic resources, which makes it more difficult to assess the *practical* feasibility of such an approach.

These experiments on the major French, English and German SHLs, Rameau, LCSH and SWD, show that we can automatically produce mappings of surprisingly good quality, even when using quite naive translation and matching methods. The lexical methods produce a relatively high coverage of the MACS manual mappings, which indicates that the use of such very simple algorithms can already support the creation of manual mappings significantly. The instance-based mapping methods provide mappings that are nearly complementary to manual ones. This is more interesting, as it indicates that, while each can be useful in different applications, the intensional and extensional semantic links are significantly different. More efforts would now be required to turn these findings into an appropriate methodology to assist manual alignment, or to evaluate to which extent imperfect mappings can still benefit to multilingual collection access for end users in the TELplus case. Our results also identify different directions to improve the performance of methods, and we will continue reporting our efforts in this area.

Acknowledgements. This work is funded by the NWO CATCH and EU eContentPlus programmes (STITCH and TELplus projects). Patrice Landry, Jeroen

Hoppenbrouwers and Geneviève Clavel provided us with MACS data. Various people at Library of Congress, DNB, BnF and TEL Office have provided or helped us with SHL and collection data, incl. Barbara Tillett, Anke Meyer, Claudia Werner, Françoise Bourdon, Michel Minguam and Sjoerd Siebinga.

References

1. http://www.theeuropeanlibrary.org/
2. http://www.theeuropeanlibrary.org/telplus/
3. Landry, P.: Multilingualism and subject heading languages: how the MACS project is providing multilingual subject access in Europe. Catalogue & Index: periodical of CILIP Cataloguing and Indexing Group 157 (to appear, 2009)
4. Day, M., Koch, T., Neuroth, H.: Searching and browsing multiple subject gateways in the Renardus service. In: Proceedings of the 6th International Conference on Social Science Methodology, Amsterdam, The Netherlands (2005)
5. Boterham, F., Hubrich, J.: Towards a comprehensive international Knowledge Organisation System. In: 7th Networked Knowledge Organization Systems (NKOS) Workshop at the 12th ECDL Conference, Aarhus, Denmark (2008)
6. Will, L.: Costs of vocabulary mapping,
   http://hilt.cdlr.strath.ac.uk/Dissemination/Presentations/
   Leonard%20Will.ppt
7. Schopman, B., Wang, S., Schlobach, S.: Deriving concept mappings through instance mappings. In: Proceedings of the 3rd Asian Semantic Web Conference, Bangkok, Thailand (2008)
8. http://www.cacaoproject.eu/
9. http://www.europeana.eu/portal/thought-lab.html
10. Zhang, S., Bodenreider, O.: Experience in aligning anatomical ontologies. International journal on Semantic Web and information systems 3(2), 1–26 (2007)
11. Lauser, B., Johannsen, G., Caracciolo, C., Keizer, J., van Hage, W.R., Mayr, P.: Comparing human and automatic thesaurus mapping approaches in the agricultural domain. In: Proceedings of the International Conference on Dublin Core and Metadata Applications, Berlin, Germany (2008)
12. Liang, A.C., Sini, M.: Mapping AGROVOC and the Chinese Agricultural Thesaurus: Definitions, tools, procedures. The New Review of Hypermedia and Multimedia 12(1), 51–62 (2006)
13. Euzenat, J., Shvaiko, P.: Ontology Matching. Springer, Heidelberg (2007)
14. http://oaei.ontologymatching.org/
15. Malaisé, V., Isaac, A., Gazendam, L., Brugman, H.: Anchoring Dutch Cultural Heritage Thesauri to WordNet: two case studies. In: ACL 2007 Workshop on Language Technology for Cultural Heritage Data (LaTeCH 2007), Prague, Czech Republic (2007)
16. http://www.ru.nl/celex/
17. Isaac, A., Summers, E.: SKOS Primer. W3C Group Note (2009)
18. http://translate.google.com/
19. Vizine-Goetz, D.: Popular LCSH with Dewey Numbers: Subject headings for everyone. Annual Review of OCLC Research (1997)

20. Svenonius, E.: The Intellectual Foundation of Information Organization. MIT Press, Cambridge (2000)
21. Isaac, A., van der Meij, L., Schlobach, S., Wang, S.: An empirical study of instance-based ontology matching. In: Aberer, K., Choi, K.-S., Noy, N., Allemang, D., Lee, K.-I., Nixon, L.J.B., Golbeck, J., Mika, P., Maynard, D., Mizoguchi, R., Schreiber, G., Cudré-Mauroux, P. (eds.) ASWC 2007 and ISWC 2007. LNCS, vol. 4825, pp. 253–266. Springer, Heidelberg (2007)
22. Salton, G., McGill, M.J.: Introduction to Modern Information Retrieval. McGraw-Hill, New York (1983)
23. Summers, E., Isaac, A., Redding, C., Krech, D.: LCSH, SKOS and Linked Data. In: Proceedings of the International Conference on Dublin Core and Metadata Applications, Berlin, Germany (2008)

An Empirical Study of User Navigation during Document Triage

Fernando Loizides and George Buchanan

Future Interaction Technologies Laboratory, Swansea University
Centre for HCI Design, City University
csfernando@swan.ac.uk, george.buchanan.1@city.ac.uk

Abstract. Document triage is the moment in the information seeking process when the user first decides the relevance of a document to their information need[17]. This paper reports a study of user behaviour during document triage. The study reveals two main findings: first, that there is a small set of common navigational patterns; second, that certain document features strongly influence users' navigation.

1 Introduction and Motivation

When seeking for information in a digital library, users repeatedly perform the task of "document triage", where they make an initial judgement of a document's potential relevance to their information need. There is a wide body of associated literature that gives indirect evidence about document triage. However, the scientific understanding of document triage specifically is limited.

The consistency and accuracy of human relevance decision making is often poor compared to nominal 'perfect' choices for known tasks on closed sets of documents [4]. Human information seekers seem to rely at least in part on search engines to 'predict' the most accurate documents for their information needs. Following a search, there is minimal tool support for the relevance decision–making of users. Observed behaviour suggests that the user's attention is narrowly focussed on a few parts of particular documents. For example, in a search result list, users mainly focus on documents presented on the first or second page [20]. Similarly, users choose only a few documents for further scrutiny, and appear to read little in detail beyond the first page [4]. This suggests that users are using only a limited pool of information to make relevance decisions.

Document triage can be viewed as a form of visual search. In computer–human interaction, the visibility principle asserts that the important aspects of a system must be visible [11]. Applying this principle to documents, one can argue that users will be influenced by what is displayed of a document. We wished to explore whether the factors suggested in the current literature (e.g. [7,19]) did indeed influence how users move and view actual documents.

Information seekers have their own mental models or "personal information infrastructures" for systematic information seeking [16] in which they follow certain visual cues. There is a healthy diversity of information seeking models

M. Agosti et al. (Eds.): ECDL 2009, LNCS 5714, pp. 138–149, 2009.

(e.g. [8,13,16]), however none of these describes document triage in detail. The current research into triage is fragmented, and this no doubt explains the lack of such a model. Available reports focus on very specific aspects or external aspects rather than document triage generally [1,2,3,5,18].

The research reported in this paper is a laboratory–based study that provides an initial assessment of the impact of common visual document features on on user behaviour during document triage. While some features focus user attention, the evidence is that conversely some relevant material is overlooked. When such issues are better understood, designers of document reader software can improve support for triage work.

We now report the study in three parts: design, results, and discussion. The paper concludes with a summary of key findings, and a view of future research.

2   Study Design

In general form our study was a laboratory–based observational experiment. Participants performed document triage on a closed corpus of documents, evaluating each for its suitability for a particular task. Log data of their interaction was captured. Pre– and post–study questionnaires plus an interview collected further data for comparison with the participant's actual behaviour during the experiment. This general structure was chosen to permit us to systematically assess the interaction between visible document content and visible document cues (e.g. headings, diagrams) and both user behaviour and relevance judgments.

2.1   Apparatus and Participant Selection

We developed a bespoke PDF reader software, that logs detailed information on the user's navigation in an XML file. Our study used a set of PDF files that were examined by the participants. To analyze the navigation log data accurately, we extracted the content from each PDF file and passed it through a custom–built parser. This permitted us to identify the exact content visible at any moment, using the view position and zooming information. The parser identifies heading, image and other visual feature information. An initial run on the document data was validated by human visual inspection to verify that the output was accurate for the set of documents we were using.

We recruited twenty participants, all having previous experience of PDF reader software. The participants were studying at postgraduate level in a computer science discipline, avoiding the specific CHI sub–disciplines that appeared in the study material. The participants' ages ranged from 21 to 38.

2.2   Study Format

Before the study, participants reported their perceived importance of common document features (title, heading, etc.). This data calibrated our findings with other work (e.g. [7,19]). Overall, our data aligned closely with earlier work.

Participant responses indicated that a few features (e.g. title and heading text) would strongly influence most users, whilst many features (e.g. body text) possessed only moderate influence. We will report individual results later, when we compare observed behaviours against subjective self–reporting. We also elicited previous experience with document reader software.

Participants were then given two information seeking tasks. The subject area of the tasks were not the speciality of the participants, but was sufficiently close to their domain to be intelligible. It was stressed to the participants that no prior knowledge of the subject was assumed, and their personal performance was not being assessed. The first information task was to find material on the interfaces of tablet PC's, the second was to find papers on specific CHI evaluation methods. The participant was then given a set of documents to evaluate against the given information need. They were instructed to give each document a rating out of 10 (1 for non–relevant documents, 10 for extremely relevant documents). The articles ranged from short papers (2 pages) to full journal papers (29 pages).

Users were given an open–ended timeframe to complete their tasks and could ask any question to clarify the task objectives. They were supplied with a dictionary should unknown words need clarification. No guidance was given regarding the papers themselves. The entire computer display was recorded using screen capturing software (BB Flashback[1]). A post–study interview was undertaken to obtain a better understanding of the participants' actions and thoughts. This interview was semi–structured with standardized questions about certain visual features (e.g. headings). Individual actions noted by the observer were also probed to elicit a better understanding of the user's intentions.

The corpus of documents was selected to permit a more detailed picture of the impact of specific document content and visual features. Whilst the majority had, broadly speaking, similar numbers of headings and other features per page, there were outliers that could reveal the impact of extreme cases. For example, one document contained text in only one typeface, size and weight.

3   Results

In this section we initally describe *navigational patterns* that we observed participants use during the document triage process. We then study what was actually visible during the triage process. Attention (i.e. viewing time) for different content is, naturally, not the same, and we report the relationship between document content and visual features and the time for which that content was visible. Finally, we report the participants' relevance judgments, and how these relate to the navigational behaviour identified in the study.

3.1   Document Navigation

From the viewing log data, we extracted a number of high–level navigation patterns that recurred in the participants' viewing of documents. Despite being

---

[1] http://www.bbsoftware.co.uk

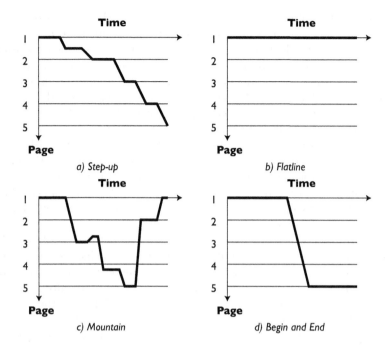

Fig. 1. Navigational Behaviours

given a free choice of action, all participants viewed documents in the same order, proceeding through the list of documents in a linear fashion.

We then studied the participants' viewing behaviours within documents. The first few seconds of every document log was linear, but from that point variations quickly emerged. These are now reported in descending frequency order. Figure 1 presents caricatured timelines to illuminate the patterns visually.

Step Navigation is our term for the predominant viewing pattern, found on 229 document passes, and in all 20 participants. As seen on Figure 1a, the initial part of the document is viewed for a prolonged period without any movement. In comparison, the other parts of the document receive periodic spans of attention, and many only viewed momentarily during rapid scrolling.

The inital view contains the title, abstract and, on most occasions, a large part of the introduction. (The precise figures of how long users remained on different parts of the document is discussed in subsection 3.2). The conclusion section of the document was often the only part of the end of the document that received any attention. Allowing for very small (<10 pixel) backtracks that are probably a result of mouse manipulation errors by the user, the user travels linearly through the document, but with attention unequally spread.

The second most common behaviour was the **flatline** pattern (see Figure 1b). Users simply scrutinize the first page or visible part of the document. In contrast to the step pattern, there is no subsequent scrolling to later document content. This was observed on 73 documents and in 18 users.

The third pattern was for the user to scrutinize the document from the beginning to the end – in the same manner as the step–up pattern – and then backtrack through the document towards the beginning, producing a **mountain** pattern(see Figure 1c). In our specific example we see the user returning to an already viewed part of the document to view it for a longer timespan (and, presumably, in greater detail). In eighteen cases, the user returned to the very beginning of the document for a final view. Overall, the mountain pattern was observed 23 times across 12 individual participants.

A last behaviour observed is the **begin–and–end** pattern (Fig. 1d). This is when the user scrutinizes the beginning of the document before directly scrolling to the conclusion of the document without any significant pause to viewing other document parts. Variations arise in backtracking to locate the actual conclusion, but in essence only two parts of the document are viewed. This behaviour was observed 33 times across all 20 users.

The models discussed above are nominal perfect models. Users often did not follow one of the models perfectly, but rather incorporate behaviours from two or more models to create hybrids. Furthermore, all the models for navigation, apart from the flatline pattern, share a common variant: where users, after reaching a specific point later on in the document, return to the beginning of the document for an extended period of time. In total, 77 of the 320 document triage runs returned to the beginning of the document, though 23 of these views were for less than one second.

3.2   Document Features and Navigation

We now report the impact of document content and visual features on the users' viewing behaviours. We commence by characterising user behaviour on pages that lack any of the features we are investigating, before turning to each feature in turn. Summarized figures are found in Figure 3.

Document Length. Document length does result in – on average – increased viewing times. The general view can be seen in Fig.2, where documents are ranked by length and with total viewing time presented. However, this picture is complicated by users using navigation tools to bypass particular pages (e.g. "go to" the penultimate page). We therefore evaluated the impact on page number on viewing time and the number of participants who viewed the page, as well as the simplified aggregate view corrected. Taking the entire viewing history of all documents, we discovered that the Pearson's correlation between the page number and the average viewing time of that page was -0.411. The correlation between the page number and the number of participants who viewed that page was -0.785. Hence, as the number of a page increases, fewer participants viewed it, and those who viewed it did so for a shorter time. This characterizes a global pattern of sharply falling attention being paid to succeeding pages.

Featureless views. Areas without any noticeable properties (i.e. plain text only) were typically scrolled past at a fast rate. This brief viewing time makes it unlikely that even superficial skimming of the page occured that could contribute

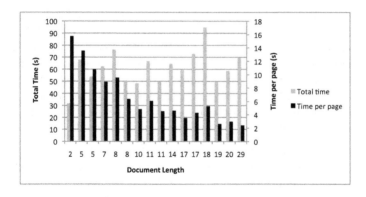

Fig. 2. Timings per document

to their relevance judgment. The average viewing time for a page view that contained plain text only was over one second (1.38s). Excluding one featureless first page with a corresponding long view time, lowers the average to below 0.5s. In contrast, the average time in view of all pages is some 5.28s. Unsurprisingly, applying Student's t–test at either the view or page level returns p<0.001 (t=4.96, df=4018). This is strong evidence that both pages and displayed views with minimal distinguishing features will receive only scant attention from the average user.

Initial Page. Abstracts and titles are considered to play a key role in user relevance judgments [19]. Both are most often located on the first page with much of the introduction. Across all documents, the first page accounted for 48.2% (SD=15.24, n=320) of all viewing time. This suggests the first–page content plays a central role in the relevance decision making of users. The average time a user spent on the first page was 23.21 seconds, in contrast with an average for all pages of 5.28s. As mentioned in Sec. 3.1, users relied solely on these three parts for making their relevance decision on 43 documents, and a further 30 documents were judged with less than 10% of the user's time spent outside the first page. Sixteen users relied solely on the first page for at least one document, and two users displayed this behaviour on 11 documents each.

Participants returned to the beginning of a document in the triage of 77 documents. In 23 instances the last view of the first page was extremely brief (< 1s). However, the average time for this second viewing for the remaining 54 cases was some 13.94s (SD=13.45), compared to an intial viewing time of 29.15s (SD=30.41). Second viewings can thus be more than cursory, though the difference between first and final view times is significant (p<0.0001,t=4.46).

Given the high impact of the first page itself, for the content features that follow, we discount the first page in all cases.

Headings. A simple correlation of the number of headings per page with the time spent with the page in view gives r=0.286 (df=4018,p<0.01). Considering the more focussed data of views, this correlation increases slightly to r=0.322.

This suggests that headings have a modest impact on navigation behaviour. The time in view of an individual heading was 6.57s (SD=2.69), while for views without a heading the average was 4.20s (SD=2.05). The t–test yields p<0.0001 (t=4.438). Furthermore, there is a correlation between the number of headings on a page and the number of participants who viewed that page for >=0.3s of r=0.354 (df=199,p<0.01). Thus, one can conclude that the presence of a heading increases the time for which user will study the page, and it also increase the likelihood that they will halt at that view rather than scroll continuously.

Pictures. In total 46 separate pictures occupied 11.9% of the total surface area of the documents, and were in view for an average of 17.73% of the participants' time. Testing the time in view for a picture (5.86s) results in Pearson's r=0.242 (p<0.05,df=720), indicating a weaker impact than headings. This is also reflected in other statistical comparisons: images fail to change the number of participants who viewed the page for >= 0.3s (r=0.042, not significant)

Tables. Tables and other statistical figures occupied 7.6% of the visible space and 9.2% of total triage time. A Pearson's coefficient of a mere 0.024 confirms that there is minimal likelihood of an effect, confirmed by other tests failing to obtain significance. It seems likely that this content plays at best a marginal role in users' information triage work.

Emphasized Text. Emphasized text includes bullet points, bold text, italic text and underlined text. When the documents included emphasized text, these areas were visible on average 7.05% of the participants' time (SD=3.56). On average participants took 3.35 seconds to triage each area containing emphasized text. In total participants spent 938 seconds of their triage time on these areas, accounting for 4.8% of their total triage time.

Conclusions Section. Of the twenty documents used in this study, 14 contained a "conclusions section". The average time on display was some 7.72 seconds when viewed (SD=2.60s), and applying the t–test against all other pages results in p<0.0001 (t=4.36), confirming an effect, and Pearson's r is 0.432 [2]. No correlation or effect is discernable when testing for either views of >= 0.3s (e.g. Pearsons=0.04) or any view. However, it must be considered that conclusions are typically late in the document, and we earlier noted a negative correlation between page numbers, viewing rates and times. Put simply, many participants fail to navigate as far as the conclusion on any given document.

Testing the conclusion page against the three previous pages, to mitigate this placement effect, strengthens all time outcomes (the average viewing time of the prior 3 pages is a low 3.56s). Applying Student's t–test to determine an effect on the number of participants who viewed the page results in p=0.051 (one–tailed,t=1.75,df=14), but the difference is small (average number of viewing participants being 11.05 for conclusions pages, and 9.82 for neighboring ones). Our observations of the study suggested, and the navigation patterns reported

---

[2] N.B. we would consider this test inappropriate on a binary value, but include its result for completeness.

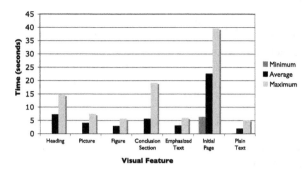

Fig. 3. Mean visible time per visual feature (in seconds)

above, both suggested that some users attempt to guess the location of the conclusions section, and this data gives some further support to that hypothesis.

Participants viewed an average of 8.5 conclusions (SD=4.73) from the 14 documents that possessed one: a rate of just over 60%. Not all were scrutinized in detail: just over 30% were seen for less than a second.

Summary. The outcome of these observations is a clear set of effects on user navigation behaviour. The initial page holds a key place in the user's triage reading. Subsequent pages receive less and less attention, with falling viewing times and declining likelihoods of a page either being viewed at all or read in detail. Headings, the conclusion section and some graphical content do result in higher display times and, in the case of headings particularly, an increasing likelihood of the page being displayed for more than a brief overview.

4   Discussion

We now recap the main impacts of the findings reported in the body of this paper. In principle, the documents presented to the participants were all rated relevant to the tasks by three subject experts. This should encourage engaged triage behaviour, as opposed to very brief rejections. Our participants reported similar subjective ratings for the importance of document features as reported in the current literature (e.g. [7]). That data is universally self–reported, and we now can compare this against actual user behaviour. In this section, we will supplement our main data with other information from the study, including subjective ratings of document features.

4.1   Navigational Behaviour

We identified four distinct types of navigational behaviour that occur during document triage. By far the most common is the *step up* pattern (see Section 3.1). In principle, this navigational behaviour allows the information seeker to skim all parts of the document, and therefore observe all of a document's features (described in Section 3.2). Variants also allow for this full document skimming. The *flatline* and *begin and end* patterns of navigation are less common than the *step up* approach, but were observed in most participants (see Section 3.1). These do not allow the user to interact with the entire content of the document and may result in the information seeker making poor relevance decisions.

4.2   Reported and Actual Behaviour

Plain text received a subjective rating of 4.8 out of 10 in the pre–study questionnaire. Plain text was, of course, visible for almost every moment of viewing time. However, when the document display contained only plain text, there was minimal time spent viewing that part of the document (Sec. 3.2). As we shall see, while content text is important, the evidence from other visual cues leads the reader's attention to a particular point. User 15 explained in the post–study interview that "If there is no text that is bold or stands out then I just scroll without reading". This seems consistent with the observed pattern of pages with no visual features being viewed for much less time than other content.

Document Length. Liu [14] has suggested that electronic reading is used less for in–depth reading. The data from our study suggests that document length has a minimal effect on users' relevance ratings. However, there is a clear effect where each succeeding page in a document receives less attention. This suggests that if a search engine determines that the relevant content for a search falls in later pages, and is absent from the first page, users may well fail to identify that material in their visual scan of the document. Byrd [6] and Harper [9] have both endeavored to make later content more visible in the user's interaction with documents. However, this earlier work failed to arrive at proven effective solutions. This topic of research appears, therefore, to be worth further study.

Main Title, Abstract and Introduction. In the pre-study questionnaire, participants viewed the main title as one of the two most important features of the document, rating it with a subjective importance of 8 out of 10. The Abstract and the Introduction ranked 3rd and 4th in importance. As all this material typically falls on the first page of a document, one would expect it to receive a considerable amount of scrutiny – and hence viewing time.

This is confirmed by the behaviour of the participants, who on average spent 48% of their triage time on the initial page of a document. Two documents had no formal abstract or introduction, and for these documents the average share of time drops to 43%. We therefore are able to add concrete data for viewing times of visual data to the findings of Cool, and others [4,12].

Subjective feedback reported that obscure and misleading titles are a negative influence in ranking, and time consuming as the reader has to work harder to establish the relevance of a document. Similarly, some abstracts were criticized for being imprecise. Participant 6 commented that he would "prefer a short abstract which doesn't over-analyze the paper". Participant 13 reinforced these points, saying "I love a short simple abstract which gets to the point, and a good main title". Due to the format of the study, we cannot discriminate between the impact of these different factors in detail.

Headings. have long been held as a significant factor in document triage[7]. The participants rated headings as the second highest visual factor when making a relevance decision (see Sec. 3). Participants repeatedly observed during interview that "headings are really important". However, we doubt that headings alone are the issue: rather, the manner in which they attract the reader's attention to body content is key. The impact of a lack of headings is often affective, comments made during the sessions included "I don't like this, there are no headings or anything" and "what is this? It doesn't even have any headings in it".

Pictures and Tables. Pictures attracted participant attention and had a mild effect on subjective relevance ratings. Participant 1 explained that "pictures give you a better idea and makes a paper more appealing to someone". As for abstracts and other content, form appears to have had an effect on the final impact of illustrations. Obscure diagrams attracted considerable criticism. Participant 8 noted that reading some of the figures was simply "too much effort". Figures had virtually no impact on user behaviour.

Conclusions Section. Conclusions received significantly more viewing time than nearby material. Only 60% of the conclusions were reached and viewing times were very brief in a third of cases. It is arguable that conclusion sections have not had, as a whole, much impact on subjective relevance ratings. The observed behaviour is consistent with reading of conclusions being informative and summative, rather than a key point in a user's triage decision.

Summary. Liu [14] characterised digital reading as hasty and incomplete. It seems that this characterisation may be found in an even more extreme form during document triage. Much of each document was simply scrolled by with minimal observation, whilst considerable viewing time was given to the first page alone. Viewing behaviours elsewhere seem influenced by highly visual content. The previous portraying of information seeking as a scent–following or berryp-icking behaviour seem relevant, but in a much higher paced form. Content that does not appear on the first page is unlikely to gain attention unless near to "honeytrap" features such as images, or a heading that reflects the user's information goal. Our data is not yet sufficiently conclusive to be certain of this simple cartoon of user behaviour. Further work is needed to finally prove or disprove this hypothesis.

5   Conclusion and Future Work

We are developing a prototype document reader that provides improved user support for triage reading. This software will provide improved within-document query support using IR methods [9] and approximate matching. Most importantly, we aim to improve the attention given to content with low visibility but a good match to the user's query. There are a number of methods for achieving such focus support; e.g. keyword highlighting, interactive feedback (e.g. non–linear srcolling) and simple visualization techniques (as already explored in [9,10,15]). We aim to compare these systematically.

User studies to gain further insight on the information triage process are also needed. Eye–tracking provides an opportunity to study visual focus in more detail. However, the centre of gaze has an imperfect correspondence with the user's centre of attention [18].

One major concern, for both user studies and system design, is that the problems of overlooked relevant material are even greater in large documents, such as PDF books of hundreds of pages. Users thus require much better support than we observe at present. It may be that user behaviour changes on longer texts to compensate for the rising cognitive demands of both visual search and precise scrolling. A related issue is that triage behaviours will certainly vary between the media concerned and the information needs and skills of the users. Further user studies with different communities (e.g. humanities researchers, lawyers) and media types are needed to provide sufficient data to provide reliable generic knowledge. Finally, as we refine and understand the process of document triage we are closer to formulating a conceptual model to represent patterns and suggest "relationships that might be fruitful to explore or test"[21].

Through a controlled laboratory–based observational study, we collected empirical data of the navigation patterns of users when viewing documents for triage. Combining self–reported significances of visual document features with the detailed navigational logs from our instrumented PDF reader, we arrived at a detailed picture of document triage as a visual search task. We uncovered a number of differences between subjective ratings and actual behaviour, and were able to characterise a set of navigational patterns that frequently occurred during triage. The study data confirms the importance of known visual cues, whilst illuminating some distinctions between them. We are also able to provide solid data to confirm the intuition of previous researchers that in digital documents little attempt is made to engage in detailed reading. This has significant repercussions for future research: e.g. user cognition may require better support to target skimming activity, and documents that lack the visual infrastructure (e.g. headings) to assist user navigation will receive low subjective relevance scores. We look forward to a more detailed assessment using eye–tracking technology to observe finer detail within the patterns we report here.

Acknowledgments

This research is supported by EPSRC Grant GR/S84798.

References

1. Badi, R., Bae, S., Moore, J.M., Meintanis, K., Zacchi, A., Hsieh, H., Shipman, F., Marshall, C.C.: Recognizing user interest and document value from reading and organizing activities in document triage. In: IUI 2006: Procs. 11th Int. Conf. on Intelligent user interfaces, pp. 218–225. ACM, New York (2006)

2. Bae, S., Badi, R., Meintanis, K., Moore, J.M., Zachhi, A., Hsieh, H., Marshall, C.C., Shipman, F.M., Costabile, M.F., Paterno, F.: Effects of display configurations on document triage. In: Costabile, M.F., Paternó, F. (eds.) INTERACT 2005. LNCS, vol. 3585, pp. 130–143. Springer, Heidelberg (2005)
3. Buchanan, G.: Rapid document navigation for information triage support. In: Procs. ACM/IEEE Joint conference on Digital libraries, p. 503. ACM Press, New York (2007)
4. Buchanan, G., Loizides, F.: Investigating document triage on paper and electronic media. In: Proceedings of the European Conference on Reasearch and Advanced Technology for Digital Libraries, vol. (35), pp. 416–427 (2007)
5. Buchanan, G., Owen, T.: Improving skim reading for document triage. In: Procs. Int. Symp. on Information Interaction in Context, pp. 83–88. ACM, New York (2008)
6. Byrd, D.: A scrollbar-based visualization for document navigation. In: Procs. ACM Conference on Digital Libraries, pp. 122–129. ACM, New York (1999)
7. Cool, C., Belkin, N.J., Kantor, P.B.: Characteristics of texts affecting relevance judgments. In: 14th National Online Meeting, pp. 77–84 (1993)
8. Ellis, D.: A behavioral approach to information retrieval system design. Journal of Documentation 45(3), 171–212 (1989)
9. Harper, D.J., Koychev, I., Sun, Y., Pirie, I.: Within-document retrieval: A user-centred evaluation of relevance profiling. Information Retrieval 7(3-4), 265–290 (2004)
10. Harper, D.J., Koychev, I., Sun, Y.: Query-based document skimming: A user-centred evaluation of relevance profiling. In: European Conference on Information Retrieval, pp. 377–392 (2003)
11. Jones, S., Jones, M., Deo, S.: Using keyphrases as search result surrogates on small screen devices. Personal Ubiquitous Comput. 8(1), 55–68 (2004)
12. Joseph, J.W.: Relevance judgments and the incremental presentation of document representations. Inforamtion Processing and Management 27(6), 629–646 (2005)
13. Kuhlthau, C.C.: Seeking Meaning. A Process Approach to Library and Information Services. Ablex Publishing, Greenwich (1996)
14. Liu, Z.: Reading behavior in the digital environment. Journal of Documentation 61(6), 700–712 (2005)
15. Loizides, F., Buchanan, G.R.: The myth of find: user behaviour and attitudes towards the basic search feature. In: Procs. ACM/IEEE-CS joint conference on Digital libraries, pp. 48–51. ACM, New York (2008)
16. Marchionini, G.: Information Seeking in Electronic Environments. Cambridge University Press, Cambridge (1995)
17. Marshall, C.C., Frank, I., Shipman, M.: Spatial hypertext and the practice of information triage. In: HYPERTEXT 1997: Proceedings of the eighth ACM conference on Hypertext, pp. 124–133. ACM Press, New York (1997)
18. Mosconi, M., Porta, M., Ravarelli, A.: On-line newspapers and multimedia content: an eye tracking study. In: SIGDOC 2008: Procs. 26th ACM international conference on Design of communication, pp. 55–64. ACM Press, New York (2008)
19. Saracevic, T.: Comparative effect of titles, abstracts and full texts on relevance judgments. In: Procs. American Society for Information Science, pp. 293–299 (1969)
20. Spink, A., Jansen, B.J., Wolfram, D., Saracevic, T.: From e-sex to e-commerce: Web search changes. Computer 35(3), 107–109 (2002)
21. Wilson, T.D.: Models in information behaviour research. Journal of Documentation 55(3), 249–270 (1999)

A Visualization Technique for Quality Control of Massive Digitization Programs

Rodrigo Andrade de Almeida, Pedro Alessio, Alexandre Topol, and Pierre Cubaud

Centre d'études et de recherches en informatique (CEDRIC)
Conservatoire National des Arts et Métiers (CNAM)
292 rue Saint-Martin, 75003 Paris, France
{rodrigo.almeida,pedro.alessio,topol,cubaud}@cnam.fr

Abstract. Massive digitization programs need massive visualization techniques for quality control. We describe the functional prototype of a 3D interactive environment enabling a rapid inspection of pages conformity for large batches of digitized books.

Keywords: visualization techniques, massive digitization, image control quality.

1 Introduction

Massive digitization programs for books collections have been proposed by a few major organizations in the past decade. In February 2008, the millionth volume has been digitized by the partnership between Google and the University of Michigan Library [1]. In Europe, similar projects are undertaken under the umbrella of the European Commission, which funds the Europeana portal. To this day, about 70 European institutions collaborate to Europeana. The French National Library, for instance, digitizes approximately 100,000 volumes per year.

These demands of public institutions have fostered the creation of specialized digitization centers that act as external contractors. The digitization process is in fact a pipeline, divided in the following tasks: (1) batch preparation, (2) scanning, (3) image processing, (4) OCR, (5) metadata production, and (6) ebook delivery. Statistical quality control is used for each step of the process. A random sample is drawn at each step. If rejected, the entire batch returns to the step before, requiring more human supervision [2]. A scanning operator can reach a speed of 700 pages per hour. So, for one single digitization center with five to ten scanning equipment, the productivity is typically around 1,000 volumes per week. Under such workload, it is clear that the digitization process must be carefully designed: a good example is given in [1]. Unfortunately, this process is currently applied with difficulty to antiquarian collections, because of their inherent heterogeneity (mix of pages and plates, fragile bindings, erratic pagination and tables, etc.).

We report in this paper some preliminary results of the DEMAT-FACTORY project, a consortium of French industrial digitization companies and research laboratories, partly funded by the State Department of Industry and the Paris region. The goal

M. Agosti et al. (Eds.): ECDL 2009, LNCS 5714, pp. 150–155, 2009.

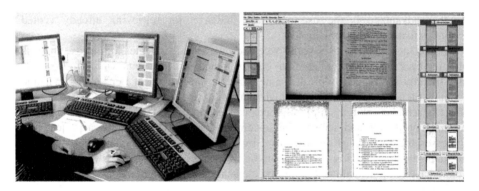

Fig. 1. Current imaging quality control in Demat-Factory: working space and interface

of this three-year project is to study how to significantly upgrade the overall quality of digitization for large, heterogeneous, antiquarian book collections, while keeping the high workload already achieved for homogenous, contemporary printed works. Among the tasks under study are the visualization methods for quality control. We focus here on the steps (2) - (3): scanning and image processing.

Step (3) is divided into the following subtasks: (a) pages separation, (b) crop and dimension control, (c) skewness removal, (d) figures identification, (e) binarization, and (f) marks removal. It is conducted automatically, usually at night. During the day, an operator validates this step. Figure 1 shows a typical working space for the operator and a screenshot of the main interface for image control. The operator processes one volume at a time on each of the three screens. The upper central part of the interface shows the scanned picture and the lower part, what is obtained after (f). The user can navigate in these views simultaneously. Thumbnails of all the scanned pictures of the volume are shown on the left side, labeled with results from automatic analysis. On the right side, the operator can restart each step from (a) to (f). All the operations are time stamped and logged.

Only a dozen pages can be checked at a time on each screen, so knowledge about the digitized document is reduced. Thumbnails ease and accelerate checking many aspects of digitized images. In some situations, however, one must inspect the images using different zoom scales or in a different zoom area. In those cases, one has to load the full view image, resulting in a time-consuming operation and hindering the fluidity of the interaction (because the visual context is abruptly changed).

We claim that massive digitizing pipelines require not only a sample browsing, but also an exhaustive browsing approach, where all images can be rapidly observed and inspected, using recent results in information visualization. In [3], authors argue that often visualizations only use one of the two following approaches.

On the one hand, the *time-based strategy*, in which discovering data depends upon time dependant actions, is commonly implemented in standard time-scrolling interfaces. Their main known issue is the abrupt change when switching from a context-browsing task to a detailed document reading. On the other hand, the *space strategy*, in which information is packed in a static view, is exploited by the Space-Filling Thumbnails technique, which tries to solve the scrolling issue [4]. This kind of display technique reveals relations between documents with little cognitive effort.

Another advantage is the use of spatial memory for retrieving already visited resources thanks to fixed positions.

The Perspective Wall uses both space and time based strategies to improve the task of going back and forth from detail to context in textual documents, with effortless and user-friendly techniques. With the same concern, the Document Lens offers dynamic zooming on page thumbnails while maintaining context awareness [5]. In [6], the same effect is achieved with a simple 3D perspective projection that offers simultaneously zooming on details and seeing near upcoming context. The Photomesa project also addresses the focus plus context issue by combining the thumbnail approach with treemap zooming [7]. However, Photomesa was developed for pictures with no obligation to have relations with neighborhood elements whereas digitization-reviewing tasks need these relations.

2   A Pan and Zoom Paper Half Pipe

The Pan&Zoom Paper Half Pipe (PZPHP) is a simplified 3D environment in which scene and navigation are reduced to an essential set of elements and operations. Pages are arranged on a wall, which has a partial pipe shape. The user controls the camera (i.e., his or her point of view in the scene), going up and down (Y), rotating (R_Y) clockwise (CW) and counter-clockwise (CCW), and zooming in or out any page (Z). The camera spatial parameters are thus mapped to a subset of three degrees-of-freedom (DOF) sensed by the input device (fig. 2, left). The goal of this interface is to provide rapid and high-resolution browsing of digitized book pages in order to detect problems in a given set of images. There are two important browsing strategies in the quality control task: *overview and detail* and *sequential browsing*. First, one must be able to see the whole set at a glance in order to identify visual anomalies or disparities (lighter or darker pages, larger or smaller margins). Then, displaying the details of "suspicious" pages or regions of pages must be done through a rapid operation. Second, one must be able to navigate through the volume following the order of the pages. One may thus identify missing pages through visual disparities between one page and its neighboring pages (e.g., a paragraph that finishes on a given page and no new paragraph starts on the succeeding one).

2.1   Scene Design and Camera Control

The pages are mapped on the inner wall of a semi-cylindrical surface. This layout affords equal length trajectories from the center of the cylinder to any of the displayed pages on a given row. On a plane wall layout, one would navigate longer to attain the pages located on the boundaries of the wall. When the camera is at the center of the scene, all the pages are visible (the radius of the cylinder varies according to the number of pages being displayed). The longitude is the main navigation axis and it is dedicated to sequential browsing. Pages are nevertheless also tiled vertically in order to take advantage of free screen area when one zooms out. Moreover, panning vertically may still serve as a "random navigation axis", along which, one examines a random sample of pages. Scroll-based interfaces like Adobe Acrobat Reader support mainly the sequential browsing strategy, displaying the pages in a column layout.

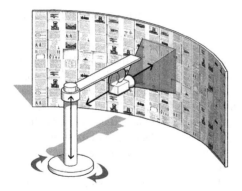 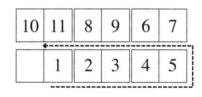

Fig. 2. On the left, 3D scene and camera movements metaphor. On the right, the layout of the displayed pages on the wall and the trajectory of a sequential navigation.

Unlike other visualization interfaces in which vanishing points distort document boundaries [3, 6], in the PZPHP, every page is always orthogonal to the camera (when zoomed out, boundary images present a light distortion). Thus, pages are neither displayed tilted nor deformed by the camera perspective. That ensures that geometrical distortions will neither confound the user nor hide hints about potential problems in the digitized pages (whether the problem comes from the physical volume or from the digitizing process). This effect is achieved thanks to the cylindrical layout and to the fact that camera rotates not around its own center but around the scene's center.

As a user may want to uninterruptedly visualize all pages of the document, from the first to the last page, we have considered other scene topologies that would be better adapted to such operation. For instance, a complete cylinder with a helical distribution of pages would permit the user to, through a same continuous movement, browse all the displayed pages in a given order. Such configuration would nevertheless make the complete overview impossible and would provide very few landmarks (retrieving an already visited page would be harder). Figure 2 (right) illustrates the order in which the pages are tiled on the wall and the ideal path for one who wants to browse pages in the sequential order. We are implementing a "pan lift" mechanism in order to compensate the discontinuity of the pan movement in the semi-cylindrical navigation. This behavior would transport the camera to the "next" line when it reaches the end of a row during the pan (i.e., a CW pan would transport the camera to the above row when panning through an even row). A condition to trigger the "pan lift" is the viewport being mainly filled by one row. This behavior would permit the user to concentrate on controlling the important variables of the task, namely the pan speed.

2.2 Implementation and Users' Feedbacks

The user controls the virtual camera through a SpaceNavigator (a six DOF isometric joystick). This device provides a tiny elastic feedback to movements along its axes. It consists in a small cylinder mounted on a heavy base. The fingertips can grasp the cylinder and it can be pushed, pulled, twisted, tilted, and rolled. Isometric devices are considered to be superior to position devices (e.g., mouse) in controlling, with higher precision and lower fatigue, trajectories of cameras in 3D scenes (and in document scrolling tasks as well)[9].

The prototype application is built using C and the OpenGL library. The master raw image of each page is cropped so as to present a 2:1 ratio and it is then sampled to seven smaller power-of-two dimensions (2048x1024 until 32x16 pixels). Power-of-two images take advantage of graphic memory space when they are loaded as textures. Moreover, they permit clear and fluid zooming thanks to the Mipmap functionality, which employs pre-calculated intermediary size images in order to reduce flickering and aliasing when the texture is scaled up or down in the viewport. During initialization of the application, the GIF images of the pages are all loaded in the graphic card's memory using texture compression functionality. As textures are all resident, there is no paging. Animation is thus completely smooth, even if the camera is rapidly moved from a 500 pages view to a high-resolution single-page detail view in less than two seconds. The final tiled wall is almost a gigapixel image – actually, 500 megapixels. When zoomed out, pages become a visual texture; illustration zones, obscure pages, distinct zones, and other critical visual information may be rapidly spotted and inspected.

Fig. 3. The PZPHP displayed on an immersive display (left) and on a 22" screen (right)

We have presented our prototype under two visual display device configurations: on a regular 22" screen and on an Elumens VisionStation, an immersive visual display device (fig. 3). The VisionStation is a hemispherical display that uses a projection-based hardware and software, for displaying images in 160 by 160 degrees field-of-view. It consists of a hemispherical surface (1.5 m diameter) attached to a small table. The user sits in front of this table where he may place the interaction input. A normal projector, coupled to a fisheye lens, is installed into the table. Such a lens provides a 180-degree horizontal and vertical projection distributing equally the same amount of pixels all over the projection surface.

A number of users (20 persons among librarians, specialists in the digitizing process, expert and novice users) have already tried our prototype; it was also presented at a digitizing hardware and services professional exhibition (FAN'2008, Paris). Some users, mainly those that are not used to 3D input devices, had a hard time in controlling the camera for zooming in and out the pages. The device's form factors and its poor elastic feedback seem to trouble most of them in performing the wanted manipulations. Moving the camera up and down, for example, is an uncomfortable manipulation - the device's top part barely moves when its pushed or pulled along that axis. Igarashi and Hinckley [8] describes the problem of the visual overflow generated by and accelerated scrolling display. Such problem is similar to the disorientation experienced by the users that tried the PZPHP when the camera is too close to the pages. A device-camera mapping that would reduce speed of yawing and vertical movements as the camera approaches the pages is right now being studied. Anyway, all users that were sensible to quality control issues were very interested in this interface and a major part of them found that it offers a value that they could not find in any other application that they had previously tried.

3   Conclusion and Future Work

We described a first prototype of a visualization interface to support high-resolution browsing of large digitized page sets. Its response time has proved to be satisfactory. At the present stage, however, the software supports neither edition nor annotation. Thus, no real world test, in which a control quality user browses a set and tags defective images in a limited time, has yet been proposed. Besides implementing such functions, we would like to conduct a usability study that helps us to find a comfortable and efficient camera mapping. Moreover, we plan to make the prototype more scalable. We would like to design a resident memory manager that would load and unload high-resolution images according to their distance from the camera viewport.

References

1. Website of the Univ. of Michigan Library,
   http://www.lib.umich.edu/news/millionth.html
2. Riley, J., Whitsel, K.: Practical Quality Control Procedures for Digital Imaging Projects. OCLC Systems & Services 21(1), 40–48 (2005)
3. Mackinlay, J.D., Robertson, G.G., Card, S.K.: The Perspective Wall: Detail and Context Smoothly Integrated. In: Proc. of CHI 1991, pp. 173–176. ACM Press, New York (1991)
4. Cockburn, A., Karlson, A., Bederson, B.B.: A Review of Overview+Detail, Zooming, and Focus+Context Interfaces. ACM CSUR 41(1), 1–31 (2008)
5. Robertson, G.G., Mackinlay, J.D.: The Document Lens. In: Proc. of UIST 1993, pp. 101–108. ACM Press, New York (1993)
6. Guiard, Y., Chapuis, O., Du, Y., Beaudouin-Lafon, M.: Allowing Camera Tilts for Document Navigation in the Standard GUI: A Discussion and an Experiment. In: Proc. of AVI 2006, pp. 241–244. ACM Press, New York (2008)
7. Bederson, B.: PhotoMesa: A Zoomable Image Browser Using Quantum Treemaps and Bubblemaps. In: Proc. of UIST 2001, pp. 71–80. ACM Press, New York (2001)
8. Igarashi, T., Hinckley, K.: Speed-dependent Automatic Zooming for Browsing Large Documents. In: Proc. of UIST 2000, pp. 139–148. ACM Press, New York (2000)

Improving OCR Accuracy
for Classical Critical Editions

Federico Boschetti, Matteo Romanello, Alison Babeu, David Bamman,
and Gregory Crane

Tufts University, Perseus Digital Library, Eaton 124, Medford MA, 02155, USA

Abstract. This paper describes a work-flow designed to populate a
digital library of ancient Greek critical editions with highly accurate
OCR scanned text. While the most recently available OCR engines are
now able after suitable training to deal with the polytonic Greek fonts
used in 19th and 20th century editions, further improvements can also
be achieved with postprocessing. In particular, the progressive multiple
alignment method applied to different OCR outputs based on the same
images is discussed in this paper.

1 Introduction

The new generation of Greek and Latin corpora that has increasingly become
available has shifted the focus from creating accurate digital texts to sophisticated digital editions. Previously prefaces, introductions, indexes, bibliographies,
notes, critical apparatus (usually at the end of the page, in footnote size), and
textual variations of different editions have either been discarded or systematically ignored in the creation of early digital collections. The ancient text that
we read in modern editions, however, is the product of editors' choices, where
they have evaluated the most probable variants attested in the manuscripts or
the best conjectures provided by previous scholars. Humanists thus need both
textual and paratextual information when they deal with ancient works.

Critical editions of classics are challenging for OCR systems in many ways.
First, the layout is divided into several text flows with different font sizes: the
author's text established by the editor, the critical apparatus where manuscript
variants and scholars' conjectures are registered and, optionally, boxes for notes
or side by side pages for the parallel translation. Second, ancient Greek utilizes
a wide set of characters to represent the combinations of accents and breathing
marks on the vowels, which are error prone for OCR systems. Third, critical
editions are typically multilingual, because the critical apparatus is usually in
Latin, names of cited scholars are spelled in English, German, French, Italian
or other modern languages, and the prefaces, introductions, translations and
indexes are also often in Latin or in modern languages. Finally, 19th century
and early 20th century editions can have many damaged text pages that present
great difficulties for conventional OCR.

M. Agosti et al. (Eds.): ECDL 2009, LNCS 5714, pp. 156–167, 2009.

2   Related Work

We can divide works related to the digitization of ancient texts into three groups: the first one concerns the analysis of manuscripts and early printed editions, the second group concerns the structure of digital critical editions (i.e. editions that register variants and conjectures to the established text) and the third group concerns OCR work performed on printed critical editions from the last two centuries.

The general approach for the first group is to provide methods and tools for computer assisted analysis and correction. Moalla et al. [17] developed a method to classify medieval manuscripts by different scripts in order to assist paleographers. Ben Jlaiel et al. [4] suggested a strategy to discriminate Arabic and Latin modern scripts that can be applied also to ancient scripts. Leydier et al. [14], [15] and Le Bourgeois et al. [13] used a method of word-spotting to retrieve similar images related to hand written words contained in manuscripts. Edwards et al. [10], on the other hand, developed a method based on a generalized Hidden Markov Model that improved accuracy on Latin manuscripts up to 75%.

The second group of studies explored recording variants and conjectures of modern authors, for instance Cervantes, such as Monroy et al. [18] or of ancient texts, for instance in Sanskrit, such as Csernel and Patte [9].

The third group of studies concerned improvements of OCR accuracy through post-processing techniques on the output of a single or multiple OCR engines. Ringlstetter et al. [27] suggested a method to discriminate character confusions in multilingual texts. Cecotti et al. [6] and Lund and Ringger [16] aligned multiple OCR outputs and illustrated strategies for selection. Namboodiri et al. [20] and Zhuang and Zhu [32] integrated multi-knowledge with the OCR output in post-processing, such as fixed poetical structures for Indian poetry or semantic lexicons for Chinese texts.

This paper further develops some guidelines first expressed in Stewart et al. [30]. In this previous research, the recognition of Greek accents in modern editions was not considered due to the technological limitations imposed by the OCR systems available.

3   Methodology

Our main interest in this research is to establish a work-flow for the massive digitization of Greek and Latin printed editions, with particular attention to the scalability of the process. The principal factors that determine the preparation of different pre- and postprocessing procedures are book collection specificities and preservation status.

3.1   Texts

Our experiments have been performed on different typologies of samples, in order to combine the aforementioned factors. Three editions of Athenaeus' *Deipnosophistae* and one of Aeschylus' tragedies have been used, by randomly extracting five pages from each exemplar. All documents have been downloaded from [12]. Athenaeus' exemplars belong to different collections and they are distributed along two centuries: Meineke's (1858) and Kaibel's (1887) editions are in the Teubner classical collection, whereas Gulick's (1951) second edition is in the Loeb classical library. Teubner and Loeb editions sensibly differ for script fonts, so that two different training sets have been created. They differ also for content organization: Meineke has no critical apparatus, Kaibel has a rich apparatus and Gulick has a minimal critical apparatus, supplementary notes and an English translation side by side.

The posthumous Hermann's (1852) edition of Aeschylus, published by Weidmann, has no critical apparatus and has a script very similar to the Teubner editions.

In this study, Greek text and critical apparatus have been separated manually, whereas English translation and notes have been ignored. In a second stage of the work, simple heuristics will be applied to classify textual areas.

Finally, in order to evaluate if and how the system could be extended to very early printed editions, an experiment has been performed on the *incunabulum* of Augustinus' *De Civitate Dei*, Venetiis 1475. In this case, even if the quality of the image is good, the irregularity of the script and the use of ligatures and abbreviations is very challenging.

3.2 OCR Engines Suitable for Ancient Greek Recognition

Three OCR engines have been employed: Ideatech Anagnostis 4.1, Abbyy FineReader 9.0 and OCRopus 0.3 in bundle with Tesseract 2.03.

Anagnostis [2] is the unique commercial OCR engine that is provided with built-in functionality for ancient Greek and it can also be trained with new fonts. Accents and breathing marks are processed separately from the character body, improving the precision of the recognition system. On the other hand, Anagnostis is not able to recognize sequences of polytonic Greek and Latin characters, such as are present in the critical apparatus. In this case, Latin characters are rendered with the Greek characters most similar in shape (for example, the Latin letter v is transformed into the Greek letter ν).

FineReader [1] is capable of complex layout analysis and multilingual recognition. Even if polytonic Greek is not implemented natively, it is possible to train FineReader with new scripts, associating the images of glyphs to their Unicode representations. For these reasons, FineReader is currently the most reliable engine to recognize texts where different character sets are mixed.

OCRopus [22] is an open source project hosted by Google Code, that can be used in bundle with Tesseract [31], illustrated by Smith [28], which is one of the most accurate open source OCR engines currently available. OCRopus/Tesseract needs to be trained in order to recognize polytonic Greek (or other new scripts, except Latin scripts) and the recognition of mixed character sets is acceptable. The output format is plain text or xhtml enriched with a microformat to register positions of words (or optionally single characters) on the page image.

3.3 Training of Single Engines

The training process is divided into two phases. First, each OCR engine has been trained with pages randomly selected from the editions used in the experiments, verifying that the training set had no overlappings with the test set. Anagnostis and FineReader have been trained with the same sets of pages, whereas OCRopus/Tesseract has been trained with a different set, in order to increase the possibility of capturing character samples ignored by the other engines. In fact, the major issue in training FineReader and OCRopus/Tesseract with ancient Greek is caused by the high number of low frequency characters (according to the Zipfian law). Unicode represents polytonic Greek both by pre-combined characters and combining diacritics, but during the training process these engines seem to analyze glyphs only as whole characters, without separation between vowels and diacritics, as Anagnostis is able to do. The entire set of pre-combined characters for ancient Greek contains more than two hundred glyphs, but some of them are employed with a very low frequency. For example, in the Athenaeus' Kaibel edition, letter ᾷ (alpha with circumflex accent, rough breathing mark and iota subscript) occurs only twice out of more than one million characters. Thus, the probability that these rare characters are sampled in the training sets is quite low. As stated above, training is based on collections and not on exemplars, for the sake of scalability. For this reason, only one training set per engine has been created for the Teubner editions, mixing pages from both Kaibel's and Meineke's exemplars.

FineReader has a good built-in training set for modern (monotonic) Greek and it is possible to use the user defined training sets either alone or in bundle with the built-in trainings. Unfortunately, while this increases the accuracy for the recognition of non-accented characters it also decreases the accuracy for the recognition of vowels with accents and breathing marks. Thus, two training sets have been created for FineReader: with and without the addition of built-in training sets.

Second, the errors produced by each engine after the first stage have been compared with the ground truth, in order to calculate the error patterns that can be corrected by the cooperation of different OCR engines. The new training sets must be identical for all the engines. For Weidmann's edition, a new set of five pages, different from both the training set and the test set, has been recognized and the hand transcription has been used as ground truth. For the other editions, a k-fold cross validation method has been performed, using all the pages but the testing one for the training.

OCR output has been post-processed with a script that adjusts encoding and formatting errors, such as Latin characters inside Greek words with the same or very similar shape (e.g. Latin character *o* and Greek character o, omicron), spaces followed by punctuation marks and other illegal sequences. A second script adjusts a small set of very frequent errors by the application of regular expressions. For example, a space followed by an accented vowel and by a consonant, an illegal sequence in ancient Greek, is transformed into space, followed by a vowel with breathing mark and a consonant.

The adjusted OCR output has been aligned to the ground truth by a dynamic programming alignment algorithm, according to the methods explained in Feng and Manmatha [11] and in van Beusekom et al. [5]. As usual, alignments are performed minimizing the costs to transform one string into the other, adding gap signs when it is necessary. In this way, n-gram alignments can be a couple of identical items (correct output), a couple of different items (error by substitution), an item aligned to a gap sign (error by insertion) or, finally, a gap sign aligned to an item (error by deletion). After the alignment, the average number of substitutions, insertions and deletions has been used to compute the average accuracy of each OCR engine. [21] offers a survey on methods to calculate approximate string matchings.

Data concerning alignments of single characters, bigrams, trigrams and tetragrams are registered in the error pattern file. For the sake of efficiency, data related to correct alignments of n-grams are registered only if the n-gram occurs at least once in a misalignment. In fact, we are particularly interested in comparing the probability that one n-gram is wrong to the probabilty that it is correct, as we will see below. The error pattern file is a table with four columns: number of characters the n-gram is constituted by, n-gram in OCR output, aligned n-gram in ground truth and a probability value, illustrated by formula (1).

$$\frac{C(a \rightarrow b)}{C(b)} * \left(\frac{C(b)}{N} \right)^{1/3} \tag{1}$$

The first factor of this value expresses the probability that, given a character (or n-gram) a in the OCR output, it represents a character (or n-gram) b in the ground truth (a is equal to b, in case of correct recognition). It is represented by the number of occurrences of the current alignment, $C(a \rightarrow b)$, divided by the total number of occurrences of the b character (or n-gram) in the ground truth, $C(b)$. The second factor of this value is the cubic root of $C(b)$ divided by the total number of characters or n-grams, N. This factor is equal for every engine, because it is based only on ground truth. The cubic root of this value is provided, according to the formula (6), which will be explained below.

3.4   Multiple Alignment and Naive Bayes Classifier

Tests have been performed on each OCR engine and the output has been adjusted with the simple post-processing scripts used also for the training samples. First of all, the two FineReader outputs (with and without the built-in trainings) have been aligned with the same methodology explained below for the alignments among different engines and we have obtained a new, more accurate FineReader output to be aligned with the other engines.

Outputs of the three engines have been aligned by a progressive multiple sequence alignment algorithm, as illustrated in Spencer [29]. The general principle of progressive alignment is that the most similar sequence pairs are aligned first, necessary gaps to align the sequences are fixed and supplementary gaps (with minimal costs) are progressively added to the previous aligned sequences, in order to perform the total alignment. In order to establish which pairs are more similar and then must be aligned first, a phylogenetic tree should be constructed, but for our triple alignment it is enough to rate each engine according to the average accuracy value established during the training process. In our tests, FineReader has scored the highest, followed by OCRopus and Anagnostis. For this reason, FineReader and Anagnostis are aligned first. The resulting OCRopus string with gap signs is aligned to Anagnostis and the new gap signs are propagated to the previously aligned FineReader string. The triple alignment is shown in Figure 1, where the gap sign is represented by underscore.

The alignment in itself is not enough to determine the most probable character: even if two engines are in agreement, but are poorly reliable for a specific character identification, the most probable character could be provided by the third engine in disagreement. Even if all the engines are in agreement, the most probable character could be another one, such as when three engines are only able to recognize Greek characters and the text is written in Latin. This situation, however, is not considered in the current study, which is limited to the selection among characters provided by at least one engine.

Formally, the probability that the current position in the original printed page e_0 contains the character x, given that the first engine e_1 provides the character c_1, the second engine e_2 provides the character c_2 and the third engine e_3 provides the character c_3, is expressed by the formula:

$$P(e_0 = x | e_1 = c_1, e_2 = c_2, e_3 = c_3) \tag{2}$$

where, in general, $P(E_0 | E_1, E_2, E_3)$, denotes the posterior probability for the event E_0, given the conjunction of the events $E_1 \cap E_2 \cap E_3$.

For example, (2) expresses the probability that the character $\breve{a}$ is in the current position on the printed page, knowing that the first engine has provided $\dot{a}$, the second engine has provided $\ddot{a}$ and the third engine has provided $\acute{a}$. These probabilities are deduced by the error pattern data recorded during the training process.

To find the highest probability among the three items provided by the engines, we have implemented a naive Bayes classifier. In virtue of the Bayes' theorem, from (2) follows:

$$[P(e_1 = c_1, e_2 = c_2, e_3 = c_3 | e_0 = x) * P(e_0 = x)] / P(e_1 = c_1, e_2 = c_2, e_3 = c_3) \tag{3}$$

Given that a naive Bayes classifier is based on the conditional independence assumption, the first factor in the numerator of (3) can be rewritten as

$$P(e_1 = c_1 | e_0 = x) * P(e_2 = c_2 | e_0 = x) * P(e_3 = c_3 | e_0 = x) \tag{4}$$

Considering that we are not interested in finding the value of the highest probability, but simply in finding the argument x_0 that provides the highest probability, we can omit the denominator of (3) and use the following formula:

$$x_0 = argmax\, P(e_1 = c_1 | e_0 = x) * P(e_2 = c_2 | e_0 = x) * P(e_3 = c_3 | e_0 = x) * P(e_0 = x) \tag{5}$$

Generalizing, we can write the equation (5) as

$$x_0 = argmax \prod_{i=1}^{n} P(e_i = c_i | e_0 = x) * P(e_0 = x)^{1/n} \tag{6}$$

where n is the number of OCR engines, e_i is a specific engine, c_i is the character provided by that engine. This equation explains why we computed the cubic root of the ground truth character probability in the equation (1). For the sake of efficiency, in this way we do not need to search for this factor and multiply it for the other factors all the times that we compute the requested term.

In our implementation, a triple agreement is unprocessed and in case of probability equal to zero, the output of the first engine (FineReader, in this case) is selected. In Figure 1 the result of the selection performed by the system is shown. In blue and red are indicated the correct characters selected from OCRopus and Anagnostis, despite the character recognized by FineReader.

Fig. 1. Multiple alignment of the three engines output

3.5   Spell-Checking Supported by Multiple Alignment Evidence

As stated above, the high number of ancient Greek pre-combined characters reduces the probability that the training sets contain some error patterns present in the test sets. In this case, the probability for a correct item is zero. On the other hand, as explained in Reynaert [25] and Stewart et al. [30], the automatic spell-checking applied to mispelled words alone is often unreliable; the first suggestion provided by the spell-checker could be wrong or, as is often the case, the word list of the spell checker does not contain proper names and morphological variants, and it thus replaces a correct word with an error. In order to reduce these issues, we have adopted a spell-checking procedure supported by the engines output evidence, filtering only the spell-checker suggestions that match a regular expression based on the triple alignment.

In order to integrate the spell-checker in our system, we have used the Aspell API [3] and we have used the word list generated by Morpheus, the ancient Greek morphological analyzer [7]. The string generated by the naive Bayes classifier is analyzed by the spell-checker. When words are rejected by the spell-checker because they are not contained in the word list, a regular expression is generated from the aligned original outputs, according to these simple rules: a) characters in agreement are written just once; b) two or three characters in disagreement are written between brackets; c) gaps are tranformed into question marks (to indicate in the regular expression that the previous character or couple of characters between brackets are optional). For example, given the aligned outputs: a) ἤλασεν, b) ἤλαστν and c) ἤλασ_ν, the regular expression generated is /[ῆῄ]λασ[ετ]?ν/. All the suggestions provided by the spell-checker are matched with this regular expression, and only the first one that matches is selected, otherwise the mispelled word is left unchanged. Further examples are shown in Figure 2. The first example, ἐξερήμωσεν, and the last example, ευφρων, merit some further consideration. The first case reflects when a correct morphological variant is not present in the spell-checker word list. No suggestion provided by the spell-checker matches the regular expression generated by aligned outputs, thus the word is correctly left unchanged. On the other hand, ευφρων is an incorrect ancient Greek word because it has neither accent nor breathing mark. In this case, none of the suggestions of the spell checker are supported by the aligned outputs evidence, thus in this case the word is incorrectly left unchanged. While the first suggestion of the spell-checker is incorrect, the third one is correct.

FineReader output	RegEx matching all OCRs	Spell-checker suggestions	Result
ἐξερήμωσεν	ἐξερή έ?[μι]ωσεν	ἐξερήμωσε, ἐξερήμωσέ, ἐξηρήμωσεν	ἐξερήμωσεν
ωπασεν	[ωοὤ]π[αο]σ[εό]ν	ὤπασεν, ὤπασέν, σπάσεν	ὤπασεν
εν᾽	[εἐ]ν᾽	ἐν, ἐν᾽ ... ἔν᾽ (34th item)	ἔν᾽
επάσης	ε?ά?πάσης	πάσης, πάσῃς ... ἁπάσης (11th item)	ἁπάσης
ἐὐθυντήριον	[εἐ][ὑυ]θυντ[ῆή]ριον	εὐθυντήριον, εὐθυντήριόν, εὐθυντῆρι	εὐθυντήριον
πρώτος	πρ[ώῶ]τος	πρῶτος, πρῶτός, πρωτὸς	πρῶτος
Κύρος	[ΚΧΗ][ύῦι]ρος	Κῦρος, Κῦρός, Κύπρος	Κῦρος
εθηκε	[εἔ]θηκε	ἔθηκε, ἔθεκέ, θῆκέ	ἔθηκε
Δυδῶν	[ΔΛ]υδῶν	Δυῶν, Διδῶν ... Λυδῶν (6th item)	Λυδῶν
λάὸν	λ[αά][ὸδ]ν	λαὸν, λαόν, Λάιόν	λαὸν
ἤλασεν	[ῆῄ]λασ[ετ]?ν	ἤλασεν, ἤλασέν, ἤασεν	ἤλασεν
ευφρων	ε?ι?[υὄ]φρωο?ν	ἐύφρων, Εὔφρων, εὔφρων (correct)	ευφρων

Fig. 2. Spell-checking supported by OCR evidence

3.6   The Last Test on a Latin *Incunabulum*

The last test has been performed using a singular engine, OCRopus, on Augustinus' *De Civitate Dei*, Venetiis 1475. We were interested in training OCRopus with Latin abbreviations and ligatures, encoded in Unicode according to the Medieval Unicode Font Initiative (MUFI) directions [19]. Images have been preprocessed with the OCRopus libraries for morphological operations, such as erosion and dilation and improvements due to preprocessing have been compared to ground truth.

4   Results

Results are evaluated comparing the accuracy of singular engines with the accuracy of the merged, spell-checked output. In order to compute the accuracy, the final output has been aligned with the ground truth. According to Reynaert [26], the accuracy has been calculated as:

$$\frac{matches}{matches + substitutions + insertions + deletions} \tag{7}$$

4.1   Accuracy of the Single Engines

Accuracy of single engines largely depends on the training sets created for each collection. Results are shown in Table 1. Both the most accurate OCR commercial application, Abbyy FineReader, and the most accurate OCR open source application, OCRopus/Tesseract are now provided with training sets that allow them to deal with polytonic Greek. In the case of Kaibel's exemplar, we have obtained better results with OCRopus/Tesseract than with Abbyy FineReader, suggesting that the open source software is currently mature enough to be applied to classical critical editions.

Results on Kaibel's and Meineke's exemplars, both Teubner editions, have been obtained using a single training set. The similarity of these results suggest that the project is scalable with pre-processing data reusable on exemplars of the same collections.

Table 1. Accuracy: single engines

Edition	FR w/o built-in training	FR with built-in training	OCRopus	Anagnostis
Gulick (Loeb)	96.44%	94.35%	92.63%	93.15%
Kaibel (Teubner)	93.11%	93.15%	95.19%	92.97%
Meineke (Teubner)	94.54%	93.79%	92.88%	91.78%
Hermann (Weidmann)	97.41%	N/A	91.84%	78.64%

4.2   Improvements Due to Alignment and Constrained Spell-Checking

Improvements due to alignment can be divided in two steps. In fact, the first gain is due to the alignment of the FineReader outputs, with and without the built-in training set, in cooperation with the user training set. In average, the improvement is +1.15% in relation to the best single engine, which is FineReader without the built-in training except in the case of Kaibel, as stated in the previous section.

The second step is the triple alignment and constrained spell-checking, which provides a gain, in average, of +2.49% in relation to the best single engine. A

Table 2. Accuracy: alignment and spell-checking

Edition	Alignment and spell-checking	Aligned FR	Best engine
Gulick (Loeb)	99.01%	98.02%	96.44%
gain	+2.57%	+1.58%	0.00%
Kaibel (Teubner)	98.17%	95.45%	95.19%
gain	+2.98%	+0.26%	0.0%
Meineke (Teubner)	97.46%	96.15%	94.54%
gain	+2.92%	+1.61%	0.00%
Hermann (Weidmann)	98.91%	N/A	97.41%
gain	+1.50%	N/A	0.00%

t-test for each exemplar demonstrates that improvements are always significant, with p<0.05. Analytical results are provided in Table 2.

The best result, as expected, concerns the most recent Loeb edition, with an accuracy rate of 99.01%. If we consider only the case insensitive text (without punctuation marks, breathing marks and accents), the accuracy arises to 99.48%. This value is especially important if we are interested in evaluating the expected recall of a text retrieval system, where ancient Greek words can be searched in upper case.

4.3   Accuracy on the Critical Apparatus

Tests on the critical apparatus of Gulick's and Kaibel's editions have been performed without a specific training for the footnote size, but with the same training sets applied to the rest of the page. Only the FineReader output with the built-in training set has been used, because the output created without it had a very low accuracy.

The average accuracy due to the triple alignment is 92.01%, with an average gain of +3.26% in relation to the best single engine, that is FineReader on Gulick's edition and OCRopus/Tesseract on Kaibel's edition. Analytical results are provided in Table 3. Also on the critical apparatus, t-test demonstrates that improvements are significant, with p<0.05.

It is important to point out that the critical apparatus, according to estimations computed in Stewart et al. [30], is approximately, on average, 5% of the page in editions with minimal information (such as Loeb editions), and 14% of the page, on average, for more informative apparatus (such Teubner editions).

Table 3. Accuracy: critical apparatus

	Alignment and spell-checking	FR with b.-in	OCRopus	Anagnostis
Gulick	90.88%	87.99%	64.79%	59.08%
gain	+2.89%	0.0%	-23.20%	-28.91%
Kaibel	93.14%	87.68%	89.54%	57.11%
gain	+3.60%	-1.86%	0.0%	-32.43%

4.4   Accuracy on the *Incunabulum*

The test performed with OCRopus on Augustinus' *De Civitate Dei* provides an accuracy of 81.05%, confirming results reached by Reddy and Crane [24].

5   Conclusion

The software developed for this study and the updated benchmarks are available on the Perseus Project website [23].

As claimed in Crane et al. [8], in order to go beyond digital incunabula it is necessary to build a digital library of classical critical editions, on which information extraction, natural language processing and corpus analysis techniques should be perfomed. A satisfactory OCR accuracy rate for the whole content of a critical edition (text and apparatus), that will allow us to lower the costs for post-corrections by hand, is one first necessary step to build the new generation of textual corpora.

Acknowledgments

This work was supported by a grant from the Mellon Foundation. We also gratefully acknowledge Marco Baroni and Tommaso Mastropasqua, of CIMeC - University of Trento (Italy), for their useful suggestions.

References

1. Abbyy FineReader Homepage, http://www.abbyy.com
2. Anagnostis Homepage, http://www.ideatech-online.com
3. Aspell Spell-checker Homepage, http://aspell.net
4. Ben Jlaiel, M., Kanoun, S., Alimi, A.M., Mullot, R.: Three decision levels strategy for Arabic and Latin texts differentiation in printed and handwritten natures. In: 9th International Conference on Document Analysis and Recognition, pp. 1103–1107 (2007)
5. van Beusekom, J., Shafait, F., Breul, T.M.: Automated OCR Ground Truth Generation. In: 9th International Conference on Document Analysis and Recognition, pp. 111–117 (2007)
6. Cecotti, H., Belaïd, A.: Hybrid OCR combination approach complemented by a specialized ICR applied on ancient documents. In: 8th International Conference on Document Analysis and Recognition, pp. 1045–1049 (2005)
7. Crane, G.: Generating and parsing classical Greek. Literary and Linguistic Computing 6(4), 243–245 (1991)
8. Crane, G., Bamman, D., Cerrato, L., Jones, A., Mimno, D., Packel, A., Sculley, D., Weaver, G.: Beyond Digital Incunabula: Modeling the Next Generation of Digital Libraries. In: Gonzalo, J., Thanos, C., Verdejo, M.F., Carrasco, R.C. (eds.) ECDL 2006. LNCS, vol. 4172, pp. 353–366. Springer, Heidelberg (2006)
9. Csernel, M., Patte, F.: Critical Edition of Sanskrit Texts. In: 1st International Sanskrit Computational Linguistics Symposium, pp. 95–113 (2007)
10. Edwards, J., Teh, Y.W., Forsyth, D., Bock, R., Maire, M., Vesom, G.: Making Latin Manuscripts Searchable using gHMM's. Advances in Neural Information Processing Systems 17, 385–392 (2004)
11. Feng, S., Manmatha, R.: A Hierarchical, HMM-based Automatic Evaluation of OCR Accuracy for a Digital Library of Books. In: JCDL 2006, pp. 109–118 (2006)
12. Internet Archive Homepage, http://www.archive.org
13. Le Bourgeois, F., Emptoz, H.: DEBORA: Digital AccEss to Books of the RenAissance. International Journal on Document Analysis and Recognition 9, 192–221 (2007)
14. Leydier, Y., Lebourgeois, F., Emptoz, H.: Text search for medieval manuscript images. Pattern Recognition 40(12), 3552–3567 (2007)
15. Leydier, Y., Le Bourgeois, F., Emptoz, H.: Textual Indexation of Ancient Documents. In: 2005 ACM symposium on Document engineering, pp. 111–117 (2005)

16. Lund, W.B., Ringger, E.K.: Improving Optical Character Recognition through Efficient Multiple System Alignment (to appear in JCDL 2009)
17. Moalla, I., Lebourgeois, F., Emptoz, H., Alimi, A.M.: Image Analysis for Paleography Inspection. In: Document Analysis Systems VII, pp. 25–37 (2006)
18. Monroy, C., Kochumman, R., Furuta, R., Urbina, E., Melgoza, E., Goenka, A.: Visualization of Variants in Textual Collations to Analyze the Evolution of Literary Works in The Cervantes Project. In: 6th European Conference on Research and Advanced Technology for Digital Libraries, pp. 638–653 (2007)
19. Medieval Unicode Font Initiative Homepage, http://www.mufi.info/fonts
20. Namboodiri, A.M., Narayanan, P.J., Jawahar, C.V.: On Using Classical Poetry Structure for Indian Language Post-Processing. In: 9th International Conference on Document Analysis and Recognition, vol. 2, pp. 1238–1242. IEEE Computer Society, Los Alamitos (2007)
21. Navarro, G.: A Guided Tour to Approximate String Matching. ACM Computing Surveys 33(1), 31–88 (2001)
22. OCRopus Homepage, code.google.com/p/ocropus
23. Perseus Project Homepage, http://www.perseus.tufts.edu/hopper/opensource
24. Reddy, S., Crane, G.: A Document Recognition System for Early Modern Latin. In: Chicago Colloquium on Digital Humanities and Computer Science: What Do You Do With A Million Books, Chicago, IL (2006)
25. Reynaert, M.: Non-interactive OCR Post-correction for Giga-Scale Digitization Projects. In: Gelbukh, A. (ed.) CICLing 2008. LNCS, vol. 4919, pp. 617–630. Springer, Heidelberg (2008)
26. Reynaert, M.: All, and only, the Errors: more Complete and Consistent Spelling and OCR-Error Correction Evaluation. In: 6th International Conference on Language Resources and Evaluation 2008, pp. 1867–1872 (2008)
27. Ringlstetter, C., Schulz, K., Mihov, S., Louka, K.: The same is not the same - postcorrection of alphabet confusion errors in mixed-alphabet OCR recognition. In: 8th International Conference on Document Analysis and Recognition, vol. 1, pp. 406–410 (2005)
28. Smith, R.: An Overview of the Tesseract OCR Engine. In: 9th International Conference on Document Analysis and Recognition, vol. 2, pp. 629–633. IEEE Computer Society, Los Alamitos (2007)
29. Spencer, M., Howe, C.: Collating texts using progressive multiple alignment. Computer and the Humanities 37(1), 97–109 (2003)
30. Stewart, G., Crane, G., Babeu, A.: A New Generation of Textual Corpora. In: JCDL 2007, pp. 356–365 (2007)
31. Tesseract Homepage, http://code.google.com/p/tesseract-ocr
32. Zhuang, L., Zhu, X.-Y.: An OCR post-processing approach based on multi-knowledge. In: Khosla, R., Howlett, R.J., Jain, L.C. (eds.) KES 2005. LNCS (LNAI), vol. 3681, pp. 346–352. Springer, Heidelberg (2005)

Using Semantic Technologies in Digital Libraries –
A Roadmap to Quality Evaluation

Sascha Tönnies[1] and Wolf-Tilo Balke[1,2]

[1] L3S Research Center, Appelstraße 9a, 30167 Hannover, Germany
[2] IFIS TU Braunschweig, Mühlenpfordstraße 23, 38106 Braunschweig, Germany
toennies@L3S.de, balke@ifis.cs.tu-bs.de

Abstract. In digital libraries semantic techniques are often deployed to reduce the expensive manual overhead for indexing documents, maintaining metadata, or caching for future search. However, using such techniques may cause a decrease in a collection's quality due to their statistical nature. Since data quality is a major concern in digital libraries, it is important to be able to measure the (loss of) quality of metadata automatically generated by semantic techniques. In this paper we present a user study based on a typical semantic technique used for automatic metadata creation, namely taxonomies of author keywords and tag clouds. We observed experts assessing typical relations between keywords and documents over a small corpus in the field of chemistry. Based on the evaluation of this experiment, we focused on communalities between the experts' perception and thus draw a first roadmap on how to evaluate semantic techniques by proposing some preliminary metrics.

Keywords: Digital Libraries, Information Quality, Semantic Technologies.

1 Introduction

Digital Libraries provide a vast amount of digitized information ranging from collections of cultural heritage to specialized topic centered portals. One of the essential differences between digital libraries and unstructured collections such as the Web, is the focus on information quality. In contrast typical Web search engines base their indexing on text-based measures from information retrieval and structural properties of the collection, e.g. link analysis, whereas digital libraries usually use indexes (manually) crafted from document metadata. Since metadata can express concepts not explicitly occurring in the document, (or leave out concepts explicitly mentioned, but not relevant for the document) the use of a metadata index generally leads to better precision and recall in information services. In addition, library indexes usually rely on controlled vocabularies providing improved retrieval features such as word sense disambiguation or cross language retrieval.

Hence, digital libraries provide an added value over unstructured document collections by offering meaningful access paths. However, given the exponential increase in newly published items even for focused collections, librarians face two serious problems. First it is increasingly costly and time consuming to properly index new items

M. Agosti et al. (Eds.): ECDL 2009, LNCS 5714, pp. 168–179, 2009.
© Springer-Verlag Berlin Heidelberg 2009

(leading to a delay in actually offering the item to customers); second in an ideal collection, the indexing has to foresee all possible (future) uses for a specific item. Moreover, the information overload for the individual customer and the increasing specialization of (research) interests force indexes to be more and more specific in the choice of appropriate indexing terms. In fact, the vision of today's digital libraries is to provide *personalized information spaces* for each individual customer.

To this end, semantic technologies have been recently proposed to bring a higher rate of automation into the indexing process. In essence semantic technologies rely on statistical methods to assess textual documents and to some degree are therefore capable of mining 'hidden' information from collections. The advantage is twofold, first document processing becomes less expensive and a higher degree of personalization is possible. Though, due to the nature of statistical methods, using these semantic techniques may not result in the same retrieval quality as manual crafted metadata. Second, for libraries, this potential decrease in quality is a serious concern; if users cannot trust in the results, the added value over simple Web searches becomes questionable. Hence, before a specific semantic technique can be adopted for use, libraries need a way to gauge the impact of the technology's use in the retrieval process.

In this paper we discuss the open problem of quality assessment for semantic techniques in digital libraries and provide a roadmap for developing quality assessment measures. We will illustrate the use of our measures specifically in the field of chemistry. The selection of chemistry is driven by the current development of the virtual topical digital library for chemistry within the ViFaChem 2 project[1]. The ViFaChem 2 project is a tight cooperation between the L3S Research Center of the University of Hannover and the German National Library of Science and Technology Hannover (TIB). The project investigates and deploys innovative value-adding services for information provisioning in the area of chemistry. To this aim chemical document corpora are annotated by bibliographic and entity-based metadata using semantic technologies. The project's vision is the creation of personal information spaces that offer a variety of relevant resources tailored to the individual user's understanding of the topic.

This paper is organized as follows: the following section will discuss related work in the field of quality assessment for (semantic) digital libraries. In Section 3 we conduct a user study in a chemical digital library and evaluate communalities in experts' interactions with automatically generated metadata in the form of related keywords. Preliminary metrics for measuring the quality of semantic technologies are then derived and discussed in section 4. We close with a short summary and outlook.

2 Related Work

In this section we will first discuss the current state of the art in assessing the quality of classical (mostly manually maintained) digital libraries and then turn to the extension to evaluating semantic technologies. A short case study shows how evaluations of such technologies are actually carried out today.

---

[1] http://www.L3S.de/vifachem

2.1 Evaluating Quality in Digital Libraries

What defines a high quality digital library? In 2000, Saracevic was one of the first authors to consider this problem [27]. He argues that any evaluation basically raises issues such as the criteria, the measures, the context and the methodology. However, his analysis shows that there is no agreement regarding the exact elements of these issues for digital library evaluation. Trying to fill some gaps in this area, Fuhr et al. developed a new description scheme using four major dimensions: collection, technology, users and uses [7]. Based on this dimensions, a questionnaire was developed and the need for an appropriate test collection was stated, similar to the TREC and CLEF initiatives. Extending this work, Gonçalves et al. [11] proposed an actual quality model for digital libraries which is deeply grounded in the formal 5S framework [12]. Exposing several digital library key concepts, several dimensions of quality were added to each concept. For each of these dimensions, the variables to measure, together with the respective S were identified.

The first comprehensive study on digital libraries evaluation frameworks is presented in [8]. The attractiveness of the collections, and the technology's ease of use are identified as key factors in assessing the quality of a digital library. Moreover, the importance of the user satisfaction is emphasized. The model presented is the interaction triptych model which defines three components of the digital library: the system, the content, and the user. In addition three axes of evaluation were provided: usability of user interaction with the system, usefulness of the content for the user, performance of managing the content by the system. Recent research is trying to adopt Web metrics, originally developed for evaluating e-commerce applications, for evaluating digital libraries [18]: preliminary results discuss, e.g., the usage of session length for evaluating the customers' satisfactions with the portal.

2.2 Extending Measures to Semantic Digital Libraries

With upcoming semantic digital libraries like JeromeDL [20] the question of quality has to be extended: what defines a high quality *semantic* digital library? Kruk et al. do not really answer this question when evaluating JeromeDL against a standard digital library measuring several traditional aspects like precision / recall and the user satisfaction [21]. The conducted user studies imply that the individual user's satisfaction seems to be higher when using semantic technologies. However, it has to be pointed out that the results shown in [21] cannot be generalized, since semantic techniques are just as good as the underlying metadata.

Particularly in the domain of collaborative tagging systems, some work investigating tag quality has been performed. According to [10] the distributions of different tags for each individual document tend to stabilize over time, i.e. more and more users add meaningful tags whereas irrelevant tags are not amplified. This result is confirmed in [13] and the authors show in addition, that tags follow a power law distribution. Considering these properties of collaborative tagging systems, it seems likely that tag data can, indeed, be a reliable source of information.

For searching and metadata creation within tagging systems, [15] proposes the exploitation of co-occurrence of users, resources, and tags. This is done using a graph model to represent the folksonomy. In [1] tag data is explored for the purpose of Web search through the use of two tag based algorithms: one exploiting similarity between tag data and search queries, and the other one utilizing tagging frequencies to determine the quality of Web pages. Chan examined a huge number of query terms posed to Powerhouse and concludes that the combined usage of folksonomies with taxonomies increases the recall of the information seeking process [3]. In contrast [25] found out that the use of only document terms yielded slightly better F-measure than using terms and tags together. The authors' results suggest that not all tags are useful descriptors for resource sharing. This leads to the question which kind of tags have a high quality: Bischoff et al. [2] showed that it is worthwhile having a common tag classification scheme for different collections – allowing tags to be compared tags used in different tagging environments. The experiments show that more than 50% of all existing tags bring new information to the resources they annotate and that a large amount of tags are accurate and reliable. A general algorithm for measuring the quality of tags is proposed in [19]. The authors decoupled the relationship between users and tag-resource pairs modeling the tag-resource pairs as nodes and co-user relationship as edges of a graph. This structure allows every two tag-resource pairs used by the same user to have different quality. The algorithm then propagates quality scores iteratively through the graph after being initialized with a set of seed nodes.

In categorization systems, especially in the ontologism field, much work has been done, and several metrics for assessing the quality of an ontology have been proposed, e.g. QOOD [9], OntoMetric [23], and OntoQA [28]. However, all these metrics remain purely on the structural level of the ontology, which is according to [29], not sufficient. In particular, the semantic quality, in terms of correctness, has to be addressed and the authors propose the development of semantically aware ontology metrics. As a first step the authors define the normalization of ontologies and introduce the term of stable metrics. The measurement of the semantic of on ontology becomes vital considering automatically generated ontologies.

2.3 Use Case Study: Evaluating the Semantic GrowBag

Let us consider a typical way of accessing digital collections. Metadata in the form of descriptive terms is often used to describe and summarize documents, and navigational access. Such terms can either be provided by the documents' authors, or be derived from controlled vocabularies, e.g. by the publisher. The collections then allow users to browse documents based on the keywords organized by some categorization system or thesaurus, i.e. searches can be broadened by choosing more general terms or focused by using more specific terms. However, creating and maintaining the underlying categorization systems is primarily done manually with very high efforts and they are often only available for specific domains.

To limit these efforts recently semantic techniques to automatically created categorization systems in the form of taxonomies have been proposed. Examples are statistical evaluation of term co-occurrences [26], language models [4], or syntactical contexts [14]. Although such techniques allow the automatic creation of taxonomies, the suitability of the resulting classification system for actually searching documents is problematic. How can the quality of such generated taxonomies be assessed? For Web search rephrasing queries in different terms is acceptable, however users of digital libraries expect clear and efficient navigation paths. Hence, the measuring of classification systems' quality becomes a vital part in the adoption of semantic technologies.

The actual measurement widely varies in semantic technology research ranging from manual inspection (of random partitions) of the taxonomy to comparison of the entire taxonomy with some kind of 'gold standard'. For instance, in the area of (bio-) medical collections the MeSH taxonomy [16] provides an often used benchmark: when putting an implementation to the test it is run over a focused collection e.g. the Medline corpus [17] and the resulting taxonomy is compared to the corresponding MeSH entries and their respective relationships. For example, in [6] a technique called Semantic GrowBag (based on term co-occurrences, for details see [5]) is used to compute more than 2000 individual taxonomies over Medline documents. It is interesting to notice that for deriving sensible topical taxonomies a minimum of about 100,000 documents was necessary, since statistical methods only provide meaningful results using a sufficiently large sample. For evaluation, the average percentage of accordance or discrepancy with respect to MeSH is presented. Still, it is not clear what these percentages mean in terms of the libraries usability when the respective taxonomies are used as classification system for navigational access.

3   Experiments over a Digital Collection of Chemical Documents

We conducted a user study by observing experts, in our case practitioners in the field of chemistry, when working over a topic restricted document collection with metadata automatically created by semantic technologies. The aim of the study was first to get a deeper understanding of the process of evaluating metadata and assessing the individual expectations second the actual helpfulness of the metadata provided.

For the experiments we used a corpus of 1000 documents randomly extracted from the Journal of Synthetic Organic Chemistry published by Thieme Publishers, Stuttgart Germany. For the metadata extraction, we focused on the author keywords which were subsequently used for automatically creating folksonomies. The actual graphs were calculated by the Semantic GrowBag technique [6] investigating higher order co-occurrences of the keywords in relation to the respective documents. A term A is considered to be 'more general' than some term B, if B usually occurs together with A, whereas A also occurs in other contexts. In that case a directed edge is added from A to B. Together with the graph structure the Semantic GrowBag technique also allows a confidence assessment for each relationship visualized by bold (strong) or dashed (weak) arrows. The Semantic GrowBag uses a biased page rank algorithm to determine this confidence. In Fig. 1 'amino acids' is considered more general than 'amino alcohols' which is indeed justified by amino alcohols being a subclass of amino acids. Note, however, that a relationship as given by the GrowBag graphs does not always express a subclass (or 'is-a') relationship, but just points out that in terms of usage as reflected by the document collection the parent term is more general than the child term.

We extracted a total of 680 graphs (e.g. Fig. 1), each representing the semantic environment for all sufficiently discriminative keywords. The page rank of each term (the number in brackets) in the graphs was also used to create the related tag clouds for the keywords (e.g. Fig. 2). The respective size of each term in the tag cloud is proportional to the page rank value of the term in the GrowBag graph. Please note that in principle the tag cloud contains all information which is available in the graph (terms and their respective page rank) just the hierarchical structure (edges) is missing.

For the actual experiments we randomly chose three query terms for each expert to evaluate the quality of the given graphs and the respective tag clouds. All experts were asked to think aloud after being exposed to the individual graph or tag cloud and provide feedback on how they assessed the quality and which metadata items were considered to be sensible for the average user of the respective collection. Moreover, after reviewing the metadata for each query term, the experts were asked about their expectations in terms of organization of the metadata and the respective correctness and completeness of the automatically created metadata vocabulary.

3.1   A Case Study

In this case study, we describe a typical expert's interaction with a generated graph (Fig. 1) / cloud (Fig. 2) for the query term '*amino alcohols*' to illustrate the conduction of our user study. A first expert was asked about the graph representation and a second about the cloud representation. The graph and cloud contain the same terms and just differ in the visualization and connections between terms.

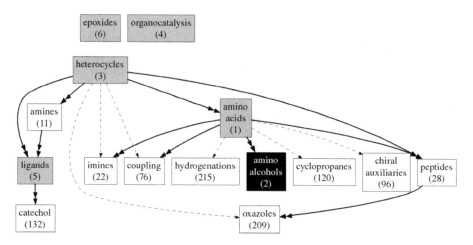

Fig. 1. The generated GrowBag graph for the keyword '*amino alcohols*'

Given the graph as shown in Fig. 1, the expert immediately pointed out that the query term represents a class of *chemical entities*; therefore, he expected to see several *attributes of this class*, typical *reaction names* where amino alcohols are used, *technical uses* and some specific terms from an *analytic* point of view. Following these expectations he clustered the elements into the following groups:

- reactions: 'coupling' and 'hydrogenations'
- classes: 'cyclopropanes', 'oxazoles', 'heterocycles', 'peptides', 'imines', 'amines', 'amino alcohols', 'amino acids', and 'epoxides'
- general concepts: 'chiral auxiliaries', 'organocatalysis', and 'ligands'
- instances: 'catechol'

In a next step, the expert noticed that there are significant differences in the generality of the terms, e.g. '*heterocycles*' has been seen as a very general term whereas '*cyclopropanes*' is a more specific term. For the last step of interaction, the relationships were analyzed: the expert considers some useful, e.g. '*peptides*' are connected via their building blocks '*amino acids*' with '*amino alcohols*' which fits better than a direct connection to 'amino alcohols' and others not useful, e.g. '*catechol*' which represents a '*hydroxyl benzene*' with no obvious connection to '*amino alcohols*'.

amines **amino acids** amino alcohols
catechol chiral auxiliaries coupling
cyclopropanes **heterocycles** hydrogenations
imines ligands organocatalysis
oxazoles **peptides**

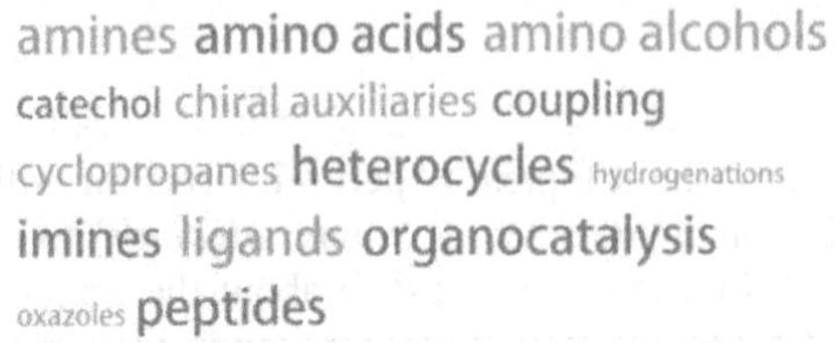

Fig. 2. The generated Tag Cloud for the keyword *amino alcohols*

After giving the equivalent tag cloud (Fig. 2) to the expert, it was interesting to note that the interaction was to a large degree identical with the graph-based representation. The expert started with predefined categories and tried to assign the terms, second the generality of the terms was judged and third the terms were linked to the query term. It has to be pointed out that the expert working on the tag cloud had much more problems during the last step, due to the way of visualization. For instance, he was surprised about the font size of '*cyclopropanes*' and '*oxazoles*'. Due to the fact that '*cyclopropanes*' is not related to the query term, he expected the font size to be much smaller than, e.g., the size of the heavily related term '*oxazoles*'.

3.2 Experimental Results

The evaluation of our observations showed that all practitioners made three major steps during the interaction with the offered metadata.

All experts started with some initial expectation for the categorization of metadata terms. First, they categorized the query term, e.g. as a substance class and then settled on semantically related subcategories based on the main category. It was interesting to see, that these subcategories varied slightly based on the background of the expert. For instance, an expert in the domain of medical chemistry also mentioned the pharmacological impact, whereas a process engineer mentioned environmental perils and toxicity. This observation leads to the conclusion that a categorization of the terms, as it is done, e.g. for the faceted browsing, is indeed useful for the customers and that the structure of a tag cloud may not always be sufficient for visualizing this kind of semantic metadata. It seems that the distribution of terms over relevant categories is one useful metric for measuring the quality of the generated metadata. In our experiments over 90% of the expected categories were indeed filled by matching keywords.

In the second step, the experts tried to understand the content of the graph / cloud. For this purpose they evaluated the terms regarding their respective generality / specifity. This was done without considering the query term. This step has been used by the experts to eliminate outliers in terms of very general or very specific keywords. In particular during our experiments the experts considered 32% of the provided keywords as being too general / specific for the respective graph / cloud.

The last step was the evaluation of the semantic closeness regarding the query term. During this evaluation step the visualization of the metadata affected the experts. Working on the graph, every term was judged individually and depicted relationships were readily taken as explanations. The experts which worked on the cloud did not have these relationships and, therefore, were confused about some terms. Even worse, the font size of the term influenced the experts far more than the confidence in the GrowBag graph. These observations imply the usage of different visualizations: using a cloud for well connected terms and using a graph for the others. In summary, the experts used their individual knowledge to understand the occurrence of the terms and if they could not make a direct connection between a keyword and the query term, they tried to connect the term via some other occurring terms in the graph. If this also failed, they considered the term as wrong or irrelevant for the query. In our experiments this happened with 12% of the occurring terms: this means that 88% have been classified correctly.

4 Towards Measuring Semantic Information Quality

The experiments in the previous section provide some ideas regarding the quality measurements for a semantic technology. Generally speaking, quality can be defined as *correctness* of information. For the field of chemistry this is especially true for data maintained in typical databases like molecular weights or boiling point of substances. However, with respective to semantic, e.g. given by author keywords, the actual correctness is somehow difficult to assess. Observing the expert we found that experts gorge the correctness rather in terms of helpfulness of a keyword and the understandability of the keywords' relationships to a query term. According to the three steps observed during the experiment, we found some communalities between experts. Based on this we will now discuss three preliminary quality metrics that of course have to be further evaluated in future work.

4.1 Degree of Category Coverage (DCC)

The evaluation of the experiments showed that all experts from the start have an implicit course topic map, together with possible classifications for entities in mind. Although the topic map differed with the individual interests of the expert, it is interesting to note that the basic entity classification was very similar (in a way, reflecting the typical cognitive instruments of a chemist). According to this implicit classification, each expert tried to categorize the metadata terms automatically created by the semantic technology. The choice of categories under consideration slightly differed according to the query term and the experts expected at least the closest categories to be filled with keywords found in the graphs, respectively clouds.

This leads to the *degree of category coverage* metric which has to measure how many of the expected categories are actually filled with terms. The more categories are filled the better the result quality is.

With $C := \{c \mid c \text{ relevant category in the topical classification}\}$ we define:

$$f(c) = \begin{cases} 1 & \text{if there is at least a single term } t \text{ in category } c \\ 0 & \text{else} \end{cases}$$

In addition, the metric also has to measure how many of the given terms do not fit to at least one of the expected categories. The more terms can be allocated, the better the result quality is.

With $T := \{t \mid t \text{ term from a given metadata subset}\}$ we define:

$$g(t) = \begin{cases} 1 & \text{if term } t \text{ belongs to some category from } C \\ 0 & \text{else} \end{cases}$$

This results in:

$$DCC = \frac{\sum_{i=1}^{|C|} f(c_i)}{|C|} + \frac{\sum_{j=1}^{|T|} g(t_j)}{|T|}$$

4.2 Semantic Word Bandwidth (SWD)

The Semantic Word Bandwidth (SWD) should reflect the results of the second interaction step: the experts estimated the overall generality / specificity of the given terms. Of course this bandwidth can only be evaluated with respect to the highest possible bandwidth. The smaller the bandwidth, the more focused is the set of related keywords.

Considering categorizations where we can rely on some ISA hierarchy, e.g. taxonomies of chemical substances, it is quite simple to determine the bandwidth. In this case, we have to identify the depth within the hierarchy for each term. Using the maximum and the minimum depth of terms normalized by the total depth of the hierarchy (*maxdepth*) the semantic word bandwidth can be defined as follows:

$$SWB = \frac{\max_{t \in T}(depth(t)) - \min_{t \in T}(depth(t))}{maxdepth}$$

In cases where no ISA hierarchy is given, it is much more complex to estimate the semantic word bandwidth. For instance, considering substances (e.g. reactants or catalysts) involved in chemical reactions could be considered more specific but in any case this would need a complex ontology describing the relationships for reactions which can currently not be found in the market place.

4.3 Relevance of Covered Terms (RCT)

The last measure used by the experts tried to determine the usefulness of a term in relation to the query term. If we consider again some ISA hierarchy or an ontology, we may express the usefulness of a term in relation to a query term as the semantic similarity between those terms. The total relevance can then be established as the average similarity of keywords to the query term.

Practically, this can be done by analyzing the underlying ontology. All keywords are associated with concepts in the hierarchy. A direct method for measuring the respective similarity is then to find the minimum length of any path connecting the two concepts [24]. However, according to [22] this may not be sufficient for more general and larger ontologies, and thus, the similarity should be a function of the attributes path length, depth and local density.

Another possibility to measure the relevance of the covered terms may be reflected by using independent semantic techniques. In our example, the Semantic GrowBag uses statistical information to compute higher order co-occurrences of keywords. Thus, the relations shown in the graphs reflect some characteristics of the underlying document collection. The naïve way of interpreting the results is that all terms covered by one graph are somehow used together with the query term. If we assume that terms which are more related to the query term are also generally used more often in relation with some document, this should also be reflected by a simple Web search query. Thus, a two term query for a query term qt and a word w_1 which are closely related should result in more hits than a query for qt and some word w_2 that are not as closely related. Preliminary experiments based on our used graphs seem to support this assumption, e.g. a Google search for the query '*amino acids* AND *amino alcohols*' yields 39,800 hits and the query '*amino alcohols* AND *cyclopropanes*' only yields 2,540 hits.

5   Conclusions and Outlook

Semantic techniques are ubiquitous in modern information systems and digital collections. In this paper we dealt with the question whether the expected loss of quality due to the use of statistical techniques can be measured. We argue that the development of such measures is especially important for their safe and sustainable application in digital libraries which generally have higher quality constrains in comparison to, e.g. Web search engines. Putting the focus on automatic metadata creation as provided by related keywords, we conducted a user study in the field of chemistry observing some experts' interaction with the created metadata. The study resulted in three major observations:

1. Domain experts always started from a (reasonably similar) cognitive classification of possible entities. They expected to find relevant terms with respect to all expected classes.
2. Considering the given metadata all experts expected to find a similar degree of generality / specificity of the keywords. The respective degree was derived relative to the general understanding of the respective domain.
3. Assessing the type of relationship between each keyword and the query term all experts tried to embed the terms in a common context. With increasing broadness of the context, the satisfaction with the keywords decreased.

Based on these observations, we proposed three measures namely degree of category coverage (DCC), semantic word bandwidth (SWB) and relevance of covered terms

(RCT). Although our preliminary results address the sensibility of the measures, a detailed investigation using several document corpora is still needed to reflect different topics and sizes. In addition, the quality of digital libraries does not only result in high precision but also in high recall. This is not faced in our metrics yet, but will be investigated in the future. Therefore, our future work will focus on the creation of suitable test corpora and will measure different semantic techniques using manual inspection together with appropriate quality measures.

References

1. Bao, S., Xue, G., Wu, X., Yu, Y., Fei, B., Su, Z.: Optimizing web search using social annotations. In: WWW 2007: Proceedings of the 16th international conference on World Wide Web. ACM Press, New York (2007)
2. Bischoff, K., Firan, C.S., Nejdl, W., Paiu, R.: Can all tags be used for search? In: CIKM 2008: Proceeding of the 17th ACM conference on Information and knowledge management. ACM Press, New York (2008)
3. Chan, S.: Tagging and Searching – Serendipity and museum collection databases. In: Proceedings of Museums and the Web 2007. Archive & Museum Informatics 2007, Toronto (2007)
4. Cimiano, P., Handschuh, S., Staab, S.: Towards the self-annotating web. In: Int. Conf. on the World Wide Web (WWW). ACM, New York (2004)
5. Diederich, J., Balke, W.-T.: The Semantic GrowBag Algorithm: Automatically Deriving Categorization Systems. In: Kovács, L., Fuhr, N., Meghini, C. (eds.) ECDL 2007. LNCS, vol. 4675, pp. 1–13. Springer, Heidelberg (2007)
6. Diederich, J., Balke, W.: Automatically Created Concept Graphs using Descriptive Keywords in the Medical Domain. In: Methods of Information in Medicine (METHODS), Schattauer, vol. 47(3) (2008)
7. Fuhr, N., Hansen, P., Mabe, M., Micsik, A., Sølvberg, I.T.: Digital Libraries: A Generic Classification and Evaluation Scheme. In: Constantopoulos, P., Sølvberg, I.T. (eds.) ECDL 2001. LNCS, vol. 2163, p. 187. Springer, Heidelberg (2001)
8. Fuhr, N., Tsakonas, G., Aalberg, T., Agosti, M., Hansen, P., Kapidakis, S., et al.: Evaluation of digital libraries. In: Int. J. on Digital Libraries, vol. 8(1) (2007)
9. Gangemi, A., Catenaccia, C., Ciaramita, M., Lehmann, J.: Qood grid: A meta-ontology-based framework for ontology evaluation and selection. In: Proc. of the 4th International Workshop on Evaluation of Ontologies for the Web (EON 2006), Edinburgh, Scotland (2006)
10. Golder, S.A., Huberman, B.A.: The structure of collaborative tagging systems (2005) CoRR abs/cs/0508082
11. Gonçalves, M.A., Moreira, B.L., Fox, E.A., Watson, L.T.: What is a good digital library? In: A quality model for digital libraries. Inf. Process Manage, vol. 43(5) (2007)
12. Gonçalves, M.A., Fox, E.A., Watson, L.T., Kipp, N.A.: Streams, structures, spaces, scenarios, societies (5s): A formal model for digital libraries. ACM Trans. Inf. Syst. 22(2) (2004)
13. Halpin, H., Robu, V., Shepherd, H.: The complex dynamics of collaborative tagging. In: WWW 2007: Proceedings of the 16th international conference on World Wide Web. ACM Press, New York (2007)
14. Hearst, M.A.: Automatic Acquisition of Hyponyms from Large Text Corpora. In: Int. Conf. on Computational Linguistics, Nantes, France (1992)

15. Hotho, A., Jäschke, R., Schmitz, C., Stumme, G.: Information Retrieval in Folksonomies: Search and Ranking. In: Sure, Y., Domingue, J. (eds.) ESWC 2006. LNCS, vol. 4011, pp. 411–426. Springer, Heidelberg (2006)
16. http://www.nlm.nih.gov/pubs/factsheets/mesh.html (last accessed on 25.03.2009)
17. http://www.nlm.nih.gov/pubs/factsheets/medline.html (last accessed on 25.03.2009)
18. Khoo, M., Pagano, J., Washington, A., Recker, M., Palmer, B., Donahue, R.A.: Using web metrics to analyze digital libraries. In: JCDL (2008)
19. Krestel, R., Chen, L.: The art of tagging: Measuring the quality of tags. In: Domingue, J., Anutariya, C. (eds.) ASWC 2008. LNCS, vol. 5367, pp. 257–271. Springer, Heidelberg (2008)
20. Kruk, S.R., Woroniecki, T., Gzella, A., Dabrowski, M.: JeromeDL - a Semantic Digital Library. In: Semantic Web Challenge (2007)
21. Kruk, S.R., Kruk, E., Stankiewicz, K.: Evaluation of Semantic and Social Technologies for Digital Libraries. In: Semantic Digital Libraries. Springer, Heidelberg (2009)
22. Li, Y., Bandar, Z.A., Mclean, D.: An approach for measuring semantic similarity between words using multiple information sources. IEEE Transactions on Knowledge and Data Engineering 15(4) (2003)
23. Lozano-Tello, A., Gómez-Pérez, A.: OntoMetric: A method to choose the appropriate ontology. Journal of Database Management, Special Issue on Ontological analysis, Evaluation, and Engineering of Business Systems Analysis Methods 15(2) (2004)
24. Rada, R., Mili, H., Bicknell, E., Blettner, M.: Development and application of a metric on semantic nets. IEEE Transactions on Systems, Man and Cybernetics 19(1) (1989)
25. Razikin, K., Goh, D.H.-L., Chua, A.Y.K., Lee, C.S.: Can social tags help you find what you want? In: Christensen-Dalsgaard, B., Castelli, D., Ammitzbøll Jurik, B., Lippincott, J. (eds.) ECDL 2008. LNCS, vol. 5173, pp. 50–61. Springer, Heidelberg (2008)
26. Sanderson, M., Croft, B.: Deriving concept hierarchies from text. In: Proc. of Int. ACM SIGIR Conf. on Research and Development in Information Retrieval, Berkeley, CA, USA. ACM, New York (1999)
27. Saracevic, T.: Digital library evaluation: toward evolution concepts. Library Trends 49(2) (2000)
28. Tartir, S., Aroinar, I.B., Moore, M., Sheth, A.P., Aleman-Meza, B.: OntoQA: Metric-based ontology analysis. In: Proceedings of IEEE Workshop on Knowledge Acquisition from Distributed, Autonomous, Semantically Heterogeneous Data and Knowledge sources (2005)
29. Vrandečić, D., Sure, Y.: How to design better ontology metrics. In: Franconi, E., Kifer, M., May, W. (eds.) ESWC 2007. LNCS, vol. 4519, pp. 311–325. Springer, Heidelberg (2007)

Supporting the Creation of Scholarly Bibliographies by Communities through Online Reputation Based Social Collaboration*

Hamed Alhoori[1], Omar Alvarez[1], Richard Furuta[1], Miguel Muñiz[2], and Eduardo Urbina[2]

[1] Center for the Study of Digital Libraries and
Department of Computer Science and Engineering
[2] Cervantes Project, Department of Hispanic Studies
Texas A&M University, USA
{alhoori,aomar,furuta,apresa,e-urbina}@tamu.edu

Abstract. Bibliographic digital libraries play a significant role in conducting research and, in the past few years, have started to move from closed to more open social platforms. However, in this, they have faced challenges (e.g., from Web spam) in maintaining the level of scholarly precision—the ratio of relevant citations retrieved by search. This paper describes a hybrid approach that uses online social collaboration and reputation based social moderation to reduce the cost and to speed up the construction of scholarly bibliographies that are comprehensive, have better quality citations and higher precision. We implemented selected social features for an established digital humanities project (the Cervantes Project) and compared the results with a number of closed and open current bibliographies. We found this can help in building scholarly bibliographies and significantly improve precision outcomes.

Keywords: Social collaboration, social moderation, social reputation, scholarly bibliography, digital libraries, digital humanities.

1 Introduction

Closed bibliographic digital libraries (BDLs), manually compiled by authorized users or automatically-generated, have existed for many years. Recently, *open social* BDLs (e.g., CiteULike[1]) have emerged. However, for specific research needs, a satisfactory level of precision and comprehensiveness is not entirely attained by either of these approaches. Current bibliographic search engines show a limited scope of coverage on literature. There is no single resource that handles the entire 2.5 million articles that emerge yearly from the 25,000 peer-reviewed journals [1], so these engines

---

* This material is based upon work supported by the National Science Foundation under Grant No. IIS-0534314.
[1] http://www.citeulike.org

M. Agosti et al. (Eds.): ECDL 2009, LNCS 5714, pp. 180–191, 2009.

access only a fraction of the literature [2]. From this limited literature, researchers concentrate further on specific groups of conferences and journals, missing other valuable related research outside of their immediate scope.

Beyond the increased information availability resulting from the increasing number of journals and conferences and their inclusion in digital libraries, there is a growing movement towards open access archives. This increases the availability of research resources in the online communities. As a consequence, papers that are not available electronically for various reasons may lose their presence in the research community.

Many digital humanities projects manually maintain online BDLs that support diverse users in locating a variety of references. In this paper we will use the example of the Cervantes Project's[2] bibliography (CIBO), which aims to represent the best resources published since 1605 about Miguel de Cervantes, the author of Don Quixote, drawn from many multilingual sources. The current CIBO bibliography gathering and filtering process is carried out by sets of contributors: the expert editors, the reviewers, and the authorized international collaborators. Consequently, delays, possibly months, can result from gathering, filtering and indexing of new publications into the CIBO.

We believe that precise social collaboration systems are a way to address each of these issues: increasing amounts of striated information, increased invisibility of off-line literature, and manually-introduced delays in filtering bibliographic information.

Most online bibliographies provide services to their users while prohibiting them from contributing. This results in a considerable loss of external knowledge. The current state of the art is moving toward two ways of interaction, where the users can benefit from the available knowledge and contribute to it. Hendry, et al. [3], mention an "amateur bibliography" that is collected by non-professionals and falls short of the standards of a professional bibliography. Although large number of references could be collected in a short span of time, this results in issues such as redundancy (repeated citations), spam, phantom author names, and phantom citations. These are not good signs of scholarly research [4] and would affect the significance of a journal (e.g., impact factor [18]) or a publication (e.g., h-index [19]). Spam also threatens social websites to undermine resource sharing, interactivity, and openness [5].

Social moderation models are elements that assist in unifying online groups to achieve consensus about common interest topics, reduce spam content, and identify members' reputations. This approach works well for social interaction and open collaboration and has been accepted in those uses. However, there is controversy about the moderation effectiveness of open environments in achieving acceptable levels of quality content and identification of users' reputations. Moderated systems have faced problems such as insufficient attention to posts, moderation delays, unfair moderations, and premature negative or positive consensus [20].

This paper's premise is that online reputation-based social collaboration (ORSC) can reach the precision level of the scholarly moderated bibliography [13] by benefiting from the "wisdom of the crowds" [6]. This approach would be more comprehensive than the regular closed bibliographies and more accurate than the open social citations websites. This would lead researchers to the required and current resources from multiple sources in less period of time. We have experimented with this issue by

---

[2] http://cervantes.tamu.edu/

implementing online social functionality for the CIBO. We have tested them on a group of CIBO users from different countries who use a variety of languages to gather, share, annotate, rank and discover academic literature. We compared our precision outcomes with a number of highly recognized, closed (e.g., WorldCat[3] and MLAIB[4]) and open (e.g. CiteULike and Bibsonomy[5]) online bibliographies.

This paper is structured as follows. We discuss the related work in Section 2. Section 3 explains the approach we used and our implementation. We present and discuss these current experiments and results in section 4. In section 5 we conclude and highlight some of the future work.

2 Related Studies

We compared main features supported by various current well-established humanities BDLs. Table 1 summarizes the main outcomes. These BDLs were initiated as long as a decade ago and most do not incorporate the social collaboration mechanisms of Web 2.0 such as social bookmarking, tagging, reviewing, ranking, etc.

Table 1. Humanities BDLs supported features

Bibliography / Features	Cervantes Project	World Shakespeare Bibliography	The Galileo Project	The Walt Whitman Archive
Developer	TAMU	Shakespeare Quarterly	Rice University	Ed Folson & Kenneth M. Price
Established	1995	1950 (physical records)	1995	1995
Searching	√	√	√	√
Browsing	√	√	√	√
Multilanguage Content	√	√	×	√
Multilanguage Interface	√	×	×	×
Review	×	√	×	√
Social Collaboration	×	×	×	×

Collaboration in bibliographies exist in several systems from areas other than the humanities. The ShaRef system [9] supports collaboration between groups of researchers. It provides authentication and access control features. Heymann, et al. [8], concluded that social bookmarking can provide search data not currently provided by other sources, though it may currently lack the size and distribution of tags necessary to make a significant impact. Santos-Neto, et al. [7], showed that the current level of collaboration in CiteULike and Connotea is consistently low. Users are adding new items much faster than they are reusing them. Only a small number of user pairs share interest over items and use the same tags, which significantly limits the potential of harnessing the social knowledge in communities. This explains the cause of the

---

[3] http://www.worldcat.org
[4] http://www.mla.org/bibliography
[5] http://www.bibsonomy.org/

relatively high spamming levels. The majority of the online social citation collections are swamped with a high level of spam [5, 10, 11]. This is a classic Web 2.0 problem: it's hard to aggregate the wisdom of the crowds without aggregating their inexperience or madness as well [12]. Bogers, et al. [11], reported, using different sizes and dates of datasets, that around 93% of BibSonomy users and 28.1% of CiteULike users are spammers, posting 84% and 31% of the spam articles and bookmarks with 88% and 53% spamming tags. [10] mentions that web spam has started targeting more specific communities, such as the scholarly world, and introduced a variety of features to fight spam in social bookmarking systems. They evaluated them with well-known machine learning methods, using the BibSonomy dataset for their experiments.

We compared the main social collaboration features of the most four popular online social citations websites (Table 2).

Table 2. Comparison of social citations features

Online Social Citations / Features	2collab[6]	BibSonomy	CiteUlike	Connotea[7]
Multilanguage interface	×	English and German	×	×
Social Bookmarking	√	√	√	√
Social Tagging	√	√	√	√
Social Reviewing	√	√	√	√
Social Ranking and Sorting	√	×	×	×
Social Filtering	√	√	×	×
Groups of interest	√	√	√	√
Reputation based social moderation	×	×	×	×

We found that most online social citations sites support the well-known social collaboration features, providing a similar set of group types in moderating the citations: *private*, *closed*, and *open*. In these three types of groups, the community is not reaching the full potential of true collaboration. In the private group the community is isolated from the world and only the previously known members can contribute. In the closed groups there is a special need to approve a member. In the open groups there is an urgent need for checking the members' contributions.

Many testing, redundant, phantom and spam citations and groups exist in these systems. All of these groups assign moderators manually, which is time consuming, and may have some influence or bias from the creators of the group. Furthermore, moderators may lose interest or be inactive for a long period of time. Moreover, in such interdisciplinary bibliographies it is hard to decide if a citation is spam or not unless it is clearly obvious or was added to a specialized group and the group members suggest that it is not related to the group's interest. None of the previous attempts that we know tried to merge the approaches in an ORSC.

---

[6] http://www.2collab.com
[7] http://www.connotea.org

3 Extending the CIBO to Support ORSC

We enhanced the existing CIBO interface to support online reputation-based social collaboration, as will be described in this section. We then compared precision results from the augmented CIBO and selected popular sites; see section 4.

3.1 Reputation Based Social Collaboration

Considering the high level of spam in social citations websites, there is a need to reflect the accuracy and quality of the users' contribution and reputation in the community when allowing them to moderate, but also there is the need to continue to benefit from the openness of self-selection.

A user's contribution can be any of these elements; citation *(C)*, tag *(T)*, rate *(R)*, review *(V)*, translation *(N)*, or filter *(F)*. Users can add new citations, tag citations, rate citations by selecting a score out of five, review citations by commenting on them, translate citations, and filter spam citations by marking them. We have three types of memberships, which are user (u), collaborator (b) and moderator (m). Users can search and share but their contributions will be moderated. We allow approval of the contributions by a moderator or by *n* collaborators; *n=(1+ceiling(JB/AB))*, where *JB* and *AB* represent the rejected and approved contributions from collaborators.

Sabater and Sierra [24] present an extensive study on a set of reputation systems considering social relationship between users. These models compute reputation based on specific elements such as ratings, levels of participation, and quality of posted information. Chen, et al. [22], present a user reputation model that is used in an user-interactive question and answer system. It combines social network analysis and user ratings. Other researchers [23] present user reputation model for a digital library and digital education community that combines individual and collaborative activity. The weights assigned to each element depend on the specific society [24].

Our model is based on a multidimensional approach. It considers the user's activity and members evaluations. The elements selection and its assignment of weights were based on CIBO moderators' experience. Our members upgrading or downgrading is done using a social reputation [21]; users obtain higher reputation in the community by having accurate contributions and receiving credits from other users. Users can be upgraded to collaborators. A collaborator can be upgraded to a moderator. Initially, we seeded the moderator list with well-known Cervantes scholars and contributors. A summary of the moderation rules and privileges are shown in Table 3.

Table 3. Moderation rules

Controls / Members	Create contribution	Approve contribution	Edit contribution
User (u)	√	×	×
Collaborator (b)	√	√ *n*b	×
Moderator (m)	√	√	√

We summarize the social reputation by using the following formulas:

If the summation of user (u) contributions $S(u)$ and the summation of users evaluations to those contributions $E(u)$, according to the element importance, time of contribution, order, and evaluator reputation (ER), exceeds a threshold value D then the user will be upgraded to collaborator. If $(S(u)+E(u))>(D\times log\ X)$, then the user will be upgraded to moderator. X is the total contributions in the system.

$S(u)$ (formula 1) is used to compute the user contributions. $S(u)$ sums the approved user contributions of C, T, R, V, N, and F after multiplying them by specified weights a to f that represent the importance of that element. $X(u)$ sums the approved user contribution of element X for a user (u), where $X \in \{C,T,R,V,N,F\}$. X_i^u represents a single user (u) contribution (i). We also multiply the sum of user contributions by reciprocal of t_i and o_i, where t_i stands for the time from the citation appearance in the literature to the time it was contributed in the CIBO, or the time from the contribution to the time of a follow up contribution such as adding new tags, rates, reviews, translations or filters. o_i stands for the order of the contribution. This will allow valid earlier contributors to gain more points that advance them to higher ranks in the community.

$$S(u) = a\sum_{i=1}^{C(u)}(\frac{C_i^u}{t_i})+b\sum_{i=1}^{T(u)}(\frac{T_i^u}{ot_i})+c\sum_{i=1}^{R(u)}(\frac{R_i^u}{ot_i})+d\sum_{i=1}^{V(u)}(\frac{V_i^u}{ot_i})+e\sum_{i=1}^{N(u)}(\frac{N_i^u}{ot_i})+f\sum_{i=1}^{F(u)}(\frac{F_i^u}{ot_i}) \quad (1)$$

To compute users evaluations we use $E(u)$ (formula 2). EX^u is a single evaluation of contribution X. $E(u)$ sums the users evaluations (EX_{ij}^u), for the user contributions after multiplying them by specified weights a' to f' that again represents the importance of that element.

$$E(u) = a'\sum_{i=1}^{C_i^u}\sum_{j=1}^{EC_j^u}(EC_{ij}^u\times ER)+b'\sum_{i=1}^{T_i^u}\sum_{j=1}^{ET_j^u}(ET_{ij}^u\times ER)+c'\sum_{i=1}^{R_i^u}\sum_{j=1}^{ER_j^u}(ER_{ij}^u\times ER)+$$
$$d'\sum_{i=1}^{V_i^u}\sum_{j=1}^{EV_j^u}(EV_{ij}^u\times ER)+e'\sum_{i=1}^{N_i^u}\sum_{j=1}^{EN_j^u}(EN_{ij}^u\times ER)+f'\sum_{i=1}^{F_i^u}\sum_{j=1}^{EF_j^u}(EF_{ij}^u\times ER) \quad (2)$$

In order to compute D, we use formula 3, where U stands for the total number of users, J the total number of rejected contributions, A the total number of approved contributions, and E the total number of evaluations.

$$D = \log(U) + \log(\frac{J}{A}\times E) \quad (3)$$

3.2   Social Technologies Applied to Bibliographies

A set of social collaboration features was implemented in CIBO to support the open social collaboration environment. Figure 1 shows the main interface as it displays a citation's details.

Fig. 1. Screenshot for a citation's details

3.2.1 Social Bookmarking

Users can participate by providing new citations using the social bookmarking feature, importing citations or manually entering them. Figure 2 shows points gained by a user after several entries.

Points for admin

Points	Approved?	Date	Operation	Category	Description
1	Approved	04/27/2008 - 16:39	vote	Uncategorized	Vote cast: node 78.
1	Approved	04/24/2008 - 19:09	vote	Uncategorized	Vote cast: node 54.
1	Approved	04/24/2008 - 14:40	vote	Uncategorized	Vote cast: node 56.
1	Approved	04/24/2008 - 14:40	vote	Uncategorized	Vote cast: node 50.
2	Approved	04/24/2008 - 14:03	insert	Uncategorized	None
2	Approved	04/24/2008 - 14:02	insert	Uncategorized	None
-2	Approved	04/23/2008 - 00:45	operation	Uncategorized	None
1	Approved	04/22/2008 - 18:24	vote	Uncategorized	Vote cast: node 41.
2	Approved	04/22/2008 - 15:01	insert	Uncategorized	None
1	Approved	04/22/2008 - 14:58	vote	Uncategorized	Vote cast: node 40.

Uncategorized points Balance: 1503

Approved points Balance: 1503
Points awaiting moderation: 0
Net points Balance: 1503

« first ‹ previous ... 72 73 74 75 76 77 78 **79** 80 next › last »

Fig. 2. Detailed view of contributors' points

3.2.2 Social Tagging

Del.icio.us and Digg are two of the popular and fastest growing social bookmarking sites that use folksonomy tagging. However, inaccurate and misleading tags are common in such open environments, which cannot be accepted in scholarly research communities. We prevent these effects by moderating the new users' tags. Users can create their own tags or reuse the previously entered tags by them or other users using auto-complete tags in real time; the implementation uses AJAX technology.

3.2.3 Social Ranking

Bibliography ranking has been used as a way to give users confident Top-N resources from the search results. A typical user only reads the first, second, or third page of results. Citations have been used as a way to rank bibliography resources. Citation-based methods deal with complex issues such as bias or self-citations, hard to detect positive or negative citations, multiple citations formats difficult to handle by computer programs, unfair consideration of new papers, venues not considered [14, 15, 16]. Other researchers [17] proposed a seed-based measure (considering top-venues and venues' authors relevance) and the browsing-based measure (considers user's behavior) to rank academic venues. However, the authors-seed needs to be updated frequently to reconsider new relevant authors. We implemented a hybrid approach, where we allowed users to rate citations on a scale of five points while benefiting from bibliographic citation details. Each user has a different weight for rating the citation according to the user reputation.

3.2.4 Social Reviewing

We implemented a feedback environment to build an active online research community. It provides reviews and comments from the users where they can interact and clarify unclear points.

3.2.5 Social Translation

As digital libraries expand their audience and content scope, there is an increasing need for resources and access tools for those resources in a variety of languages [14]. The Cervantes Project's international scope requires the inclusion of content and system functionalities in multiple languages, since Cervantes literature has been translated to various languages and bridges between cultures need to be established.

Users can choose the preferred available language at any moment while using the system. This choice will automatically translate the interface to that language and will select only the content with that language. Using the Google Translate API [25], we provided a translation capability for the comments. Bibliographic data can be entered in a language and then manually translated to a new language or linked to existing bibliographic data or publications in other languages (Figure 3).

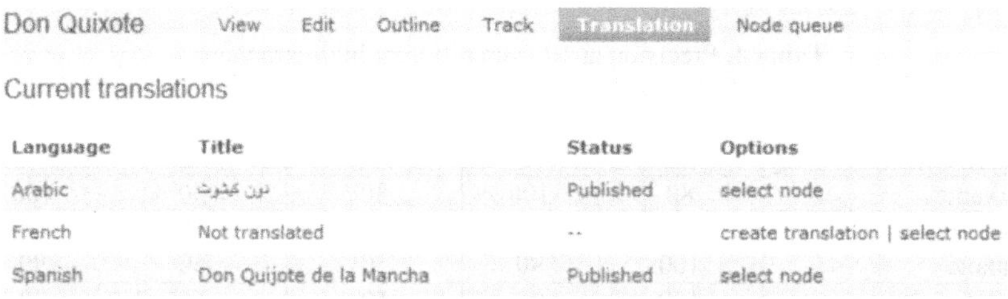

| Don Quixote | View | Edit | Outline | Track | Translation | Node queue |

Current translations

Language	Title	Status	Options
Arabic	دون كيشوت	Published	select node
French	Not translated	--	create translation | select node
Spanish	Don Quijote de la Mancha	Published	select node

Fig. 3. A publication available translations

3.2.6 Social Filtering

Retrieving citations that are irrelevant, incorrect or spam frustrates the researchers and affects their productivity. We tried to mitigate this scenario by empowering the users to discover and filter any such results, spam, or spammers by reporting them for moderation. It is more as a social encouragement, since first users who discovered and reported these results would be given higher weights compared with subsequent users who report them. A moderator or *n* Collaborators (see 3.1) can approve the requests by editing or hiding contributions or banning a spammer. Moderators will be able to view these changes for any follow up request and future statistics.

3.2.7 Social Discovery and Networking

By providing the previous social facilities, we allowed the researchers to share and discover latest academic literature without worrying about inaccurate bibliographic data. They can search and browse the citations contributed by the users, collaborators, moderators, or combination of them. They can discover what the hot topics are in the research field and what is significant to other like-minded researchers by viewing what they read, cited, tagged, ranked, or reviewed. Therefore, they can identify the related researchers with similar interest that they can network with.

4 Evaluation and Discussion

From the set of online citations websites available on the Internet, we selected the most reliable closed sites to digital humanities and the most popular open social citations sites. We used two closed BDLs, WorldCat and MLA International Bibliography (MLAIB), and four open social citations websites, CiteULike, Connotea, 2collab and BibSonomy, that contain millions of citations. We compared their precision outcomes with the augmented CIBO. Precision in our experiments was calculated as the number of relevant citations retrieved by a search divided by the total number of citations retrieved by that search at several milestones. Cervantes Project experts decided the common keywords and tags that are used in Cervantes literature, and we used those as search terms. The experts also evaluated the relevancy of retrieved documents. After gathering the results from the different resources, we found that Connotea and 2collab contain only few citations about Cervantes. Therefore, we removed them from the comparison. Table 4 shows a sample of precision to CIBO at the first 10 retrieved citations compared with CiteULike, BibSonomy, WorldCat and MLAIB. We used different lengths of keywords and tags combinations to search the bibliographies. Table 5 shows the average precision percentage % at 10 (P10), 20, 30, 40, and 50 for the bibliographies.

Table 4. Precision at 10 from different bibliographies

BDLs / Search terms	WorldCat	MLAIB	CiteULike	BibSonomy	CIBO
Cervantes	80	100	30	30	100
سيريفانتس	0	0	0	0	40
Quixote	100	90	50	50	90
Quijote	100	90	50	50	90
Cervantes plays	90	40	30	00	80
Miguel de Cervantes Poetry	30	10	0	0	100
Cervantes Windmills	80	100	30	10	80
Sancho panza	100	100	20	0	100
Dulcinea	80	80	10	0	50
Cervantes Blanket	10	30	10	0	0
Cervantes Island	30	30	0.0	0	90
Cervantes Persiles	80	70	10	0	90

Table 5. Average of precision from 10 to 50

Precision% BDLs	P10	P20	P30	P40	P50
CIBO	91	38	23	15	11
WorldCat	78	36	24	17	13
MLAIB	74	31	20	15	11
CiteULike	24	8	5	4	3
BibSonomy	14	4	2	1	1

Figure 5 shows that CIBO performs better than all the compared BDLs at precision 10. At precision 20 it is still ahead with 2% from WorldCat. At precision 30, World-Cat goes ahead with 1%.

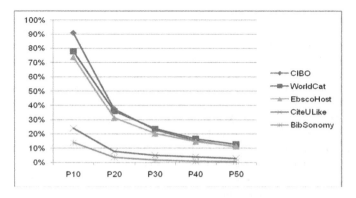

Fig. 5. Precision of the compared BDLs with CIBO

While CIBO achieved higher precision at 10 and 20, its precision started to decrease later on. This pattern occurs mainly because the users' rate and filter the initial results while neglecting the subsequent outcomes.

Our findings show how closed BDLs have considerably enhanced precision performance over the open social citation systems. This seemingly justifies the argument of scholarly communities to keep using closed environments but also increases the limited scope of coverage on literature. However, using the ORSC approach produces a precision performance competitive to general and closed bibliographies on searches for Cervantes-related topics. It also supports the personalization of the information and shows who are the active researchers. This visibility helps identify researchers for future collaborations.

5   Conclusion and Future Work

The open bibliography environments were originally conceived as websites for exchanging citations and reviews of global publications, taking advantage of the large communities available on the Internet. These sites offer a variety of benefits, but the lack of moderation brings high levels of spam. In addition, many contributors are more enthusiastic than experienced. A lack of moderation may be acceptable for social sites but regarding scholarly communities, the content quality is a priority.

In this paper we have investigated the precision outcomes of a hybrid bibliography system created by an online digital humanities community. Our current experimental results indicate that using ORSC would improve the quantity and usage of scholarly bibliography and improve the quality and creditability of social citations sites.

We intend to automate more portions of the moderation process by checking the contributed citations to the closed and open online citations websites. We will evaluate the reviews and comments (positive or negative) by identifying and interpreting annotation patterns and semantics to give a relevance weight to each source, which would help also in the ranking. We plan also to investigate the existing work identifying hidden spam to get statistics to automate the process of filtering.

References

1. Harnad, S., Brody, T., Vallieres, F., Carr, L., Hitchcock, S., Gingras, Y., Oppenheim, C., Hajjem, C., Hilf, E.R.: The Access/Impact Problem and the Green and Gold Roads to Open Access: An Update. Serials review 34(1), 36–40 (2008)
2. Hull, D., Pettifer, S.R., Kell, D.B.: Defrosting the Digital Library: Bibliographic Tools for the Next Generation Web. PLoS Comput. Biol. 4(10), 10–1000204 (2008)
3. Hendry, D.G., Jenkins, J.R., McCarthy, J.F.: Collaborative bibliography. Information Processing and Management: an International Journal 42(3), 805–825 (2006)
4. Jacso, P.: Testing the Calculation of a Realistic h-index in Google Scholar, Scopus, and Web of Science for F. W. Lancaster. Library Trends 56(4), 784–815 (2008)
5. Heymann, P., Koutrika, G., Garcia-Molina, H.: Fighting Spam on Social Web Sites: A Survey of Approaches and Future Challenges. IEEE Internet Computing 11(6), 36–45 (2007)
6. Sorowiecki, J.: The Wisdom of the Crowds: Why the Many Are Smarter Than the Few and How Collective Wisdom Shapes Business, Economies, Societies and Nations, 1st edn., New York (2004)
7. Santos-Neto, E.: Ripeanu, M., Iamnitchi, A.: Content Reuse and Interest Sharing in Tagging Communities. Technical Notes of the AAAI 2008 Spring Symposia - Social Information Processing, pp. 81-86. Stanford, CA, USA (March 2008)
8. Heymann, P., Koutrika, G., Garcia-Molina, H.: Can Social Bookmarking Improve Web Search. In: WSDM 2008 (2008)
9. Wilde, E., Anand, S., Bücheler, T., Jörg, M., Nabholz, N., Zimmermann, P.: Collaboration Support for Bibliographic Data. International Journal of Web Based Communities 4(1), 98–109 (2008)
10. Krause, B., Hotho, A., Stumme, G.: The Anti-Social Tagger - Detecting Spam in Social Bookmarking Systems. In: AIRWeb 2008: Proceedings of the 4th International Workshop on Adversarial Information Retrieval on the Web (2008)
11. Bogers, T., van den Bosch, A.: Using Language Modeling for Spam Detection in Social Reference Manager Websites. In: 9th Dutch-Belgian Information Retrieval Workshop (DIR 2009), pp. 87–94. Enschede, The Netherlands (2009)
12. Torkington, N.: Digging the Madness of Crowds (2006),
    http://radar.oreilly.com/archives/2006/01/
    digging-the-madness-of-crowds.html

13. Hendry, D.G., Carlyle, A.: Hotlist or Bibliography? A Case of Genre on the Web. In: 39th Annual Hawaii International Conference on System Sciences (HICSS), vol. 3(04-07), p. 51b (2006)
14. Larsen, B., Ingwersen, P.: The Boomerang Effect: Retrieving Scientific Documents via the Network of References and Citations. In: 25th Annual international ACM conference on Research and development in information retrieval (SIGIR), Tampere, Finland (2002)
15. Larsen, B., Ingwersen, P.: Using Citations for Ranking in Digital Libraries. In: 6th ACM/IEEE-CS joint conference on Digital libraries. Chapel Hill, NC (2006)
16. Yang, K., Meho, L.: CiteSearch: Next-generation Citation Analysis. In: 7th ACM/IEEE-CS joint conference on Digital libraries, Vancouver, British Columbia, Canada (2007)
17. Yan, S., Lee, D.: Toward Alternative Measures for Ranking Venues: A Case of Database Research Community. In: 7th ACM/IEEE-CS joint conference on Digital libraries, Vancouver, British Columbia, Canada (2007)
18. Garfield, E.: Citation analysis as a tool in journal evaluation. Science 178, 471–479 (1972)
19. Hirsch, J.E.: An index to quantify an individual's scientific research output. Proceedings of the National Academy of Sciences 102(46), 16569–16572 (2005)
20. Lampe, C., Resnick, P.: Slash(dot) and burn: distributed moderation in a large online conversation space. In: SIGCHI Conference on Human Factors in Computing Systems, Vienna, Austria (2004)
21. Windley, P.J., Daley, D., Cutler, B., Tew, K.: Using reputation to augment explicit authorization. In: ACM Workshop on Digital Identity Management, Fairfax, Virginia, USA (2007)
22. Chen, W., Zeng, Q., Wenyin, L., Hao, T.: A user reputation model for a user-interactive question answering system: Research Articles. In: Concurrency and Computation: Practice and Experience, Hong Kong, vol. 19-15, pp. 2091–2103 (2007)
23. Jin, F., Niu, Z., Zhang, Q., Lang, H., Qin, K.: A User Reputation Model for DLDE Learning 2.0 Community. In: 11th international Conference on Asian Digital Libraries: Universal and Ubiquitous Access To information, Bali, Indonesia (2008)
24. Sabater, J., Sierra, C.: Review on Computational Trust and Reputation Models. Artifitial Intelligence Review 24(1), 33–60 (2005)
25. Google AJAX Language API, http://code.google.com/apis/ajaxlanguage/

Chance Encounters in the Digital Library

Elaine G. Toms and Lori McCay-Peet

Centre for Management Informatics
Dalhousie University
6100 University Ave
Halifax, Nova Scotia, Canada
etoms@dal.ca, mccay@dal.ca

Abstract. While many digital libraries focus on supporting defined tasks that require targeted searching, there is potential for enabling serendipitous discovery that can serve multiple purposes from aiding with the targeted search to suggesting new approaches, methods and ideas. In this research we embedded a tool in a novel interface to suggest other pages to examine in order to assess how that tool might be used while doing focused searching. While only 40% of the participants used the tool, all assessed its usefulness or perceived usefulness. Most participants used it as a source of new terms and concepts to support their current tasks; a few noted the novelty and perceived its potential value in serving as a stimulant.

Keywords: Digital libraries; serendipitous discovery.

1 Introduction

Introducing the potential for serendipitous discovery within digital libraries brings also the possibility for diverting attention away from the task, leading to unproductiveness. Like many tangible fields of endeavour, e.g., medicine, biology and engineering, "finding information without seeking it through accidental, incidental or serendipitous discoveries" [20] may lead to a fresh approach to a problem, novel information, or a fruitful departure leading to solutions to other problems. The challenge from a design perspective is in how to enable serendipity without also causing a non-productive distraction or interruption.

Serendipitous information retrieval, which also has been called *information encounters* [3], *chance encounters* [19] and *incidental information acquisition* [23] "occurs when a user acquires useful information while interacting with a node of information for which there were no explicit *a priori* intentions" [19]. In each case, the retrieval or viewing of the information object occurred when the user made an accidental and often perceptive discovery. The connection between user and object is likely influenced by the person's prior knowledge and experience within a particular problem space and by the person's recognition of the 'affordances' within that information object [19].

In this research, we introduced and tested "Suggested Pages," a list of items that are somewhat related to the currently viewed webpage to assess how users interact

M. Agosti et al. (Eds.): ECDL 2009, LNCS 5714, pp. 192–202, 2009.

with it while in the process of doing other work. Because the Suggestions were dynamically created and based on the current page viewed, we had little control over what might actually appear. Potentially the Suggestions were directly related to the task, but given the nature of information retrieval algorithms, they might also be semantically dissimilar.

2   Previous Work

Evidence for serendipitous encounters is unmistakable in prior research in both physical and virtual information environments. Ross [15] and Williamson [23] in separate studies found incidental information, respectively, through reading and conversation pertinent to daily lives. More recently, Rimmer and colleagues [14] observed serendipitous interactions in the use of text in a library speculating on the advantage that the printed work provided.

While many acknowledge the challenge of designing a system that would enable serendipity, but not inhibit productivity [16, 22], triggering a serendipitous encounter has been proposed and tested in a number of ways. Back in 1968, Grose and Line [8] proposed that books be arbitrarily shelved to facilitate browsing. More recently, Campos and de Figueiredo [2], developed a software agent called 'Max' that wandered the web selecting links to follow, at times random, in order to induce serendipitous discovery. Twidale and colleagues [22] found that spurious results were perceived as serendipitous, offering potential opportunities for new directions. Erdelez [5] found in preliminary studies that it is possible to trigger information encountering episodes by embedding information known to relate to a participant's secondary information needs in a list of search results.

In a newspaper reader, Toms [21] introduced an "Items-to-Browse" interface tool that suggested other pages to read based on the page currently being read. She noted that participants found the most interesting articles from the list of articles suggested by the system while people were reading the daily newspaper. While ordinarily following a standard menu-based system, participants used these suggestions which subsequently invoked a chance encounter that they may not otherwise have experienced had they read the newspaper according to their usual practices. Each identified the reason for selecting an article from the Items-to-Browse; their motivations varied from the esoteric and sensational, to connections with a thought or event held in memory [18].

As well, serendipitous encounters cannot solely be enabled through the design of information systems and individual differences have been found to play a role in the degree to which people make or are receptive to making a serendipitous discovery [4, 9]. While some people, for example, are not easily distracted from the information task at hand, others are very sensitive to noticing the information objects that surround them [4]. Franken [6] describes human ability to process information as limited, motivating us to be selective about the information we process and tending to process information that is consistent with information we already have. McBirnie [11] suggests information literacy education should highlight the value of serendipity in the information search so that users will be more open to chance discoveries rather than aiming solely for the controlled, efficient search.

Conceptually, each incident of serendipitous encounter is embedded in a primary information seeking and retrieval episode. One generally does not seek serendipity; it occurs usually in the process of some other activity. Erdelez [5] proposed that a typical information encountering (IE) episode consists of some or all of the following elements: Noticing, Stopping, Examining, Capturing, and Returning. While some information seekers may simply perceive (Notice) encountered information, perception may develop into an interruption of the original information seeking task (Stopping), followed by evaluation (Examining) and mining (Capturing) of the encountered information, and finally Returning to the original task. These elements are similar to the stages of the interruption management stage model [10, 12]: interruption, detection, interpretation, integration, and resumption. Latorella [10] goes further by mapping out the behaviours associated with these stages: oblivious dismissal, unintentional dismissal, intentional dismissal, pre-emptive integration, and intentional integration.

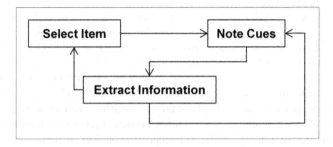

Fig. 1. A Digital Browsing Instant

In her study of browsing, Toms [17] referred to these episodes as "Digital Browsing Instants" as illustrated in Figure 1. A person may select an item to support a primary task, and in the context of examining that item may notice cues, e.g., word, phrase or concept, that acts "as a stimulus influencing the user's focus" [17]. When information is extracted, a serendipitous acquisition of information – a chance encounter – occurs. Not all cues result in serendipity; serendipity only occurs when a connection is made. "The serendipitous discovery is not a matter of blind luck, rather it is the recognition of a valuable document attribute/connection discovered by means outside established access system rules and relying on a user's self knowledge" [13].

This research follows from Toms' [17, 18, 21] initial study which pre-dates the Web and was used in the context of everyday use – news reading. In the work reported here, a list of suggested webpages was embedded within each page display and modified with each page viewed. This concept was deployed in a novel interface (see http://www.interactiveIR.org/public) used to access a version of the Wikipedia. The objective of our research was to assess how such a tool is used in the context of a work task operationalized as a set of pre-defined tasks. While news-reading is a browsing process, one with few pre-ordained notions, the task here had defined objectives, and we examine how users would use such a tool in this context.

3   Methods

This research is part of a larger study that is looking at the role of task in search as well as the interface elements to be introduced to search interfaces. In the research reported here, we have focused specifically on how the tool meant to induce serendipity, Suggested Pages, was used. Part of the challenge of studying serendipity is how to artificially induce it. In this case, by imposing a device in the context of a second study we were able to observe, but not control the action. Until the data was collected we had no idea of whether in fact the tool was used at all. Post session questions enabled us to ask about use without interfering with the conduct of the larger study. The methods describe how the data was collected, but we have provided only the results specifically related to this feature.

3.1   System – WikiSearch

The system used was an experimental system that enabled searching the contents of the Wikipedia. Each page selected from a search results list for viewing contained ordinary hypertext links as well as links to search within the wiki. As illustrated in Figure 2, each page also displayed a box of *Suggested Pages* in the upper right corner. This set of page links was created by entering the first paragraph of the currently displayed page as a search string. The top five results not including the current page were displayed. Not unlike the Suggestions described earlier, this list had the potential to provide more specific pages about the topic, or be distracting. wikiSearch runs on Lucene 2.2, an open source search engine using the vector space model. The Wikipedia XML documents are indexed using the Lucene standard analyzer and its default stemming and stop word filtering.

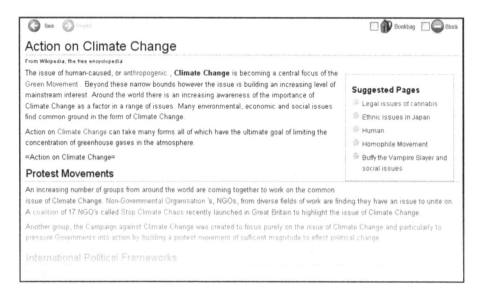

Fig. 2. WikiSearch Page view

3.2 Tasks

Over the course of the study, participants completed three tasks from a set of 12 tasks developed for INEX 2006. The tasks were intended to represent work tasks and varied significantly according to topic. Topics varied from recommending certain types of bridge structure to choosing between venues on a trip to Paris, and sorting out the difference between fortresses and castles. One example of the 12 tasks is:

> As a member of a local environmental group who is starting a campaign to save a large local nature reserve, you want to find some information about the impact of removing the trees (logging) for the local pulp and paper industry and mining the coal that lies beneath it. Your group has had a major discussion about whether logging or mining is more ecologically devastating. To add to the debate, you do your own research to determine which side you will support.

To respond to each task, participants were required to add wikipages to a Bookbag, a tool at the interface used to collect relevant documents, and to rate the relevance of each page. Over the course of responding to these tasks, participants were exposed to the Suggested Pages.

3.3 Metrics

To assess the Suggested Pages, we examined each use to qualitatively identify how selected suggested pages related to the original task, and using a series of closed and open-ended questions that asked participants to assess the tool.

3.4 Participants

The 96 participants (M=49, F=47) were primarily (90%) students from the university community, and from mixed disciplines. 25% held undergraduate degrees and 12% graduate or other degrees. 84.4% were under 27. They were an experienced search group with 86.5% searching for something one or more times a day, and also relatively frequent users of the Wikipedia (54% use it at least weekly).

3.5 Procedure

Data collection took place in a laboratory setting, a seminar room where 5 to 7 people were processed simultaneously using laptops with wired network connections scattered around the room. A research assistant was always present for any questions or interventions. Participants were simply told that we were assessing how people search, the intent of the larger project, but were not briefed on the precise objectives – assessing their use of Suggested Pages – of the work reported here.

Participants were presented with the following steps in a series of self-directed webpages: 1) Introduction which introduced the study, 2) Consent Form that outlined the details of participation, 3) Demographics and Use Questionnaire to identify prior knowledge and experience, 4) Tutorial and practice time using the wikiSearch system, 5) Pre-Task Questionnaire, 6) Assigned Task to be completed using wikiSearch integrated into the interface, 7) Post-Task Questionnaire, 8) Steps 5 to 7 were repeated for the other two tasks, 9) Post-Session Questionnaires to identify user perception of aspects of the system, and 11) Thank-you for participating page.

3.6 Data Analysis

First, data regarding the use of the Suggested Pages was extracted from the log files which recorded the use of all interface tools provided by this system. The suggested pages viewed by any participant were then manually coded, by comparing each to the original assigned task. Second, the data from questionnaires were loaded into SPSS to calculate means and standard deviations (SD). Third, the responses to open-ended questions were loaded into Excel and manually coded for users' perception of the use and usefulness of the Suggested Pages.

4 Results

Three types of data used to assess the use and perception of Suggested Pages include a) how the Suggested Pages tool was used by participants, b) the participants' assessment of the tool, and c) their perceived usefulness of the tool.

4.1 Use of Suggestions

Thirty-eight (40%) of the 96 participants used the Suggested Pages a total of 92 times. While the Suggested Pages were used with all tasks, not all participants used the tool for each task. To assess how the tool was used we considered two aspects: a) the relationship of the Suggested Page to the task, and b) the location within the search process.

Of the 92 instances, 54 were pages directly related to responding to the task, of which all were declared relevant by the participant who assigned an average relevancy of 3.96 out of a possible 5.0. Twenty-nine pages were considered somewhat related which means that pages had aspects of the topic at a significantly broader or narrower level. Of these pages, eight were declared relevant by the participant.

The remaining nine pages bore no logical connection to the topic of the task. Of those, one was declared relevant by the participant. The topics of these pages varied considerably:

Unclean animals	Suburb
Aphid	Domestic water system
Value added	D
Works which reference MIT	.km
Tourist guy	

Some of these topics could be classified as curiosity, defined as "had no concept of what might be found; title was esoteric, eccentric or sensational" [18].

4.2 User Perception of the Tool

At the end of the study, participants indicated the extent to which they agreed with a series of statements about the interface. Three of those questions related to the

Suggested Pages. The seven-point scale varied from strongly disagree to strongly agree. As indicated in Table 1, participants were ambivalent in their response. They found the Suggested Pages pertinent to their task, but also found the pages diverted them from their task. They were in agreement about the surprise element.

Table 1. User Perception of *Suggested Pages*

Statements	Mean	SD
I found pages listed in *Suggested Pages* that were pertinent to the task I was working on	4.19	1.50
I found unexpected pages listed in *Suggested Pages*	5.00	1.26
I found that the pages listed in *Suggested Pages* lead me away from my search task	4.40	1.48

4.3 Perceived Usefulness of Suggested Pages

In two open-ended questions, participants indicated when they thought they would find the Suggested Pages useful, and when they would not use them. In responding to both questions, 21 (20%) people indicated that they did not notice the feature, did not use it, and would not use it in future. Some of these responses related to visibility: "I didn't really pay much attention to it to be honest" (P562), and because of generalized practices: "Because of the way it is placed on the outside of the screen I felt it was a distraction and I thought it was sponsored links, a bit like you have on a google page" (P582).

Although 38 people directly clicked on a page listed in Suggested Pages, others examined the list without clicking on a link and used that information subsequently in the search process. Overwhelmingly, the comments about the pertinence of Suggested Pages related to its perceived relevance to the search. The Suggested Pages were perceived to be most useful "when the suggestions were close to the topic" (P558) or "related to the page I was currently looking at" (P563), and perceived as less useful "when the suggested pages are unrelated or not specific enough to my original topic" (P553).

Participants amplified that perspective indicating when the Suggested Pages were likely to be used such as when "I didn't find what I was looking for on the page and it had other suggested pages that were relevant" (P548), or "when the initial search didn't provide much" (P618). Participants also found it "useful to have topics worded differently so you can see other ways of describing your search" (P580), or to assist with "finding the 'correct' search words to get relevant information" (P560). In addition to giving different perspectives on direction of the search, the Suggested Pages were perceived as helping when "looking for a broad sense of the topic" (P574), "to get more variance on a topic" (P575, or "when the topic I was covering had many areas of interest" (P610). But they found them useless when "the suggestions obviously had nothing to do with what I was searching for" (P609).

In addition to assisting directly with the task at hand, participants also found that the Suggested Pages gave them perspective on the topic which appeared to be of two types. On the one hand, seeing related topics helped "with your general understanding of a topic, even though it may not help you with your particular question" (P621). While on the other hand, the Suggested Pages seemed to be a low cost learning device; they could "look at the suggested list, click on the link to check, and then decide which information was relevant that way" (P596).

The Suggested Pages served as stimulants, particularly "when the pages suggested things that I hadn't thought to search for" (P589), or "made me think of something else to search for" (P594). But not all participants saw this as value. Said one, "Obviously if the suggested page is related to my search than I would find it useful but a lot of the time the suggested pages were unrelated or did not contain any information about the specific topic I was looking for" (P564). There seemed to be a tension not only among participants but within the responses of individuals between the value of being lead astray and staying on topic. Some were concerned that "it would take me way further away from where I wanted to be" (P573).

But not everyone considered this to be negative. Some found the Suggested Pages were more useful when "something interesting popped up" (P577), or "...did not take me too far away from my research task" (P583) or "mainly useful for interest, not information" (P634). "I found it could be distracting. It tempted me to look at information that was either slightly related or totally unrelated to my original search" (P633).

Because they were in a laboratory situation, their behaviour was not always typical in how they would use the Suggested Pages: "I looked at it once and saw a suggested page that had nothing to do with my search and would have clicked on it if I was at home (P622).

In other cases, participants had a strategy for how to use the Suggested Pages. As P602 noted "When there are multiple pages to a search result and the first half of the results have nothing to do with the topic, then the suggested pages would be most useful." However, participants would tend not to use them when "I am doing a quick search and need some quick and general information" (P700).

Participants were equally adamant about when to ignore the suggestions indicating they would ignore them "if I really wanted to stay on the topic that I searched" (P545), when "you have found all the information you are looking for" (P569), when I "have no need or interest to expand my research on a topic" (P617) or "if I already had a general understanding of the topic I was looking up" (P598).

Time was a factor in the perceived usefulness. Said one, "When I am pressed for time and cannot explore extremely tangential suggestions" (P583). On the other hand, "perhaps if I were searching a topic to find out as much as I could about it, without limits, I would use this feature more" (P629).

Others had their own ideas about how to make the Suggested Pages feature more useful. Rather than providing you with "more general information about your topic and related ones, instead... [provide] more depth or information" (P621). "The selections that it offered were just not that relevant, which is important for academic searches. If you know the subject well, you don't want to be bothered - and if you don't, you don't want to be led astray. There should be some sort of obvious link between your search term and the suggested pages, if possible. Suggestion for the sake of suggestion should be minimized" (P638). For one this was very much about control: "I don't think I would ever select sites from the selected pages unless it was really really pertinent to my search. I don't think that a computer can think for you and tell you what you think you are looking for. I would rather critically assess the information I am getting and change my search accordingly" (P582).

5 Analysis and Discussion

In summary, 96 participants were exposed to a tool, Suggested Pages, which identified a set of possible relevant pages, some of which were very relevant and some of which were diversions as it depended on how the search algorithm processed the query which used the contents of first paragraph of the displayed pay. The participants, as expected, used this tool significantly less than any other interface tool provided by this system. Some did not see it, while others avoided it, perceiving that it would lead them astray, or that the suggestions were not very useful. When used, it was mostly used to aid with the task at hand which is what would be expected given the nature of the task. Some however, viewed the tool as a potential for digression, but not always a distracting one.

When used, the Suggested Pages helped to broaden or narrow search and to generate ideas for keywords to use in further searching. But the feature was not perceived useful when participants wanted to avoid being distracted and stay on topic due to time constraints or to finish the tasks.

Some participants indicated that they did not need to use this feature because they had other strategies. In some cases the information was already found and they had no need to look further or they were confident in their knowledge of the topic at hand and/or in their personal search abilities.

Our participants expressed elements of Toms' [17] Digital Browsing Instant, Erdelez's [5] five typical information encountering elements, and McFarlane and Latorella's interruption management stage model [12]. Interestingly, while many commented on the Suggested Pages' value in providing other connections, very few used the tool for branching out into other topics which may have been due to the primary experimental scenario in which they were immersed.

This study highlights the tendency of an information seeker to focus on one problem at a time [5, 6]. Many participants appeared unwilling to give up the reins of the search, underlining McBirnie's [11] suggestion of the *paradox of control* where the potential for novel, useful information is sacrificed for the sake of the efficiency of the information search. The willingness to allow the search to take information seekers on a serendipitous rather than straight path may be another key individual difference between participants' varied uses and perceptions of the Suggested Pages. Would, for example, those more inclined to curiosity be also more inclined to follow potential serendipitous paths? More research needs to be done to gain a better understanding of the types of individual differences that lend themselves to openness to serendipitous discovery and how this can be better facilitated in information systems.

Clearly participants view the Suggested Pages as serving two purposes. Some expected it to support the task, serving a purpose much like the Google's "similar pages" and were annoyed when taken astray. But some were clearly intrigued by the suggestions. Perhaps the presentation of suggestions needs to be differentiated so that the support for the task is separated from items that are intended to lead one astray. A core challenge may be embedded in the way we work, and the need to stick to the task. Non-task related topics may indeed be disruptive. On the other hand, they may also serve to disorient in order to re-orient, as well as introduce approaches and ideas related to the topic that had not previously been considered.

At the same time, the way in which participants used the Suggested Pages may be limited by the system. We disabled the browser's bookmarking capability for example. While we added a tool for collecting useful pages related to the topic, we did not provide the capability for multiple instantiations of it so that items not related to the task, but of interest to the participant could be tracked.

Our implementation of the Suggested Pages list was a blunt force product based strictly on similarly to first paragraph of the displayed page. In the previous test [e.g., 17, 21], the list was created using a weighted Boolean model with no control for length resulting in the unlikeliest of similarities among the list in many cases. Other examples have included randomness [2, 22] or known ancillary problems of users [5], but what makes the best triggering device is an outstanding research question. Even once we know how to develop that "list," we additionally will need to examine how best to present the list, i.e., how to provide the cues. As important will be when to introduce a serendipitous device, as there is a fine-line between enabling serendipity and being distracting.

6 Conclusion

In this research we exposed 96 users to a serendipitous inducing tool while asking them to complete a set of goal-oriented tasks. While the tool was used by about half in highly selected ways a few elected to view items not related to their assigned task. While participants readily made the connection between suggested pages and their task, few noted the value of disconnected findings which may be more related to selected individual differences. There remains much to be done to more fully understand how to trigger an effective serendipitous encounter and this may reside in further user modeling.

Acknowledgments. Chris Jordan implemented the search engine. Alexandra MacNutt, Emilie Dawe, Heather O'Brien, and Sandra Toze participated in the design and implementation of the study in which the data was collected. Research was supported by the Canada Foundation for Innovation, Natural Science and Engineering Research Council Canada (NSERC), and the Canada Research Chairs Program.

References

1. Adams, A., Blandford, A.: Digital Libraries' Support for the User's 'Information Journey'. In: Proceedings of the 5th ACM/IEEECS (JCDL 2005), pp. 160–169. ACM Press, New York (2005)
2. Campos, J., Figueiredo, A.D.: Searching the Unsearchable: Inducing Serendipitious Insights. In: Proceedings of the workshop program at the fourth international Conference on Case-Based Reasoning, Washington, D.C. (2001)
3. Erdelez, S.: Information Encountering: A Conceptual Framework for Accidental Information Discovery. In: Proceedings of ISIC 1996, Information Seeking in Context, Tampere, Finland. Taylor Graham, London (1997)
4. Erdelez, S.: Information Encountering: It's More Than Just Bumping Into Information. B. Am. Soc. Inform. Sci. 25(3), 25–29 (1999)

5. Erdelez, S.: Investigation of Information Encountering in the Controlled Research Environment. Inform. Process. Manag. 40(6), 1013–1025 (2004)
6. Franken, R.E.: Human Motivation. Brooks/Cole, Monterey (1982)
7. Gibson, J.J.: The Theory of Affordances. In: Shaw, R., Brandsford, J. (eds.) Perceiving, Acting and Knowing: Toward an Ecological Psychology, pp. 67–82. Lawrence Erlbaum, Hillsdale (1977)
8. Grose, M.W., Line, M.B.: On the Construction of White Elephants: Some Fundamental Questions Concerning the Catalogue. Lib. Assn. Rec. 70, 2–5 (1968)
9. Heinström, J.: Psychological Factors Behind Incidental Information Acquisition. Libr. Inform. Sci. Res. 28(4), 579–594 (2006)
10. Latorella, K.A.: Investigating Interruptions: Implications for Flightdeck Performance NASA/TM-1999-209707. National Aviation and Space Administration, Washington (1999)
11. McBirnie, A.: Seeking Serendipity: The Paradox of Control. Aslib Proceedings 60(6), 600–618 (2008)
12. McFarlane, D.C., Latorella, K.A.: The Scope and Importance of Human Interruption in Human-Computer Interaction Design. Hum.-Comput. Interact. 17(1), 1–61 (2002)
13. O'Connor, B.: Fostering Creativity: Enhancing the Browsing Environment. Int. J. Info. Mgmt. 8(3), 203–210 (1988)
14. Rimmer, J., Warwick, C., Blandford, A., Gow, J., Buchanan, G.: An Examination of the Physical and the Digital Qualities of Humanities Research. Inform. Process. Manag. 44(3), 1374–1392 (2008)
15. Ross, C.S.: Finding Without Seeking: The Information Encounter in the Context of Reading for Pleasure. Inform. Process. Manag. 35, 783–799 (1999)
16. Stelmaszewska, H., Blandford, A.: From Physical to Digital: A Case Study of Computer Scientists' Behaviour in Physical Libraries. Int. J. Digital. Lib. 4(2), 82–92 (2004)
17. Toms, E.G.: Browsing Digital Information: Examining the "Affordance" in the Interaction of User and Text. Unpublished Ph.D. Dissertation. University of Western Ontario (1997)
18. Toms, E.G.: What Motivates the Browser? In: Exploring the Contexts of Information Behaviour: Proceedings of the Second International Conference on Research in Information Needs, Seeking and Use in Different Contexts, pp. 191–208. Taylor Graham (1998)
19. Toms, E.G.: Information Exploration of the Third Kind: The Concept of Chance Encounters. In: A position paper for the CHI 1998 Workshop on Innovation and Evaluation in Information Exploration Interfaces (1998),
    http://www.fxpal.com/chi98ie/submissions/long/toms/index.htm
20. Toms, E.G.: Serendipitous Information Retrieval. In: Proceedings, First DELOS Network of Excellence Workshop Information Seeking, Searching and Querying in Digital Libraries, Zurich, Switzerland, December 11-12, pp. 17–20 (2000), ERCIM-01-W01
21. Toms, E.G.: Understanding and Facilitating the Browsing of Electronic Text. Int. J. Hum.-Comput. Int. 52, 423–452 (2000)
22. Twidale, M.B., Gruzd, A.A., Nichols, D.M.: Writing in the Library: Exploring Tighter Integration of Digital Library Use with the Writing Process. Inform. Process. Manag. 44(2), 558–580 (2008)
23. Williamson, K.: Discovered by Chance: The Role of Incidental Information Acquisition in an Ecological Model of Information Use. Libr. Inform. Sci. Res. 20(1), 23–40 (1998)

Stress-Testing General Purpose Digital Library Software

David Bainbridge[1], Ian H. Witten[1], Stefan Boddie[2], and John Thompson[2]

[1] Department of Computer Science
University of Waikato
Hamilton, New Zealand
{davidb,ihw}@cs.waikato.ac.nz
[2] DL Consulting Ltd
Innovation Park
Hamilton, NZ
{stefan,john}@dlconsulting.com

Abstract. DSpace, Fedora, and Greenstone are three widely used open source digital library systems. In this paper we report on scalability tests performed on these tools by ourselves and others. These range from repositories populated with synthetically produced data to real world deployment with content measured in millions of items. A case study is presented that details how one of the systems performed when used to produce fully-searchable newspaper collections containing in excess of 20 GB of raw text (2 billion words, with 60 million unique terms), 50 GB of metadata, and 570 GB of images.

1 Introduction

Today we are witnessing a great upsurge in nationally funded digital library projects putting content on the web. As the volume of data to be stored in these repositories increases, the question of the scalability of the software used becomes crucial. The focus of this paper is to detail and assess what is known about three widely used open source general purpose digital library systems: DSpace, Fedora and Greenstone.

The structure of the paper is as follows. First we briefly describe the three pieces of software under review, with emphasis on the core technologies they are based on, before discussing in turn known results (from a variety of sources) of scalability testing. Then we take, as a case study, the work undertaken in building newspaper-based digital library systems for the National Libraries of New Zealand and Singapore. These collections currently contain over half a million OCR'd items each, and are on track to be scaled up to twice and four times their size, respectively.

2 Background

DSpace is a collaborative venture between Hewlett Packard and MIT's Library, heavily optimized for institutional repository use [6]. Ready to use out of the box,

M. Agosti et al. (Eds.): ECDL 2009, LNCS 5714, pp. 203–214, 2009.

it has been widely adopted for this purpose. It is written in Java, and is servlet based, making extensive use of JSP. Tomcat is recommended by its developers with either PostgreSQL or Oracle as the relational database management system. Other RDBMS can be used with it (such as MySQL) through JDBC.

It can be configured to support full-text indexing (utilizing Lucene) for Word, PDF, HTML and plain text. Documents are either ingested individually through the web (intended for author submission), or locally (i.e. on the server where DSpace is installed) through batch processing by command-line scripts into a collection. A DSpace repository is represented as a hierarchy of communities and sub-communities, with collections forming the leaves to this hierarchy.

Fedora is founded upon a powerful digital object model, and is extremely flexible and configurable [1,4]. A repository stores all kinds of objects, not just the documents placed in it for presentation to the end-user. For example a Service Definition (or SDef) object defines a set of abstract services and the arguments they take, which can then be associated with a document in the repository to augment its behavior; a Service Deployment (or SDep) object provides a concrete implementation of an SDef. Ingesting, modifying and presenting information from this rich repository of objects is accomplished through a set of web services. These are grouped by function into two APIs, one for access and the other for management. Both "lite" (RESTful) and a full (SOAP-based) versions of the APIs are available.

Documents are ingested either using command-line scripts or the Fedora Administration tool, which is a Java-based application. No matter which option is used, ultimately the data is passed through the management API (which includes authentication), and so these tools function equally well remotely as they do locally to the Fedora installation. Documents must be encoded in XML in either FOXML, the Fedora extension to METS, or ATOM.

Like DSpace, the core system is written in Java and makes use of Servlets. Again a relational database is part of the mix. Fedora ships with McKoi—a light-weight pure Java implementation—and it can also be set up to work with MySQL, DB Derby, Oracle 9 and PostgreSQL. Full-text indexing is possible through a third-party package (GSearch) into which Lucene, Zebra or Solr indexing packages can be plugged.

Unlike DSpace, Fedora is not a turn-key software solution, and software development is necessary to shape it into the desired end product. In the case where the shaping is of interest to others, it is of course possible to package this up and make it available to give a more ready-to-run solution. For the main part, these additions sit on top of the Fedora architecture are of secondary importance in terms scalability issues. Of primary importance is how well the core repository handles large volumes of data, typically in the form of documents and metadata, but also the expressed relationships between objects. In Fedora, such relationships are handled by a triple-store.

In terms of a design philosophy, Greenstone sits between DSpace and Fedora. The standard download is a turn-key solution, however it includes several alternative software components (with different strengths and weaknesses) that are the software equivalent of being "hot swappable." The ingest phase to Greenstone is written in Perl, with the delivery mechanism implemented in C++ and executed as a CGI program using a standard web server.

A system of document parsing plugins allows for a extensive range of file formats to be full-text indexed, along with the automatic extraction (and manual assignment) of metadata: plain text, e-mail messages, Postscript, PDF, HTML, Word, PowerPoint, Excel, and OpenOffice documents are just some of the formats supported. Records from metadata-only formats can be treated as documents or bound to their source document counterpart if present: Greenstone processes metadata in MARC, Bibtex, Refer (EndNote), Dublin Core and LOM (Learning Object Metadata) formats, amongst others.

As mentioned above, the database and indexing systems are swappable components. MG, MG++ and Lucene are available for full-text indexing; GDBM and SQL-lite for the metadata database. MG's strength is that compression is built into the indexing technique, but it is not incremental [8]. This contrasts with Lucene, which can support incremental building but does not include compression. In designing a collection, a digital library can choose (literally with the click of a button) to trade the amount of space needed to represent the built collection against speed of rebuilding, or *vice versa*, as the needs of the project dictate.

The principal requirement for the database system used by Greenstone is that it supports *get-record* and *set-record* operations. The most basic of flat-file database management systems provides this, and they are trivial operations for a relational database system to support. Greenstone ships with both the GDBM (flat-file) and SQL-lite (relational) database systems.

Greenstone also supports web-based submission of documents (aka DSpace) and a graphical tool called the Librarian Interface that can be configured for local or remote access (aka Fedora). All three systems provide the option of running command-line scripts.

2.1 Discussion

The different aims of these three software tools manifest themselves in the decisions made in developing the underlying software architecture, and how it is deployed. It is also the case that there are significant differences seen in the web technologies used by Greenstone and the other two. This stems from the fact that one of Greenstone's important design requirements is to run on early versions of Windows (right down to Windows 3.1) for deployment in developing countries [7].

Greenstone in fact comes in two flavors: Greenstone 2 is the production version that is described above, and is what is assessed here; Greenstone 3 is a more recent development that is structured as a research framework. It is written in Java, makes use of web services and servlets, is backwards compatible with Greenstone 2, and has optional support for a relational database system. In terms of applicability of this paper, scalability findings from DSpace and

Fedora that stress-testing these shared web technologies are broadly applicable to Greenstone 3; furthermore, despite the different web technologies used between Greenstone 2 and the other two DL systems, there are many parallels as to how the ingest process is performed, particularly with the command-line scripts. Many of the lessons learnt in the case study of building a newspaper DL using Greenstone are applicable. We return to this point in our conclusion.

3   Stress-Testing

We now review known results from testing the scalability of Fedora, DSpace and Greenstone.

3.1   Fedora

The first known stress-testing of Fedora was by the University of Virginia Library in 2001. Still in an embryonic form compared to the Fedora Project as we know it today, Virginia worked from the reference implementation Cornell had produced, and over a period of two years modified it for use in a web environment. For performance reasons, it was at this point that the decision to incorporate a relational database was made. A testbed of half a million heterogeneous digital objects—consisting of images, scientific data and a variety of XML encoded documents, such as e-texts—was established and performance of the digital library evaluated.

Using a Sun Ultra80 dual-processor as the server, a laptop in another location of the campus was used to simulate 20 simultaneous users accessing the repository. The average response time of the server was approximately half a second. By synthetically replicating objects the repository was grown to 1 million and then 10 million items, with the same experiment repeated at each stage. By the time there were 10 million objects stored in the repository the server response time was found to be between 1–2 seconds.

In 2004 The National Science Digital Library (NSDL) took the decision to base its content management system on Fedora. A factor in it being chosen was its established credentials with regard to scalability through the work of Cornell and Virginia. By 2007 NSDL had grown to the point where the repository stored over 4.7 million objects with 250 million triples representing the various relationships expressed between objects. The latter was causing problems with the component they had selected to support this aspect of the architecture, Kowari, and it motivated the development of MPTStore[1] as an alternative solution that had better scalability characteristics.

A recent and more comprehensive project to study the scalability of Fedora is underway at FIZ Karlsruhe in Germany. This project is assessing a variety of aspects of configuring a high-volume Fedora repository. Before scaling up to 14 million items, a fine-grained analysis was conducted on a test set of 50,000 documents. Parameters studied included:

---

[1] http://mptstore.sourceforge.net

- Different versions of Java, and tuning components of the JRE to control garbage collection, heap size, etc.
- Different database and triple-store backends (such as MPTStore and Mulgura), including examples of using them with default values and fine-tuning their settings.
- Remote versus local access, and internally versus externally managed content.

Ingesting 14 million items took three weeks, using a dual core 2.4 GHz Intel processor with 2 GB of RAM, and resulted in 750 million triple-store items. A small number of documents failed to be ingested; the researchers are still investigating why. Some noteworthy recommendations from the testing to date are:

- There are efficiency advantages to using Java 1.6 over 1.5.
- Mass ingest using a remote connection to the relational database is significantly faster (by a factor of 2) than the same operation performed locally.
- It is worth fine-tuning the configuration of the database and triple-store, but unfortunately most adjustments lower the ability to recover from errors.

It is perhaps initially surprising that a remote database was faster than a local one. This is attributed to the streamlining of disk IO when undertaken remotely. Database updates no longer have to compete with disk activity associated with reading the source documents.

3.2   DSpace

The U.S. National Library of Medicine has developed a digital repository called SPER (System for the Preservation of Electronic Resources), based on DSpace. It uses MySQL for the relational database and includes an enhanced command-line ingest facility. In 2007 they undertook a detailed study of DSpace designed to stress test it in this configuration [3].

To form a testbed, they combined two collections on Food and Drug-related Notices Of Judgment to form a new collection containing nearly 18,000 documents. Each document comprises TIFF image data, OCR'd text and descriptive metadata. The majority of documents (75%) contain 1–2 pages; the longest one has 100 pages. These files were then replicated to form a collection of over a million (1,041,790) items, and located in a hierarchy of 32 communities, consisting of 109 collections.

Ingesting the testbed took slightly over 10 days using a Sun Microsystems X4500 server with a dual core 2.8 GHz processor. The version of Java used was 1.4. The files were processed without error, and the experiment repeated a second time to gather more statistics. A prominent result found was that the ingest time per document increased roughly linearly with the size of the collection. Ingesting their sample documents into an empty repository took approximately 0.2 seconds on average. With 1 million items in it, the average ingest time had risen by a factor of 7 to around 1.4 sec. The overall conclusion was that performance was satisfactory at this scale of operation.

3.3 Greenstone

Initial scalability testing on Greenstone was conducted by its developers (also authors of this paper) during 1998–1999 under Linux. Using real-world data, in one case a collection was formed from 7 GB of text, and in another (a union catalog), 11 million short documents were used (about 3 GB text). Both examples used MG for indexing (this was the only indexing tool supported at that point) and built (first time) without any problems.

In these tests, we did not attempt to use a larger volume of data as there was nothing appropriate available at the time. Using 7 GB twice over to make 14 GB does not really test the scalability of the indexing tool because the vocabulary has not grown accordingly, as it would with a real collection. The issue of vocabulary size is particular acute when working with OCR'd text, a point we return to in the case study below.

Independent stress testing of Greenstone was carried out in 2004 by Archivo Digital, an office associated with the Archivo Nacional de la Memoria (National Memory Archive), in Argentina. Its test collection contained sequences of page images with associated OCR text. They used an early version of Greenstone (2.52) on an 1.8 GHz Pentium IV server with 512 MB RAM running Windows XP Professional. There were 17,655 indexed documents, totaling 3.2 GB of text files, along with 980,000 images in TIFF format. Full-text indexes were built of both the text and the titles.

They found that almost a week was spent collecting documents and importing them. (No image conversion was done.) Once the import phase was complete, it took almost 24 hours to build the collection. The archives and the indexes were kept on separate hard disks to reduce the overhead that reading and writing from the same disk would cause.

3.4 Discussion

Both Greenstone and Fedora undertook scalability testing early on, and it is fair to say that, given the available technology at the time, both performed well. Since DSpace is offered as a turn-key solution, there is evidence of greater expectations within the community that deploys this software that is should simply just continue to work as the volume grows.[2] This contrasts with Fedora's design philosophy, where an organization is expected to undertake their own IT development to get going with it, and therefore (as with the case of MPTStore) there is a greater willingness to adjust the installation when things do not work out as expected.

Greenstone sits somewhere between these two projects. It has a pluggable indexer and database infrastructure, and includes several options as part of the standard install. If one particular technology hits a limit it is easy to switch to an alternative. For example, MG's Huffman coding calculations overflow at around 16 GB of text, however if this poses a difficulty, it is trivial to switch the

---

[2] http://wiki.dspace.org/index.php/ArchReviewIssues

collection to use Lucene instead (which does not have this restriction). There are always trade-offs, of course: a Greenstone collection based on Lucene does not support case-sensitive matching, whereas ones based on MG and MG++ do.

3.5 Comparable Testing

Scalability tests have been conducted by different groups at different times on different content, and so it is difficult to compare the results produced by the different systems directly. Consequently, we installed DSpace (version 2.1), Fedora (version 3.1) and Greenstone (version 2.81) on the same computer and compared them ingesting the same set of files: 120,000 newspaper articles comprising 145 GB of images and OCR'd text. (The files in question are an excerpt that comprises about 2% of the scanned newspaper collection described in the next section.)

The workstation used was a dual-processor 3 GHz quad-core Xeon processor with 4 GB of memory. Rounding to the nearest minute, the processing time for the three systems was surprisingly close: Greenstone (using Perl 5.8) took 35 minutes; DSpace (using Java 1.5) took 34 minutes; and Fedora (same version of Java) took 78 minutes. In all cases the digital library server was run locally to where the source documents were located. With a server that was remote from the source content, the processing time for Fedora could perhaps be halved to around 37 minutes due to the effect noticed by FIZ Karlsruhe (noted above).

4 Case Study

We now present, as a case study, the construction of a large collection in Greenstone. This contains approximately 1.1 million digitized pages from national and regional newspaper and periodicals spanning the years 1839–1920, and was undertaken by DL Consulting for the National Library of New Zealand. Over half (69%) of the images in the collection have been OCR'd, comprising nearly 20 GB of raw text: 2 billion words, with 60 million unique terms that is full-text searchable. This number will increase over the years until all images are fully searchable.

Resulting from the OCR'd data, the collection consists of 8.3 million searchable newspaper articles, each with its own metadata (much of it automatically generated). The total volume of metadata is 50 GB—three times as much as the raw text! A large part of this is information that records the location of each word in the source image, along with coordinate information for each article. This is used in the digital library interface to highlight search terms in the images and to clip individual articles out of the newspaper pages. Before being built into a digital library collection the metadata is stored in an XML format, which occupies around 600 GB, slightly less than 1 MB per newspaper page. The result of this work, known as Papers Past, is available on-line and can be can be viewed at *http://paperspast.natlib.govt.nz*.

4.1   Building the Papers Past Collection

Text. About 820,000 pages have been processed by optical character recognition software to date (early 2009) for the Papers Past collection. These contain 2.1 billion words; 17.25 GB of raw text. The images were digitized from microfilm, but the paper originals from which the microfilm was obtained were of poor quality, which inevitably resulted in poor performance of the OCR software.

The high incidence of recognition errors yields a much larger number of unique terms than would normally be expected. The 2.1 billion words of running text include 59 million unique terms (2.8%). Our tests show that this will continue to increase linearly as further content is added. By contrast, clean English text typically contains no more than a few hundred thousand terms, and even dramatic increases in size add a relatively small number of new terms. As a point of comparison, the Google collection of n-grams on the English web[3] is drawn from 500 times as many words (1,025 billion) but contains less than a quarter the number of different words (13.6 million, or 0.0013%). However, words that appear less than 40 times were omitted from this collection, a luxury that we did not have with Papers Past because of the importance of rarely-used place and personal names for information retrieval.

Enormous vocabularies challenge search engine performance [8]. Moreover, the high incidence of errors makes it desirable to offer an approximate search capability. Neither MG nor MG++ provide approximate searching, so it was decided to use the Lucene indexer because of its fuzzy search feature and proven scalability—it has been tested on collections of more than 100 GB of raw text. Consequently we worked on improving and extending the support for Lucene that Greenstone provides.

Metadata. Papers Past involves a massive amount of metadata. Its specification demanded that newspaper articles be viewable individually as well as in their original context on the page. The 820,000 OCR'd pages comprise 8.3 million individual articles, each with its own metadata. To meet the specification, the physical coordinates that specify an article's position on the page were stored as metadata, which is used to clip the article-level images from pages at runtime.

A further requirement was that search terms be highlighted within page images. In order to do so, the bounding-box coordinates of each and every word in the collection must be stored. These word coordinates represent 49 GB of metadata. Putting together all the article, page, and issue information yielded a total of 52.3 GB of metadata.

The collection's source files are bi-tonal digital images in TIFF format. From these, the OCR process generates METS/ALTO XML representation [2]. This includes all the word and article bounding-box coordinates, as well as the full text, article-level metadata and issue-level metadata. The resulting source data includes 137,616 METS files, one per newspaper issue, and 8,381,923 ALTO files, one per newspaper page. Together these amount to a total of 570 GB of

---

[3] The Google n-gram collection is available on six DVDs from
http://www.ldc.upenn.edu/

XML, slightly under 1 MB per page. All this XML is imported into Greenstone, which indexes the text with Lucene and stored the metadata and bounding-box coordinates in a database.

From the very beginning, Greenstone has used the GNU database management system (GDBM) for storing and retrieving metadata. Alternative database backends were added later. GDBM is fast and reliable, and does this simple job very well. Crucially, and of particular importance for librarian end-users, GDBM can be installed on Windows, Linux and Macintosh computers without requiring any special configuration. However, the design of Papers Past exposed limitations. GDBM files are restricted to 2 GB; moreover, look-up performance degrades noticeably once the database exceeds 500 MB.

A simple extension was to modify Greenstone to make it automatically spawn new databases as soon as the existing one exceeds 400 MB. Currently, Papers Past uses 188 of them. With this multiple database system Greenstone can retrieve word coordinates and other metadata very quickly, even when running on modest hardware.

Images. The archival master files for Papers Past are compressed bi-tonal TIFF images averaging 770 KB each. The full collection of 1.1 million images occupies 830 GB.

A goal of the project is that no special viewer software is required other than a modern web browser: users should not need browser plug-ins or downloads. However, TIFF is not supported natively by all contemporary browsers. Consequently the source images are converted to a web friendly format before being delivered. Also, processing is required to reduce the image file size for downloading, and in the case of individual articles, to clip the images from their surrounding context.

Greenstone normally pre-processes all images when the collection is built, stores the processed versions, and serves them to the user as required. However, it would take nearly two weeks to pre-process the 1.1 million pages of Papers Past, at a conservative estimate of 1 page/sec. To clip out all 8.3 million articles and save them as pre-prepared web images would take over 3 months, assuming the same rate. The preprocessed images would consume a great deal of additional storage. Furthermore, if in future it became necessary to change the size or resolution of the web images they would all need to be re-processed. Hence it was decided to build an image server that converts archival source images to web accessible versions on demand, and maintains a cache of these for frequently viewed items.

A major strength of Greenstone is its ability to perform well on modest hardware. The core software was designed run on everything from powerful servers to elderly Windows 95/98 and even obsolete Windows 3.1/3.11 machines, which were still prevalent in developing countries. This design philosophy has proven immensely valuable in supporting large collections under heavy load. Greenstone is fast and responsive. On modern hardware, it can service a large number of concurrent users. However, the image server is computation-intensive and consumes significant system resources, though the overall system can cope with moderately heavy loads on a single modern quad-core server. (At the New Zealand National Library it is run on a Sun cluster.)

Building the collection. It takes significant time to ingest these large collections into Greenstone, even without the need to pre-process the source images. Greenstone's ingest procedure consists of two phases. The first, called importing, converts all the METS/ALTO data into Greenstone's own internal XML format. The second, called building, parses the XML data and creates the Lucene search index and the GDBM metadata databases.

The import phase processes approximately 100 pages/min on a dual quad-core Xeon processor with 4 GB of main memory, taking just over four days to import 820,000 pages. This stage of the ingest procedure may be run in batches and spread over multiple servers if necessary. The building phase processes approximately 300 pages/min on the same kind of processor, taking around 33 hours to build the collection. The latter step is incremental, so when new content is added (or existing content is altered) only the affected items need processing.

Future improvements. The problem of large-scale searching is exacerbated by the presence of OCR errors. An obvious solution is to remove errors prior to indexing [5]. However, we decided not to attempt automatic correction at this stage, in order to avoid introducing yet more errors. In our environment this solution is workable so long as Lucene continues to perform adequately on uncorrected text. However, we do plan to investigate the possibility of eliminating the worst of the OCR errors.

Singapore National Library has embarked on a comparable newspaper project using Greenstone. This currently contains approximately 600,000 newspaper pages, and will grow to more than two million (twice the size of Papers Past) The Singapore collection uses grayscale JPEG-2000 master source files, which average 4.5 MB per page; the existing 600,000 pages consume 2.6 TB of storage.

4.2 Apache Solr

We have conducted some indicative experiments with Apache Solr. Solr is based on Lucene, crafted to optimize performance in a web environment. It features faceted search, caching and a mechanism for distributed indexes, along with convenient API access from Java, JavaScript, PHP, Ruby and other web-integration-friendly programming languages.

To trial this indexing technology we combined the OCR'd text and metadata from the New Zealand and Singapore National Libraries to produce a total of nearly 17 million articles (16,901,237). Running on a 2.33 GHz dual-core Xeon processor, it took 30 hours (wall-clock time) to index the files. The server is shared by others; however the build was done over the weekend and consequently the machine was lightly loaded.

Comparing query times between Lucene and Solr, on this size of collection Lucene took between 7–10 seconds (processor time) to produce the first 500 matching documents. This range factors in a caching effect of files through the operating system that was observed, which resulted in a saving of approximately

2 seconds after the initial query. For the equivalent queries using Solr, query times were in the range 0.24–0.33 seconds—over 20 times faster. This stark difference is all the more impressive when it is remembered that Solr is using essentially the same indexing technique, only better tuned. For example, one thing it does is run a daemon process to respond to queries, rather than perform each query in isolation.

5   Summary and Conclusions

This article has described a range of scalability tests applied to the general-purpose open source digital library software solutions DSpace, Fedora, and Greenstone. Indicative scalability tests where conducted by the developers of Greenstone and Fedora early in their development history. Other than taking a long time to ingest, these tests worked flawlessly on test sets that exceeded 10 million items.

The Fedora project synthesized a 10 million item collection from a starting set of documents that numbered 500,000. The test looked at the server response time from simulated user requests for content. As the size of the repository grew, response time slowed from 0.5 to 1–2 seconds, still within acceptable bounds. There was no mention of full-text indexing being tested at that point, but that is only to be expected, because—to the best of the authors' knowledge—support for this was not added until some years later. Eight years on, with the software in a considerably more mature form, FIZ Karlsruhe has undertaken a comprehensive program of stress testing this software. Again, replicated data was used to produce a high volume testbed (14 million items).

DSpace has been stress-tested with 1 million documents; again replication was used to produced the test set. With full-text indexing becoming a standard feature of digital library software, some caution is necessary when interpreting results from scalability tests when replicated data is used. Vocabulary size does not grow the way it would in a real collection, and consequently the indexing component is not placed under as much pressure as it would under normal conditions.

With Greenstone, which has supported full-text search from the outset, testing has focused on high-volume real-world data. As with Fedora, testing was undertaken early in the development process, and tests with 11 million items and 7 GB of text helped confirm the viability of the software infrastructure. Ten years on, Greenstone's scalability has recently been assessed through substantial national newspaper projects in New Zealand and Singapore. Each contains over half a million OCR'd pages (text and images) and are on track to be scaled into the multi-million range.

As a case study, this paper has presented details of building the Papers Past collection for the New Zealand National Library. The advertised accuracy of OCR software is around 99.9%, but this is at the character level and assumes high quality images. The error rate at the word level is much higher. The Papers Past images were a century old and were digitized from microfilm and microfiche. The image quality is poor, which compounds the problem of errors. Our experience with this OCR text is that vocabulary size grows linearly with collection size.

This is potentially a severe stress point in any digital library system that offers full-text indexing.. However, for the current (and projected) scale of operation the indexing software coped admirably.

To operate at this scale, Greenstone software developments were modest. First, the existing multi-indexer framework was fine-tuned to increase its support for Lucene. Second, the standard flat-file database (GDBM) was extended to support multiple instances, to overcome the performance degradation that occurred when database files exceeded 500 KB. Testing showed that this arrangement more than adequately satisfied the needs of these projects.

The two newspaper projects helped identify some bottlenecks in the building (and rebuilding) of collections, and these were addressed as part of this work. Greenstone's importing phase is particularly amenable to being distributed across several servers. Moreover, Greenstone supports a system of "plugouts" that allows this phase of the operation to output to other formats such as FedoraMETS and DSpace. Greenstone's importing can be fed directly into the ingest processes of DSpace and Fedora, thereby passing the benefits of the improvements made to Greenstone on to these projects as well.

References

1. Lagoze, C., Payette, S., Shin, E., Wilper, C.: Fedora: an architecture for complex objects and their relationships. International Journal on Digital Libraries 6(2), 124–138 (2006)
2. Littman, J.: Technical approach and distributed model for validation of digital objects. D-Lib Magazine 12(5) (2006)
3. Misr, D., Seamans, J., Thoma, G.R.: Testing the scalability of a DSpace-based archive. Technical report, National Library of Medicine, Bethesda, Maryland, USA (2007)
4. Payette, S., Lagoze, C.: Flexible and extensible digital object and repository architecture (FEDORA). In: Nikolaou, C., Stephanidis, C. (eds.) ECDL 1998. LNCS, vol. 1513, pp. 41–59. Springer, Heidelberg (1998)
5. Reynaert, M.: Non-interactive OCR post-correction for giga-scale digitization projects. In: Gelbukh, A. (ed.) CICLing 2008. LNCS, vol. 4919, pp. 617–630. Springer, Heidelberg (2008)
6. Smith, M., Bass, M., McClella, G., Tansley, R., Barton, M., Branschofsky, M., Stuve, D., Walker, J.: DSpace: An open source dynamic digital repository. D-Lib Magazine 9(1) (2003), doi:10.1045/january2003-smith
7. Witten, I.H., Bainbridge, D.: A retrospective look at greenstone: lessons from the first decade. In: JCDL 2007: Proceedings of the 2007 conference on Digital libraries, pp. 147–156. ACM Press, New York (2007)
8. Witten, I.H., Moffat, A., Bell, T.C.: Managing gigabytes: compressing and indexing documents and images, 2nd edn. Morgan Kaufmann, San Francisco (1999)

The NESTOR Framework: How to Handle Hierarchical Data Structures

Nicola Ferro and Gianmaria Silvello

Department of Information Engineering, University of Padua, Italy
{ferro,silvello}@dei.unipd.it

Abstract. In this paper we study the problem of representing, managing and exchanging hierarchically structured data in the context of a *Digital Library (DL)*. We present the *NEsted SeTs for Object hieRarchies (NESTOR)* framework defining two set data models that we call: the "Nested Set Model (NS-M)" and the "Inverse Nested Set Model (INS-M)" based on the organization of nested sets which enable the representation of hierarchical data structures. We present the mapping between the tree data structure to NS-M and to INS-M. Furthermore, we shall show how these set data models can be used in conjunction with *Open Archives Initiative Protocol for Metadata Harvesting (OAI-PMH)* adding new functionalities to the protocol without any change to its basic functioning. At the end we shall present how the couple OAI-PMH and the set data models can be used to represent and exchange archival metadata in a distributed environment.

1 Motivations

In *Digital Library Systems (DLSs)* objects are often organized in hierarchies to help in representing, managing or browsing them. For instance, books in a library can be classified by author and then by subject and then by publishing house. Documents in an archive are organized in a hierarchy divided into fonds, sub-fonds, series, sub-series and so on. In the same way the internal structure of an object can be hierarchical; for example the structure of a book organized in chapters, sections and subsections or a web page composed by nested elements such as body, titles, subtitles, paragraphs and subparagraphs. One very important tool extensively adopted to represent digital objects such as metadata, text documents and multimedia contents — the *eXtensible Markup Language (XML)* — has an intrinsically hierarchical structure.

Representing, managing, preserving and sharing efficiently and effectively the hierarchical structures is a key point for the development and the consolidation of DLS technology and services. In this paper we propose the *NEsted SeTs for Object hieRarchies (NESTOR)*[1] framework defining two set data models that we call: the "Nested Set Model (NS-M)" and the "Inverse Nested Set Model (INS-M)". These models are defined in the context of the ZFC (Zermelo-Fraenkel with

---

[1] Nestor is a Greek myth [1]; a king of Pylos in Peloponnesus, who in old age led his subjects to the Trojan War. His wisdom and eloquence were proverbial.

M. Agosti et al. (Eds.): ECDL 2009, LNCS 5714, pp. 215–226, 2009.

the axiom of Choice) axiomatic set theory [7], exploiting the advantages of the use of sets in place of a tree structure. The foundational idea behind these set data models is that an opportune set organization can maintain all the features of a tree data structure with the addition of some new relevant functionalities. We define these functionalities in terms of flexibility of the model, rapid selection and isolation of easily specified subsets of data and extraction of only those data necessary to satisfy specific needs.

Furthermore, these set data models can work in conjunction with the *Open Archives Initiative Protocol for Metadata Harvesting (OAI-PMH)* [12] that is the standard *de-facto* for metadata sharing between DLSs in distributed environments. The extension of OAI-PMH with these set data models allows the protocol to manage and exchange complex hierarchical data structure in a flexible way. The extension of OAI-PMH shall permit the exchange of data belonging to a hierarchy with a variable granularity without losing the relationships with the other data in the hierarchy. Furthermore, the OAI-set which is a constituent part of the protocol will be used also to organize the data and not only to enable the selective harvesting. A concrete use case is the archival data that are organized in a hierarchy which preserve the meaningful relationships between the data. When an archival object is shared it has to preserve all the relationships with the preservation context and with the other objects in the archive; since the use of tree data structure in this context turns out to be problematic in terms of accessibility and flexibility, we shall show that the use of the proposed data models in conjunction with OAI-PMH overcomes many of these issues.

The paper is organized as follows: Section 2 briefly defines the tree data structure. Section 3 defines the two proposed set data models and presents the mapping functions between the tree data structure and the set data models. Section 4 describes how OAI-PMH can extend its functionalities by exploiting the NS-M or the INS-M; moreover this section presents a use case in which the set data models and OAI-PMH can be used together to exchange full expressive archival metadata. Section 5 draws some conclusions.

2   The Tree Data Structure

The most common and diffuse way to represent a hierarchy is the tree data structure, which is one of the most important non-linear data structures in computer science [8]. We define a tree as $T(V, E)$ where V is the set of nodes and E the set of edges connecting the nodes. V is composed by n nodes $V = \{v_1, \ldots, v_n\}$ and E is composed by $n - 1$ edges. If $v_i, v_j \in V$ and if $e_{ij} \in E$ then e_{ij} is the edge connecting v_i to v_j, thus v_i is the parent of v_j. We indicate with $d_V^-(v_i)$ the **inbound degree** of node $v_i \in V$ representing the number of its inbound edges; $d_V^+(v_i)$ is the **outbound degree** of $v_i \in V$ representing the number of its outbound edges. $v_r \in V$ is defined to be the **root** of $T(V, E)$ if and only if $d_V^-(v_r) = 0$; $\forall v_i \in V \setminus \{v_r\}$, $d_V^-(v_i) = 1$. The set of all **external nodes** is $V_{ext} = \{v_i : d_V^+(v_i) = 0\}$ and the set of all the **internal nodes** is $V_{int} = \{v_i : d_V^+(v_i) > 0\}$.

We define with $\Gamma_V^+(v_i)$ the set of **all the descendants** of v_i in V (including v_i ifself); vice versa $\Gamma_V^-(v_i)$ is the set of **all the ancestors** of v_i in V (including v_i ifself). We shall use the set Γ in the following of this work, so it is worth underlining a couple of recurrent cases. Let $v_r \in V$ be the root of a tree $T(V, E)$ then $\Gamma_V^-(v_r) = \{v_r\}$ and $\Gamma_V^+(v_r) = V$. Furthermore, let v_i an external node of $T(V, E)$, then $\Gamma_V^+(v_i) = \{v_i\}$.

3   The Set Data Models

We propose two set data models called *Nested Set Model* (NS-M) and *Inverse Nested Set Model* (INS-M) based on an organization of nested sets. The most intuitive way to understand how these models work is to relate them to the well-know tree data structure. Thus, we informally present the two data models by means of examples of mapping between them and a sample tree.

The first model we present is the **Nested Set Model** (NS-M). The intuitive graphic representation of a tree as an organization of nested sets was used in [8] to show different ways to represent tree data structure and in [3] to explain an alternative way to solve recursive queries over trees in SQL language. An organization of sets in the NS-M is a collection of sets in which any pair of sets is either disjoint or one contains the other. In Figure 1 we can see how a sample tree is mapped into an organization of nested sets based on the NS-M.

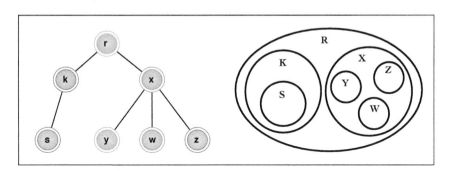

Fig. 1. The mapping between a tree data structure and the Nested Set Model

From Figure 1 we can see that each node of the tree is mapped into a set, where child nodes become *proper subsets* of the set created from the parent node. Every set is subset of at least of one set; the set corresponding to the tree root is the only set without any supersets and every set in the hierarchy is subset of the root set. The external nodes are sets with no subsets. The tree structure is maintained thanks to the nested organization and the relationships between the sets are expressed by the set inclusion order. Even the disjunction between two sets brings information; indeed, the disjunction of two sets means that these belong to two different branches of the same tree.

The second data model is the **Inverse Nested Set Model** (INS-M). We can say that a tree is mapped into the INS-M transforming each node into a set, where each parent node becomes a subset of the sets created from its children. The set created from the tree's root is the only set with no subsets and the root set is a proper subset of all the sets in the hierarchy. The leaves are the sets with no supersets and they are sets containing all the sets created from the nodes composing tree path from a leaf to the root. An important aspect of INS-M is that the intersection of every couple of sets obtained from two nodes is always a set representing a node in the tree. The intersection of all the sets in the INS-M is the set mapped from the root of the tree. Fig. 2 shows the organization of nested sets created from the branch of a tree.

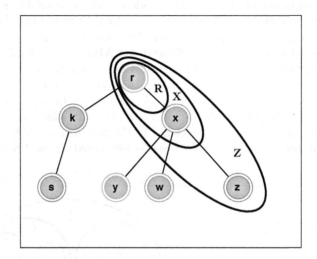

Fig. 2. How the branch of a tree can be mapped into the INS-Model

It is worthwhile for the rest of the work to define some basic concepts of set theory: the family of subsets and the subfamily of subsets, with reference to [4] for their treatment. However, we assume the reader is confident with the basic concepts of ZFC axiomatic set theory, which we cannot extensively treat here for space reasons.

Definition 1. *Let A be a set, I a non-empty set and C a collection of subsets of A. Then a bijective function $A : I \longrightarrow C$ is a **family** of subsets of A. We call I the **index** set and we say that the collection C is **indexed** by I.*

We use the following notation $\{A_i\}_{i \in I}$ to indicate the family A; the notation $A_i \in \{A_i\}_{i \in I}$ means that $\exists\, i \in I \mid A(i) = A_i$. We call **subfamily** of $\{A_i\}_{i \in I}$ the **restriction** of A to $J \subseteq I$ and we denote this with $\{B_i\}_{j \in J} \subseteq \{A_i\}_{i \in I}$.

Definition 2. *Let A be a set and let $\{A_i\}_{i \in I}$ be a family. Then $\{A_i\}_{i \in I}$ is a* ***Nested Set*** *family if:*

$$A \in \{A_i\}_{i \in I}, \tag{3.1}$$

$$\emptyset \notin \{A_i\}_{i \in I}, \tag{3.2}$$

$$\forall A_h, A_k \in \{A_i\}_{i \in I}, h \neq k \mid A_h \cap A_k \neq \emptyset \Rightarrow A_h \subset A_k \vee A_k \subset A_h. \tag{3.3}$$

Thus, we define a Nested Set family (NS-F) as a family where three conditions must hold. The first condition (3.1) states that set A which contains all the sets in the family must belong to the NS-F. The second condition states that the empty-set does not belong to the NS-F and the last condition (3.3) states that the intersection of every couple of distinct sets in the NS-F is not the empty-set only if one set is a proper subset of the other one [6,2].

Theorem 1. *Let $T(V, E)$ be a tree and let Φ be a family where $I = V$ and $\forall v_i \in V$, $V_{v_i} = \Gamma_V^+(v_i)$. Then $\{V_{v_i}\}_{v_i \in V}$ is a Nested Set family.*

Proof. Let $v_r \in V$ be the root of the tree then $V_{v_r} = \Gamma_V^+(v_r) = V$ and thus $V \in \{V_{v_i}\}_{v_i}$ (condition 3.1). By definition of descendant set of a node, $\forall v_i \in V$, $|V_{v_i}| = |\Gamma_V^+(v_i)| \geq 1$ and so $\emptyset \notin \{V_{v_i}\}_{v_i \in V}$ (condition 3.2).

Now, we prove condition 3.3. Let $v_h, v_k \in V$, $h \neq k$ such that $V_{v_h} \cap V_{v_k} = \Gamma_V^+(v_h) \cap \Gamma_V^+(v_k) \neq \emptyset$, ab absurdo suppose that $\Gamma_V^+(v_h) \not\subseteq \Gamma_V^+(v_k) \wedge \Gamma_V^+(v_k) \not\subseteq \Gamma_V^+(v_k)$. This means that the descendants of v_h share at least a node with the descendants of v_k but they do not belong to the same subtree. This means that $\exists\, v_z \in V \mid d_V^-(v_z) = 2$ but then $T(V, E)$ is not a tree. $\qquad\square$

Example 1. *Let $T(V, E)$ be a tree where $V = \{v_0, v_1, v_2, v_3\}$ and $E = \{e_{01}, e_{02}, e_{23}\}$, thus $\Gamma_V^+(v_0) = \{v_0, v_1, v_2, v_3\}$, $\Gamma_V^+(v_1) = \{v_1\}$, $\Gamma_V^+(v_2) = \{v_2, v_3\}$ and $\Gamma_V^+(v_3) = \{v_3\}$. Let $\{V_{v_i}\}_{v_i \in V}$ be a family, where $V_{v_0} = \{v_0, v_1, v_2, v_3\}$, $V_{v_1} = \{v_1\}$, $V_{v_2} = \{v_2, v_3\}$ and $V_{v_3} = \{v_3\}$. Then, from theorem 1 it follows that $\{V_{v_i}\}_{v_i \in V}$ is a NS-F.*

In the same way we can define the Inverse Nested Set Model (INS-M):

Definition 3. *Let A be a set and let $\{A_i\}_{i \in I}$ be a family. Then $\{A_i\}_{i \in I}$ is an* ***Inverse Nested Set*** *family if:*

$$\emptyset \notin \{A_i\}_{i \in I}, \tag{3.4}$$

$$\forall \{B_j\}_{j \in J} \subseteq \{A_i\}_{i \in I} \Rightarrow \bigcap_{j \in J} B_j \in \{A_i\}_{i \in I}. \tag{3.5}$$

Thus, we define an Inverse Nested Set family (INS-F) as a family where two conditions must hold. The first condition (3.4) states that the empty-set does not belong to the INS-F. The second condition states that the intersection of every subfamily of the INS-F belongs to the INS-F itself.

Theorem 2. *Let $T(V, E)$ be a tree and let Ψ be a family where $I = V$ and $\forall v_i \in V$, $V_{v_i} = \Gamma_V^-(v_i)$. Then $\{V_{v_i}\}_{v_i \in V}$ is an Inverse Nested Set family.*

Proof. By definition of the set of the ancestors of a node, $\forall v_i \in V$, $|V_{v_i}| = |\Gamma_V^-(v_i)| \geq 1$ and so $\emptyset \notin \{V_{v_i}\}_{v_i \in V}$ (condition 3.4).

Let $\{B_{v_j}\}_{v_j \in J}$ be a subfamily of $\{V_{v_i}\}_{v_i \in V}$. We prove condition 3.5 by induction on the cardinality of J. $|J| = 1$ is the base case and it means that every subfamily $\{B_{v_j}\}_{v_j \in J} \subseteq \{V_{v_i}\}_{v_i \in V}$ is composed only by one set B_{v_1} whose intersection is the set itself and belongs to the family $\{V_{v_i}\}_{v_i \in V}$ by definition.

For $|J| = n-1$ we assume that $\exists\, v_{n-1} \in V \mid \bigcap_{v_j \in J} B_{v_j} = B_{v_{n-1}} \in \{V_{v_i}\}_{v_i \in V}$; equivalently we can say that $\exists\, v_{n-1} \in V \mid \bigcap_{v_j \in J} \Gamma_V^-(v_j) = \Gamma_V^-(v_{n-1})$, thus, $\Gamma_V^-(v_{n-1})$ is a set of nodes that is composed of common ancestors of the $n-1$ considered nodes.

For $|J| = n$, we have to show that $\exists\, v_t \in V \mid \forall\, v_n \in J$, $B_{v_{n-1}} \cap B_{v_n} = B_{v_t} \in \{V_{v_i}\}_{v_i \in V}$. This is equivalent to show that $\exists\, v_t \in V \mid \forall\, v_n \in J$, $\Gamma_V^-(v_{n-1}) \cap \Gamma_V^-(v_n) = \Gamma_V^-(v_t)$.

Ab absurdo suppose that $\exists\, v_n \in J \mid \forall\, v_t \in V$, $\Gamma_V^-(v_{n-1}) \cap \Gamma_V^-(v_n) \neq \Gamma_V^-(v_t)$. This would mean that v_n has no ancestors in J and, consequently, in V; at the same time, this would mean that v_n is an ancestor of no node in J and, consequently, in V. But this means that V is the set of nodes of a forest and not of a tree. □

Example 2. *Let* $T(V, E)$ *be a tree where* $V = \{v_0, v_1, v_2, v_3\}$ *and* $E = \{e_{01}, e_{02}, e_{23}\}$, *thus* $\Gamma_V^-(v_0) = \{v_0\}$, $\Gamma_V^-(v_1) = \{v_0, v_1\}$, $\Gamma_V^-(v_2) = \{v_0, v_2\}$ *and* $\Gamma_V^-(v_3) = \{v_0, v_2, v_3\}$. *Let* $\{V_{v_i}\}_{v_i \in V}$ *be a family where* $V_{v_0} = \{v_0\}$, $V_{v_1} = \{v_0, v_1\}$, $V_{v_2} = \{v_1, v_2\}$ *and* $V_{v_3} = \{v_0, v_2, v_3\}$. *Then, from theorem 2 it follows that* $\{V_{v_i}\}_{v_i \in V}$ *is a INS-F.*

4   Set-Theoretic Extensions of OAI-PMH

The defined set data models can be exploited to improve the data exchange between DLSs in a distributed environment. In this context the standard *de-facto* for metadata exchange between DLSs is the couple OAI-PMH and XML. The main reason is the flexibility of both the protocol and the XML that foster interoperability between DLSs managing different kinds of metadata coming from different kinds of cultural organizations. It is worthwhile to describe the functioning of OAI-PMH to understand how it can take advantage of the NS-M and INS-M and how it can be extended to cope with the exchange of hierarchical structures. OAI-PMH is so widely diffuse in the field of DL that we can assume the reader is familiar with its underlying functioning. The main feature of OAI-PMH we exploit is selective harvesting; this is based on the concept of *OAI-set*, which enables logical data partitioning by defining groups of records. Selective harvesting is the procedure that permits the harvesting only of metadata owned by a specified OAI-set. In OAI-PMH a set is defined by three components: `setSpec` which is mandatory and a unique identifier for the set within the repository, `setName` which is a mandatory short human-readable string naming the set, and `setDesc` which may hold community-specific XML-encoded data about the set.

OAI-set organization may be hierarchical, where hierarchy is expressed in the `setSpec` field by the use of a colon [:] separated list indicating the path from the root of the set hierarchy to the respective node. For example if we define an OAI-set whose `setSpec` is "A", its subset "B" would have "A:B" as `setSpec`. In this case "B" is a proper subset of "A": $B \subset A$. When a repository defines a set organization it must *include set membership information in the headers of the records* returned to the harvester requests. Harvesting from a set which has sub-sets will cause the repository to return the records in the specified set and recursively to return the records from all the sub-sets. In our example, if we harvest set A, we also obtain the records in sub-set B [13].

In OAI-PMH it is possible to define an OAI-set organization based on the NS-M or INS-M. This means that we can treat the OAI-sets as a Nested Set Family (NS-F) or as an Inverse Nested Set Family (INS-F). The inclusion order between the OAI-sets is given by its identifier which is a `<setspec>` value. This `<setspec>` value is also added in the header of every record belonging to an OAI-set. In the following we describe how it is possible to create a Nested Set family of OAI-Set and afterward how the same thing can be done with an Inverse Nested Set family.

Let $\mathcal{O}$ be a Nested Set family and let I be the set of the `<setspec>` values where $i \in I = \{s_0 : s_1 : \ldots : s_j\}$ means that $\exists \ O_j \in \{O_i\}_{i \in I} \ | \ O_j \subset \ldots \subset O_1 \subset O_0$. Every $O_i \in \{O_i\}_{i \in I}$ is an OAI-set uniquely identified by a `<setspec>` value in I. The `<setspec>` values for the $O_k \in \{O_i\}_{i \in I}$ are settled in such a way to maintain the inclusion order between the sets. If an O_k has no superset its `setspec` value is composed only by a single value (`<setspec>`s_k`</setspec>`). Instead if a set O_h has supersets, e.g. O_a and O_b where $O_b \subset O_a$, its `setspec` value must be the combination of the name of its supersets and itself separated by the colon [:] (e.g. `<setspec>`$s_a : s_b : s_h$`</setspec>`). Furthermore, let $R = \{r_0, \ldots, r_n\}$ be a set of records, then each $r_i \in O_j$ must contain the setspec of O_j in its header.

Throughout $\{O_i\}_{i \in I}$ it is possible to represent a hierarchical data structure, such as a tree, in OAI-PMH providing a granularity access to the items in the hierarchy and at the same time enabling the exchange of a single part of the hierarchy with the possibility of reconstructing the whole hierarchy whenever it is necessary. The next section presents a concrete use case of the NS-F mapping a tree data structure enabling its representation and exchange by means of OAI-PMH; a visual idea of this procedure can be seen in Fig. 4.

In the same way we can apply the INS-M to OAI-PMH; Let $\mathcal{U}$ be an Inverse Nested Set family and let J be the set of the `<setspec>` values where $j \in J = \{s_0 : s_1 : \ldots : s_k\}$ means that $\exists \ U_k \in \{U_j\}_{j \in J} = U_k \subset \ldots \subset U_1 \subset U_0$. In $\{U_j\}_{j \in J}$ differently that in $\{O_i\}_{i \in I}$ the following case may happen: Let $U_i, U_k, U_w \in \{U_j\}_{j \in J}$ then it is possible that $U_w \subset U_i$ and $U_w \subset U_k$ but either $U_i \not\subseteq U_k$ and $U_k \not\subseteq U_i$. If we consider $\{U_j\}_{j \in J}$ composed only of U_i, U_k and U_w, the identifier of U_i is `<setspec>`s_i`</setspec>` and the identifier of U_k is `<setspec>`s_k`</setspec>`. Instead, the identifier of U_w must be `<setspec>`$s_i : s_w$`</setspec>` and `<setspec>`$s_k : s_w$`</setspec>` at the same time; this means

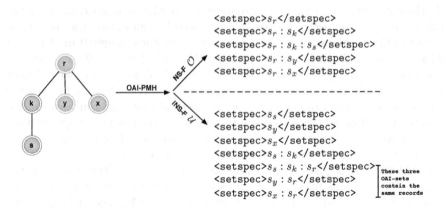

Fig. 3. The *setspec* values of the OAI-sets belonging to the NS-F $\{O_i\}_{i \in I}$ and the INS-F $\{U_j\}_{j \in J}$ obtained from a sample tree

that in $\{U_j\}_{j \in J}$ there are two distinct OAI-sets, one identified by `<setspec>`$s_i : s_w$`</setspec>` and the other identified by `<setspec>`$s_k : s_w$`</setspec>`. This is due to the fact that the intersection between OAI-sets in OAI-PMH is not defined set-theoretically; indeed, the only way to get an intersection of two OAI-sets is enumerating the records. This means that we can know if an OAI-record belongs to two or more sets just by seeing whether there are two or more `<setspec>` entries in the header of the record. In this case the records belonging to U_w will contain two `<setspec>` entries in their header: `<setspec>`$s_i : s_w$`</setspec>` and `<setspec>`$s_k : s_w$`</setspec>`; note that only the `<setspec>` value is duplicated and not the records themselves.

In Figure 3 we can see how a sample tree can be represented in OAI-PMH exploiting the OAI-sets organization. In the upper part we reported the `<setspec>` values of the OAI-sets organized in a NS-F $\{O_i\}_{i \in I}$, instead in the lower part we reported the `<setspec>` values of the OAI-sets organized in INS-F $\{U_j\}_{j \in J}$.

With this view of OAI-PMH we can set a hierarchical structure of items as a well-defined nested set organization that maintains the relationships between the items just as a tree data structure does and moreover we can exploit the flexibility of the sets exchanging a specific subset while maintaining the integrity of the data. Indeed, in the header of the items there is the set membership information which, if necessary, enables the reconstruction of the hierarchy or part of it. Throughout the NS-M and INS-M it is possible to handle hierarchical structures in OAI-PMH simply by exploiting the inner functionalities of the protocol; indeed, no change of OAI-PMH is required to cope with the presented set data models.

The choice between NS-M and INS-M is based on the application context: NS-M fosters the reconstruction of the lower levels of a hierarchy starting from a node, vice versa INS-M fosters the reconstruction of the upper levels. This difference between the models should become clearer if we consider a relevant example of how OAI-PMH and the presented set data models can be successfully used to overcome well-known problems in data exchange.

4.1   The Set Data Models and OAI-PMH Applied to the Archives

This subsection describes how we can exchange archival metadata in a distributed environment and it is a continuation of the work presented in [5]. A brief introduction regarding the archive peculiarities is worthwhile for a better understanding of the proposed solutions. An archive is a complex cultural organization which is not simply constituted by a series of objects that have been accumulated and filed with the passing of time. Archives have to keep the context in which their documents have been created and the network of relationships among them in order to preserve their informative content and provide understandable and useful information over time. The context and the relationships between the documents are preserved thanks to the strongly hierarchical organization of the documents inside the archive. Indeed, an archive is divided by fonds and then by sub-fonds and then by series and then by sub-series and so on; at every level we can find documents belonging to a particular division of the archive or documents describing the nature of the considered level of the archive (e.g. a fond, a sub-fonds, etc.). The union of all these documents, the relationships and the context information permits the full informational power of the archival documents to be maintained.

In the digital environment an archive and its components are described by the use of metadata; these need to be able to express and maintain such structure and relationships. The standard format of metadata for representing the complex hierarchical structure of the archive is *Encoded Archival Description (EAD)* [9], which reflects the archival structure and holds relations between documents in the archive. On the other hand to maintain all this information an EAD file turns out to be a very large XML file with a deep hierarchical internal structure. Thus, accessing, searching and sharing individual items in the EAD might be difficult without taking into consideration the whole hierarchy. On the other hand, users are often interested in the information described at the item level, which is typically buried very deeply in the hierarchy and might be difficult to reach [11]. These issues can be overcome by describing the hierarchical organization of an archive as a family of sets in the NS-M, where the documents belonging to a specific division of the archive become metadata belonging to a specific set.

In Fig. 4 we can see two approaches to representing the archival organization and documents. The first approach is the EAD-like one in which the whole archive is mapped inside a unique XML file which is potentially very large and deeply hierarchical. All information about fonds, sub-fonds or series as well as the documents belonging to a specific archival division are mapped into several XML elements in the same XML file. With this approach we cannot exchange precise metadata through OAI-PMH, rather we have to exchange the whole archive. At the same time it is not possible to access a specific piece of information without accessing the whole hierarchy [10].

The second approach is based on the NS-M. The archival hierarchy is mapped into a family Φ that for theorem 1 is a NS-F. In Φ the documents are represented as items belonging to the opportune set. In this way the context information and the relationships between the documents are preserved thanks to the nested set

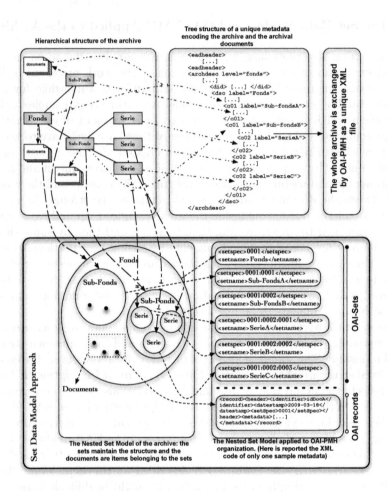

Fig. 4. The hierarchical structure of an archive mapped into a metadata with a tree data structure, the alternative mapping in the NS-M and in OAI-PMH

organization and at the same time they are not bound to a rigid structure. Then, Φ is represented in OAI-PMH throughout the family $\{O_i\}_{i \in I}$ of OAI-sets obtained setting of the `<setspec>` values as described in the previous subsection. For instance, the set obtained from the root has a "0001" identifier, the set mapped from the children of the root are identified by "0001:0001", "0001:0002" and so on. Thus, from the identifier of an OAI-set we can reconstruct the hierarchy through the ancestors to the root. By means of OAI-PMH it is possible to exchange a specific part of the archive while at the same time maintaining the relationships with the other parts of it. The NS-M fosters the reconstruction of the lower levels of a hierarchy; thus, with the couple NS-M and OAI-PMH applied to the archive, if a harvester asks for an OAI-set representing for instance a sub-fond it recursively obtains all the OAI-subsets and items in the subtree rooted in the selected sub-fonds.

This approach can also be applied with the INS-M mapping the archival hierarchy into a INS-F $\{U_j\}_{J \in J}$ following the procedure illustrated in the previous section. In this case there is a big difference in the harvesting procedure; indeed, if a harvester asks for an OAI-set representing for instance an archival series it recursively obtains all the OAI-subsets and records in the path from the archival series to the principal fond that is the root of the archival tree. The choice between a NS-M or INS-M should be done on the basis of the application context. In the archival context the application of the INS-M would be more significant than the NS-M. Indeed, often the information required by a user stored in the external nodes of the archival tree [11]. If we represent the archival tree by means of the INS-F, when a harvester requires an external node of the tree it will receive all the archival information contained in the nodes up to the root of the tree. This means that a Service Provider can offer a potential user the required information stored in the external node and also all the information stored in its ancestors nodes. This information is very useful for inferring the context of an archival metadata which is contained in the required external node; indeed, the ancestor nodes represent and contain the information related to the series, sub-fonds and fonds in which the archival metadata are classified. The INS-M fosters the reconstruction of the upper levels of a hierarchy that in the archival case often contain contextual information which permit the relationships of the archival documents to be inferred with the other documents in the archive and with the production and preservation environment.

5   Conclusions

We have discussed the relevance of the hierarchical structures in computer science with a specific examination of the DLSs. We have presented the tree data structure and highlighted the more relevant aspects to our treatment of hierarchical structures. We have also presented the NESTOR framework defining two set-theoretical data models called Nested Set Model and Inverse Nested Set Model as alternatives of the tree data structure. Furthermore, we have shown how a tree can be mapped in one model or the other. These models maintain the peculiarities of the tree with the flexibility and accessibility of sets. We have shown how the protocol OAI-PMH can be extended by exploiting the NS-M or the INS-M. Lastly we have presented a significant application of the presented set data models in conjunction with OAI-PMH represented by the archives. Indeed, we have shown how the hierarchical archive organization can be represented and exchanged in OAI-PMH and thus between different DLSs in a distributed environment.

Acknowledgments

The authors wish to thank Maristella Agosti for her support and collaboration to bring forth this work. The work reported has been supported by a grant from the Italian Veneto Region. The study is also partially supported by the TELplus

Targeted Project for Digital Libraries, as part of the eContentplus Program of the European Commission (Contract ECP-2006-DILI- 510003).

References

1. Collins Dictionary of The English Language. William Collins Sons & Co.Ltd (1979)
2. Anderson, K.W., Hall, D.W.: Sets, Sequences, and Mappings: The Basic Concepts of Analysis. John Wiley & Sons, Inc., New York (1963)
3. Celko, J.: Joe Celko's SQL for Smarties: Advanced SQL Programming. Morgan Kaufmann, San Francisco (2000)
4. Davey, B.A., Priestley, H.A.: Introduction to Lattices and Order, 2nd edn. Cambridge University Press, Cambridge (2002)
5. Ferro, N., Silvello, G.: A Methodology for Sharing Archival Descriptive Metadata in a Distributed Environment. In: Christensen-Dalsgaard, B., Castelli, D., Ammitzbøll Jurik, B., Lippincott, J. (eds.) ECDL 2008. LNCS, vol. 5173, pp. 268–279. Springer, Heidelberg (2008)
6. Halmos, P.R.: Naive Set Theory. D. Van Nostrand Company, Inc., New York (1960)
7. Jech, T.: Set Theory - The Third Millenium Edition. Springer, Heidelberg (2003)
8. Knuth, D.E.: The Art of Computer Programming, 3rd edn., vol. 1. Addison Wesley, Reading (1997)
9. Pitti, D.V.: Encoded Archival Description. An Introduction and Overview. D-Lib Magazine 5(11) (1999)
10. Prom, C.J., Rishel, C.A., Schwartz, S.W., Fox, K.J.: A Unified Platform for Archival Description and Access. In: Rasmussen, E.M., Larson, R.R., Toms, E., Sugimoto, S. (eds.) Proc. 7th ACM/IEEE Joint Conference on Digital Libraries (JCDL 2007), pp. 157–166. ACM Press, New York (2007)
11. Shreeves, S.L., Kaczmarek, J.S., Cole, T.W.: Harvesting Cultural Heritage Metadata Using the OAI Protocol. Library Hi Tech. 21(2), 159–169 (2003)
12. Van de Sompel, H., Lagoze, C., Nelson, M., Warner, S.: Implementation Guidelines for the Open Archive Initiative Protocol for Metadata Harvesting. Technical report, Open Archive Initiative (2002)
13. Van de Sompel, H., Lagoze, C., Nelson, M., Warner, S.: Implementation Guidelines for the Open Archive Initiative Protocol for Metadata Harvesting - Guidelines for Harvester Implementers. Technical report, Open Archive Initiative, p. 6 (2002)

eSciDoc Infrastructure:
A Fedora-Based e-Research Framework

Matthias Razum, Frank Schwichtenberg, Steffen Wagner, and Michael Hoppe

FIZ Karlsruhe, Hermann-von-Helmholtz-Platz 1,
76344 Eggenstein-Leopoldshafen, Germany
firstname.surname@fiz-karlsruhe.de

Abstract. eSciDoc is the open-source e-Research framework jointly developed by the German Max Planck Society and FIZ Karlsruhe. It consists of a generic set of basic services ("eSciDoc Infrastructure") and various applications built on top of this infrastructure ("eSciDoc Solutions"). This paper focuses on the eSciDoc Infrastructure, highlights the differences to the underlying Fedora repository, and demonstrates its powerful und application-centric programming model. Further on, we discuss challenges for e-Research Infrastructures and how we addressed them with the eSciDoc Infrastructure.

1 Introduction

Digital Repositories undergo yet again a substantial change of paradigm. While they started several years ago with a library perspective, mainly focusing on publications, they are now becoming more and more a commodity tool for the workaday life of researchers. Quite often the repository itself is just a background service, providing storage, persistent identification, preservation, and discovery of the content. It is hidden from the end-user by means of specialized applications or services. Fedora's approach of providing a repository architecture rather than an end-user tool matches this evolution. eSciDoc (Dreyer, Bulatovic, Tschida, & Razum, 2007), from the start of the project nearly five years ago, has always been in line with this development by separating backend services (eSciDoc Infrastructure) and front-end applications (eSciDoc Solutions)[1].

E-Science and e-Research trigger several new challenges (Hey & Trefethen, 2005). Whereas many e-Science applications concentrate on massive amounts of data and how to store, manipulate, and analyze them, eSciDoc focuses more on a more holistic approach to knowledge management in the research process. If not only the final results of the research process, but all intermediate steps from the first idea over experimentation, analysis, and aggregation should be represented within a repository, the digital library becomes a 'e-Research Infrastructure'. Figure 1 depicts the process and eSciDoc's approach with a generic infrastructure and specialized solutions supporting the various steps of the research process.

---

[1] http://www.escidoc.org/

M. Agosti et al. (Eds.): ECDL 2009, LNCS 5714, pp. 227–238, 2009.

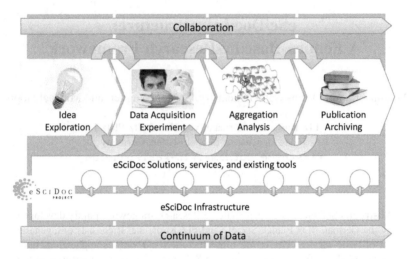

Fig. 1. eSciDoc supports the whole research process with a generic infrastructure and specialized applications, services, and integration of existing tools

2   Challenges for e-Research Infrastructures

A whole new set of requirements have to be taken into consideration for e-Research infrastructures. In the following, we will name a few of them and show how eSciDoc currently supports researchers, data curators, librarians, and developers with powerful, yet simple-to-use features. Intentionally left out are challenges like the data deluge (Hey & Trefethen, 2003) and massive parallel computing/grid computing, which we don't address with eSciDoc.

Maintain both data and publications. Primary data differs a lot from traditional publications. It comes in various and sometimes exotic file formats, metadata profiles are specific to disciplines or even projects, and it quite often includes datasets consisting of several files. eSciDoc's flexible content models and compound objects allow to store these complex data objects. The support for arbitrary metadata profiles for each object, both on the logical and the physical (file) level allows for proper description and discovery.

Reliable citations. Citations have always been the backbone of scholarly communication and a challenge for the web, digital libraries, and electronic archives. Much of the value of digital resources for scholarly communication lies in enabling resources to be cited reliably with resolvable and actionable links over long periods. Therefore, libraries, archives, academic institutions, and publishers have an interest in the persistence of resource identification (Powell & Lyon, 2001). New publication types like datasets, simulations, etc. in combination with increasing numbers of born-digital materials requires the adoption of a digital equivalent of the traditional paper-based citations. eSciDoc supports a wide range of persistent identifiers on the object and file level. Versioning ensures that once cited, objects will always appear in the same way as originally perceived by the citing author. To avoid 'broken' references, eSciDoc will never delete a once published object. Instead, objects may be withdrawn, which means that access to the files is inhibited, but the metadata and a note explaining the reason for the withdrawal are still accessible.

Focus on researchers. E-Research infrastructures are built for researchers, so their needs and working attitudes should be the main focus. Research is a dynamic process, and tools should support this process, not block it. eSciDoc allows researchers to build their own solutions easily because of much basic functionality (storage, search, authentication, access rights) being provided by the infrastructure. This separation of concerns allows them to focus on their scholarly 'business logic'. Open programming interfaces support a wide range of programming and scripting languages when building your own solution. More important, researchers only reluctantly change their working habits. eSciDoc's APIs comply with the web architecture, thus facilitating mash-ups, integration with existing tools and scientific software packages, and data exchange with other scientific repositories.

Veracity and fidelity of research and re-use of data. Supporting the whole research process with tools and maintaining all (digital) artifacts that are created or modified in the course of this process may help to enrich traditional publications with supplementary materials. Such data may help others in reproducing and validating results. Provenance metadata, tagging, semantic links between objects, and multiple metadata records of arbitrary profiles provide context not only for long-term preservation, but as well for discovery and re-use of objects.

Collaboration. Collaboration has always been an important aspect of scholarly work. E-Research infrastructures can support and even amplify the collaborative aspect by technical means. Cross-institutional teams can benefit from distributed authentication (Shibboleth), which supports virtual working groups. Flexible access policies allow for fine-grained rights management, thus giving researchers the option to exactly control the dissemination of their artifacts created in the course of the research process at any given time. Team work, especially when geographically distributed, requires versioning and audit trails, so that team members can track changes and eventually roll back unwanted or unintentional modifications of any artifact.

Mixture of open access and private material. Research data is not always freely accessible. Researchers often dislike the idea of releasing intermediate results before the final publication. Some datasets might not be publishable at all (e.g., due to privacy issues). Even within a project team, fine-grained access rights might be necessary. eSciDoc implements a single point of policy enforcement as part of the infrastructure, which leads to a lightweight solution-side implementation. Even badly designed solutions cannot compromise the overall security.

Preservation. The manifold file formats, the complex compound objects in e-Research, and the lack of standards in describing datasets appropriately impede the long-term preservation of the data stored or referenced in e-Research infrastructures. eSciDoc doesn't provide a comprehensive solution for this problem, especially as preservation has many organizational (i.e., non-technical) aspects to be looked at. However, eSciDoc supports several metadata records of arbitrary profiles per record, which not only caters for descriptive, technical, and administrative metadata records, but allows describing the same object out of different perspectives. This is especially important for discovery and re-use of objects across disciplines. An Audit Trail and a PREMIS-based event history[2] keep track of changes to the object. Fedora's standardized METS[3]-based XML object container ensures a software-independent storage format, can be validated, and supports checksums for the digital content. The file-based storage model eases backup and restore operations, accompanied with Fedora's unique rebuild capabilities.

3 Differences between Fedora and eSciDoc

Fedora (Lagoze, Payette, Shin, & Wilper, 2005), the Flexible Extensible Digital Object Repository, is a renowned solution for all kinds of applications, including e-Research use cases. Fedora is the repository underneath the eSciDoc Infrastructure, and it already provides many features to address the above mentioned challenges. So why did we chose to encapsulate Fedora in an extensive middleware approach (which the eSciDoc Infrastructure actually is) and what are the main differences on a functional level between an out-of-the-box Fedora installation and eSciDoc?

Fedora provides a very generic set of functionalities, addressing the needs of various communities and use cases. On the other hand, this means that it only provides low-level functionality, requiring developers to spend time implementing high-level services. eSciDoc tries to fill that gap by adding these high-level services on top of Fedora while hiding some of the more complex aspects of Fedora, thus increasing the productivity of developers. However, this advantage contrasts with a reduction of flexibility.

1. Datastreams and Object Patterns
A Fedora object consists of several datastreams, which contain either XML or binary content. The repository developer has to define the layout and allowed contents for each datastream – the 'content model'. Fedora 3.0 implements the Content Model Architecture (CMA)[4], which codifies the previous implicit content model approaches. CMA assures that objects always comply with the model, but it does not help to find the best possible model. Content modeling is one of the challenges to start with Fedora, even for simpler use cases. Therefore, eSciDoc introduced 'object patterns' for basic object types: Items and Containers (see figure 2). For both object patterns, the layout, naming, and allowed content is predefined. eSciDoc objects are therefore less flexible, but simplify data modeling. Still, they provide flexibility where needed (e.g. storing metadata records). Our experience shows that this model accommodates for most use cases.

eSciDoc provides several more basic objects types, which are more fixed in nature and therefore are not seen as 'object patterns'. The most relevant ones are the Organizational Unit and the Context. Organizational Units are used to represent hierarchical structures of organizations with its institutes, departments, working groups, and so on. These hierarchies can be used to identify owners of objects, but at the same time can

---

[2] http://www.loc.gov/standards/premis/
[3] http://www.loc.gov/standards/mets/
[4] http://www.fedora-commons.org/documentation/3.0b1/userdocs/digitalobjects/cmda.html

be referenced from within metadata records to state the affiliation of authors. Contexts are administrative entities, which on the one hand side state the legal and organizational responsibility, on the other hand side they are used to store configuration information.

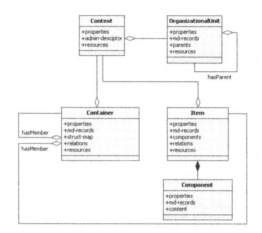

Fig. 2. eSciDoc Data Model, showing the two object patterns 'Item' and 'Container'. Items may have one or more components, which hold the different manifestations of the content.

The **Item pattern** represents a single entity with possibly multiple manifestations. Each manifestation is included in the Item as separate part, the **Component**. A Component includes the manifestation-related metadata, some properties (e.g., mime-type, file name), and the content itself. The **Container pattern** aggregates Items or other Containers. Object patterns support multiple metadata records, object relations (within and outside of eSciDoc), and some logistic and lifecycle properties. All the information is stored in one or more Fedora objects, following the 'atomistic' content model paradigm. However, the user will only work with a single eSciDoc object. All the complex work is done behind the scenes by the eSciDoc Infrastructure.

2. Object Lifecycle

Items and Containers both implement a basic lifecycle in the form of a simple workflow. Each eSciDoc object derived from one of the object patterns is created in status 'pending'. Submitting the object forwards it to the quality assurance stage, from where it can be either sent back to the creator for revision or released (i.e., made publicly accessible). In rare cases, released Items need to be withdrawn (e.g., because of copyright infringements), which is the last status in the lifecycle. For each status, different access rights (based on policies, roles, and scopes) may be defined. Moving objects from one status to the next is as easy as invoking a single method. Figure 3 depicts the available states of the object lifecycle and their succession.

Objects that are in status 'released' are not fixed. Authors may decide to further work on them. In this case, a new version of the object will be created, and a version status will be set to 'pending', whereas the object status will remain 'released'. This means that the author has a working version accessible only to her and maybe invited collaborators, whereas the rest of the world still sees the released version of the object. As soon as the new working version has been submitted and released, it will be presented to non-privileged users as the latest version, complementing the previously released version.

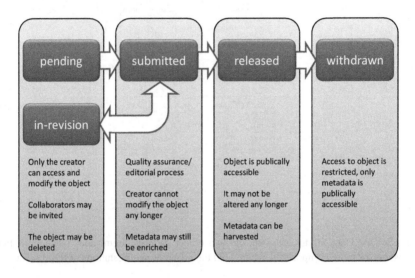

Fig. 3. eSciDoc's object lifecycle with different states and default access policies

3. Versioning

eSciDoc extends Fedora's versioning approach. Because of the atomistic content model, a single eSciDoc object may in fact be a graph-oriented composition of Fedora objects. However, the user conceives eSciDoc objects as one entity and expects what we call 'whole-object versioning' (WOV). Therefore, the eSciDoc Infrastructure maintains a special datastream to facilitate that view by keeping track of modifications upon all digital objects that, together, form a graph-oriented composition, such as multi-object content models with whole/part (Item resource with Component parts) or parent/child (Container resource with member objects) relationships.

eSciDoc differentiates between 'working versions' and 'releases', which is tightly coupled with the object lifecycle. Working versions are only accessible to authors and collaborators, whereas releases are publicly visible. For releases, eSciDoc implements a graph-aware versioning. If a part changes, there must be a new version of the whole, i.e., if a Component is updated, the Item itself is updated as well. If a child is either added or removed, there must be a new version of the parent. If an existing child is just modified, the parent object is not versioned. This is a simplification to avoid too many versions of parents with many children. However, if a Container is released, the Infrastructure ensures that all relationships to its members are updated first, which means that the simplification only affects working versions and not releases (which are the only ones to be cited or referenced).

Keeping track of changes within a compositional network can be done without actually creating separate copies of all its objects, which could raise scalability issues when there are many changes. Instead, eSciDoc versioning re-uses the existing Fedora content versioning scheme, augmented with the additional WOV metadata stream carried in "parent" and "whole" digital objects. This "whole-object versioning" metadata is a simple XML tree with a node for every version of the resource or the parent/child graph. The Object Manager updates the WOV metadata whenever a write operation is completed upon a part or child object. So, the eSciDoc Infrastructure will generate a single intellectual version of the whole graph-oriented composition of Fedora objects even for actions that involve modifications of several datastreams or several Fedora objects. However, not all method invocations create new versions, e.g. forwarding an object to the next state in its lifecycle. For operations that create no new versions eSciDoc maintains an additional event log that tracks all actions using PREMIS.

4. Application-oriented Representation

The fact that an eSciDoc object actually is a graph of Fedora objects with multiple datastreams is completely transparent to the user. The eSciDoc Infrastructure exposes its contents as XML representations, which contain all relevant information for typical application scenarios. That includes version information, metadata, and references to other objects or parts within the same object. References are expressed as XLink simple links (DeRose, Maler, & Orchard, 2001). Making object relations explicit by means of XLink simple links allow for easy navigation through the object graph. Rarely requested parts of an eSciDoc object, like the event log, are not included in the standard representation, but can easily be retrieved by means of 'virtual resources'. Virtual resources again are represented as XLink simple links with an 'href' attribute, so access to this additional information is just a 'mouse click away'.

'Parts' of an object may be the object properties, one of the metadata records, or a component. Each part is retrievable via its XLink 'href' attribute and thus may be seen as sub-resource of the entire resource. If, for example, a user is interested in just one metadata record, there is no need to retrieve the representation of the entire object.

5. Authentication and Authorization

eSciDoc relies on distributed authentication systems like directory servers with an LDAP interface and Shibboleth (Scavo & Cantor, 2005). In distributed authentication systems like Shibboleth, users are maintained in the identity management system (or 'identity provider', IdP) of their home institution, where their credentials like usernames and passwords are kept. Consequently, eSciDoc includes no user management. However, in order to be able to associate users with roles, the infrastructure creates and maintains proxy objects for users that have accessed at least once an eSciDoc Solution. Each user proxy object consists of a unique name and a set of attributes. The attributes are initially populated during the first login and are updated with each subsequent login, based on the unique name. The attribute set depends on either the Shibboleth federation or the configuration of the directory server. The attributes are mainly used for personalization features.

eSciDoc replaces Fedora's built-in authorization mechanism, mainly because eSciDoc secures access to resources stored both within and outside of Fedora, and the need to handle object graphs as single entities. All authentication and authorization functionality is encapsulated in the eSciDoc Infrastructure, therefore the application programmer has not to consider the complexity of its implementation – this is especially important when several applications run on the same infrastructure, eventually sharing data. Every request to the infrastructure has to pass through the included authentication and authorization layer to access a method of the business logic. The eSciDoc authorization relies on five concepts: users (represented by the above mentioned user proxy objects), policies, roles, scopes, and groups:

Policies define access rights of users for resources, based on a set of rules. eSciDoc uses XACML to express these policies and rules. During the evaluation of policies, both standard XACML and eSciDoc-specific attributes are matched against conditions expressed in these rules. Additionally, policies may define target actions. In this case, the policy only is evaluated if the requested action matches one of the actions defined in the target of the policy. Typically, actions are directly mapped to method invocations of an eSciDoc API. If the action matches, rules are applied. Rules again may define target actions. In such a case, the rule will only get evaluated if the requested action matches one of the rule's target actions.

Roles define a set of rights, expressed in one or more XACML policies (Moses, 2004). Typically, administrators grant roles instead of explicit policies to users. Roles quite often describe a real-world responsibility like 'author', 'metadata editor', or 'collaborator'. Any non-authenticated, anonymous user gets granted with the 'default user' role, i.e. the role encompasses all actions that are allowed for everyone (e.g., retrieving Items in status 'released').

Scopes further constrain roles. They correspond to eSciDoc resources (e.g., Containers, Contexts, or Organizational Units) and are expressed by eSciDoc-specific resource attributes. An example might help to understand the concept: Ann works for the 'Foo' project. The system administrator created a new Context for her project and assigned her the role 'author'. But Ann shall only be allowed to create and modify objects within the Context of her own project. So if a second Context for Bob's project 'Bar' exists, the system administrator will grant Ann the generic role 'author' with the scope 'Context Foo'. Bob will have the same role, but with a different scope 'Context Bar'.

Groups are an additional concept to simplify the management of access rights. Groups may contain users or other groups. The member definition may be static (i.e. explicitly defined by the system administrator) or dynamic (i.e. based on user attributes). Groups can be associated with roles. If a user belongs to a group, he or she automatically inherits the roles of the group. If a user belongs to several groups at the same time, all associated group roles are inherited, complementing the roles that are directly associated with the user. Groups allow for the automatic assignment of roles (and thus policies) to users that have never logged in before. A common scenario is to give deposit access to the Context of an institute based on the attribute that contains the name of the institute (Organizational Unit).

With this powerful and fine-grained authentication and authorization mechanism, eSciDoc is well equipped to fulfill all kinds of access right requirements in common e-Research scenarios.

6. Persistent Identification

A persistent identifier identifies a resource independently from the storage system or location. There are many competing persistent identification systems, like PURL[5], Handle System[6], and DOI[7]. The general approach to make an identifier persistent is by separating the identifier from the locator and provide a mapping mechanism between both. If a resource is moved to a new location, only the mapping needs to be updated with new location, whereas all references stay stable. This, persistent identifiers provide ongoing access and reference to a resource and can be used whenever a permanent link is required. In science and humanities, stable references (e.g., for citations) are of great importance. With the advent of e-Research, not only publications, but all relevant objects should become citable (Klump, Bertelmann, Brase, Diepenbroek, Grobe, & Höck, 2006).

A persistent identifier typically consists of two parts: a prefix or namespace, which uniquely identifies the organization that is responsible for the persistently identified content, and a suffix, which uniquely identifies a resource within the scope of the organization. The process to create the suffix of PIDs is generally called minting. Different opinions exist on how much semantics a suffix should include (Kunze, 2003). Keeping identifiers semantic-free is a widely accepted approach (Sollins & Masinter, 1994). A minter should be configurable to address the varying requirements of different organizations.

Fedora comes with a default, database-based Identifier module, which only generates local identifiers. It has no built-in support for minting and assignment of PIDs. OhioLINK has extended Fedora with their HandlePIDGeneratorModule[8] for the Handle System. It can automatically assign a handle instead of a local identifier for each Fedora object, but constraints the length of the handle – similar to the default identifier – to a length of 64 characters.

eSciDoc persistent identifiers are not a replacement for the local identifier, but additional attributes of an object. A resource is retrievable either via its PID or its local identifier. Length and structure of a PID is not limited. Different PID generators (minters) are configurable, including simple serial numbers, checksums, and advanced rules based on the NOID minter[9].

eSciDoc differentiates between three kinds of PIDs: Object PID, Version PID, and Content PID. Each resource has an Object PID, which identifies the whole resource, including all of its versions. Using the Object PID will always retrieve the latest version of a resource. Each version of a resource can be identified with a Version PID, which will always retrieve the exact version of a resource. Additionally, all binary content of an object (e.g., PDF, image, etc.) is persistently identifiable by its Content PID.

The default configuration of eSciDoc's object lifecycle ensures that each resource is assigned with an Object and Version PID before a version is 'released' (i.e., made publicly available). The assignment of PIDs is not triggered automatically by a status change in the object lifecycle. Instead, it is the responsibility of each eSciDoc solution

---

[5] http://purl.org/
[6] http://www.handle.net/
[7] http://www.doi.org/
[8] http://drc-dev.ohiolink.edu/wiki/HandleGenerator
[9] http://www.cdlib.org/inside/diglib/ark/noid.pdf

to trigger the minting and assignment of PIDs. So it is up to the application designer when and how to assign PIDs, with respect to differing content models and file types. eSciDoc Solutions may even provide suffix values for the PID handler.

Minting and registering of PIDs in eSciDoc is managed by the PID Manager, one of the services of the eSciDoc Infrastructure. In its default configuration, the PID Manager currently supports only the Handle System. The PID Manager Service can be extended to support other systems as well (e.g. PURL). A concurrent use of several PID systems is possible, as well as including externally managed PIDs (i.e. PIDs that are assigned to an object before it is ingested into eSciDoc).

4 Application-Oriented Programming Model

The eSciDoc e-Research environment is built as a service-oriented architecture (SOA). The infrastructure consists of several independent services. Each service implements both a REST (Fielding, 2000) and a SOAP API. The APIs support simple CRUD (Create, Retrieve, Update, and Delete) and task-oriented methods. These APIs together with object patterns, their XML representations, versioning, and the powerful authentication and authorization form eSciDoc's application-oriented programming model, which focuses on the mindset of application developers and hides the technical details of the implementation as much as possible.

As already mentioned earlier, we see the ability to integrate existing tools and software packages with the eSciDoc Infrastructure as an important design goal. Another one is to account for various programming languages for solution development, including scripting languages for fast prototyping, mash-ups, and thin-client development. As a proof of concept, we have integrated the *Schema Driven Metadata Editor for eResearch* developed by ARCHER[10] and MAENAD[11] in order to allow comfortable editing of metadata records of eSciDoc objects. Another example is the integration of DigiLib, the versatile image viewing environment for the internet[12]. The most simplistic solution can be built based on XSLT transformations of object representations delivered by the eSciDoc Infrastructure. All necessary transformations are delegated to the browser of the user.

The API is representation-oriented, i.e. changing an object typically means retrieving the representation, changing it, and sending it back to the eSciDoc Infrastructure, thus following a *load-edit-save* paradigm. It is possible to just modify parts or single values in the XML representation of a resource and send it back for update. Some properties or parts of the representation are purely informational and therefore not updateable. These are ignored by the eSciDoc Infrastructure when a representation is sent back in order to store it. So it is actually possible to use a retrieved object representation as template for creating a new one. A developer has just to care for the essential properties.

In contrast to the representation-oriented approach, some actions like forwarding a resource in workflow are executed via task-oriented methods, which encapsulate these operations and strictly separate them from others.

---

[10] http://www.archer.edu.au/
[11] http://metadata.net/sfprojects/mde.html
[12] http://digilib.berlios.de/

The XML representations of eSciDoc objects and their exposure as web resources, together with the built-in methods of the HTT Protocol (GET, PUT, DELETE, etc.), form the CRUD-based web programming interface of the eSciDoc infrastructure. That can easily be used from existing applications or even websites. It fits well with AJAX web development techniques and supports easy integration of repository features into applications. Because of the built-in simple workflow, versioning, and authentication and authorization, already the simplest possible client solution supports basic repository features. The communication between client and infrastructure relies on well-known standards for data representation and distributed communication. Thus, the eSciDoc Infrastructure is a developer-friendly application framework.

5 Conclusion and Outlook

The eSciDoc Infrastructure encapsulates Fedora as its core component, but adds a wide range of higher-level services and its application-oriented programming model. It allows building various types of solutions, from light-weight Javascript hacks to fully-blown Java applications. It fulfills the vision of creating an efficient, flexible, programmer-friendly e-Research framework supporting web-based publication, collaboration and communication for research environments assembled with the repository capabilities of Fedora Digital Repository.

Acknowledgements

eSciDoc has been funded by the German Ministry for Education and Research (BMBF) from 2004-2009. Both Max Planck Society and FIZ Karlsruhe have substantially added additional resources to the project to meet the very ambitious project goals. We are very grateful for this financial support.

The design and implementation of the eSciDoc Infrastructure as well as the underlying concepts of eSciDoc have been conceived by a much larger team than represented by the authors of this paper, both at FIZ Karlsruhe and at the Max Planck Digital Library. We received a lot of very valuable input from researchers and librarians from various institutes of the Max Planck Society and other organizations. Finally, we would like to thank Fedora Commons for their great software and support.

References

DeRose, S., Maler, E., Orchard, D.: XML Linking Language (XLink) Version 1.0. W3C Recommendation (2001), http://www.w3.org/TR/2000/REC-xlink-20010627/

Dreyer, M., Bulatovic, N., Tschida, U., Razum, M.: eSciDoc - a Scholarly Information and Communication Platform for the Max Planck Society. In: German e-Science Conference, Baden-Baden (2007)

Fielding, R.T.: Architectural Styles and the Design of Network-based Software Architectures. Retrieved from University of California (2000),
http://www.ics.uci.edu/~fielding/pubs/dissertation/
fielding_dissertation.pdf

Hey, T., Trefethen, A.: Cyberinfrastructure for e-Science. Science 308(5723), 817–821 (2005)

Hey, T., Trefethen, A.: The Data Deluge: An e-Science Perspective. Grid Computing, 809–824 (May 30, 2003)

Klump, J., Bertelmann, R., Brase, J., Diepenbroek, M., Grobe, H., Höck, H.: Data Publication in the Open Access Initiative. Data Science Journal 5, 79–83 (2006)

Kunze, J.A.: Towards Electronic Persistence Using ARK Identifiers (July 2003),
http://www.cdlib.org/inside/diglib/ark/arkcdl.pdf

Lagoze, C., Payette, S., Shin, E., Wilper, C.: Fedora: an architecture for complex objects and their relationships. International Journal on Digital Libraries 6(2), 124–138 (2005)

Moses, T.: eXtensible Access Control Markup Language Version 2.0. OASIS Committee draft 04: access_control-xacml-2.0-core-spec-cd-04 (2004)

Powell, A., Lyon, L.: The DNER Technical Architecture: scoping the information environment. JISC Information Environment Architecture (May 18, 2001)

Scavo, T., Cantor, S.: Shibboleth Architecture – Technical Overview. Working Draft: draft-mace-shibboleth-tech-overview-02 (2005)

Sollins, K., Masinter, C.: Functional Requirements for Uniform Resource Names. Request for Comment 1737, IETF Network Working Group (1994)

Collaborative Ownership in Cross-Cultural Educational Digital Library Design

Pauline Ngimwa[1], Anne Adams[1], and Josh Underwood[2]

[1] Institute of Educational Technology (IET), Jenny Lee Building,
Open University, Milton Keynes, UK
{p.g.ngimwa,a.adams}@open.ac.uk
[2] London Knowledge Lab, Institute of Education
University of London UK
j.underwood@ioe.ac.uk

Abstract. This paper details research into building a Collaborative Educational Resource Design model by investigating two contrasting Kenyan / UK design case-studies and an evaluation of end-users and designers' perceptions of digital libraries and their usage patterns. The two case-studies compared are; case study 1 based on formal learning in an African university digital library. Case study 2 is centered on informal learning in an ongoing rural community digital library system which has a collaborative design model that is being designed, developed and reviewed within the UK and Africa. A small scale in-depth evaluation was done with 21 participants in case-study 1 but related to and with implications for the second case-study. In-depth user issues of access, ownership, control and collaboration are detailed and reviewed in relation to design implications. Adams & Blandford's 'information journey' framework is used to evaluate high-level design effects on usage patterns. Digital library design support roles and cultural issues are discussed further.

Keywords. Digital library design, Educational digital libraries, African Context of Use, Cross-cultural usability.

1 Introduction

Digital library end-users increasingly want control of what they use, how and when they use it as well as how it is designed. They are also becoming content creators because of democratization of content creation via the Internet [1]. Participatory design [2] and user centered design [3] approaches have for a long time supported designs of systems according to users needs. These approaches can also support the increased ownership felt by end-users in a system. However, applying these approaches can be complicated, time-consuming and expensive. Further complications can also ensue when resources, end-users and designers are separated by distance and culture. In a recent policy report for the 'American Library Association' [4], however, participatory design within digital libraries has been highlighted as an imperative. The report recommends that social networking and similar participatory tools must be tested and utilized at the core of the library.

M. Agosti et al. (Eds.): ECDL 2009, LNCS 5714, pp. 239–249, 2009.

Increasingly digital library design research has looked at the role of Web 2 applications [5, 6] in developing end-user control and ownership [7] to avoid digital libraries being 'passive warehouses' [8]. However, there needs to be an understanding of the underlying ownership and roles that end-users and information experts have in the design and use of these resources. In addition, only a small number of research findings have reviewed what these issues of end-user control, personalization and ownership mean within different cultures [9] and how this may change on-going support and management from information professionals [10]. There is therefore need for a model that can guide development of collaborative digital library designs that bring together end users, information experts and designers. Such a model should consider issues of ownership and roles of key players and their contexts. For educational digital libraries, such a model can provide support in merging digital library capabilities with the learning design needs.

The focus of the research reported here is on development of a collaborative educational resource model. We do this in reference to case-studies and an evaluation of two contrasting digital libraries for different Kenyan end-user communities. Both projects are at different stages in the development process thus providing some interesting insights into cultural similarities with differences in end-user engagement and control rationale. In order to understand end-users usage patterns over time in relation to the design of the digital libraries, we have used Adams & Blandford's [11] information journey framework. This framework identifies an information journey for end-users who interact with information temporally, traveling through a personal or a team-based information journey and using different resources through the stages of their information journeys.

2   Digital Library Case-Studies

We have used two contrasting case-studies with different types of end-users in formal and informal learning environments to research on the development of a Collaborative Resource Design (CERD) model. The first case study is the traditional digital library system common in most African universities. This usually takes up the form of a hybrid system of databases integrated into existing print resources. Normally this is a controlled system which follows laid-down organizational procedures i.e. recognized classification, cataloguing and metadata schemes. Collection usually consists of licensed resources. The second case-study is a community-based information system whose design is collaborative and co-owned with an element of academic control and co-ordination.

Although the two case studies may appear to contradict each other, they help us to present a case for a collaborative digital library design model by drawing on the comparison. In addition, the two systems represent four common elements applicable for a digital library design which help us to understand the similarities and differences between the two approaches. These elements are: the end-users who in both case studies are involved in some form of e-learning (formal and informal). The second element is the information resources. The third element is the digital library experts or facilitators. The last element is the context; both case studies share a common cultural context.

As will be seen in the following sections, using these contrasting case studies allowed us to identify differences between collaborative and non collaborative digital library designs and what this implies to the success of a digital library system.

2.1 Traditional Digital Library Design

This case study is a Kenyan university library whose electronic resources comprise of over 100 databases of licensed resources, organised and integrated into the library online catalogue which is hosted in the library website (figure 1). Users are the university academic community of around 37,000 students and 1,400 academics. Access is mainly through IP authentication which means that remote access is not supported; users must be within the university network system in order to access the resources. Unfortunately, technological challenges faced by the university such as expensive and inadequate bandwidth and limited computers affect effective usage of these resources.

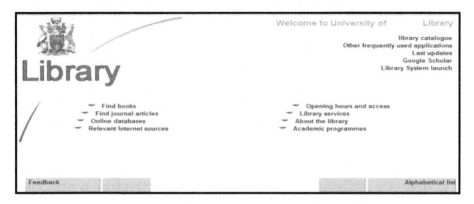

Fig. 1. University library website

Due to limited library budget, decision of what to acquire is usually made by the library. Usage of the collection is mainly polarized around academics, graduate students and final year undergraduate students. Publicity of new subscriptions is usually done through the university intranet or emails sent to the lecturers with the hope that they will alert their students. Users have limited participation in the interface design, organisation and personalisation of the collection.

2.2 Collaborative Co-ownership Digital Library Design

VeSeL (Village e-Science for Life) [12], is a research project funded by the UK's Engineering and Physical Sciences Research Council and focusing on designing technologies that are appropriate and sustainable for rural development. A group of UK based researchers partnering with the Kenyan University (case study 1) initially visited around 5 farming communities in two regions along with key players (e.g. agricultural extension workers, local middle men and private advisors). After a risk-benefit analysis, two communities were selected to work with. From the onset, the project set up to have a mutual understanding of each other's contexts and needs.

Thus the two communities were involved in the identification of their needs and technological solutions, and in the design process of the technology. The communities expressed a desire not just for acquisition and delivery of agricultural information but also for sharing this information with others. There was also the desire for local ownership of any technological solution developed.

In the following field trip the team employed ethnographic methods and interviews on the two communities' ways of living, how they interacted with existing technology and their reaction to new technologies as they were introduced to them. All the researchers involved were of African descent in order to boost cultural understanding in the whole process. Both field visits were conducted in partnership with the Kenyan University partner. The research team also explored a number of scenarios for activities with the communities, assisted by the University students. These included community blogging, water management, community mapping, agricultural and community podcasting, agricultural trails and group-based activities for community schools. However, some of these activities like the agricultural trails and community mapping were discarded almost immediately after finding them to be inappropriate in the field. Since then mobile resource kits have been deployed to communities. These support information access, capture and dissemination. The kit comprises of a Macbook laptop, solar charger and GPRS modem for internet access, digital cameras and audio recorders for capturing data. Initial training was provided around using email, blogging and posting photos and searching for information both online and locally in organic farming resources preloaded on the laptops. Follow-up training has been provided both face-to-face and remotely over Skype in response to breakdowns and requirements identified by users.

The Vesel design process was informed by guidelines derived from the concept of 'fluidity' described in De Laet and Mol's [13] analysis of the Zimbabwe Bush pump, an example of a particularly successful technology in a development context. Four principles are worth some emphasis for the purpose of this paper:

1. *Clear and Present Need*: the design process and technology to be designed should have clear value to the community, e.g. by addressing current needs.
2. *The place of the community*: Providing a role for the community in the design, on-going use and maintenance of the technology is vital to its sustainability.
3. *Ownership and access*: The community must feel that they own the technology and that it is freely accessible to them and adaptable by them.
4. *Distributed action*: Implementation of technology requires that methods and insights of the local community are paramount.

The design approach taken on by the team is one that trains end users to "take on design roles and self report their progress with the technology as participant ethnographer" [14]. For example farmers are encouraged to collect and post data from their farms as a simple blog posting, as depicted in initial sketches (figure 2). This user-generated information is later linked to an online Knowledge Management System (KMS), a kind of a Content Management System which has the basics of a digital library infrastructure.

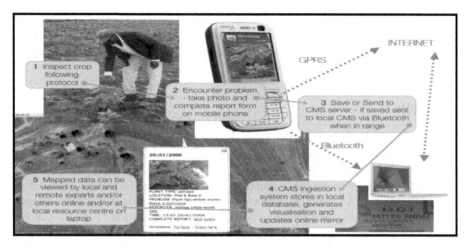

Fig. 2. Data flow and technology for resource management

3   Evaluation Research Method

A small scale in-depth qualitative study was done involving participants from Faculties of Agriculture and Computer Science in case study 1 university. 21 participants (13 students, four lecturers and four librarians) were purposefully sampled. These participants were also involved directly or indirectly in the VeSeL project described in case study 2 above. Participants' experience in the VeSeL project and use of the University library meant they had a clear concept of what was meant by digital resources within both these contexts. Their participation in the two case studies also meant that perceptions could be gathered regarding their different roles as end-users and content developers. For case-study 1 the in-depth qualitative analysis was supplemented by an ethnographic evaluation [15] with field notes on the University and its library. Relevant field documents were gathered to give a complete picture of participants' physical and temporal context and related needs. Further data were collected from VeSel project meetings. Much of this data was fed into the background case-study details whilst some related to key evaluation issues.

3.1   Data Collection and Analysis

Formal interviews were used to get data from the students in order to understand their perceptions of digital resources. These were supplemented with informal interviews with lecturers and librarians as a way of validating the students' responses. Because of their involvement in the VeSeL project, they provided data some of which related directly to case study two as will be seen in the result presentation section. Ethnographic field notes and VeSeL project meeting notes were integrated into the analysis. Collected data was analyzed thematically in line with the first stages of a grounded theory analysis with open coding completed and synthesis of all the data into common themes. This analytic approach was preferred because it allows themes to emerge from the data, thus uncovering previously unknown issues.

Some quantitative data was also collected (Table 2) and used to further triangulate and verify aspects of the in-depth analysis. However, as the concepts, design process and research questions were predominately perceptual a qualitative nature was kept to research methodology.

4   Results

The findings from perceptions of current information and digital library usage identified interesting design issues. In general, the study established that the traditional model of the digital library design and management service provided is one-way, non-collaborative and non-user centred compared to the collaborative design approach in the VeSeL project. An overview of the impact of the design on Adam's and Blandford's end-users information journey (discussed further in section 5) shows that for the traditional model (non-collaborative), the three stages were distinct. However, in the collaborative model the stages were inter-related and intertwined with no distinction in resources between the three stages. This can be seen more clearly from an analysis of the resources utilised in the information journey of the two design approaches (See Table 1).

An in-depth thematic analysis of the interviews, ethnographic field data and quantitative data collected revealed 3 key issues: access and ownership, control, and collaboration. These issues were clearly inter-related and often derived from the design approach of the digital library. A review of this interaction is presented in the discussion section.

Table 1. Comparison of resources needed between traditional and VeSeL digital libraries in the information journey

	DL end-users	Initiation	Facilitation	Interpretation
Traditional DL end-users	Students	Lecturers, Peers, Coursework & books	Lecturer, Peers, books, Web Resources, Digital Libraries	Lecturers, Peers, web-resources
	Librarians	Colleague, DL email alert & bulletin, Student queries,	Digital library, Book, web-resources	Colleagues
	Lecturers	Colleague Course development, Research	Book, Colleague, web-resources, Digital Libraries	Colleagues, web-resources
VESSEL end-users	community end-users	Family, Neighbours, Experts		
	resource developers	Colleagues, Content Developers, Community end-users.		
	content developers	Community end-users, Colleagues, books, web-resources		

4.1 Access and Ownership

One of the most important barriers to the traditional libraries usage related to access issues. These constraints were identified as overshadowing the users' needs. For instance, the use of an IP authentication method meant that users had no control over where they wanted to access the resources from; they were forced to be on the university premises in order to access the resources. One participant noted the following when she was asked where she accessed the resources from:

> "Mostly at the university, because the university has some certain membership. So you are able to get into some libraries free of charge. Anything I have tried like from my work place, am normally forced to pay something..." Computer Science Postgraduate student

The fact that the users were rarely involved in the acquisition of resources meant a clash between the end-user demand and what was eventually provided. Most students indicated that they were not able to access the resources they needed because they did not have the right authentication. The reason for this was that the university had not subscribed full-text although students could see the abstracts from the publishers' databases. Students were not aware that this was a subscription issue:

> "...we were given a print out of passwords of some resources but most of the times they were not useful to my topic... the few that were close to my topic I tried accessing them but I could not get to them." Agricultural student

This issue was growing in importance which was verified by the usage statistics gathered for popular journals not provided (See Table 4).

Table 2. 10 most popular journals denied access due to lack of subscription license *(Source: data obtained from Wiley Interscience Publishers, 2008)*

Month (2008)	No. of cases denied
January	55
February	109
March	113
April	77
May	66

Table 2 shows that there were a large number of popular journals that students wished to access and that the library had not subscribed to. However, the study further established that this mismatch between what the students require and what is provided by the library did not deter them from using other information and digital resources. Students looked for alternative sources such as their lecturers, free resources from the internet or visited other institutions that had better access to these resources. This seemed to suggest that students have owned or have the desire for ownership of the usage process. This contrasts the approach taken by the VeSeL project whereby the information resources are open access but the communities have ownership because they have participated in developing it.

4.2 Control Issues

The study also showed that those students and academics whose IT skills were superior tended to take charge of the acquisition and usage of the resources. These participants were found to have had limited engagement with the librarians who they perceived as not collaborating with them to provide a demand driven service:

> "What we have is supply driven... the demand has never spoken...We keep saying we have so many thousands of journals, how useful are they? That's why I subscribe to mine ..."Academic

In addition, these users were self-directed and explored a multitude of different approaches that would make their information usage richer. For instance, upon discovering that there was limited local information, two students decided to create a website (http://www.try-african-food.com/) that would host local content and use Web 2.0 tools to share it with the rest of the world. This resource was and is constructed, maintained and supported without the aid of the library.

> "It is the students who came up with this idea and said "why don't we build our own site?" Lecturer

This accentuates the end-users' desire for ownership, control and participation in development of digital resources.

4.3 Collaboration Issues

The initiative to design the Try-African-Food website with social networking tools highlights issues of collaboration in the design process. Users in the traditional model desired a stake in the design and usage of the digital library and because this was not available, they looked for a way out by taking charge of the process such as designing the Try-African-Food website. The presence of a blog in the website advances the end-users desire for a more collaborative and engaging process in information resources.

In contrast to the traditional digital library model, the Vesel project was identified as a more collaborative model. As was conceived at the design stage, every stake holder participated in the design of the system. Community needs were identified at the start of the project. Although these initial needs were not always captured or related accurately, an ongoing iterative approach to requirements gathering means an organic nature to the design which is continually being developed. They were involved in the design of the technology which has been identified as appropriate for solving their problems. Online spaces i.e. community and school blogs and websites were developed where the communities and school children network and exchange knowledge. This was seen as leading to the creation of an informal e-learning environment, where the communities can access knowledge that will improve their livelihood and at the same time participate in knowledge exchange with other interested parties. This entire process is highly collaborative and user-focused. However, what this evaluation has also identified is that it provides an increased sense of ownership amongst all the parties involved.

5 Discussion

Brewer [8] argues that digital libraries must be pro-active and dynamic in their support of users' changing information needs so as not to become 'passive warehouses' of navigable information. However, for digital libraries to effectively support end-users, there is a need to understand them within their context. Looking at the findings from this study in relation to the previous findings for the users' information journey helps us to understand some interesting issues around users changing needs for information control and changing roles. Adams & Blandford [11] identified a users 'information journey'; from the initiation of information requirements, through the facilitation of information to the user and finally to the interpretation and application of that information. It is interesting to see from the two case-studies that there are two completely different approaches to user needs in these different stages of the information journey. It is also helpful to use this framework as applied to an evaluation of the design of these digital libraries.

The traditional digital library model (case study 1) highlights the end-user utilizing the library with fixed needs that may develop slightly through their searching and browsing activities. This research identified problems here around issues of access and authentication. This was noted as leading to poor perceptions of control and ownership of the digital library. The collaborative co-ownership model (case study 2), however, highlights not only changing information requirements but a deeper level of control on how to formulate the information requirements (in the information initiation stage) through data collection in the field (e.g. sensors, mobile devices) and the user controlled format of the information questions (e.g. a photo of a problem with a crop).

Traditional libraries can be great at facilitating information but they provide poor support on interpreting the information received. Often the users are not supported in understanding the information given and just left to swim in it. This study established that there is limited interaction between students and librarians. Students frequently turned to their lecturers or peers to try and understand the information acquired. Lankes et al [4] argue for participatory librarianship where the librarian is at the centre of all information process roles.

However, fundamentally the librarian has still remained the facilitator of the information with no interpretation role. The support role must match the end-user's changing needs, an important consideration for the development of the CERD model.

Collaborative design (case study 2) merges facilitation and interpretation and closely relates it to information initiation through an iterative process. For example the farmer initiates an information need which could be a problem about her crop. She checks with her neighbor or an expert for facilitation of the information. At the interpretation stage, the farmers and the experts are learning from each other. The experts are checking with the farmers to see if they are meeting their information needs and designing appropriate technologies for meeting these needs. Farmers are feeding back to the process and also sharing their new experiences through use of multimedia technologies and resources. Farmers are also learning new information facilitation techniques by being introduced to social networking tools such as blogs. The experts and the communities (end users) are engaged with each other at all the stages in a collaborative process. What this study has identified is the empowering way that this in turn creates ownership of the process, the resources and the technologies. This is a vital contribution for the development of the CERD model. The three key players in the design process must collaborate through an iterative process along all the stages of the user's information journey.

Finally, it is interesting to note that case study 2 included designers, content developers and community end-users across great distances and from different cultures. Initially there were problems with communication between these parties, but as these continued to be highlighted and dealt with they have diminished. End-users ownership has been surprisingly high regardless of these issues. Case study 1 is co-located with the end-users from a similar culture. However, the students & academics noted several poor design issues (e.g. problems with authentication) along with a lack of ownership and control. There are some clear cultural implications of these two design approaches which will need further investigation for the purpose of the development of the CERD model.

6 Conclusion

What does this new wave of digital libraries that are collaboratively designed and end-user controlled mean to the future of the information professionals, their practice and training? What are the implications for the designs of digital libraries? How does this relate to the end-users, particularly students who are self directed in their learning and are already benefitting from Web 2.0 social networking tools, and demanding a stake in the design process? Ultimately, how does this feed in to the development of the CERD model? First of all, information professionals have to step out of the traditional practice of meeting end-users' needs within the confines of current library practice. Both information professionals and educators need to refocus their services by working more collaboratively in order to make clear connections between digital resources and learners' needs as emphasized by Littlejohn et al. [16]. As seen in the study, this can best be achieved by bringing all the three players on board in the design process. A model that makes this possible and one that integrates the collaborative capabilities of the emerging Web 2.0 technologies is the recommendation of our research. Furthermore, such a model may also help to reduce cultural barriers in the usage of educational digital resources by making them more cultural specific. However, further research will be required to review this further.

Acknowledgments. This work is funded by a CREET scholarship and the VeSeL project is funded by an EPSRC grant. We are grateful for the help of the Kenyan university and community along with Digital Library designers and managers for each case-study.

References

1. Giersch, S., Leary, H., Palmer, B., Recker, M.: Developing a review process for online resources. In: Proceedings of the JCDL, Pittsburgh, PA, p. 457 (2008)
2. Muller, M.J.: Participatory Design: The Third Space in HCI. In: The human-computer interaction handbook: fundamentals, evolving technologies and emerging applications, pp. 1051–1068 (2002)

3. Vredenburg, K., Mao, J., Smith, P.W., Carey, T.: A survey of user-centered design practice. In: Proceedings of SIGCHI conference, pp. 471–478. ACM Press, New York (2002)
4. Lankes, D.R., Silverstein, J., Nicholson, S.: Participatory Networks: the library as conversation, http://iis.syr.edu/projects/PNOpen/ParticiaptoryNetworks.pdf
5. Lund, B., Hammond, T., Flack, M., Hannay, T.: Social Bookmarking Tools (II): A Case Study – Connotea. D-Lib Magazine 11(4) (2005)
6. Hunter, J., Khan, I., Gerber, A.: HarvANA-Harvesting Community Tags to Enrich Collection Metadata. In: Proceedings of the JCDL, Pittsburgh, PA, pp. 147–156 (2008)
7. Davis, L., Dawe, M.: Collaborative design with use case scenarios. In: Proceedings of the 1st JCDL, Roanoke, VA, USA, June 24-28, pp. 146–147 (2001)
8. Brewer, A., Ding, W., Hahn, K., Komlodi, A.: The role of intermediary services in emerging digital libraries. In: Proceedings of DL 1996, pp. 29–35. ACM Press, Bethusda (1996)
9. Tibenderana, P., Ogao, P.: Acceptance and use of electronic library services in Ugandan universities. In: Proceedings of the JCDL, Pittsburgh, PA, pp. 323–332 (2008)
10. Heinrichs, J.H., Lim, K., Lim, J., Spangenberg, M.A.: Determining Factors of Academic Library Web Site Usage. Journal of the American Society for Information Science and Technology 58(14), 2325–2334 (2007)
11. Adams, A., Blandford, A.: Digital libraries' support for the user's 'information journey'. In: Proceedings of the 5th JCDL, Denver, CO, USA, pp. 160–169 (2005)
12. VeSeL project (2009), http://www.veselproject.net/
13. De Laet, M., Mol, A.: The Zimbabwe Bush Pump: Mechanics of a Fluid Technology. Social Studies of Science 30(2), 225–263 (2000)
14. Walker, K., Underwood, J., Waema, T.M., Dunckley, L., Abdelnour-Nocera, J., Luckin, R., Oyugi, C., Camara, S.: A resource kit for participatory socio-technical design in rural Kenya. In: Proceedings of the SIGCHI conference on Human factors in computing systems, Florence, IT (2008)
15. Hughes, J., King, V., Rodden, T., Andersen, H.: Moving out from the control room: ethnography in system design. In: Proceedings of the CSCW Chapel Hill, pp. 429–439. ACM Press, New York (1994)
16. Littlejohn, A., Cook, J., Campbell, L., Sclater, N., Currier, S., Davis, H.: Managing educational resources. In: Conole, G., Oliver, M. (eds.) Contemporary perspectives in e-learning research, pp. 134–146. Routledge, London (2006)

A Hybrid Distributed Architecture for Indexing

Ndapandula Nakashole[1,*] and Hussein Suleman[2]

[1] Max-Planck Institute for Computer Science
Saarbruecken, Germany
nnakasho@mpi-inf.mpg.de
[2] Department of Computer Science, University of Cape Town
Private Bag, Rondebosch, 7701, South Africa
hussein@cs.uct.ac.za

Abstract. This paper presents a hybrid scavenger grid as an underlying hardware architecture for search services within digital libraries. The hybrid scavenger grid consists of both dedicated servers and dynamic resources in the form of idle workstations to handle medium- to large-scale search engine workloads. The dedicated resources are expected to have reliable and predictable behaviour. The dynamic resources are used opportunistically without any guarantees of availability. Test results confirmed that indexing performance is directly related to the size of the hybrid grid and intranet networking does not play a major role. A system-efficiency and cost-effectiveness comparison of a grid and a multiprocessor machine showed that for workloads of modest to large sizes, the grid architecture delivers better throughput per unit cost than the multiprocessor, at a system efficiency that is comparable to that of the multiprocessor.

1 Introduction

Distributed architectures are de facto data scalability platforms as evidenced by the scale of data handled by service providers on the Web such as those that provide search and storage services. With ever-expanding digital library collections, scalable services are needed to provide efficient access to data.

In recent years, Web search engines have enabled users on the Web to efficiently search for documents of interest. Results are returned in a few seconds, with potentially relevant documents ranked ahead of irrelevant ones. These technology companies compete with one another to provide high quality search services requiring complex algorithms and vast computer resources, at no direct financial cost to the user. However Web search engine spiders often do not completely index data stored in digital libraries so search has to be provided as part of the digital library software suite.

Many large-scale Web service providers — such as Amazon, AOL, Google, Hotmail and Yahoo! — use large data centres, consisting of thousands of commodity computers to deal with their computational needs. In 2003, Google's search engine architecture had more than 15,000 commodity class PCs with fault-tolerant software [3]. The key advantage of this architectural approach is its ability to scale to large data collections and millions of user requests. For example, Web search engines respond to millions

---

* Work done while a student at the University of Cape Town.

M. Agosti et al. (Eds.): ECDL 2009, LNCS 5714, pp. 250–260, 2009.

of queries per day at a low latency. Clusters of commodity computers are known for their better cost/performance ratio in comparison to high-end supercomputers. However, there is still a high cost involved in operating large data centres. Such data centres require investment in a large number of dedicated commodity PCs. In addition, they need adequate floor space, cooling and electrical supplies. IDC[1] reported that in 2007 businesses spent approximately $1.3 billion to cool and power spinning disk drives in corporate data centres and this spending is forecasted to reach $2 billion in 2009 [6].

For an organisation hosting a digital library, whose primary focus is not information retrieval, it may be difficult to justify expenditure on a data centre. In addition, if the computers will not be used for other tasks, they may not be highly utilised at all times. Furthermore, it is not clear how much more data centres can be scaled up at a reasonable cost if both data and workload continue to grow in the coming years. Baeza-Yates et al.[5] estimate that, given the current amount of Web data and the rate at which the Web is growing, Web search engines will need 1 million computers in 2010. It is therefore important to consider other approaches that can cope with current and future growth in data collections and be able to do so in a cost-effective manner.

This paper proposes an alternative architecture — a hybrid scavenger grid consisting of both dedicated servers and dynamic resources in the form of idle workstations to handle medium- to large-scale search engine workloads. The dedicated resources are expected to have reliable and predictable behaviour. The dynamic resources are used opportunistically without any guarantees of availability. These dynamic resources are a result of unused capacity of computers, networks and storage within organisations, exploiting work patterns of the people within an organisation. Dedicated nodes are needed to provide search services that are reliable and have a predictable uptime. Due to the limited number of dedicated nodes, they cannot provide the scalability required to handle indexing of large data collections. Thus the dedicated nodes should be augmented with the dynamic nodes that become available during non-working hours. From the dedicated nodes, the architecture gets reliability; from the dynamic nodes it gets scalability.

The rest of the paper is organised as follows: Section 2 discusses related work; Section 3 discusses the design and implementation details of the search engine; Sections 4 to 6 present evaluation details; and Section 7 provides concluding remarks.

2   Related Work

Scavenger Grids. A "scavenger" or "cycle-scavenger" grid is a distributed computing environment made up of under-utilised computing resources in the form of desktop workstations, and in some cases even servers, that are present in most organisations. Cycle-scavenging provides a framework for exploiting these under-utilised resources, and in so doing providing the possibility of substantially increasing the efficiency of resource usage within an organisation. Global computing projects such as FightAIDS@Home [10] and SETI@Home [21] have already shown the potential of cycle-scavenging. The idea

---

[1] International Data Corporation (IDC) is a market research and analysis firm specialising in information technology, telecommunications and consumer technology. http://www.idc.com

of a hybrid scavenger grid is an extension of cycle-scavenging — it adds the notion of dedicated and dynamic resources.

Hybrid Grid Architectures. Although there has been work done on information retrieval for cluster(for example Hadoop [14]), grid (for example GridLucene [17]) and peer-to-peer(for example Minerva [18]) architectures, there has been virtually no published work that proposes the use of a hybrid scavenger grid for information retrieval. However a few works have investigated the use of hybrid scavenger grid architectures for other applications. A recent doctoral dissertation [1] investigated the use of a combination of dedicated and public resources in service hosting platforms. It observed that by designing appropriate resource management policies, the two types of resources can combined be utilised to increase the overall resource utilisation and throughput of the system. The BitTorrent[20] peer-to-peer video streaming platform relies on unused uplink bandwidth of end-user computers. Das et al.[7] have proposed the use of dedicated streaming servers along with BitTorrent, to provide streaming services with commercially valuable quality assurances while maintaining the scalability of the BitTorrent platform.

Enterprise Search Toolkits. A number of commercial systems dedicated to organisational search have been developed. FAST [9], OmniFind [19] and other traditional enterprise search engines are software toolkits. These toolkits do not mention the hardware infrastructure required to handle large scale intranet search. It is up to the organisation to determine the hardware infrastructure with the storage and computational power to deliver the desired scalability. The Google Search Appliance and the Google Mini Search Appliance [13] are hardware devices that provide intranet search able to handle up to millions of pages. These devices have initial acquisition costs as opposed to using resources already at the organisation's disposal in conjunction with a small number of dedicated resources.

3   Design and Architecture

The focus of this work is not on developing new information retrieval algorithms but rather on a different distributed architecture. Therefore, the developed prototype uses the Lucene [16] open source search engine as the underlying information retrieval engine. In addition, it uses the Condor job scheduler [15] for job submission and tracking. For distributed data management, the system employs the Storage Resource Broker (SRB)[4] data grid middleware which provides a layer of abstraction over data stored in various distributed storage resources, allowing uniform access to distributed data.

The architecture of the experimental search engine has five main components (see Fig. 1). The User Interface provides an access point through which queries enter the system. The Scheduler performs job allocation and uses Condor to distribute jobs to the dedicated and dynamic nodes which run the Worker Agents. Worker Agents refer to the software that executes on the nodes of the grid. The characteristics of the nodes dictate the types of jobs they can perform. The dedicated nodes are dedicated to search engine operations and thus as long as they are up and running they are available to execute jobs and provide services that are required for the search engine to operate. For

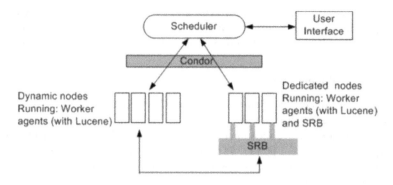

Fig. 1. High level components of the experimental search engine architecture. These are: User Interface, Scheduler, Condor, Worker Agents and SRB.

this reason, the dedicated nodes are responsible for providing persistent storage for the indices via SRB and also for responding to queries. Because availability of the dynamic nodes cannot be guaranteed they only perform text indexing.

The Scheduler has the task of splitting the data collection into chunks and ingesting the chunks into SRB. The Scheduler also starts the Worker Agent software on the dedicated and dynamic nodes. It does this by first contacting Condor to get a list of the available nodes. The Scheduler then creates Condor jobs that instruct the nodes to run the Worker Agent software. Worker Agents request data to index. Upon receiving a request for a data chunk, the Scheduler allocates a new chunk to the requesting machine. The Scheduler specifies the data chunk allocated to the machine by indicating the location of the chunk on SRB. The Worker Agents run a Lucene indexer on the chunk and ingest the resulting sub-index on SRB. Once all the chunks are indexed, all the sub-indices located on a single SRB server are merged into a single index.

When a query is posed to the system via the User Interface, it is passed on to the Scheduler which routes the query to all the SRB storage servers that store indices. The SRB servers independently search their indices and return their results to the Scheduler. Finally, the Scheduler merges the results before returning them to the user. The next sections present the results of system performance evaluation.

4 Experimental Setup

4.1 Hardware

Four computing systems were used for the experiments. The first is a set of machines called the **Dynamic nodes** — they are 66 desktop machines within a 100 Mbps Ethernet network. Each machine is equipped with a 3 GHz Pentium 4 processor, 512 MB of RAM and a 40 GB hard disk. The second is a set of machines called **Dedicated nodes** which are 13 desktop class computers interconnected by a Gigabit Ethernet network. Each machine is equipped with a 3 GHz Pentium 4 processor, 512 MB of RAM and an 80 GB hard disk. The third is a desktop class computer with a 2.33 GHz Intel Core 2

Duo processor, 2 GB of RAM and a 250 GB SATA hard disk — this is the **Scheduler**. The fourth system is a **multi-core machine** (server) with a 3GHz Intel Quad-Core Xeon processor, 8 GB of RAM and a 500 GB SATA hard disk.

4.2   Data set and Query Set

The system was evaluated on a data collection crawled from the .ac.uk domain, which is the domain of academic institutions in the United Kingdom. The collection is 70.27 G of 825,547 documents. The collection has various file types(PDF 87,189; DOC 21,779; TXT 2,569; RTF 2,042 and HMTL 711,968). In order to test for data scalability, the collection was duplicated in cases where the data collection needed for the experiments is larger than the actual size of the collection. Query performance experiments did use duplicated data to simulate larger collections. The reason behind this is that duplicating the data collection only changes the index in one dimension. This can affect querying performance. It does not however affect indexing performance since in distributed indexing the data is indexed in small jobs and there are no duplicates within each indexing job. Each partial index is independent of subsequent partial indices and thus the index building process is not affected by data duplication.

Typical query logs from the domains crawled were not available. Instead, test queries used are top queries of the Web search volume made accessible via the Google Insights for Search service [12]. Google Insights for Search provides the most popular queries across specific regions, categories and time frames. The categories chosen are those that are typically covered by academic institutions, namely: Science, Sports, Telecommunications, Society, Health, Arts and Humanities. All the queries within the query set return a non-empty result set on the index of the AC.UK collection. The total number of queries in the set is 1008, with an average number of terms per query of 1.5 and the longest query contains 3 terms.

The first set of experiments, as shown in the next section, focused on how the dynamic nodes of the grid can be best organised and utilised to deliver the best indexing performance.

5   Varying Dynamic Indexers

Within the hybrid scavenger grid, the number of dynamic indexers plays a major role in indexing time. Ideally, as the number of dynamic indexers increases, indexing time decreases linearly. This experiment aimed to find the number of dynamic indexers that delivers the best performance. Best in this sense means that indexing time is reduced and also that the indexers involved are well utilised.

Distributed indexing can be organised in one of two ways. With Local Data distributed indexing, machines index data that is stored on their local disks and transfer the partial index to one of the index SRB storage servers. With Non-local Data distributed indexing, machines download source data that is stored on the storage servers on SRB and also store the resulting indices on SRB storage servers. Intuitively, the Local Data indexing approach achieves superior performance because of data locality and thus it incurs less network transfer time. The Local Data indexing approach was used in the experiments reported here.

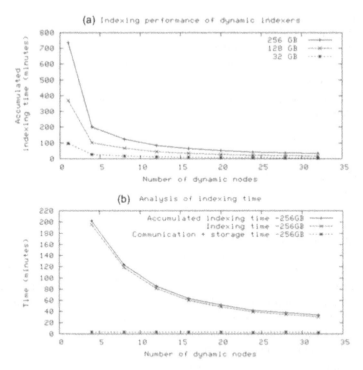

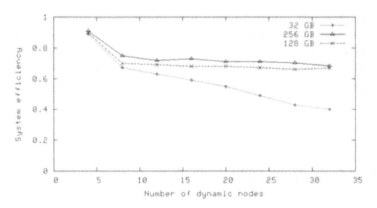

Fig. 2. Indexing performance for dynamic indexers, with 6 SRB storage servers

Fig. 3. System efficiency during indexing

Indexing performance for various data sizes was analysed. The accumulated indexing time is the total time spent on all tasks performed to index a collection of a given size, including job scheduling and communication. Indexing time is the time spent on actual indexing of data as opposed to other tasks such as file transfer or job scheduling. Communication and storage time is the time to transfer indices and to ingest the indices into SRB.

From Fig. 2 it is clear that as the number of dynamic nodes increases, indexing time decreases and that a large part of indexing time is spent on actual indexing. Communication and storage time for transmitting and storing indices on storage servers remains more or less the same even as the number of dynamic indexers increases. What has not been shown is how resource utilisation is affected as more dynamic nodes are added to the grid. Fig. 3 shows the grid system efficiency for varying numbers of dynamic nodes. System efficiency measures utilisation of resources — how busy the resources are kept.

Parallel system performance of a system consisting of n processors is often characterised using speedup — the serial execution time divided by parallel execution time: $speedup(n) = time(1)/time(n)$.

System efficiency is defined as the speedup divided by the number of processors:

$$system\ efficiency(n) = speedup(n)/n$$

Thus efficiency is maximised at 1.0 when n=1. From Fig. 3, it can be seen that for the case of 32 GB, when more than 24 nodes are used, system efficiency goes down to below 50% of full efficiency. Therefore, at the 24 node point adding more nodes decreases indexing time but utilisation per machine decreases to levels where each machine does little work, with for example each machine doing under 60 seconds of indexing. For the 128 GB and 256 GB cases, system efficiency also declines with increasing numbers of dynamic nodes. However, due to the increased workload the system remains relatively efficient, reaching a minimum of 67% and 68% efficiency respectively.

This experiment has shown that for a given workload, the number of dynamic nodes can be increased to index the collection in the shortest possible time. However, adding more nodes to the grid in order to achieve the shortest indexing time can result is poor utilisation of resources with system efficiency falling to levels below 50%. Therefore, the optimal number of dynamic nodes is the one that results in lowest indexing time at a system efficiency above a defined threshold.

The experiment reported thus far has shown indexing performance of the hybrid scavenger grid. The question to ask at this stage is how performance of the hybrid scavenger grid comparecompares to other architectures and whether it is worth investing in dedicated nodes and maintaining a grid, if the cost is the same as that of a middle or high end multi-processor server which has comparable performance.

6 Hybrid Scavenger Grid versus Multi-core

While cost-effective scalability is one of the advantages of a hybrid scavenger grid-based search engine, the challenge is the process of designing, implementing and running a search engine on such a distributed system. Limitations of the grid such as the unpredictable nature of dynamic nodes and job failure rate can hinder performance. Thus it can be argued that with the advent of multi-core technology, powerful servers can be purchased for low prices and thus the centralised architecture should be the architecture for workloads of certain magnitudes.

Experiments were carried out to compare the cost-effectiveness and system efficiency of the quad-core machine to that of the hybrid scavenger grid.

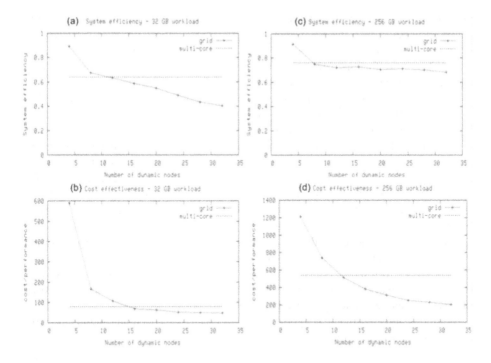

Fig. 4. System efficiency and cost-effectiveness: 32 GB and 256 GB

To determine the cost-effectiveness of a system with n processors which cost $cost(n)$, performance and cost are combined to obtain cost/performance [22]:

$$costperf(n) = \frac{cost(n)}{1/time(n)}$$

A system is more cost-effective than the other when its costperf value is smaller than the other system's. The cost of a system depends on one's point of view. It can be hardware cost for processors, memory, I/O or power supplies. For the purpose of this experiment, the cost only includes processor cost. The prices used are list prices in US dollars (as of 7 December, 2008)[11]. The processor (Intel Quad-Core Xeon X5472/3 GHz) in the quad-core machine costs $1,022 and a typical desktop processor (Intel Core 2 Duo E8400/3 GHz)[2] costs $163.

Fig. 4 (a) and (b) show system efficiency and cost-effectiveness of both systems, for the workload of 32 GB. The system efficiency of the multi-core is constant at 0.64 since the number of cores are fixed, whereas that of the grid varies with the number of dynamic nodes. It can be seen that for more than 12 dynamic nodes, the efficiency of the grid is lower than that of the multi-core, and continues to decline as more nodes

---

[2] The experiments used Pentium 4 machines, however these are no longer listed in the price list from Intel — currently new desktop computers typically have an Intel Core 2 Duo processor and thus the price of a Core 2 Duo was used.

are added. It can also be seen that the cost-effectiveness (Fig. 4 (b)) of the grid is only significantly better than the multi-core when 24 or more nodes are used. However, at this point the efficiency (Fig. 4 (a)) of the grid is 0.49 whereas that of the multi-core is 0.64. Therefore for this particular workload it can be concluded that multi-core is a better choice since employing the grid leads to poorly utilised resources.

Fig. 4 (c) and (d) show system efficiency and cost-effectiveness of both systems, for the workload of 256 GB. The system efficiency of the multi-core is 0.76. The efficiency of the grid is lower than that of the multi-core when more than 8 dynamic nodes are used — it remains relatively high and reaches a minimum of 0.68 for 32 dynamic nodes. It can be seen in Fig. 4 that the grid performs better and is more cost-effective when 12 or more dynamic nodes are used. At that point the grid has a system efficiency of 0.72 which is 4% less than that of the multi-core. For this workload, it can be concluded that the grid is more cost-effective and at the same time utilisation of the grid resources is relatively high.

This experiment has shown that for small workloads, although the grid provides better performance and cost-effectiveness for large numbers of dynamic nodes, the system efficiency goes to low levels that render the usefulness of the grid questionable. For modest to large workloads, the grid is a more beneficial approach achieving better cost-effectiveness and maintaining relatively high system utilisation.

Having established that the hybrid scavenger grid is a beneficial architecture for search engine indexing, it important to also evaluate its performance for searching.

7   Querying Performance Analysis

In a scalable search engine, query response time should remain more or less constant even as the size of the searched index increases. Moreover, the index distribution among the index storage servers should enable query response times to be more or less the same for different queries — the time to respond to individual queries should not be substantially longer for some queries while it is shorter for others.

From Fig. 5(a) it can be seen that the average query response time remains fairly constant even as the data size is increased. Query throughput is determined by the performance of the query servers and also by the arrival rate of queries at the scheduler [2] . The attained average throughput is 10.27 queries per second. This means that the 6 storage servers used can process up to 36,972 queries per hour or roughly close to a million queries per day. With the average throughput of 10.27, the average time per query is 0.10 seconds. The query throughput attained is comparable to that of other distributed systems. Badue et al [2] reported that with a cluster of 8 machines, they observed a query throughput of 12 queries per second, with an average time per query of 0.12 seconds.

Fig. 5(b) shows that response times for all the queries is below one second, with an average query response time of 0.13 seconds, a minimum of 0.034 seconds and maximum of 0.39 seconds. This experiment has shown that the average query response time remains fairly constant, that query response times are below one second and that the variance in query response times is not substantial.

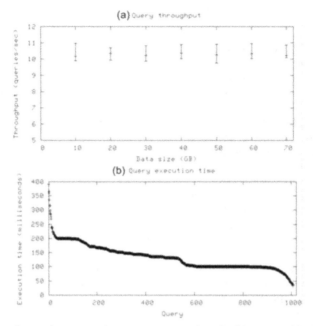

Fig. 5. Querying performance. The query response times in (b) are sorted in descending.

8 Conclusions

The hybrid scavenger grid proves to be a feasible architecture for a search engine that supports medium- to large-scale data collections within an intranet. The system reduces indexing time and responds to queries within sub-seconds. The resources of the system can be organised in a way that delivers the best performance by using the right number of nodes. The desired levels of performance and system efficiency determine the optimal number of dynamic/static nodes to index a collection.

The scalability of the architecture comes from the fact that more dynamic nodes can be added as required. Data scalability is vital as collections within digital libraries continue to grow at fast rates. For example, digital libraries of scholarly publications have become increasingly large as academic and research institutions adopt open access institutional repositories, using tools such as EPrints[8], in order to maximise their research impact. As an institution grows, so does the amount of the data it produces but also human resources increase. Assuming the normal practice of one computer per person, there will always be enough dynamic nodes to deal with the increasing data within an institution.

One possible future work direction is to evaluate the system's ease of maintenance. Maintaining a distributed system requires significant human effort. Large data collections of several terabytes of data require a large grid consisting of large numbers of dynamic nodes. As the size of the grid grows, the effort required to operate and maintain the grid also becomes greater. Therefore, it would also be of interest to know the human cost of a hybrid scavenger grid operation in comparison with the other architectures while taking into account performance, hardware cost-effectiveness and resource efficiency.

References

1. Asaduzzaman, S.: Managing Opportunistic and Dedicated Resources in a Bi-modal Service Deployment Architecture. PhD thesis. McGill University (2007)
2. Badue, C., Golgher, P., Barbosa, R., Ribeiro-Neto, B., Ziviani, N.: Distributed processing of conjunctive queries. In: Heterogeneous and Distributed IR workshop at the 28th ACM SIGIR Salvador,Brazil (2005)
3. Barroso, L.A., Dean, J., Hölzle, U.: Web search for a planet: The Google Cluster Architecture. IEEE Micro. 23(2), 22–28 (2003)
4. Baru, C.K., Moore, R.W., Rajasekar, A., Wan, M.: The SDSC storage resource broker. In: Proceedings of the 1998 conference of the Centre for Advanced Studies on Collaborative Research,Toronto, Canada (1998)
5. Baeza-Yates, R., Castillo, C., Junqueira, F., Plachouras, V., Silvestri, F.: Challenges on distributed web retrieval. In: ICDE, Istanbul, Turkey, pp. 6–20. IEEE, Los Alamitos (2007)
6. Computerworld Inc. Storage power costs to approach $2B this year (2009),
   http://www.computerworld.com
7. Das, S., Tewari, S., Kleinrock, L.: The case for servers in a peer-to-peer world. In: Proceedings of IEEE International Conference on Communications, Istanbul, Turkey (2006)
8. EPrints. Open access and institutional repositories with EPrints (2009),
   http://www.eprints.org/
9. FAST. FAST enterprise search (2008),
   http://www.fastsearch.com
10. FightAIDS@Home. Fight AIDS at Home (2008),
    http://fightaidsathome.scripps.edu/
11. Intel Cooporation. Intel processor pricing (2009),
    http://www.intc.com/priceList.cfm
12. Google. The Google Insights for Search (2008),
    http://www.google.com/insights/search/
13. Google. The Google search appliance (2008),
    http://www.google.com/enterprise/index.html
14. Hadoop. Apache Hadoop (2008),
    http://hadoop.apache.org/
15. Litzkow, M., Livny, M.: Experience with the condor distributed batch system. In: Proceedings of the IEEE Workshop on Experimental Distributed Systems (1990)
16. Lucene. Lucence search engine (2008), http://lucene.apache.org/
17. Meij, E., Rijke, M.: Deploying Lucene on the grid. In: Open Source Information Retrieval Workshop at the 29th ACM Conference on Research and Development on Information Retrieval, Seattle, Washington (2006)
18. Michel, S., Triantafillou, P., Weikum, G.: MINERVA: a scalable efficient peer-to-peer search engine. In: Proceedings of the ACM/IFIP/USENIX 2005 International Conference on Middleware. Grenoble, Greece (2005)
19. OmniFind. OmniFind search engine (2008), http://www-306.ibm.com/software/data/enterprise-search/omnifind-yahoo
20. Pouwelse, J.A., Garbacki, P., Epema, D.H.J., Sips, H.J.: The bittorrent p2p file-sharing system: Measurements and analysis. In: Castro, M., van Renesse, R. (eds.) IPTPS 2005. LNCS, vol. 3640, pp. 205–216. Springer, Heidelberg (2005)
21. SETI@Home. Search for extraterrestrial intelligence at home (2007),
    http://setiathome.berkeley.edu/
22. Wood, D.A., Hill, M.D.: Cost-effective parallel computing. IEEE Computer 28, 69–72 (1995)

A Concept for Using Combined Multimodal Queries in Digital Music Libraries*

David Damm[1], Frank Kurth[2], Christian Fremerey[1], and Michael Clausen[1]

[1] University of Bonn, Department of Computer Science III
Römerstraße 164, 53117 Bonn, Germany
{damm,fremerey,clausen}@iai.uni-bonn.de
[2] Research Establishment for Applied Science (FGAN), FKIE-KOM
Neuenahrer Straße 20, 53343 Wachtberg, Germany
kurth@fgan.de

Abstract. In this paper, we propose a concept for using combined multimodal queries in the context of digital music libraries. Whereas usual mechanisms for content-based music retrieval only consider a single query mode, such as query-by-humming, full-text lyrics-search or query-by-example using short audio snippets, our proposed concept allows to combine those different modalities into one integrated query. Our particular contributions consist of concepts for query formulation, combined content-based retrieval and presentation of a suitably ranked result list. The proposed concepts have been realized within the context of the PROBADO Music Repository and allow for music retrieval based on combining full-text lyrics search and score-based query-by-example search.

1 Introduction

Increasing amounts of digitized musical content result in the need for managing them automatically. Especially for libraries holding a vast amount of musical content that is steadily increasing due to ongoing digitization there is a high demand on automatisms to cope with the large number of documents. At the same time, technology from the field of Music Information Retrieval (MIR) has been developed for content-based querying of music collections using different modalities, e.g. by entering text or score-fragments, by cropping audio snippets from audio recordings, by whistling, singing or humming melodies, or by tapping a rhythm. Several of those query types have been realized for existing music collections [1,2]. In this paper, we consider the music repository set up within the PROBADO Digital Library Initiative [3].

The retrieval functionalities commonly provided by existing query engines, consisting of the steps of query formulation, query processing, visualization of query results and user interaction are, in a sense, *unimodal*. That is, there is no

---

* This work was supported in part by Deutsche Forschungsgemeinschaft (DFG) under grant INST 11925/1-1.

M. Agosti et al. (Eds.): ECDL 2009, LNCS 5714, pp. 261–272, 2009.

general concept on how to integrate and jointly process multiple query types in this processing chain, hence affording *multimodal* retrieval. Exceptions to this are some straight-forward approaches which process different query modalities separately and then use unions or intersections of the individual query results [4,5,6]. Especially in the case that a user has only sparse or fuzzy knowledge regarding the different modalities, the combination of different, multimodal queries might compensate this issue and eventually result in improved search results. Think for example of a user remembering a rough harmonic progression as well as some of the lyrics of a piece of music he listened to on the radio.

As currently available search engines are limited to unimodal queries, queries may be formulated very quickly, which is very convenient for the user. On the other hand, those unimodal queries frequently result in a lot of matches in case the query is not specific enough. Consequently, this kind of searching is likely to end up in the time-consuming process of reformulating the query in order to constrain the high number of matches. To overcome the limitations of unimodal retrieval, this paper presents a retrieval concept for realizing combined, multimodal querying in the context of music libraries. Particularly, we consider the task of integrating full-text (lyrics-) queries and audio retrieval using the query-by-example paradigm. We present concepts for the full processing chain of (a) formulating multimodal queries within a suitable user interface, (b) content-based retrieval including an adequate ranking strategy and subsequent generation of a result list, (c) suitably visualizing the list of query results providing a user-friendly and content-based overview representation, and (d) enabling possible user-feedback and interaction. The approach has been motivated in the real-life context of setting up the PROBADO Music Repository [3]. The special case of combining the textual and acoustic modalities serves as a starting point for integrating several different query modalities that will be of interest within the PROBADO Music Repository.

The rest of this paper is organized as follows. The subsequent Sect. 2 contains an overview on the current status of the Digital Music Repository within the PROBADO Digital Library Initiative. Sect. 3 contains our proposed concept of combining multimodal queries and presents a retrieval mechanism for the example of combined full-text- and audio retrieval. The paper concludes in Sect. 4 with prospects on future work.

2   The PROBADO Music Repository

The PROBADO Digital Library Initiative is a research effort to develop next generation Digital Library support for non-textual documents. It aims to contribute to all parts of the digital library workflow from content acquisition to semi-automatic indexing, search, and presentation of results. Currently, two different repositories are set up, PROBADO-3D containing 3D objects from architecture, and the PROBADO Music Repository considered in this paper.

Sections 2.1 and 2.2 contain an overview on the PROBADO Music Repository which is currently set up at the Bavarian State Library (BSB), Munich. While 2.1

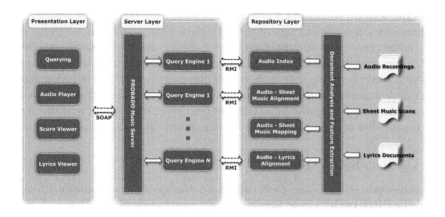

Fig. 1. Overview of the structure of the PROBADO Music Repository

contains a general system overview, 2.2 puts a focus on the generic functionalities for content-based retrieval.

2.1 System Overview

The PROBADO Music Repository incorporates the storage of and the access to digital music documents. For preservation purposes, digital copies of available musical content held by the BSB such as audio recordings, sheet music and other music-related material are made. For indexing purposes, these digital copies are analyzed and annotated by recent state-of-the-art MIR techniques.

One key task is to build up content-based search indexes in order to search for, e.g., lyrics phrases or score fragments. Another key task is to consolidate all available documents (particularly audio recordings and scanned sheet music) for the same piece of music and relate them among each other. Mapping- and synchronization-techniques are used to create alignments between meaningful entities within (a) sheet music pages and time segments within audio recordings (score-audio-synchronization), (b) words and time segments of audio recordings (text-audio-synchronization), and (c) time segments of different audio recordings of the same piece of music. A more detailed view on the topic of extracting meaningful entities from scanned sheet music and its mapping to audio recordings is given in [7,8,9]. The resulting benefits are content-based and cross-modal searching capabilities, synchronous multimodal playback and visualization of pieces of music, and advanced cross-modal browsing capabilities.

All these functionalities are realized and implemented within a modular system as depicted in Fig. 1. The system is organized as classical 3-tier architecture and consists of the *presentation layer*, the *server layer* and the *repository layer* (left to right). Search indexes, annotations and linking structures between different modalities are obtained in a preprocessing step which is carried out offline in the repository layer. The access to index structures and synchronization data as well as the delivery of musical content to the user takes place in the server layer. The presentation layer consists of user interface components for accessing musical content, especially content-based searching for musical content, navigation and browsing within search results, as well as synchronized playback of audio and sheet music or lyrics. For each system interaction such as retrieving search results and accessing musical content, there is a dedicated module, referred to as *Query Engine*. The communication between the presentation and the server layer is provided by a service-oriented architecture (SOA) [10].

2.2 Content-Based Querying

This subsection gives some more detail on the query functionality provided within the PROBADO Music Repository. As mentioned before, a key task in the context of the PROBADO Music Repository is to enable content-based search using lyrics phrases or score fragments as queries. Due to the consolidation of all musical content belonging to the same piece of music, each content-based search may also be viewed as cross-modal. This is, one can use either of the visual or textual modalities as queries, while aiming to find matches in the other modality. Up to now, two distinct options for content-based querying are available, *lyrics-based retrieval* as proposed in [11] and *score-based retrieval* using audio matching as proposed in [12]. Both approaches use indexing techniques to achieve a high retrieval efficiency.

The lyrics-based retrieval allows for formulating a query in the textual modality by entering a few words in order to find the positions within audio recordings where the words are sung. The mapping of positions within the lyrics text document to time segments within an audio recording is performed using lyrics-enriched MIDI files as described in [11]. Here, onset times of individual words or syllables are explicitly given within a musical context. This information, in turn, is then used to synchronize the lyrics to the audio recording. The subsequently used indexing technique is based on inverted files which are well known from classical full-text retrieval [13] and enhanced for the special case of a lyrics search. The search is fault-tolerant w.r.t. misspelled or omitted words in both the query as well as the lyrics, see [11].

The score-based retrieval follows the query-by-example paradigm. A query is formulated in the visual modality by selecting a portion of a sheet music page, particularly a few consecutive measures. The system retrieves all occurrences of the selected music excerpt within the indexed audio recordings. Note that the sheets of music are images obtained from scanned analogue pages and thus the actual musical content or semantic is expressed in the visual modality. For the purpose of synchronization with audio recordings, the images have to be transformed to a symbolical representation. This is done by optical music recognition (OMR) techniques [14,15]. Although the output of the OMR processing is quite error-prone it is sufficient for synchronization with the audio recordings. Exploiting this synchronization, instead of querying the selected score excerpt, the according snippet of an associated (synchronized) audio recording is used for the search process. Here, a sequence of audio features is extracted from the snippet and subsequently a feature-based search on an audio features index is performed. Due to the extraction of consecutive features that reflect the chromatic harmonic progression of the underlying audio snippet at a coarse level, the audio retrieval system is robust against changes in timbre, instrumentation, loudness and transposition and therefore musically similar snippets can be found regardless of a particular performance [7,8]. For a more detailed view, we refer to [16] and the references therein.

3   Combining Multimodal Queries

In this section, we propose how to incorporate a concept for combined multimodal queries into the typical stages of a query-retrieval chain, particularly query formulation (Sect. 3.1), content-based retrieval and ranking (Sect. 3.2), presentation of query results (Subsect. 3.3) and mechanisms for user feedback and navigation (Sect. 3.4).

One of our key contributions is that our system enables the user to formulate a query that may consist of different modalities, including the textual, visual and auditory modality. In particular, he can query a combination of metadata, lyrics and audio fragments. For this, he formulates single, unimodal queries and adds them successively to his search. Single queries are gathered in a special structure

Fig. 2. User interface for query formulation tabs (top left), Query Bag (top right), display of the results (bottom left) and the document viewer (bottom right)

for representing sets of queries, referred to as *Query Bag*. After the Query Bag is submitted to the retrieval system, the user is presented a list of pieces of music that match in at least one modality regarding his query. To organize the result list we introduce a new ranking approach based on a combination of multiple result lists ensuring that pieces of music containing more matching modalities are given a higher rank, see 3.2. The results, i.e. pieces of music, may be viewed in detail and played back in the document viewer module. Additionally, both the result list and the player can be used for querying, query refinement and document navigation, see 3.4.

To give a user-friendly and intuitively operable interaction environment, our approach was to incorporate the look-and-feel of well-tried Internet search platforms that a wide range of users is familiar with. The user interface for the retrieval, browsing, playback and navigation in musical content is completely web-based and runs in every state-of-the-art Internet browser. Fig. 2 shows a snapshot of a typical system configuration. Similar to popular existing query engines, the top part of the GUI contains the query formulation area while the result view area is located below. Both areas are further subdivided to facilitate the subsequently described functionalities. The query formulation area is split into both a tab cards region and the Query Bag region. The result area is divided into the result list pane and the document viewer.

In the following, a further in-detail look at each particular stage of the query-retrieval chain is given. Throughout the whole section, the piece of music "Gefrorne Tränen" belonging to the song cycle "Winterreise" by Franz Schubert will serve as our running example.

3.1   Query Formulation and Interface

In this subsection, we give a detailed look at the query formulation and its interface. The query formulation area, shown in Fig. 2 (top), consists of various query formulation forms per modality which are organized as tab cards (top left). It further contains the Query Bag (top right), where single queries can be added to, viewed, revised or removed. Currently, the user is enabled to formulate metadata, lyrics and audio fragment queries within the appropriate, designated tab card. Additional query formulation types such as entering a melody by a virtual piano, humming a melody or tapping a rhythm are planned to be integrated in the future.

From within any tab card the user has the choice to either perform an immediate, unimodal search using the just formulated query (classical query) or to add the latter to the Query Bag and continue with the formulation of another query in order to gather a couple of unimodal queries. The Query Bag stores all queries and offers an overview representation of all gathered queries. So, the user at any time is informed about which queries he has collected so far. Each single query inside the Query Bag can be examined more precisely by clicking the plus-sign icon left to the query. To the right of each query there are icons for reformulating the query and for removing it from the Query Bag as well, by clicking either the pencil- or the "x" icon, respectively. By clicking the pencil icon, the corresponding query formulation tab opens for editing. Once the user has finished assembling the individual queries, the search button at the bottom of the Query Bag can be clicked in order to submit them to the search engine as one integrated, multimodal query. Subsequently, a multimodal search is performed.

3.2   Content-Based Retrieval and Ranking

Once the Query Bag is submitted, the system disassembles it and delegates each contained single query to an appropriate query engine which is capable of handling the particular type of query. The query engines act independently from each other and for each modality a homogeneous list of matches is returned. In this, each match consists of a document ID, the position of the matching segment, and a ranking value $r \in [0,1]$. In case of content-based queries, the latter segments are generally short parts of the document. If, however, a document matches due to its metadata description, the document is said to match at every position; i.e., a matching segment ranges from the beginning to the end of the document.

Due to the synchronization of different document types such as audio recordings, sheets of music and lyrics documents, all matching segment boundaries can be expressed in the time domain, i.e. translated to a start- and an end-timestamp. Thus, all segments are directly comparable, which will be exploited in the subsequent combined ranking and merging. For merging and ranking of multiple result lists returned by the different query engines into a single, integrated result list we use a straight-forward bottom-up approach explained in the following.

Each result list returned by a query engine consists of document IDs and for each document ID there exists a list of matching segments. These segment lists are inserted into a hashtable, where a single data entry stores a piece of music's ID in together with related segment lists. For each inserted segment list, the respective modality is stored as well. With this, all inserted segment lists associated to the same piece of music are clustered and stored within a single hashtable data entry. Subsequently, for each entry of the hashtable a merging of the contained segment lists is performed. This step is now described in detail. Let M be the global number of queried modalities and m the local number of non-empty segment lists stored in a currently considered hashtable entry. We now consider the merging step of two segment lists, $L^1 := \{(b_1^1, e_1^1, r_1^1), \ldots, (b_{|L^1|}^1, e_{|L^1|}^1, r_{|L^1|}^1)\}$ and $L^2 := \{(b_1^2, e_1^2, r_1^2), \ldots, (b_{|L^2|}^2, e_{|L^2|}^2, r_{|L^2|}^2)\}$, where the i-th entry of list k is a segment $s_i^k = (b_i^k, e_i^k, r_i^k)$ consisting of a beginning- and an ending timestamp as well as a ranking value. In this, we assume that each segment list corresponding to a single modality does only contain non-overlapping segments. Let L be the merged, integrated segment list. For merging two lists L^1 and L^2 into L, we consider two cases. If a segment s_i^k does not overlap in time with any segment s_j^l of the other list, s_i^k is simply copied to L. If there is a temporal overlapping of a segment $s_i^k := (b_i^k, e_i^k, r_i^k) \in L^k$ and a segment $s_j^l := (b_j^l, e_j^l, r_j^l) \in L^l$, s_i^k and s_j^l are merged into a new segment $s := (\min(b_i^k, b_j^l), \max(e_i^k, e_j^l), r)$ which is inserted

Fig. 3. Document viewer in the score visualization mode. Multimodal content of selected measures can be queried.

into L. Note that overlaps do reflect simultaneously arising hits and for this reason, we want them to get higher ranked in general. To additionally promote small individual ranking values r_i^k, r_j^l in the latter case of segmental overlap, the assigned ranking value is defined as $r := (r_i^k + r_j^l) \cdot f_{boost}$, where $1 \leq f_{boost} \leq M$ is a constant global boosting factor. The merging of the m segment lists is done iteratively until no residual segment list remains. Note that the factor f_{boost} is applied only once during the processing of the segment lists. When all m segment lists are merged into a single, integrated segment list, all of the segments' ranking values are normalized by applying the factor $1/(M \cdot f_{boost})$ resulting in a final ranking value in the interval $[0, 1]$. This algorithm can be implemented in a straight-forward manner with a time complexity linear in list lengths, as long as each list L^k is sorted ascending w.r.t. the beginning timestamps b_i^k of its matching segments $s_i^k := (b_i^k, e_i^k, r_i^k)$.

In the end, for every piece of music there results an individual, integrated list of multimodal matching segments along with assigned ranking values. The overall ranking value for a piece of music is determined by the maximum ranking value of its integrated segment list. Finally, the pieces of music are put into a new result list and sorted in descending order of their respective ranking values. This means that the final result list is organized such that the more modalities within pieces of music do match, the higher their assigned ranking values are. Therefore they occur at earlier positions in the list. In turn, pieces of music matching in less modalities occur at later positions in the list.

3.3   Integrated Presentation of Query Results

Typically, available search engines provide the user with a flatly organized result list only where the list entries commonly consist of single documents. However, in the case of the music domain, there are multiple document types (in our case audio recordings, sheet music and lyrics documents) representing a piece of music using different modalities. As in our applications we have multiple documents of the different types available for a piece of music, we believe that it is of special interest to present all those documents in a collective manner, even if some of them do not match a user's query. Therefore, we took this consideration into account concerning the presentation of query results.

The bottom area of Fig. 2 shows the result list (left) and the document viewer (right). While the result list shows the matching pieces of music regarding the query, the document viewer offers access to all content belonging to the currently selected piece of music. It furthermore gives a more detailed view on matching regions within its multimodal content and is also responsible for playing back the latter. As mentioned before, the resulting matches are presented to the user not at document level. Instead the user is offered every piece of music where at least one document representing that piece contains one or more matches to the current query. All documents belonging to the same piece of music that match the user's query are summarized within a single list entry. The entry shows the artist's name and the title of the piece of music, a lyrics excerpt as well as the matching documents along with their number (in brackets). Additionally, at the bottom there are links to show more titles of the artist and to save the result (see also Sect. 3.4). A more detailed view of the single matching documents as well as the exact matching positions therein, is given in the document viewer.

One key feature of the document viewer is the integrated display of matching segments along the timeline bar at the bottom. Besides the adjustment of the current playback position by using the slider knob, it is used to show all matching positions within the currently selected multimodal contents used for playback. The matching positions are indicated by colored boxes along the timeline bar, where the color and brightness of the boxes encode modality and ranking value, respectively. Additionally, matching segments within "inactive" documents, i.e. others than those ones used for playback, are displayed as gray boxes.

Below the timeline bar there are further buttons to control the playback state (start/pause, stop) as well as the playback position. While the control buttons retain their positions, the labels are exchanged depending on the currently selected visualization type, i.e. scores or lyrics, in the center.

Another key feature of the document viewer is to simultaneously playback multimodal content associated to the currently selected piece of music. Here, the content might be audio recordings, sheet music and lyrics as well, where the last two ones are displayed in the center area. From within a pop-up menu that is reachable by clicking the album cover in the top left corner, the user can choose which contents are used for playback. Here, all available content associated to the piece of music are shown.

While an audio recording is played back, available sheet music or lyrics are also displayed synchronously; i.e., the user can visually track the currently played measure or the currently sung words, respectively (cf. Fig. 3 and 4). Due to this style of enjoying music in a multimodal way, the document viewer may be

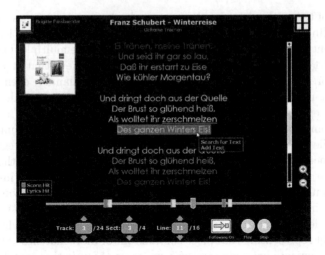

Fig. 4. Document viewer in the lyrics visualization mode. Selected text can be queried.

thought of as being a video player, but in addition it provides sophisticated user interaction options such as navigation and query refinement, which are examined in Sect. 3.4.

3.4   Query Refinement and Navigation

From within the result list, for each retrieved piece of music, the user is enabled to request more titles of the same artist by choosing the appropriate "get more titles from artist..."-link which is available from the context menu. Once the user selects this option, the Query Bag is flushed, rebuilt with only a simple metadata query consisting of the artist's name and a subsequent new search is performed, what finally results in an updated list that displays all pieces of music by this artist contained in the database.

From within the document viewer, the user can select which audio-visual content is used for playback in case that more than one document per modality is available. For example, if different audio recordings of a piece of music are available, the user has the choice to decide which specific performance he wants to listen to. With this functionality, one is also allowed to switch between different performances while retaining the actual musical playback position. Thus, the user can additionally draw local comparisons between different interpretations of a piece of music. Sheet music books may also be exchanged, if more than one is available.

Furthermore, the user may thumb easily through the sheet music book or the lyrics text, by clicking on a measure or a word, respectively, whereupon the playback position is changed accordingly.

Moreover, the user can utilize content-based searching capabilities from within visual content following the query-by-example paradigm. When the user selects a portion of either a sheet music page or the lyrics text, he can use this excerpt for a new query, see Fig. 3 and 4. He has the option to start either a completely new search based only on the selected portion, or to add the query as an additional partial query to the Query Bag. In the case of sheet music, a portion may consist of two modalities, score and text (cf. Fig. 3). Here, the user can choose, whether he queries both modalities together or separate from each other.

As matching segments within multimodal contents are displayed as boxes along the timeline bar at the bottom of the document viewer, they can be simultaneously utilized for navigation purposes. By clicking on a box, the playback is started or resumed at the corresponding time position. This functionality enables the user to jump directly to the found segments matching the user's query.

4   Conclusions and Future Work

In this paper, we presented a concept for integrated multimodal access to digital music libraries. Especially, we considered the integration of multimodality into all stages of the query-retrieval chain, aiming at facilitating web-based user access to pieces of music by means of the various available kinds of music documents.

The concept of combining multiple queries has not to be restricted to using different modalities as used in this paper. Naturally, multiple queries of the same modality (like multiple score-based queries in one Query Bag) might be used. Using the Query Bag approach, even intentionally different queries might be combined into a single query resulting in an integrated result list, readily available for browsing.

The future work on the topic of this paper as well as in the context of the PROBADO Music Repository, on the one hand consists of improving the retrieval performance of the query engines described in Sect. 2.2. On the other hand, we want to investigate the idea of a *direct* alignment of lyrics from within sheet music, recognized by OMR software, and "clear" text documents, containing the lyrics as well. This allows for a direct mapping between word positions within the text documents and image regions within the scanned sheet music, which, in turn, allows for a mapping to time segments due to the synchronization of sheet music and audio recordings. Thus, the need for an intermediate representation like MIDI as proposed in [17] is not applicable. In the context of the PROBADO Music library project, further studies will be conducted to evaluate and improve the system design and usability.

References

1. Musipedia: http://www.musipedia.org/
2. Typke, R.: Melodyhound: Search within the music (2006),
   http://melodyhound.com/
3. Kurth, F., Damm, D., Fremerey, C., Müller, M., Clausen, M.: A framework for managing multimodal digitized music collections. In: Christensen-Dalsgaard, B., Castelli, D., Ammitzbøll Jurik, B., Lippincott, J. (eds.) ECDL 2008. LNCS, vol. 5173, pp. 334–345. Springer, Heidelberg (2008)

4. de Kretser, O., Moffat, A., Shimmin, T., Zobel, J.: Methodologies for distributed information retrieval. In: Proceedings of the 18th International Conference on Distributed Computing Systems (ICDCS 1998), pp. 66–73 (1998)
5. Henrich, A., Robbert, G.: Combining multimedia retrieval and text retrieval to search structured documents in digital libraries. In: DELOS Workshop: Information Seeking, Searching and Querying in Digital Libraries (2000)
6. Kailing, K., peter Kriegel, H., Schönauer, S.: Content-based image retrieval using multiple representations (2008)
7. Hu, N., Dannenberg, R., Tzanetakis, G.: Polyphonic audio matching and alignment for music retrieval. In: Proceedings of the IEEE Workshop on Applications of Signal Processing to Audio and Acoustics (October 2003)
8. Bartsch, M.A., Wakefield, G.H.: Audio thumbnailing of popular music using chroma-based representations. IEEE Transactions on Multimedia 7(1), 96–104 (2005)
9. Müller, M., Fremerey, C., Kurth, F., Damm, D.: Mapping sheet music to audio recordings. In: Proceedings of the 9th International Conference on Music Information Retrieval, ISMIR 2008 (2008)
10. W3C: Web services, http://www.w3.org/2002/ws/
11. Müller, M., Kurth, F., Damm, D., Fremerey, C., Clausen, M.: Lyrics-based audio retrieval and multimodal navigation in music collections. In: Kovács, L., Fuhr, N., Meghini, C. (eds.) ECDL 2007. LNCS, vol. 4675, pp. 112–123. Springer, Heidelberg (2007)
12. Kurth, F., Müller, M.: Efficient Index-based Audio Matching. IEEE Transactions on Audio, Speech and Language Processing 16(2), 382–395 (2008)
13. Witten, I.H., Moffat, A., Bell, T.C.: Managing Gigabytes, 2nd edn. Van Nostrand Reinhold (1999)
14. Byrd, D., Schindele, M.: Prospects for improving OMR with multiple recognizers. In: Proceedings of the 7th International Conference on Music Information Retrieval (ISMIR 2006), pp. 41–46 (2006)
15. Jones, G.: Sharpeye music reader (2008), http://www.visiv.co.uk/
16. Kurth, F., Müller, M., Fremerey, C., Chang, Y., Clausen, M.: Automated synchronization of scanned sheet music with audio recordings. In: Proceedings of the 8th International Conference on Music Information Retrieval (ISMIR 2007), September 2007, pp. 261–266 (2007)
17. Krottmaier, H., Kurth, F., Steenweg, T., Appelrath, H.J., Fellner, D.: Probado — a generic repository integration framework. In: Kovács, L., Fuhr, N., Meghini, C. (eds.) ECDL 2007. LNCS, vol. 4675, pp. 518–521. Springer, Heidelberg (2007)

A Compressed Self-indexed Representation of XML Documents*

Nieves R. Brisaboa[1], Ana Cerdeira-Pena[1], and Gonzalo Navarro[2]

Database Lab., Univ. da Coruña, Spain
{brisaboa,acerdeira}@udc.es
Dept. of Computer Science, Univ. of Chile
gnavarro@dcc.uchile.cl

Abstract. This paper presents a structure we call XML Wavelet Tree (XWT) to represent any XML document in a compressed and self-indexed form. Therefore, any query or procedure that could be performed over the original document can be performed more efficiently over the XWT representation because it is shorter and has some indexing properties. In fact, XWT permits to answer XPath queries more efficiently than using the uncompressed version of the documents. XWT is also competitive when comparing it with inverted indexes over the XML document (if both structures use the same space).

1 Introduction

XML[1] has long ago become the standard for representing semi-structured documents and W3C has defined the language XPath[2] for querying XML documents allowing constraints on both structure and content. Recently, several works have been devoted to the problem of modelling and querying XML documents and new query languages or XPath extensions have been proposed [10,9,3].

On the other hand, the research in text compression has experimented a big advance in the last years. Different compression methods have been proposed, demonstrating beyond doubts that the use of word-based statistical semi-static compressors, such as Plain and Tagged Hufman, ETDC, *(s,c)*-DC or RPBC [12,6,8], perfectly fulfil IR requirements because those compressors allow querying the compressed version of the text up to 8 times faster than the uncompressed version. That is, the text is compressed to about 30%-35% of its original size and can be kept in that compressed form all the time, because direct search of words and phrases can be performed over that compressed version. Therefore the text only need to be uncompressed to be shown to a human user, but any process, for IR or any other purpose, can be done over the compressed text. In this way, not only storage space is saved, but also time. Time is the critical factor in efficiency and processing a compressed version of a document saves time when we need to access to disk looking for a document, when it is transmitted through a network, or more importantly, when it is processed.

---

* Funded in part by MEC grant TIN2006-15071-C03-03, for the Spanish group; and for the third author by Fondecyt grant 1-080019 (Chile).

M. Agosti et al. (Eds.): ECDL 2009, LNCS 5714, pp. 273–284, 2009.

More recently, compression techniques have become even more sophisticated allowing not only a compressed representation of the text, but also self-indexed representations using the same compressed space (about 35% of the original size). Those compressed and self-indexed representations of the text successfully compete with the classical inverted indexes, even if they use compression strategies. Among those compressed and self-indexed document representations, the Word Suffix Arrays [11,7] and the Wavelet Trees [4] are some of the most powerful.

In this paper we present a modified wavelet tree, based on a *(s,c)*-DC compressor, to create a self-indexed and compressed version of XML documents. Our representation, which we call XML Wavelet Tree (XWT), uses only about 30%-40% of the space of a XML document and provides some self-indexing properties that can be successfully used in answering XPath queries.

Notice that any XML document can be represented, using our XWT, in a compressed and self-indexed form, therefore any processing or query that could be performed over the original XML document can also be performed over the XWT representation. Moreover, due to the fact that the XWT representation is smaller and has some indexing properties, any processing will be more efficient over the XWT representation than over the original uncompressed document.

2   Previous Work

Among the different word-based byte-oriented semi-static statistical compression methods available in the state of the art, we use *(s,c)*-DC [5,6] as basis of our representation because it provides flexibility to compress with different models the *tags* of a XML document and the rest of its words. On the other hand, among the self-indexing structures available we chose to work with the WT presented in [4] because it is the only one that could be adapted in order to represent, in a compact way, the structure of the document (that is, the XML *tags*) separated from the rest of the words.

2.1   *(s,c)*-Dense Code

(s,c)-Dense Code is a word-based semi-static statistical prefix-free encoder. In a first pass over the source text the different words and their frequencies are obtained (the *model*). Then, the vocabulary is sorted by frequency and a codeword is assigned to each word (shorter codewords to more frequent words). In a second pass, the compressor replaces each word by its codeword leading to a compressed representation of the text.

As other compressors, *(s,c)*-DC distinguishes between bytes[1] that do not end a codeword, called *continuers*, and bytes that only can appear as the last byte of a codeword, *stoppers*. In this case, where s is the number of *stoppers* and c indicates the number of *continuers* $(s+c = 256)$, *stoppers* are the bytes between 0 and $s-1$ and *continuers*, are those between s and $s+c-1 = 255$. To minimize compression ratios, optimal values for s and c are computed for the specific

---

[1] For simplicity, we focus on the byte oriented version.

word frequency distribution of the text [5]. Then given source symbols sorted by decreasing frequencies, the corresponding *(s,c)*-DC encoding process gives one-byte codewords to the words in positions from 0 to $s - 1$. Words ranked from s to $s + sc - 1$ are sequentially assigned two-byte codewords. The first byte of each codeword has a value in the range $[s, s + c - 1]$, that is, a *continuer*. The second byte, the *stopper*, has a value in range $[0, s - 1]$. Words from $s + sc$ to $s + sc + sc^2 - 1$ are assigned three byte codewords, and so on.

Example 1. The codes assigned to symbols $i \in 0 \dots 15$ by a (2,3)-DC are as follows: $\langle 0 \rangle$, $\langle 1 \rangle$, $\langle 2,0 \rangle$, $\langle 2,1 \rangle$, $\langle 3,0 \rangle$, $\langle 3,1 \rangle$, $\langle 4,0 \rangle$, $\langle 4,1 \rangle$, $\langle 2,2,0 \rangle$, $\langle 2,2,1 \rangle$, $\langle 2,3,0 \rangle$, $\langle 2,3,1 \rangle$, $\langle 2,4,0 \rangle$, $\langle 2,4,1 \rangle$, $\langle 3,2,0 \rangle$ and $\langle 3,2,1 \rangle$.

2.2 Byte-Oriented Wavelet Tree (WT)

In [4] we presented a novel reorganization of the codewords bytes of a text compressed with any word-based byte-oriented semi-static statistical prefix-free compression technique. This reorganization, called *Wavelet Tree*, consists basically on placing the different bytes of each codeword at different WT nodes instead of sequentially concatenating them, as in a typical compressed text.

The root of the WT is represented by all the first bytes of the codewords, following the same order as the words they encode in the original text. That is, let assume we have the text words $\langle w_1, w_2 \dots w_n \rangle$, whose codewords are $cw_1, cw_2 \dots cw_n$, respectively, and let us denote the bytes of a codeword cw_i as $\langle c_i^1 \dots c_i^m \rangle$ where m is the size of the codeword cw_i in bytes. Then the root is formed by the sequence of bytes $\langle c_1^1, c_2^1, c_3^1 \dots c_n^1 \rangle$. At position i, we place the first byte of the codeword that encodes the ith word in the source text, so notice that the root node has as many bytes as words has the text. We consider the root of the WT as the first level. Therefore, second bytes of the codewords longer than one byte are placed in nodes of a second level. The root has as many children as different bytes can be the first byte of a codeword of two or more bytes. That is, in a (190, 66)-DC encoding scheme, the root will have always 66 children, because there are 66 bytes that are *continuers*. Each node X in this second level contains all the second bytes of the codewords whose first byte is x, following again the same order of the source. That is, the second byte corresponding to the jth occurrence of byte x in the root, is placed at position j in node X. Formally, let suppose there are t words coded by codewords $cw_{i_1} \dots cw_{i_t}$ (longer than one byte) whose first byte is x. Then, the second bytes of those codewords, $\langle c_{i_1}^2, c_{i_2}^2, c_{i_3}^2 \dots c_{i_t}^2 \rangle$, form the node X. The same idea is used to create the lower levels of the WT. Looking into the example, and supposing that there are d words whose first byte codewords is x and whose second one is y, then node XY is a child of node X and it stores the byte sequence $\langle c_{j_1}^3, c_{j_2}^3, c_{j_3}^3 \dots c_{j_d}^3 \rangle$ given by all the third bytes of that codewords. Those bytes are again in the original text order. Therefore, the resulting WT has as many levels as bytes have the longest codewords.

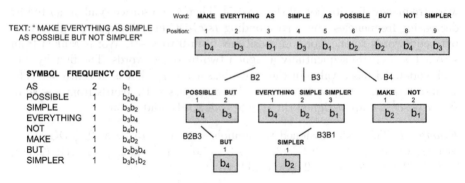

Fig. 1. Example of WT

In Fig. 1[2], a WT is built from the text MAKE EVERYTHING AS SIMPLE AS POSSIBLE BUT NOT SIMPLER. Once codewords are assigned to all the different words in the text, their bytes are spread in a WT following the reorganization of bytes explained. For example, b_3 is the 9th byte of the root because it is the first byte of the codeword assigned to 'SIMPLER', which is the 9th word in the text. In turn, its second byte, b_1, is placed in the third position of the child node $B3$ because 'SIMPLER' is the third word in the root having b_3 as first byte. Likewise, its third byte, b_2, is placed at the third level in the child node $B3B1$, since the first and second byte of the codeword are b_3 and b_1, respectively.

Original codewords can be rebuilt from the bytes spread along the different WT nodes using *rank* and *select* operations. Let be B a sequence of bytes, b_1, $b_2 \ldots b_n$. Then, *rank* and *select* are defined as:

- $rank_b(B,p) = i$ if the number of occurrences of the byte b from the beginning of B up to position p is i.
- $select_b(B,j) = p$ if the jth occurrence of the byte b in the sequence B is at position p

The two basic procedures using the WT are *locating* a word in the text and *decoding* the word placed at certain position. Both are easily solved using *select* and *rank* operations, respectively.

To find the first occurrence of 'SIMPLER', we will start at the bottom of the tree and go up. As we can see in Fig. 1, the codeword of 'SIMPLER' is $b_3b_1b_2$, therefore, we start at node $B3B1$, in the third level, and search for the first occurrence of the byte b_2 computing $select_{b_2}(B3B1, 1) = 1$. In this way, we obtain that the first position of that node ($B3B1$) corresponds to the first occurrence of 'SIMPLER'. Now, we need to locate in node $B3$ the position of the first occurrence of byte b_1. Again, this is obtained by $select_{b_1}(B3, 1) = 3$, that newly indicates our codeword is the third one starting by b_3 in the root node. Finally, by calculating $select_{b_3}(root, 3) = 9$, we can answer that the first occurrence of 'SIMPLER' is at 9th position in the source text.

---

[2] Note that only the shaded byte sequences are stored in tree nodes; the text is shown only for clarity.

To decode a word we use rank operations. To know which is the 7th word in the source text we start reading $root[7] = b_2$. According to the encoding scheme we know that the code is not complete, so we will have to read a second byte in the second level of the WT, more precisely, in the node $B2$. To find out which position of that sequence we have to read, we use $rank_{b_2}(root, 7) = 2$. Therefore, $B2[2] = b_3$, gives us the second byte of the codeword. Again b_3 is not a *stopper*, so we need to continue the procedure. In the child node $B2B3$, that corresponding to the two first bytes of the codeword we have just read (b_2b_3), we have to read the byte which is at position $rank_{b_3}(B2, 2) = 1$. Finally, we obtain $B2B3[1] = b_4$, which marks the end of the searched codeword. As a result, we have the codeword $b_2b_3b_4$, that corresponding to 'BUT', which is precisely the 7th word in the source text, as expected.

The performance of the WT depends on the implementation of the *rank* and *select* operations, because they are the base for any procedure over this structure. A detailed description of their implementation can be found in [4]. It is based on a structure of partial counters to avoid counting the number of occurrences of a searched byte from the beginning of a WT node. There is a tradeoff between space and time. If we use more partial counters, we need more space, but *rank* and *select* operations will be more efficient.

3   XML Wavelet Tree (XWT)

Phase I: Parsing the XML document and assigning codewords. The first step to obtain the XWT is to parse the input XML document to create the vocabulary and compute the frequencies distribution (the *model*). The *parsing* process distinguishes different kind of words depending on whether a word is[3]: *i*) a name of a *start-tag* or an *end-tag*, *ii*) the name of an attribute, *iii*) an attribute value, *iv*) a word appeared inside a *comment*, *v*) a word appeared inside a *processing instruction*, or *vi*) a word of the XML document content.

That is, when compressing, a same word will be assigned different codewords depending on the category it belongs to. For example, if the word *book* appears as content (e.g. ... *the great book* ...), but also as an attribute value (e.g. *category= "book"*) and inside a comment (e.g. ⟨! − − ... *this book is* ... −−⟩) it will be stored as three different entries in the vocabulary, one for each different category.

Keeping this difference between same words according to its function in the XML document structure increases the vocabulary size, however it is translated into efficiency and flexibility when querying.

It is also in the *parsing* that some *normalization* operations take place (all according to [1]). For instance, *empty-element tags* are translated into its corresponding pair of *start-end tag* (e.g. ⟨*tag_name/*⟩ → ⟨*tag_name*⟩ ⟨*/tag_name*⟩) and redundant spaces and spaces inside tags are eliminated (e.g. ⟨*tag_name*   ⟩ → ⟨*tag_name*⟩), etc.

---

[3] Division implicitly given by the different kind of XPath queries[2].

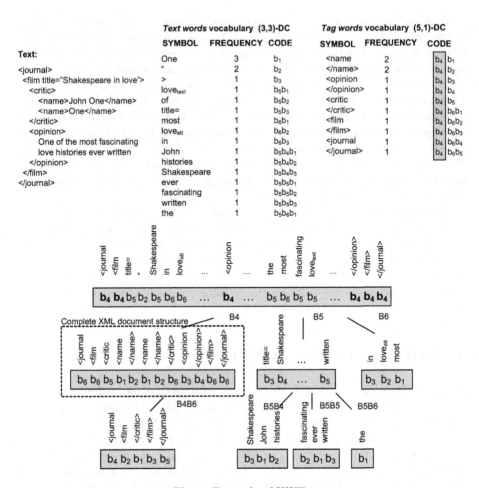

Fig. 2. Example of XWT

While parsing the XML document, two vocabularies are created. One stores the different *start-* and *end-tags* and therefore the structure of the document. The other stores the rest of the words. We call them, *tag words* and *text words* vocabulary, respectively (see Fig. 2).

As it was explained in Section 2, *(s,c)*-Dense Code uses different bytes for *continuers* and for *stoppers*. So it is easy to see how reserving a *continuer* to be the first byte of the codewords assigned to *tag words* (in Fig. 2, see the bytes shaded in the CODE column of the *tag words* vocabulary) it is possible to keep them all located in the same branch of the XWT (see the branch $B4$ in Fig. 2). Remember that they follow the document order and hence maintain their relationships like in the original XML document. But what is even more striking is that this feature implicitly provides an efficient way to solve structural queries. To do this, we only need to deal with those nodes of the XWT storing the structure of the document, and omit the rest of the compressed text.

Therefore, all the words of the *text words* vocabulary are assigned a codeword following a *(s,c)*-Dense Code encoding scheme, keeping aside one *continuer*. In Fig. 2, where a $(3,3)$-DC encoding scheme is used to encode *text words*, the first of the *continuers*, b_4, has been selected as the one reserved. Notice that it is not used as a first byte of any of the codewords assigned to the *text words*.

Because of this, compression could be affected, so to minimize this loss, *tag words* are also coded according to another optimal values of s and c. That is, on the one hand we keep the selected *continuer* as the first byte of the *tag* codewords to store the XML structure isolated. On the other hand, the remaining bytes of the *tag* codewords will be given following their own *(s,c)*-DC scheme. In the example of the Fig. 2, codewords assigned to *tags* follow a $(5,1)$-DC encoding scheme, after the first byte, that is always the *continuer* b_4 (shaded column).

Phase II: Compressing and creating the XWT. Once codewords are assigned to words, we do a second pass over the text replacing each word by its codeword and storing these codeword bytes along the different nodes of the XWT (it is possible to precalculate the number of nodes as well as their sizes in advance). So, by keeping an array of markers indicating the next writing position for each node, they are filled sequentially following the order of the words in the text.

4   Using XWT

4.1   Decompression

To decompress from a random text word j (*random decompression*), we follow the procedure explained in Section 2.2. But now, we take into account the use of two vocabularies with different encoding schemes. That is, we first access to the jth byte of the root node of the XWT to get the first byte of the codeword and then we check if the byte read, b_i, matches or not with the *continuer* used to mark *tag words*. Depending on this, going down in the XWT to obtain the remaining bytes is done by using the corresponding s and c values.

If we want to decompress the whole text from the beginning (*full decompression*), we can follow a more efficient procedure. Given that the sequences of bytes of all the XWT nodes follow the original order of the words in the source text, *full decompression* can be efficiently implemented using pointers to the next positions to be read in each node. That is, when going to a child node to read the following byte of an uncomplete codeword, we do not need to compute any *rank* operation to find out what byte of this child node sequence we have to read. It always will be the next one to process in that child node.

4.2   Answering XPath Queries

Since the XWT structure is an exact representation of the XML document, any operation over the original text can be done over such representation. Therefore, all XPath queries can be answered using our representation. Indeed, some of them take benefit of the implicit indexing properties provided by the own XWT structure and are efficiently answered.

Counting. To *count()* the number of occurrences of a word (e.g. *tag*, name of an attribute, attribute value, word inside a comment or node content, etc.) we just compute how many times the last byte of the codeword assigned to that word appears in its corresponding XWT node. Therefore, if a word is encoded with a codeword xyz (being x and y, *continuers* and z, a *stopper*), it is only necessary to count the number of bytes z in the node XY. That is, we only do $rank_z(node_{XY}, i)$, where i is the size of the node XY. In turn, if the codeword has just one byte, z, we will do $rank_z(root, n)$, where n is the number of words in the text, that is, the number of bytes in the root.

Locating. We can locate the position in the text of any occurrence of a word (typical XPath queries as *//book*, *//@title*, etc.) searching its last byte in the corresponding XWT node and performing consecutive *select* operations up to the root. If we want to locate all the occurrences of a word, this process is repeated for each one. Since the traversed XWT nodes are the same for each occurrence and these will be processed consecutively, select operations and thus the whole process, can be sped up by using pointers to the already found positions in the WT nodes.

Locating phrases. To locate a *phrase* pattern we start locating the first occurrence of the least frequent word of the pattern in the root node. Then we check if all the first bytes of the codewords of each word of the phrase pattern match with the previous and next bytes of the root node. If those matches happen, we follow validating the rest of the bytes of the corresponding codewords. But if it is not the case, we save going down into the XWT and we simply locate the next occurrence of the least frequent word to be processed in a same way.

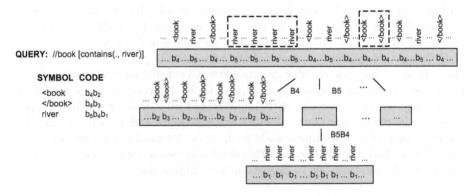

Fig. 3. Example of searching pairs of start-end tags containing a word

Searching pairs of start-end tags containing a word. In XPath, a *predicate* is a filter applied to a set of XML nodes. For simplicity, here we have chosen predicates over text: $//tag\_name\ [contains(., word_{text})]$. That is, we are interested in reporting the pairs of start-end tags that fulfill $\langle tag\_name \rangle \ldots word_{text}$ $\ldots \langle /tag\_name \rangle$. For example: $//book\ [contains(., river)]$ (see Fig. 3).

We begin locating the first occurrence of the desired word (*river*, in the example) in the root node. Then, by counting the number of occurrences of the desired *start-tag* ($\langle book$ in Fig. 3) placed before that position and that of the desired *end-tag* ($\langle /book \rangle$) we will know how many of the element nodes we are looking for contain that occurrence of the word. We can easily figure out those number of occurrences dealing only with the branches of the XWT storing the *tags*. In the example, we locate the first occurrence of *river* which is surrounded by the first occurrences of $\langle book$ and $\langle /book \rangle$. Therefore they are reported as a hit.

Now, instead of performing the same process with the next occurrence of the word (in the example, the 2nd occurrence of *river*), we can skip some text looking for the first occurrence of the desired *start-tag* placed after the position of the just located occurrence of the word. That is, in the example, we locate the second occurrence of $\langle book$ and its corresponding *end-tag*, and then we look for an occurrence of *river* between their positions. Given that there is one occurrence (the 6th occurrence of *river*), the 2nd occurrence of the element node *book* is also reported as a hit. By doing this we skip the occurrences of *river* that could be before the second occurrence of $\langle book$, and which are not interesting for the search (those occurrences of *river* surrounded by a striped rectangle in Fig. 3). After that, we proceed in one of the two ways. If the XML element node we are searching allows *self nesting*, we take the first occurrence of the word placed after the position of the desired *start-tag* just located (the 2nd occurrence of $\langle book$). If not, we take the next occurrence of the word after its corresponding *end-tag* (it is the case of the example, so we take the 7th occurrence of *river*). In both cases, we repeat the whole procedure. Again, this allows skipping those occurrences of the element node that could not contain any occurrence of the word searched (in Fig. 3 we skip the 3th occurrence of *book*).

Although we have explained the algorithm for the particular case of XML element nodes and words being part of their content, it can be generalized to predicates over other element nodes. That is, queries like $//tag\_name_1[//tag\_name_2]$ but also $//tag\_name_1//tag\_name_2$.

Searching attributes values. Another important query in XPath is to find all the occurrences of an attribute having a given value, being it a simple word or a phrase. That is, queries like $@att\_name = "att\_value"$.

Whatever the case of the value, the algorithm to find out those attributes is that aimed at searching *phrase patterns*. That is, we will find all the phrase patterns given by the phrase built from the name of the attribute and its value: `att_name="att_value"`.

Other queries. We have just explained a common subset of the XPath queries. However, any other one can be answered using the representation we have presented. Some other queries like, for example, //tag_name [position() = i] or //tag_name [position() <= n] can be solved by simple locating the ith occurrence or the n-first occurrences of the *tag*, respectively, instead of locating all as we have seen. If we want to cope with queries involving *parent* XPath axis, /, it is not hard to imagine how to incorporate it from the discussion about //tag_name₁[//tag_name₂].

5   Experimental Results

An isolated Intel®Pentium®Core 2 Duo 2.13 GHz system, with 2 GB dual-channel DDR-667Mhz RAM was used in our tests. It ran Ubuntu 8.04 GNU/Linux (kernel version 2.6.24.23). The compiler used was gcc version 4.2.4 and -09 compiler optimizations were set. Time results measure CPU user time in seconds.

The four different XML documents used to run our experiments are:

- 0.5d and 9d : files generated with *xmlgen*, an XML data generator developed inside *XMark Project* (http://monetdb.cwi.nl/xml/).
- dblp : file corresponding to the revision of April 16, 2008.
- psd7003 : file of the public proteins database, *Integrated Protein Informatics Resource for Genomic and Proteomic Research* (http://pir.georgetown.edu/).

Table 1. Description of the documents used and compression properties

XML doc.	size	EN	MD	VT	VNT	#T	#NT	R1	R2	CT	DT
0.5d	55,32	832	12	148	85	1,665	9,468	31.82	29.06	4.16	0.66
dblp	282,42	6,928	6	70	1,750	13,856	61,649	41.50	37.32	28.84	3.68
psd7003	683,64	21,305	7	128	3,142	42,611	106,621	41.29	40.35	60.43	6.84
9d	1007,12	15,040	12	148	743	30,080	171,595	31.28	28.57	69.61	12.24

On the one hand, Table 1 presents the name of the XML documents used, their size in MBytes, their number of XML element nodes (EN)(x10³), their maximum depth level (MD), the number of different words in *tag words* (VT) and *text words* (VNT)(x10³) vocabularies, and the number of *tag words* (#T) and *text words* (#NT) that compose each document (x10³). On the other hand, the last four columns of Table 1 also show, respectively, the compression ratios (in %) obtained by XWT (R1) and the *(s,c)*-DC compressor (R2) over each XML document, as well as the compression (CT) and decompression (DT) times (in seconds) using XWT. Notice that XWT represents each XML file using about 30%-40% of its original size and, which is more striking, XWT only uses 3% more space than the needed to compress the documents with *(s,c)*-DC. That is, the powerful indexing capabilities of XWT only need 3% of extra space over the compressed text.

In Table 2 we can see the times obtained to answer the different common XPath operations explained in Section 4.2. The results presented are obtained using a XWT implementation with a waste of 3% of extra space for the structures of the partial counters used to speed up *rank* and *select* operations. From column 1 through column 12 we present the times obtained for *count* all the

Table 2. Searching operations

	1000 < f ≤ 10000				100 < f ≤ 1000				1 ≤ f ≤ 100				TCW	ATT
	count (μs)	first (μs)	all (ms)	snip. (ms)	count (μs)	first (μs)	all (ms)	snip. (ms)	count (μs)	first (μs)	all (ms)	snip. (ms)	(ms)	(ms)
0.5d	3.33	4.44	3.39	65.39	3.59	4.16	3.30	19.77	0.74	8.19	0.04	0.23	8.78	0.04
dblp	3.55	5.42	5.09	68.31	2.92	10.84	2.58	11.04	0.22	20.71	0.03	0.06	21.83	0.13
psd7003	3.04	6.64	6.06	55.65	3.40	6.33	2.06	7.27	0.41	15.80	0.04	0.10	62.08	0.02
9d	3.47	4.41	13.84	214.94	3.19	8.26	2.97	11.38	0.52	8.98	0.04	0.12	80.13	0.04

occurrences, locate the *first* position, locate *all* the positions, and extract all the 10-words *snippets* of a word. We distinguish 3 groups of words depending on their frequency f: *i)* $1000 < f \leq 10000$, *ii)* $100 < f \leq 1000$ and *iii)* $1 \leq f \leq 100$ and show the average time of searching for 100 distinct words (skipping *stopwords*) randomly chosen from the two vocabularies in each group.

In turn, columns 13 and 14 of Table 2 show, respectively, the average times obtained to locate all the occurrences of a certain pair of *start-end tags* containing a word (TCW) and to locate all the occurrences of an *att_name = "att_value"* pattern (ATT). In the first case, we have randomly chosen 100 *tags* and 100 *text words* from their respective vocabularies and have performed the algorithm. For the second operation, we used 100 randomly chosen pairs of the different *att_name = "att_value"* pairs with frequency between 1 and 100 existing in each XML document. Notice that here the search times of locate all the occurrences of a certain *att_name = "att_value"* pattern depend also on the number of words that form the *"att_value"*. The greater the number of words, the fewer the number of false positives we find in the root of the XWT that will spend time being processed down in the XWT. Moreover, it also depends on the frequency of the least frequent word of the *att_name = "att_value"* pattern. The greater the frequency, the greater the number of possible candidates we will check.

To properly valuate these data we need to take into account that, long ago [12], it has been clearly established beyond doubts that any kind of word or phrase search over the uncompressed text takes up to 8 times more time than to perform the same search over the compressed text, due to the fact that processing the uncompressed text imply to process around three times more bytes. Therefore, it only makes sense to compare our data, about searches to answer the different XPath queries, with those that could be obtained using a compressed version of the text. But in [4] it was experimental tested the performance of WT against compressed text. Different compressors were used to create the compressed text and to obtain the codewords for the WT. In all the cases the WT was dramatically faster to perform any kind of search, thanks to the self-indexing properties.

As consequence, it is more interesting to compare our results against those that can be obtained using a compressed and indexed version of the XML documents. In [4] the performance of WT was compared against inverted indexes to blocks of text (not to individual words occurrences). Different block sizes in the inverted index and different number of partial counters in the WT were used in order to compare the efficiency of both approaches when using different amounts of space. Results clearly proved the superior efficiency of the WT in searching words and phrases. The WT was superior even in recovering snippets when the amount of used space was inferior to the 40% of the original text size.

In our case, where not only we need to find a word, but also to process the text around it to know if that word is between some specific pair of *start-end tags*, or if it is an attribute value, etc. the use of our XWT will be even more advantageous.

6   Conclusions and Future Work

In this paper we introduce a strategy for compressing XML documents to about 35% of their size giving them, furthermore, self-indexing properties. This strategy is based on the use of a data structure we called XML Wavelet Tree (XWT). XWT is a new approach to the problem of storing, processing and querying XML documents in a time and space efficient way. Although our results are promising, more research must be done to improve the self-indexing properties. This is especially important for the tags representation, because most XPath queries imply the use of the document structure that the tags provide. On the other hand a systematic experimental evaluation of our XWT must be done comparing its performance with some of the other efficient XML representations.

References

1. Xml 1.0, W3C Recommendation of Extensible Markup Language (XML) Version 1.0, 5th edn., http://www.w3.org/TR/REC-xml
2. Xpath 2.0, W3C Recommendation of XML Path Language (XPath) Version 2.0, http://www.w3.org/TR/xpath20
3. Bordogna, G., Pasi, G.: Personalised indexing and retrieval of heterogeneous structured documents. Inf. Retr. 8(2), 301–318 (2005)
4. Brisaboa, N.R., Fariña, A., Ladra, S., Navarro, G.: Reorganizing compressed text. In: SIGIR 2008, pp. 139–146 (2008)
5. Brisaboa, N.R., Fariña, A., Navarro, G., Paramá, J.R.: (s, c)-dense coding: An optimized compression code for natural language text databases. In: Nascimento, M.A., de Moura, E.S., Oliveira, A.L. (eds.) SPIRE 2003. LNCS, vol. 2857, pp. 122–136. Springer, Heidelberg (2003)
6. Brisaboa, N.R., Fariña, A., Navarro, G., Paramá, J.R.: Lightweight natural language text compression. Inf. Retr. 10, 1–33 (2007)
7. Brisaboa, N.R., Fariña, A., Navarro, G., Places, A.S., López, E.R.: Self-indexing natural language. In: Amir, A., Turpin, A., Moffat, A. (eds.) SPIRE 2008. LNCS, vol. 5280, pp. 121–132. Springer, Heidelberg (2008)
8. Culpepper, J.S., Moffat, A.: Enhanced byte codes with restricted prefix properties. In: Consens, M.P., Navarro, G. (eds.) SPIRE 2005. LNCS, vol. 3772, pp. 1–12. Springer, Heidelberg (2005)
9. Fuhr, N., Grobjohann, K.: Xirql: A query language for information retrieval in XML documents. In: SIGIR 2001, pp. 172–180 (2001)
10. Li, H.-G., Aghili, S.A., Agrawal, D., Abbadi, A.E.: Flux: fuzzy content and structure matching of XML range queries. In: WWW 2006, pp. 1081–1082 (2006)
11. Manber, U., Myers, G.: Suffix arrays: a new method for on-line string searches. SIAM Journal on Computing 22(5), 935–948 (1993)
12. Moura, E., Navarro, G., Ziviani, N., Baeza-Yates, R.: Fast and flexible word searching on compressed text. TOIS 18(2), 113–139 (2000)

Superimposed Image Description and Retrieval for Fish Species Identification

Uma Murthy[1], Edward A. Fox[1], Yinlin Chen[1], Eric Hallerman[2],
Ricardo da Silva Torres[3], Evandro J. Ramos[3], and Tiago R.C. Falcão[3]

[1] Department of Computer Science, Virginia Tech, Blacksburg, VA 24061, USA
[2] Department of Fisheries and Wildlife Sciences, Virginia Tech, Blacksburg,
VA 24061, USA
[3] Institute of Computing, University of Campinas, Campinas, SP, Brazil
umurthy,fox,ylchen@vt.edu, ehallerm@vt.edu, rtorres@ic.unicamp.br

Abstract. Fish species identification is critical to the study of fish ecology and management of fisheries. Traditionally, *dichotomous keys* are
used for fish identification. The keys consist of questions about the observed specimen. Answers to these questions lead to more questions till
the reader identifies the specimen. However, such keys are incapable of
adapting or changing to meet different fish identification approaches, and
often do not focus upon distinguishing characteristics favored by many
field ecologists and more user-friendly field guides. This makes learning
to identify fish difficult for Ichthyology students. Students usually supplement the use of the key with other methods such as making personal
notes, drawings, annotated fish images, and more recently, fish information websites, such as Fishbase. Although these approaches provide useful
additional content, it is dispersed across heterogeneous sources and can
be tedious to access. Also, most of the existing electronic tools have limited support to manage user created content, especially that related to
parts of images such as markings on drawings and images and associated
notes. We present SuperIDR, a superimposed image description and retrieval tool, developed to address some of these issues. It allows users to
associate parts of images with text annotations. Later, they can retrieve
images, parts of images, annotations, and image descriptions through
text- and content-based image retrieval. We evaluated SuperIDR in an
undergraduate Ichthyology class as an aid to fish species identification
and found that the use of SuperIDR yielded a higher likelihood of success in species identification than using traditional methods, including
the dichotomous key, fish web sites, notes, etc.

Keywords: superimposed information, image annotation, image retrieval,
fish, species identification, biodiversity, user study.

1 Introduction

Identification of fish species is critical to study of fish ecology and management of fisheries, and follows from precise observation of external morphology,

M. Agosti et al. (Eds.): ECDL 2009, LNCS 5714, pp. 285–296, 2009.

coloration, and internal characters. Fish species identification is used in many important tasks such as stream water assessment, where the presence (or absence) of species and their number determine the quality of the stream water. However, learning to correctly identify fishes is difficult for Ichthyology students, who often find traditional dichotomous keys intimidating. Dichotomous keys (Figure 1-a) consist of questions about morphological features of a specimen. Questions are presented in pairs called couplets, and one member of a couplet should describe the specimen in question. Depending on which member of the pair is appropriate, the user will be directed to further questions until, after traversing through a series of couplets, the specimen is identified. For example, if the specimen lacks paired fins and a jaw and has seven gill openings on each side, then it is a lamprey (a family of fishes), and the user is directed to questions to identify the species of that lamprey. If it has paired fins, a jaw, and a single gill opening, then it will belong to any of a number of families, and the user will be asked other questions to identify the family.

However, such keys are incapable of adapting or changing to meet different fish identification approaches, and do not accommodate the range of learning styles actually utilized by students. In particular, keys often do not focus upon distinguishing characteristics favored by many field ecologists and more user-friendly field guides (e.g., Page and Burr 1991 [1]).

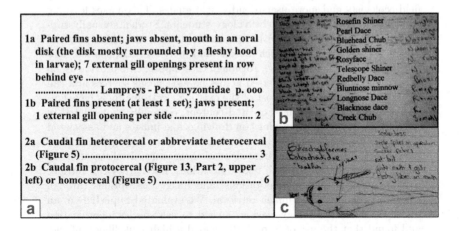

Fig. 1. Fish species identification methods. a) Snippet of a dichotomous key; b) Printed list of species superimposed with notes by student on distinguishing characteristics; c) Marked diagram of a fish, indicating distinguishing characteristics, along with other details of the fish, on a notecard.

In addition to the key, students use a variety of other artifacts and tools to study species identification, such as:

- **Paper-based artifacts and tools**: dichotomous keys, personal notes, notecards, textbooks, field guides, marked drawings, annotated printed pictures of fish, printed lists of families, genera, species, etc.

– **Electronic artifacts and tools**: images from Google, Yahoo! search, images on a PowerPoint presentation or Word document, FISHBASE [2], EFISH [3], EKEY [4][1], etc.

Almost always a student uses a combination of these artifacts and tools. For example, a student may use digital pictures of fish specimens organized in a PowerPoint presentation. He might then make notes in writing (or typed notes) corresponding to the fish pictures on the slides. Another example is the use of notecards (Figure 1-c). On one side of the notecard, a student might sketch fish diagrams, annotate them with the distinguishing characteristics, and on the other side write a description of the fish. While studying fish species, students typically move back and forth among various artifacts and tools and a preserved specimen, all the while trying to memorize distinguishing characteristics, scientific names, and other information about the fish.

Although the aforementioned methods work for students, they tend to be tedious, time-consuming, and sometimes unsuccessful. Some practical problems are challenging technical terms used in dichotomous keys, insufficient visual or descriptive information for definitively answering questions posed in a dichotomous key, absence or limited variety of reference specimens, inability to share information or identification problems with others not physically present, and the tediousness of accessing information (images, descriptions, markings, notes, etc.).

1.1   Better Information Management and Access

The above overview on fish species identification indicates that electronic support for learning to identify fish species is limited and can be improved considerably. There are many issues to consider, including better support for:

– describing and accessing fish images
– managing and accessing user created content, especially that related to parts of images such as markings on drawings and images and associated notes
– sharing this information with others
– providing all of the above listed capabilities in a well-integrated solution

We chose to focus on providing better support to manage and access parts of images and related information. Several domains require scholars to work with images with a significant number of details. In the past, paper-based tools and techniques to work with such information have been used with a fair amount of ease and success. Although the electronic world has provided us with several advantages over paper, such as ease of editing, ease of sharing, and better ability to store, organize, and access information (searching, browsing, etc.), yet the electronic tools used to support working with parts of images are usually not well integrated or do not interoperate well, leading to ineffective and inefficient task execution.

---

[1] Details of these systems are in Section 4.

We developed SuperIDR, an image description and retrieval tool, which addresses some of these questions by enabling the user to add content in the form of image markings and annotations, and providing improved image search using text- and content-based image retrieval. Evaluation of SuperIDR in an undergraduate Ichthyology class showed that students identified more unknown fish specimens correctly with SuperIDR than with traditional methods (dichotomous keys, notecards, fish web sites, etc.).

2   A Superimposed Image Description and Retrieval Tool (SuperIDR)

We developed SuperIDR, a superimposed image description and retrieval tool, with the aim of helping users to work with parts of images *in situ* - i.e., being able to select, annotate, retrieve, and share parts of images in the context of the original image. The basis for the functionality was a result of combining *Superimposed Information* on images along with *Content-Based Image Retrieval*.

Superimposed information (SI) refers to new information laid over existing information (such as bookmarks, annotations, etc.) [5]. Superimposed applications (SAs) allow users to lay new interpretations over existing or base information. SAs employ "marks", which are references to selected regions within base information. SAs enable users to (a) deal with information of varying granularity, and (b) select or work with information elements at sub-document level while retaining the original context. In SuperIDR, we worked with *image marks*, or references to parts of fish images such as fin, mouth, body, tail, etc., in the context of the entire fish image.

Content-Based Image Retrieval (CBIR) systems aim to retrieve images similar to a user-defined specification or pattern based on content properties (e.g., shape, color or texture), usually encoded into feature vectors [6]. We use the Content-Based Image Search Component (CBISC), an Open Archives Initiative (OAI)-compliant component that provides an easy-to-install search engine to query images by content [7]. It can be readily tailored for a particular collection by a domain expert, who carries out a clearly defined set of pilot experiments. It supports the use of different types of vector-based image descriptors, and then easily combines them to yield improved effectiveness. CBISC encapsulates a metric index structure to speed up the search process, which can be easily configured for different image collections. For SuperIDR, we used the .NET version of CBISC [8] to index and retrieve complete images as well as parts of images (defined by image marks).

SuperIDR is an extension of a PC-only version that we developed earlier [9]. It has been developed in C# and uses MySQL as the database. We developed SuperIDR to work with tablet PCs, taking advantage of pen-based input. We felt that this would emulate, as close as possible, the use of a pen on paper, which many biologists are used to since they make markings often. Also, we felt that when used in the field, it would be more convenient to work with pen input versus using a touchpad or keypad. For example, a fisheries scholar could use the pen-input to mark a feature on a fish image and then write notes describing that mark.

SuperIDR is seeded with details of 207 species of freshwater fishes of Virginia, taken from [10]. Each species has a representative image as shown in Figure 2-b. In addition to making annotations, SuperIDR allows searching and browsing of species descriptions, images, image marks, and annotations. A user can search in one of two ways: 1) perform text-based search (full-text and field-wise search, powered by Lucene.NET[2]) on species descriptions and annotations, where the query may include terms, phrases, or their boolean combination; 2) perform content-based image search on images and annotated-image-marks, where the query could be a complete image or part of an image. Finally, in SuperIDR, a user can browse through species information either through a taxonomic organization of species based on family and genera or through an electronic version of the dichotomous key. The scenario described below illustrates how an Ichthyology student would use SuperIDR.

Scenario: Matt is a junior, majoring in Fisheries, enrolled in the Ichthyology class. He has some experience with species identification but still is quite intimidated by dichotomous keys. In the past, he has supplemented the use of the keys with personal notes, pictures from the web, textbooks, etc. The Ichthyology teacher has decided to use a new software tool in class, SuperIDR. Matt is moderately familiar with computers and is enthusiastic about using SuperIDR, as it will help him with species identification.

Matt walks through the various features of SuperIDR, beginning with a taxonomy browser, where all species have been organized according to families and then genera. He browses to one of this week's species, the *redear sunfish* (Figure 2-a). The species description screen provides him details about the fish, including physical description, habitat, food habits, etc. (Figure 2-b). However, it does not have distinguishing characteristics explained in detail, which are an important part in identifying the fish species. Matt adds a new annotation using a pen input. He marks the fish picture and associates the marks with text explanations, notes in Matt's own words, making it easier for Matt to remember and learn about this species. He makes several such annotations on the different fish images about which he is learning (Figure 2-c).

In two weeks, Matt's teacher holds a practice specimen identification test. Some of the unknown specimens are in jars and some are in the form of images posted on the course web site. Matt examines the fish specimens for specific characteristics and uses text search on annotations and species descriptions (Figure 2-d, e). Browsing through the result list and occasionally clicking on a result for details, he is able to narrow down the options to the species of the unknown specimen. Sometimes he uses the electronic key available in SuperIDR. For some images of unknown fishes, Matt uses the content-based image search feature. He vaguely has some idea about the fish, but does not remember the exact keywords. He can search on all or part of the fish image content and browse through the result list as he would do with the text search results (Figure 2-f, g). Using SuperIDR not only makes it faster for Matt to identify the species of the unknown specimens but also reinforces descriptions of those species through distinguishing characteristics.

---

[2] http://incubator.apache.org/lucene.net/

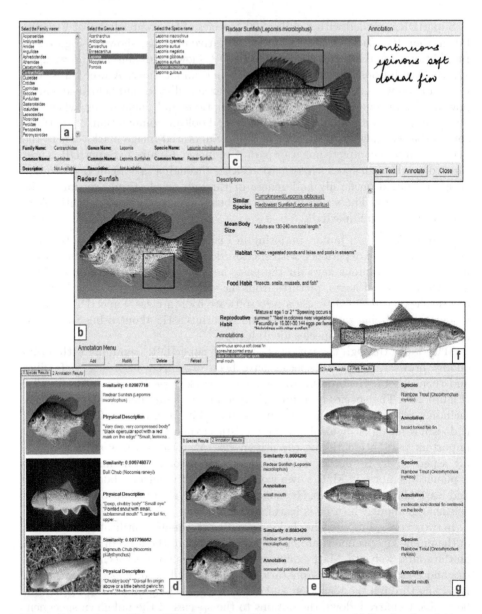

Fig. 2. Screenshots of SuperIDR features: a) Taxonomy browser; b) Species description screen shows details of species and annotations - the highlighted annotation (bottom right) is associated with a marked region in the image; c) Annotation screen – pen input is used to mark the fish image and "write" the annotation, which gets automatically recognized; d) Eight species description results for the text query '"red mark" "small mouth" "pointed snout" "no spots"'; e) Two annotation results for the same text query; f) Content-based image search, where the query is the marked region that covers black dots on the body of a rainbow trout; g) Image search results, which can be annotated image marks (shown in the figure) and/or complete images.

3   Classroom-Based Evaluation

We had several interactions with fisheries students, faculty, and researchers to help with development and improvement of SuperIDR. In a two-month-long longitudinal formative evaluation with five fisheries scholars (three students, one faculty member, and one researcher), we made improvements and added new features to SuperIDR, including replacing database indexing with full text indexing of species descriptions and annotations, adding user controls for better management of annotations, and fixing bugs.

In order to assess the effectiveness of SuperIDR as an aid to species identification, we conducted an experiment in an undergraduate Ichthyology class. In the experiment, we compared the use of SuperIDR with traditional methods of species identification (either by using the key, personal notes, markings on images, use of websites, notecards, etc., or a combination of the aforementioned methods). Most students in that class were juniors in Fisheries Science. Participation in the evaluation study was voluntary and students' performance on the quiz (which was part of the experiment) was not included in their final grades.

There were 28 students in the class and all were present in the first meeting of the experiment. However, we considered data from only 18 of them since the remaining either did not show up for the remaining meetings or we had incomplete data from them. Seventeen of the 18 students rated themselves as moderately or very familiar with computers although 13 of them indicated that they had very low expertise with tablet PCs[3]. Sixteen students had previously taken a course on species identification (not necessarily fish species), so most of them were familiar with the task. Most students used annotations of some form on paper for learning, but most of them did not use digital annotation systems.

We conducted the experiment in April 2008, in the 12th week of a 15-week-long class. SuperIDR was set up on tablet PCs. Due to limited availability of tablet PCs, we asked the students to form teams of two (based on their seating in class that day) and share a tablet PC. Students were given a tutorial on the tool as well as a user manual for later reference. We asked the students to use the tool for the week, explore its features and make several annotations on images of the species they had studied, in order to prepare for a test the following week.

In a lab period in the following week, we had two sessions - session 1 and session 2. We divided the class (consisting of teams) into two parts - A and B. In each session, students working in teams had to identify 20 unknown specimens. In session 1, teams in part A used only SuperIDR as an aid to species identification, and teams in part B used traditional methods. In session 2, teams in part A used traditional methods and teams in part B used only SuperIDR.

In the experiment, we collected the following data:

- Species identification responses for the 40 unknown specimens: For a specimen to be correctly identified, its family, genus, and scientific name had to be correctly identified. We computed a team's score, i.e., the number of correctly identified specimens, for each session.

---

[3] For these questions, students rated themselves on a scale of 1 to 5 with 1 being least familiar and 5 being very familiar.

- Entry (demographic data and data about prior experience with species identification and software tools) and exit questionnaire responses (qualitative feedback on the use of SuperIDR)
- Log data of user interaction with SuperIDR

3.1   Experiment Data Analysis

Species Identification Responses: Table 1 provides a summary of the scores for teams using different methods across the two sessions. In the experiment, each team worked with one specimen at a time, and three main factors impacted the outcome – correct or incorrect – of a species identification task: 1) the nature of the specimen; 2) the team (of two students) working on it; and 3) the method used to identify it - the tablet PC tool method or traditional method. Keeping this in mind, we used the generalized linear model with a binomial (logit) link function to analyze the species identification responses. Generalized linear models, an extension of the linear modeling process, allow models to be fit to data that follow probability distributions other than the Normal distribution (such as the Poisson or the Binomial distribution) [11].

Statistical analysis using R[4] showed that the the team (p-value=0.015) and the method (p-value=0.011) had a significant impact on the outcome of the species identification task, while the session (hence, the nature of specimen) did not impact the outcome significantly. The mean values in Table 1 show that using SuperIDR yielded a higher likelihood of identifying a specimen correctly than using traditional methods.

Exit Questionnaire Responses[5]: Students gave their feedback on the tool through an exit questionnaire. In general, students' knowledge of computer-assisted fish identification improved during the course of the study. Students still preferred the use of the key (10) versus SuperIDR (6) (two students didn't respond to this question). One reason for this could be the timing of the study, which was towards the end of the semester. At that time, students already had established practices for species identification. Most were busy practicing their known skills for upcoming exams, and did not spend too much time on the tool (also indicated in the log data). Another reason for the students' preference was that many were frustrated with the pen input. Although we felt that it might have been useful, students were frustrated because of the poor ink response (slow and distorted sometimes) and the recognition of handwriting. Students also felt that the number and quality of images was low. This was a serious drawback since there was no support for students to add their fish pictures or diagrams to the existing image collection. Some of these reactions are reflected in the students' comments:

- "Very neat but may take a while to master all the key concepts."
- "It was very helpful."

---

[4] http://www.r-project.org/
[5] The entry questionnaire responses have been summarized above, where information about the participants is given.

Table 1. Number of correct responses, from 20 specimens, of different teams using traditional methods and using SuperIDR

Traditional Methods			SuperIDR		
Team ID	Session	Correct	Team ID	Session	Correct
2	1	15	2	2	18
4	1	16	4	2	17
6	1	13	6	2	17
11	1	12	11	2	16
3	2	8	3	1	14
5	2	13	5	1	15
9	2	12	9	1	10
10	2	10	10	1	14
13	2	11	13	1	11
Mean		12.2	**Mean**		14.67

- "Very helpful for taxonomy, still needs better photos."
- "If you had started the program at the beginning of the semester, there would be higher success and better likelihood that we may use it."

Log Data: Log data showed that students used the Browse feature the most (logged 2450 times) and made 500 search requests with 99 being image searches and 451 being text searches. We were very interested in knowing how students make use of image marks and of superimposed information in species identification, through both annotation and searching. However, there was no record of use of the Annotation feature in the log data. The reason for this (as also mentioned above) is the timing of the experiment. Most students used it only on the two days they came to class for the experiment.

3.2 Evaluation Summary

Overall, we believe that SuperIDR was well received by Fisheries scholars as an aid to species identification. Students identified more specimens correctly with SuperIDR than with traditional methods, with just a week of use. The questionnaire responses indicated a preference for traditional methods versus SuperIDR and the log data did not record activity on the annotation front. As mentioned earlier, we think this was because of the timing and duration of the experiment. One more indication of student interest in the tool is that six out of the fourteen teams chose to keep SuperIDR for three more weeks till the end of the semester. Students said that they wanted to explore the tool further. In addition to the results mentioned, we received several suggestions for extensions and improvements to the tool.

4   Related Work

Recently, there have been systems and tools developed to support biodiversity researchers and scholars. For example, Lyons et al. [12] described a photo-based computer system for identifying Wisconsin fishes that features multiple images of each species which can be accessed by dichotomous key, a query tool, and a slide show. EcoPod is a PDA-based application, which replaces traditional paper field guides with a mobile computing platform [13]. It focuses on enabling users to work with keys to identify species in the field. It has limited support for adding user content and also does not have any CBIR capabilities. In [14], the authors achieve a high level of accuracy in automatically identifying moth species using data mining techniques. However, their system does not provide any text annotation or retrieval capabilities.

Other systems include the popular online FishBase [2], an information system providing information to fisheries professionals on the 31,000 known species of fishes. Information access in FishBase is limited to browsing and field-wise searching. FishBase also provides forums to discuss problems with other scholars. EFish [3] is another system providing species identification and life history information for 200 species of freshwater fishes of Virginia. Information access in EFish is limited to browsing and there is no facility to add user content. EKEY [4], a predecessor to SuperIDR, is a web-based system with an electronic dichotomous key, a taxonomy browser and provides text- and shape-based search. SuperIDR builds upon the features of these systems, while enabling the user to add content to the existing information base. Also, it provides support for working with specific parts of images and performing content-based image description and retrieval. In addition, it has pen-input capabilities, mimicking free-hand drawing and writing on paper.

Many digital libraries have annotation capabilities, usually focusing on annotations of complete documents. There has been work done to provide annotation support for image digital libraries such as [15], [16]. Also, there are several photo annotation systems, including online tools such as Flickr[6] and Fototagger[7]. However, most of these systems do not combine the capabilities to work with text- and content-based description and retrieval of parts of images. With regard to integrating subdocuments with digital libraries, we developed an architecture for representing SI in DSpace[8] [17]. The focus of that work was mainly on text documents, with limited support for images.

5   Discussion and Future Work

We presented SuperIDR, a tool that combines text- and content-based image description and retrieval, as an aid to fish species identification. Students performed well with SuperIDR and it was generally well received by fisheries students, faculty, and researchers.

Our initial goal was to provide support for scholars to work with images with significant numbers of details such as those in Fisheries Sciences. Through field trips, in-class observations, and interactions with fisheries scholars, we found

---

[6] http://flickr.com

[7] http://www.fototagger.com/

[8] http://dspace.org

that working with parts of images is important for fish species identification. Some reasons for this are: 1) distinguishing characteristics usually focus on part of the fish (and fish image); 2) students work with (annotate, browse, and search) parts of images in their notes, online information, and keys, while studying and while identifying fishes; 3) they frequently go back and forth between notes and marked fish images; and 4) they compare parts of different fish images to study differences between species, genera, and families. However, we did not find any significant evidence of such use from our experiments (log data). As mentioned earlier, we believe this was due to the timing and duration of our experiment.

We feel that we need to understand and provide better digital support for scholars to work with parts of images in the digital world. Towards this, we are conducting in-depth interviews with the students who participated in the classroom experiment. We have conducted four such interviews and have received more details on use of SuperIDR and on working with parts of images. For example, one student said that SuperIDR would be useful in any field with an "–ology" suffix. Another student said that the most useful features in SuperIDR with regard to species identification are working with images and adding personal notes to images.

We have made further improvements to SuperIDR based on feedback received from the user studies, and will be making it available to download to fisheries scholars. A natural extension of this system is to make it web-enabled, thus supporting content-sharing across image management systems and social networks.

We used *fish species identification* as the specific scholarly task to situate our research ideas. However, we believe that our work is applicable to any scholarly task/domain involving images with a significant number of details, such as analyzing paintings in art history, examining a building style in architecture, understanding trees in dendrology, etc.

Acknowledgments

Thanks go to Microsoft (under a tablet PC grant) for funding this project. We also thank our previous sponsors who helped initiate this work including NSF (DUE-0435059), CAPES, FAPESP, FAEPEX, and CNPq. We especially thank Dr. Donald Orth, Ryan McManamay, Lindsey Pierce, and all the participants in the evaluations. Many thanks go to Jason Lockhart and the College of Engineering at Virginia Tech for loaning us tablet PCs to conduct experiments. We are grateful to the American Fisheries Society for use of copyrighted material from Jenkins and Burkhead's *Freshwater Fishes of Virginia*.

References

1. Page, L., Burr, B.: A Field Guide to Freshwater Fishes: North America North of Mexico (Peterson Field Guides). Houghton Mifflin Co., New York (1991)
2. WorldFishCenter: Fishbase: A global information system on fishes (1997), http://www.fishbase.org/home.htm

3. Helfrich, L., Newcomb, T., Hallerman, E., Stein, K.: Efish: The virtual aquarium (2001),
   http://www.cnr.vt.edu/efish/
4. da, S., Torres, R., Hallerman, E., Jenkins, R.E., Burkhead, N.M., Herrington, B.: Ekey - the electronic key for identifying freshwater fishes (2004),
   http://fwie.fw.vt.edu:8080/ekey/
5. Maier, D., Delcambre, L.M.L.: Superimposed information for the internet. In: WebDB (Informal Proceedings), pp. 1–9 (1999)
6. da, S., Torres, R., Falcão, A.X.: Content-based image retrieval: Theory and applications. Revista de Informática Teórica e Aplicada 13(2), 161–185 (2006)
7. da, S., Torres, R., Medeiros, C.B., Goncalves, M.A., Fox, E.A.: A digital library framework for biodiversity information systems. International Journal on Digital Libraries 6(1), 3–17 (2006)
8. Kozievitch, N.P., Falcão, T.R.C., da, S., Torres, R.: A .Net implementation of a content-based image search component. In: Demo Session, 23th Brazilian Symposium on Databases, Campinas, Brazil (2008)
9. Murthy, U., da Silva Torres, R., Fox, E.A.: Sierra - a superimposed application for enhanced image description and retrieval. In: Gonzalo, J., Thanos, C., Verdejo, M.F., Carrasco, R.C. (eds.) ECDL 2006. LNCS, vol. 4172, pp. 540–543. Springer, Heidelberg (2006)
10. Jenkins, R.E., Burkhead, N.M.: Freshwater Fishes of Virginia. American Fisheries Society, Bethesda (1994)
11. McCullagh, P., Nelder, J.A.: Generalized linear models. CRC Press, Boca Raton (1989)
12. Lyons, J., Hanson, P., White, E.: A photo-based computer system for identifying wisconsin fishes. Fisheries 31(6), 269–275 (2006)
13. Yu, Y., Stamberger, J.A., Manoharan, A., Paepcke, A.: EcoPod: a mobile tool for community based biodiversity collection building. In: JCDL 2006: Proceedings of the 6th ACM/IEEE-CS Joint Conference on Digital Libraries, pp. 244–253. ACM Press, New York (2006)
14. Mayo, M., Watson, A.T.: Automatic species identification of live moths. Knowledge-Based Systems 20(2), 195–202 (2007)
15. Jochum, W., Kaiser, M., Schellner, K., Wirl, F.: Living memory annotation tool — image annotations for digital libraries. In: Kovács, L., Fuhr, N., Meghini, C. (eds.) ECDL 2007. LNCS, vol. 4675, pp. 549–550. Springer, Heidelberg (2007)
16. Stein, A., Thiel, U., Brocks, H.: COLLATE – collaboratory for annotation, indexing and retrieval of digitized historical archive material (2004),
    http://www.collate.de/
17. Archer, D.W., Delcambre, L.M., Corubolo, F., Cassel, L., Price, S., Murthy, U., Maier, D., Fox, E.A., Murthy, S., Mccall, J., Kuchibhotla, K., Suryavanshi, R.: Superimposed information architecture for digital libraries. In: Christensen-Dalsgaard, B., Castelli, D., Ammitzbøll Jurik, B., Lippincott, J. (eds.) ECDL 2008. LNCS, vol. 5173, pp. 88–99. Springer, Heidelberg (2008)

Significance Is in the Eye of the Stakeholder

Angela Dappert and Adam Farquhar

The British Library
Wetherby, West Yorkshire LS23 7BQ
{Angela.Dappert,Adam.Farquhar}@bl.uk

Abstract. Custodians of digital content take action when the material that they are responsible for is threatened by, for example, obsolescence or deterioration. At first glance, ideal preservation actions retain every aspect of the original objects with the highest level of fidelity. Achieving this goal can, however, be costly, infeasible, and sometimes even undesirable. As a result, custodians must focus their attention on preserving the most significant characteristics of the content, even at the cost of sacrificing less important ones. The concept of significant characteristics has become prominent within the digital preservation community to capture this key goal. As is often the case in an emerging field, however, the term has become over-loaded and remains ill-defined. In this paper, we unpack the meaning that lies behind the phrase, analyze the domain, and introduce clear terminology.

Keywords: Digital preservation, properties, characteristics, significant properties, significant characteristics, applicable properties, requirements.

1 Introduction

Custodians of digital content take action when the material that they are responsible for is threatened by, for example, obsolescence or deterioration. At first glance, ideal preservation actions retain every aspect of the original objects with the highest level of fidelity. Unfortunately, achieving this goal can be costly, infeasible, and sometimes even undesirable. As a result, custodians must focus their attention on preserving the most significant characteristics of the content, even at the cost of sacrificing less important ones. Furthermore, we must verify that the preservation actions we apply actually preserve these characteristics. The concept of significant characteristics has become prominent within the digital preservation community to capture this key goal [9].

The term *significant characteristic* has become over-loaded and remains ill-defined. This has some unfortunate consequences. First, communication is hampered, because the term is used in substantially different ways by different authors. Second, based on an extensive analysis of policy and strategy documents related to digital preservation [7], the current definitions do not actually meet the needs of content custodians. Content custodians need to express priorities, as well as requirements that go beyond the significance of properties and values. Third, implementations based on existing definitions fail to meet the needs of content custodians because they focus too

M. Agosti et al. (Eds.): ECDL 2009, LNCS 5714, pp. 297–308, 2009.

tightly on characteristics of content and format, and do not take account of the context in which preservation actions take place.

1.1 Related Work

Chris Rusbridge [18] eloquently states why the quest for faithfulness to the original in all respects is both excessive and impractical in most preservation situations. Original work on significant characteristics comes out of the Cedars project [5], work at the Australian National Archives [14], the InSPECT project [12], PLANETS [1, 2, 7, 9, 19] and others. Surveys of related work are provided by Knight [13] and Wilson [21].

Terminology is used inconsistently and includes significant properties [e.g., 10, 12, 13], significant characteristics [1], essence [14], aspects [8], and others. Nonetheless, a widely accepted definition for significant properties is Andrew Wilson's [21]:

"The characteristics of digital objects that must be preserved over time in order to ensure the continued accessibility, usability, and meaning of the objects, and their capacity to be accepted as evidence of what they purport to record."

The term "characteristics", which describes what must be preserved in this definition, is interpreted in two conflicting ways. Some interpret it to refer to the abstract properties of file formats [e.g., 1, 12], whereas others interpret it to refer to the values of properties of specific digital objects [2].

We also find different interpretations of the term "digital objects", which describes whose characteristics need to be preserved. In 2002, an OCLC/RLG working group[16] stated that the properties of data objects need to be preserved; Brown [3] applies it to information objects as opposed to data objects in the OAIS sense of the terms [4]; Becker [1] applies it to the characteristics of specific file formats. Knight hints that the characteristics of the environments in which digital objects are rendered may also have to be preserved [12], but this idea is not fully articulated.

The need to clarify the difference between significant characteristics and representation information has repeatedly been voiced [e.g., 10, 13], but not yet addressed.

1.2 Contributions

In this paper, we probe into the meaning of Wilson's definition. The exploration has led us to shift focus from a priori significance of characteristics in files or file formats to a new model in which stakeholders state requirements expressing significance. In contrast with previous work, we

- distinguish "properties" and "characteristics" (Section 2.1);
- provide a conceptual model, identify the types of objects which may have properties and characteristics, and unify the treatment of properties and characteristics across preservation objects, preservation actions, and their environments (Section 2.2);
- clarify who and what determines significance (Section 3);
- list observations about practical uses of significant characteristics. They justify why we treat significant characteristics as first class concept that is a subtype of requirement (Section 3);
- clarify the difference between significant characteristics, applicable properties and representation information (Section 4).

2 Foundations

2.1 Modelling Language – What Must Be Preserved?

In order to write with a reasonable level of precision, we need to introduce a basic vocabulary to talk about entities, properties, values, and so on. We use an object-oriented model with roots in [6]. The core terms in this vocabulary are:

Entity – Anything whatsoever.

Class – A class is a set of entities. Each of the entities in a class is said to be an instance of the class.

Individual – Entities that are not classes are referred to as individuals.

Property – A property is an individual that names a relationship.

Characteristic – A property / value pair associated with an entity. The value is an entity. This relationship is illustrated in Figure 1.

Facet – A facet is a property / value pair associated with a characteristic. The value is an entity.

Constraint – A Boolean condition involving expressions on entities.

Unless otherwise specified, a characteristic is directly associated with an entity. It is sometimes useful to associate a characteristic with all of the instances of a class. We refer to this as a *class characteristic*. Furthermore, we say that a property *applies* to a class if it can be meaningfully associated with some instances of the class.

Fig. 1. Properties and characteristics

We can use this language in the domain of digital objects and preservation. For example, *file* is a class; *f1.txt* is an instance of the class *file*; *fileSize* is a property; the property *fileSize* applies to *file*; the file *f1.txt* has the characteristic *fileSize* = 131342.

If every instance of *myDigitalSoundObject* has been virus-scanned, then it has the class characteristic *isVirusScanned* = "yes".

Important additional information about a characteristic, such as how a value is encoded, the unit of measure, or the algorithm or tool used to compute it can be specified using facets.

Under this terminology, it is clear that a characteristic (property / value pair) may be preserved by a preservation action, but that the abstract property cannot be. It is therefore not sensible to speak about preserving a "significant property."

2.2 Conceptual Model - Whose Characteristics Are Captured?

A key aspect of our model is that each of the classes *preservation object*, *environment*, and *preservation action* illustrated in Figure 2 may have properties and characteristics. It is important to distinguish the types of entity which are characterized. They play different roles during preservation processes and have different applicable

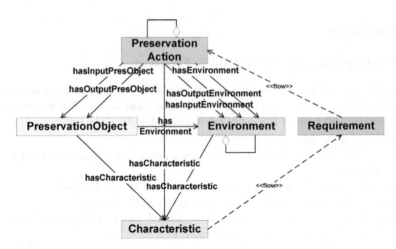

Fig. 2. Conceptual model: Characteristics of preservation objects, preservation actions and environments

properties. The labelled arrows summarise some of the properties that apply to the class' instances. This section discusses each of these concepts in more detail.

Preservation Object

The *preservation object* concept corresponds to those objects in need of preservation. It has subclasses on three tiers, as illustrated in Figure 3. The top two tiers are associated with specific physical representations of digital objects. The top tier comprises physical objects, such as bitstreams and its subclasses including *bytestreams* and *files*. The middle tier comprises *representations* of logical objects consisting of *representation bitstreams* that are needed to create a single rendition of a logical object (e.g., the set of *html* and *gif* files[1] needed to render the web version of a journal article). The bottom tier comprises logical objects such as *intellectual entities* and *components*.

These concepts are explained in detail in [8] and [9]. This presentation is somewhat modified to align terminology with PREMIS [15] and FRBR [11].

An *intellectual entity* is a distinct intellectual or artistic creation. PREMIS [15] defines it as a set of content that is considered a single intellectual unit for purposes of management and description. The *intellectual entity* can be extended in ways to meet the needs of stakeholders. For example, in the library setting, common subclasses include *collection*, *work*, and *expression*. In an archival setting, subclasses such as *fonds* and *series* are also relevant. Most repositories support discovery and delivery of *intellectual entities* such as books, videos, and articles. They may augment these with *work* and *expression* subclasses to capture useful FRBR distinctions [11]. *Intellectual entities* may also correspond to larger structures, such as *collections*, which may not be of interest to the end-user, but may be significant in preservation decisions.

---

[1] The formal definition of such a statement would of course contain a persistent unique identifier of the exact version of the file formats. For improved readability of examples we casually refer to file formats by their file extension.

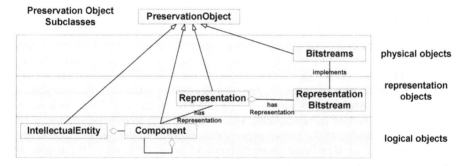

Fig. 3. Preservation Object Subclasses

During preservation, it is often necessary to consider fine-grained *components* of an *intellectual entity*. Examples include *table, image, title, substring,* or even an individual *character*. The *component* entity can be decomposed in several ways, such as by the type of content (e.g., *textComponent, imageComponent*), or by structure (e.g., *headerComponent* or *tableOfContentsComponent*). Values for characteristics of components can be measured from their associated representations (e.g. the *font* of a character component can be extracted from its representation bitstream.).

Properties can be applicable to objects in every tier. For example:

- *fileSize* or *encoding* are applicable to files.
- *numberOfFilesInTheRepresentation, totalRepresentationSize, resolution,* or *preservationLevel* are applicable to representations.
- *pageCount* or *frameRate* are properties applicable to intellectual entities such as a journal article or video. *Alignment* is a property applicable to a *textComponent*. *SemanticInterpretation* can be a characteristic of any component.

Environments

Preservation objects don't exist in isolation. A user or system interacts with an object in an *environment*. Therefore, every preservation object is associated with one or more environments that support different purposes or functions. Examples of environment purposes include delivery (remote or local), creation, ingest, and preservation. Examples of environment functions include rendering, editing, executing, and printing.

Every environment may be broken down into sub-environments that are needed for the interpretation and representation of the preservation object. Examples include hardware and software environments, the community, budgetary factors, the legal system, and other internal and external factors. They correspond to an extended notion of the environment description of *representation information* [4] and are enumerated in [8].

Environments have characteristics. For example:

- *memoryUsage* = "low" is a characteristic of a software tool environment that renders the preservation object.
- *numberOfIntermediateCopies* <= 3 and *preservesColourDepth* = "yes" are characteristics of a preservation service which is part of a preservation action's environment. They can be captured in a preservation services registry.

Preservation Actions

Custodians of digital content take actions to mitigate the risks that they identify. A preservation action event takes place when a preservation service is invoked. A preservation action is applied to an initial, or input, preservation object and environment. The result of the action is either a new output preservation object and/or a new environment. Together they mitigate the risk that the action addresses. For example, a Microsoft Word bytestream is migrated to a *pdf* bytestream in order to lock in the desired look-and-feel of the document. The output environment must support a *pdf* viewer. Characteristics of the output preservation object and the output environment are validated against significant characteristics in order to quantify the degree of compliance. This approach to describing preservation actions works for migration, emulation, hardware replacement, and other solutions.

Every preservation action is associated with the environment required for its own execution. The hardware on which the action is executed and the preservation service that is invoked are parts of this environment.

Preservation actions may have characteristics. For example, *numberOfIntermediateCopiesProduced* = 2 is a characteristic of a preservation action. This might be used to identify preservation actions that violate copyright regulations or license agreements that limit the number of intermediate copies created.

3 Observations about Significance in Digital Preservation

Observation 1

An idea, concept, act, or thing is not inherently significant. A stakeholder attributes significance to something, typically in a context relevant to some purpose or goal. In the digital preservation context, significance is determined by the stakeholders involved in the preservation process. These include the producer of the digital object, the custodian who holds it, and the consumer who will access it. The stakeholder's priorities may be captured as requirements ("business rules") by the custodian, who needs to ensure that preservation actions satisfy them. Requirements are an explicit statement of a stakeholder's values. These requirements influence the preservation process, and are often captured in preservation guiding documents, such as strategy or business documents. The conceptual model must have a requirement concept for capturing significance explicitly.

There is a notion that significant characteristics refer to the intellectual content - the essence of the digital object. In contrast, other characteristics are merely circumstantial, not significant, and can be ignored in preservation actions. Unfortunately, it is not possible to determine out of context which properties reflect content and which reflect circumstance. Consider a number that is formatted with the colour red. In some settings, the colour may be for a visual effect - simply pretty, circumstantial, and insignificant; in another setting, the colour may be to indicate that it is to be understood as negative and therefore has a significant semantic impact. This can only be determined by the stakeholder capturing significance explicitly.

Observation 2

Stakeholders specify constraints on both preservation objects and environments. Jeff Rothberg introduced widely used criteria to evaluate authenticity [17]: *content, context, appearance, structure,* and *behaviour.* These are sometimes misinterpreted as exhaustive categories for significant characteristics [e.g. 12]. The consequence is to limit significant characteristics to "informational entities" - the logical preservation object itself - and exclude bytestreams, representations, or environments.

In contrast, the characteristics of preservation actions constrain the context in which significant characteristics apply, but are not themselves significant.

Observation 3

Significant characteristics are not simple property/value pairs which a stakeholder declares to be significant. Our analysis of policy and strategy documents [8] shows that stakeholders need to state more complex requirements that can be expressed as constraints, using a constraint language such as OCL [20]. They often need to include specifications such as contexts, invariants, pre-conditions and post-conditions.

In many cases, for example, a stakeholder considers characteristics to be significant only when some additional conditions are met - that is, a context is specified. As a result, the language that we use to define significant characteristics must be expressive enough to include a context.

Sometimes the conditions involve preservation object or environment characteristics:

- If *componentType* = "text" then *fontSize* is significant.
- If *environmentType* = "preservation" then *resolution* is significant.
  At other times the conditions involve preservation action characteristics:
- If *preservationActionType* = "bitPreservation" then *fileSize* is significant.

Observation 4

Significance is not absolute and binary. We can not only choose which characteristics should be significant, but would like to specify an importance factor which is a measure of the relative significance of the characteristic for the stakeholders. I may consider each of two conflicting things significant and prioritise one as more significant than another. This prioritisation is essential for both decision making and planning.

Finally, requirements may tolerate some deviation or error. For example, an office document migration that produced a result with different hyphenation or pagination might be acceptable in many situations. We can allow for a tolerance factor which specifies to what degree deviation from the required value can be tolerated. During evaluation of a preservation action the importance and tolerance factors can be combined into a weighted measure of the significant characteristic.

Observation 5

In many cases, we wish to include the possibility of capturing improvements to an object. A common preservation action is normalization of digital objects upon ingest. This may be done to reduce the variety of formats held, but may also be done to improve characteristics in the original. For example, we might migrate files which are in formats that are susceptible to degradation to files in a more resilient format, or move static tables to spreadsheets which enable pivot tables. In this case the characteristics *fileFormatResilience* = "high" or *enablesPivotTables* = "yes" are significant characteristics which were not found in the original. Another preservation action which improves upon the original is the manual restoration of a file by a curator to the state it was presumed to have had before a corruption. Another common example can be found in CAD drawings or data sets. As technology improves, consumers desire to perform new functions on old data in ways that were previously not possible.

Observation 6
While characteristics capture values at a given moment in time, significant characteristics capture constraints on characteristics across time – before and after a preservation action.

As a result of Observations 2, 3, 4, 5, and 6, the language that we use to define significant characteristics must be able to express relationships other than the simple preservation of a value.

The above observations illustrate that significant characteristics are a subclass of preservation guiding requirements [8, 9]. Ideally, we would rename them to "significance requirements", but were reluctant to break too radically with current terminology. We recommend that significant characteristics which express requirements or business rules should in the general case be represented as explicit first class objects in a data model. Figure 2 introduces this separate concept.

We define significant characteristics as:

Requirements in a specific context, represented as constraints, expressing a combination of characteristics of preservation objects or environments that must be preserved or attained in order to ensure the continued accessibility, usability, and meaning of preservation objects, and their capacity to be accepted as evidence of what they purport to record.

4 Discussion

Using the conceptual model and the definition of significant characteristics, we can now investigate some implications of the definition and the relationship of significant characteristics to related digital preservation concepts.

4.1 Implications of the Conceptual Model

The conceptual model was motivated by our findings during the analysis of preservation policy and strategy documents [8]. It suggests the need for developing approaches that allow stakeholders to express constraints with prioritisation and tolerances.

It supports a wide array of preservation activities found in real organisations. Characteristics of different entities are used to express requirements for different preservation activities or purposes. For example, bit-preservation actions such as media refresh preserve characteristics at the file or representation level such as *fileSize*, *encoding*, or the *numberOfFilesInTheRepresentation*. In contrast, migration actions can be expected to change these characteristics.

Significant characteristics at the representation level can express requirements associated with the representations' different purposes, such as preservation versus access copies. *Resolution* = "high" and *preservationLevel* = "9" may be significant characteristics of a representation that is aimed at preserving archival quality.

A significant characteristic that is considered an inherent requirement of a logical component and does not vary from representation to representation should be captured on the logical level. These requirements need to be satisfied by all migration or emulation actions applied to this logical component. For example the requirement *sematicInterpretation* = "negative number" may be declared significant for all representations of a numberComponent. Different representations of the numberComponent can satisfy it by rendering it as a red number, adding a minus sign or surrounding it by parenthesis, but the logical requirement must be satisfied for all of them.

Significant characteristics of intellectual entities can model high level policy and strategy requirements, such as legal or fiscal requirements that must be satisfied after any preservation action.

Significant characteristics of environments make it possible to express requirements whose aim is preserving the look-and-feel of an information object, since the look-and-feel is determined by the combination of the data object and its environment. These significant requirements support emulation and migration activities equally. Environmental factors can also be external or internal policy factors which permit the expression of policy constraints.

4.2 File Formats and Properties

The basic consequence of this analysis is that significance is not inherent in or determined by the file formats of digital objects – but by the needs and requirements of stakeholders in the preservation activities. This enables us to make sense of common preservation activities, such as migration to less expressive file formats. For example, some stakeholders will be satisfied by migration from a *word* document to a simple text file when the original contains only simple text components (i.e., no formatting, headers, tables, and so on). A radio station might be satisfied by a migration that only preserves the audio stream of a video object. The analysis also shows why there can be disagreement about the significance of a property between stakeholders. Disagreement reflects different requirements and priorities among stakeholders. For example, the rotational frequency of a shape in a piece of online art may be significant to the artist, but not for many viewers.

The analysis also clarifies the role of archival subsets of file formats, such as *pdf/a*. The well-designed archival format profile will support properties that are of interest to a substantial community of stakeholders and appear in a substantial subset of content in the full file format.

The preservation community is establishing registries of file formats and properties that apply to them [12, 19]. These are registries of applicable properties[2] rather than of significant characteristics. A stakeholder may indicate that some of the applicable properties are not significant in certain contexts. This increases the set of preservation actions that are appropriate. Conversely, a stakeholder may indicate preconditions which rule out preservation actions that would have been appropriate considering only the file format's applicable properties.

4.3 Significant Characteristics and Representation Information

How do the significant characteristics of this conceptual model relate to representation information, as defined in OAIS [4, 16]? Representation information is "the

---

[2] There are also properties which describe a file format itself rather than the objects that are represented in files. They often appear in stakeholder requirements and enable stakeholders to choose formats that suit their business needs. For example, a custodian might require files to be represented in formats defined by an open standard, or in common use, or with high resilience to degradation damage.

information that maps a Data Object into more meaningful concepts. An example is the ASCII definition that describes how a sequence of bits (i.e., a Data Object) is mapped into a symbol."

Representation information is a set of characteristics describing the preservation object and its environment. Furthermore, representation information is specified for a specific context, namely for a given "designated community". It will vary for different designated communities. Additionally, the purpose of representation information is to guarantee the accessibility, usability, and meaning of preservation objects. All these characteristics of representation information agree with the definition of significant characteristics. It becomes obvious, that representation information is NOT a form of significant characteristics when we realize that it does not specify characteristics that need to be preserved or attained, nor does it specify requirements for preservation actions. Representation information is the set of important characteristics of a data object that are needed to make sense of it for a given designated community at a given time. It does not specify constraints for transformations over time, and it does not specify characteristics of an acceptable derived data object.

A piece of representation information, for example, may be the fact that a given data object requires a certain software package for its proper rendering. This does not imply that the corresponding information object after a migration must use this same software package.

Some pieces of representation information may, however, be declared to be significant for preservation purposes. For example, the semantic interpretation of a data object, such as the characteristic that a given *numberComponent* is to be interpreted as "body weight", is likely to be considered significant in most contexts.

5 Conclusion

This article has examined the concept of significance in digital preservation and presented a new model that places significance in the hands of stakeholders. The model has extended the domain of significant characteristics beyond digital objects to include environments. The model has consequences for implementations of preservation metadata dictionaries, property registries, and preservation services.

This work has been conducted within the larger context of defining a conceptual model and specific vocabulary for supporting preservation processes [8] within the PLANETS project. Significant characteristics can be considered one specific form of preservation guiding requirements which are discussed in [8].

This work has been presented within the digital preservation framework, but may apply to other transformation applications such as rendering accessible versions of digital objects for disabled users.

Acknowledgement

Work presented in this paper is partially supported by the European Community under the Information Society Technologies (IST) Programme of the 6th FP for RTD - Project IST-033789. The author is solely responsible for the content of this paper.

References

1. Becker, C., et al.: A generic XML language for characterising objects to support digital preservation. In: SAC 2008: Proceedings of the 2008 ACM symposium on Applied computing, pp. 402–406 (2008)
2. Becker, C., et al.: Plato: A Service Oriented Decision Support System for Preservation Planning. In: JCDL 2008, Pittsburgh, Pennsylvania, USA (June 2008)
3. Brown, A.: White Paper: Representation Information Registries. PLANETS report PC/3-D7 (2008), http://www.planets-project.eu/docs/reports/ Planets_PC3-D7_RepInformationRegistries.pdf
4. CCSDS. Reference Model for an Open Archival Information System (OAIS). CCSDS 650.0-B-1, Blue Book (the full ISO standard) (January 2002), http://public.ccsds.org/publications/archive/650x0b1.pdf
5. CEDARS Project, http://www.leeds.ac.uk/cedars/
6. Chaudhri, V., Farquhar, A., Fikes, R., Karp, P., Rice, J.: OKBC: A programmatic foundation for knowledge base interoperability. In: Proceedings of the 1998 National Conference on Artificial Intelligence (1998)
7. Clausen, L.: Opening Schroedinger's Library: Semi-automatic QA reduces uncertainty in object-transformation. In: Kovács, L., Fuhr, N., Meghini, C. (eds.) ECDL 2007. LNCS, vol. 4675, pp. 186–197. Springer, Heidelberg (2007)
8. Dappert, A., Ballaux, B., Mayr, M., van Bussel, S.: Report on policy and strategy models for libraries, archives and data centres. PLANETS report PP2-D2 (2008), http://www.planets-project.eu/docs/reports/ Planets_PP2_D2_ReportOnPolicyAndStrategyModelsM24_Ext.pdf
9. Dappert, A., Farquhar, A.: Modelling Organisational Goals to Guide Preservation. In: iPRES 2008: The Fifth International Conference on Preservation of Digital Objects (2008), http://www.bl.uk/ipres2008/ipres2008-proceedings.pdf
10. Hockx-Yu, H., Knight, G.: What to Preserve?: Significant Properties of Digital Objects. In: Report on the JISC/BL/DPC Workshop of April 7, 2008, British Library Conference Centre (2008); The International Journal of Digital Curation 3(1) (2008), http://www.ijdc.net/./ijdc/article/view/70/70
11. IFLA Study Group: Functional Requirements for Bibliographic Records (1998), http://www.ifla.org/VII/s13/frbr/frbr.pdf (Retrieved on December 1, 2007)
12. Knight, G.: Framework for the definition of significant properties (2008), http://www.significantproperties.org.uk/documents/ wp33-propertiesreport-v1.pdf
13. Knight, G., Pennock, M.: Data Without Meaning: Establishing the Significant Properties of Digital Research. In: iPRES 2008: The Fifth International Conference on Preservation of Digital Objects (2008), http://www.bl.uk/ipres2008/ipres2008-proceedings.pdf
14. National Archives of Australia: An Approach to the Preservation of Digital Records (December 2002), http://www.naa.gov.au/Images/ An-approach-Green-Paper_tcm2-888.pdf
15. PREMIS Editorial Committee: PREMIS Data Dictionary for Preservation Metadata, Version 2 (March 2008), http://www.loc.gov/standards/premis/v2/ premis-2-0.pdf

16. The OCLC/RLG Working Group on Preservation Metadata: Preservation Metadata and the OAIS Information Model. A Metadata Framework to Support the Preservation of Digital Objects (2002),
http://www.oclc.org/research/projects/pmwg/pm_framework.pdf
17. Rothenberg, J.: Preserving Authentic Digital Information. In: Authenticity in a Digital Environment. The Council on Library and Information Resources (2000),
http://www.clir.org/pubs/reports/pub92/rothenberg.html
18. Rusbridge, C.: Excuse me...some digital preservation fallacies? Ariadne, 46 (February 2006), http://www.ariadne.ac.uk/issue46/rusbridge/
ISSN 1361-3200
19. The National Archives: PRONOM,
http://www.nationalarchives.gov.uk/pronom/
20. Warmer, A., Kleppe, A.: The Object Constraint Language. In: Getting Your Models Ready for MDA, Addison-Wesley Longman Publishing Co., Boston (2003)
21. Wilson, A.: Significant Properties Report, InSPECT Work Package 2.2, Draft/Version 2 (2007), http://www.significantproperties.org.uk/documents/
wp22_significant_properties.pdf

User Engagement in Research Data Curation

Luis Martinez-Uribe[1] and Stuart Macdonald[2]

[1] University of Oxford, e-Research Centre, 7 Keble Road, Oxford OX1 3QG, UK
`luis.martinez-uribe@oerc.ox.ac.uk`
[2] University of Edinburgh, EDINA, 160 Causewayside, Edinburgh, EH9 1PR, Scotland
`stuart.macdonald@ed.ac.uk`

Abstract. In recent years information systems such as digital repositories, built to support research practice, have struggled to encourage participation partly due to inadequate analysis of the requirements of the user communities. This paper argues that engagement of users in research data curation through an understanding of their processes, constraints and culture is a key component in the development of the data repositories that will ultimately serve them. In order to maximize the effectiveness of such technologies curation activities need to start early in the research lifecycle and therefore strong links with researchers are necessary. Moreover, this paper promotes the adoption of a pragmatic approach with the result that the use of open data as a mechanism to engage researchers may not be appropriate for all disciplinary research environments.

Keywords: digital curation, research data management, open data, digital repository services, user engagement.

1 Introduction

Research methods and practice, including scholarly communication, are experiencing a radical transformation in the digital age. New tools and infrastructures make possible the generation of digital research data outputs as well as new ways to use, share and reuse them. There is a growing acceptance of the importance of curating research data in order to preserve them and make them re-usable for future generations with libraries, computing services and other service units within academic institutions working together to develop digital repositories to curate this type of research output.

We believe that engagement with researchers, the user communities in this case, is crucial in order to develop systems that will meet their needs. Whilst some argue that open data is the way forward, it is not clear that it will help engage researchers with digital curation activities. Thus this paper will attempt to answer the following research questions:

Is open data the correct concept to engage the research community?
What other methods can be used to facilitate engagement in data curation?

2 Open Access Repositories and Researchers' Requirements – A Balancing Act

Open Access (OA) enthusiasts have written about the inevitability of 24 hours a day and 7 days a week access to all research papers and their citations *"for free, for all*

M. Agosti et al. (Eds.): ECDL 2009, LNCS 5714, pp. 309–314, 2009.
© Springer-Verlag Berlin Heidelberg 2009

and forever." [1] Primarily led by technological developments, the increase in the overall volume of research, the increasing uncertainty about content preservation and by the strong dissatisfaction of academic libraries subjected to constant increase of journal subscription prices, digital repositories were built and employed within research institutions far and wide [2]. Being content provider as well as user and re-user of these information systems, researchers can be regarded as the key user community. Nonetheless, it has been argued no formal detailed requirements analysis has taken place in order to identify and address researchers' needs and concerns related to such scholarly communication systems [3]. As a result the user community has been overlooked in the developmental phases of technology design and implementation of the information system ultimately meant to serve them. Arguably the repository infrastructure developed was not, in most cases, built to address researchers' needs but those of libraries' and librarians. This has led to a struggle to find ways to populate repositories with researchers' output. In recent times there have been an increase in the number of institutional and research funders OA mandates, the knock-on effect from which will see the requirement of significant investment in awareness raising activities in order to highlight the benefits to researchers of using and depositing research materials in such repository systems. Such a process may have been expedited had the library and research worlds been more closely involved in a more agile digital repository design and development with an iterative requirements phase.

3   Research Data Repositories - Learning From Experience

When it comes to the research data setting we have to approach the problem from a new perspective. We have to evolve and learn from previous experiences in order to develop repository services capable of dealing with the management and curation of research data by addressing researchers' needs.

Although open data is becoming a widely used term, there is not a consistent formalisation of the concept. Murray-Rust [4] suggests that the concept of open source software can be extended to that of open data in that data should be freely available for re-use and modification without restriction. The virtues of 'Open Data' have been praised and evangelized by many since OECD's declaration back in 2003 [5] however many research communities are currently not in a position to make their data available on those terms.

The JISC-funded DISC-UK DataShare project explored a number of technical, legal and cultural issues surrounding research data in repository environments. It built on the existing collaboration of data librarians and data managers from the Universities of Edinburgh, Oxford, Southampton and LSE and investigated mechanisms for ingesting and sharing research data in existing institutional repository systems for those researchers willing to openly share them. Project partners identified a number of barriers pertaining to the researcher and the research setting that would impact on data sharing [6], including:

- Reluctance to forfeit valuable research time to prepare datasets for deposit, e.g. anonymisation, codebook creation, formatting
- Concerns over making data available to others before it has been fully exploited

- Concerns that data might be misused or misinterpreted, e.g. by non-academic users such as journalists
- Concerns over loss of ownership, commercial or competitive advantage
- Concerns that repositories will not continue to exist over time
- Unwillingness to change working practices
- Uncertainty about ownership of IPR
- Concerns over confidentiality and data protection

It may be argued that open data is a reality in some disciplines: in crystallography there are a number of established repositories including the Cambridge Crystallographic Data Centre and Crystallography.net with further discussions taking place about a federation of crystallography data repositories [7]. Molecular biologists have been sharing data through repositories like the Protein Data Bank (PDB) since the seventies [8] and geneticists have also been sharing nucleotide sequences through GenBank from the early eighties [9]. These are great examples of communities embracing the benefits of open data but it is important to highlight that these initiatives were lead by visionary researchers in those fields. There is an interesting analogy with domain specific publication repositories like arXiv or RePEc. They represented successful examples of author self-archiving repositories but this didn't translate in wide acceptance and use of open access repositories.

4   Different Approaches - Researchers' Needs Connecting Data Management and Curation

A more research inclusive and bottom up approach has been taken by data and information management activities in the Universities of Edinburgh and Oxford in order to understand better how researchers work, what are the drivers behind their information management and sharing activities and what services they require.

At Edinburgh a team of social scientists and information service specialists (respectively, from the Institute for the Study of Science, Technology and Innovation, and from Information Services at the University of Edinburgh and the Digital Curation Centre) are carrying out a RIN-funded study designed to enhance understanding of how researchers in the life sciences locate, evaluate, manage, transform and communicate information as part their research processes, in order to identify how information-related policy and practice might be improved to better meet the needs of researchers. Information diaries were completed by over 50 life-science researchers from eight sub-disciplinary research groups. An interview schedule was constructed in order to investigate further the findings from the diaries. 24 interviews were conducted across the groups followed by focus-group discussions. An in-depth study was also employed on one of the groups. Interim findings suggests that some disciplines lend themselves more than others to open data and that there's much variety, specificity and complexity in terms of research data within the examined groups. Research data created via models/simulations, observations, and experiments are intrinsically linked with the data collection methodologies and instrumentation and as such may be better placed within a Virtual Research Environment (VRE) and/or a staging repository-type environment [10] as there are often issues surrounding the unraveling of data content when sophisticated and domain-specific proprietary systems are used. In addition, certain data cannot be considered conventionally open for example: data controversial by nature (stem cell data, brain scans); data received from industrial partners, licensed data products and the ensuing derived data products, data leading to development of patents or commercial products. Other findings include:

- Most life science researchers spend much of their time searching for and organising data however data curation and/or sharing only becomes crucial to them at certain stages of the research process.
- The groups investigated lack any obvious or explicitly appointed data/information managers, leaving individuals to manage their own information/data in a non-formal fashion.
- There is an implicit feeling across the groups surveyed that only the researchers themselves have the subject knowledge necessary to curate their own research data.
- Researchers in the life sciences express a keen sense of 'ownership' and protectiveness towards their data. However there is confusion or uncertainty about their rights with respect to data ownership

In Oxford an internal scoping study [11] on research data curation took place throughout 2008 and involved the Office of the Director of IT, Computing Services, the Library and the Oxford e-Research Centre. The aim of the study was to capture the requirements for digital repository services to manage and curate research data. A requirements gathering exercise took place and around 40 researchers across disciplines were interviewed to find out about their data management practices and capture their requirements for services to help them manage data. The findings from this exercise showed that the vast majority of researchers felt that there were potential services that could help them. The following scenarios present some of the challenges, found during the scoping study, that researchers are facing with their data and represent the types of needs that data repository services should be trying to address:

- In some cases, researchers had generated data several years ago and now could not make sense of them as they had not kept enough information on how the data was created in the first place;
- In scientific disciplines research groups require secure storage for their large volume of data generated by instruments such as electronic microscopes or by computing simulations run in GRID systems;
- Many clinical research centres compiling data for decades and spending months to migrate data formats in order to avoid format obsolescence;
- In many cases researchers want to make their articles' accompanying data available online in a sustainable way and they do not have the institutional infrastructure to do this so they published the data on their departmental website.

The scoping study is now being followed up by the JISC funded Embedding Institutional Data Curation Services in Research (EIDCSR) project. This project will attempt to address the data management and curation requirements of two research groups who produce and share data. EIDCSR involves the partners of the scoping study with Research Services and IBM to integrate research workflows with the

Fedora Digital Asset Management System, long-term file storage and underpinning these efforts with policy development and economic models. A key aspect of this project will be the possibility to work with researchers from the moment they generate the data, this will ensure that the necessary and appropriate curation actions are taken early in the research lifecycle.

Oxford and Edinburgh are also both involved in the development of the Data Audit Framework[1] (DAF) which helps to establish relationships with research communities around the issues of data curation. This methodology provides organisations with the means to identify, locate, describe and assess how they are managing their research data. The methodology goes some way to enabling data auditors to identify and engage researchers regarding their research data holdings. It also provides information professionals who wish to extend their support for research data within the university community with a vehicle for engaging with researchers in addition to a focus for discussion of data curation practices. This may manifest itself through local data management training exercises to equip researchers with the skills and tools to deal with funders' data management and sharing policies. Indeed, the Edinburgh Data Audit Implementation Project [12] states that 'staff had numerous comments and suggestions for improvement of data management at different levels indicating an awareness of the issues, even where it has not been made a priority to address.'

Engagement with researchers through the activities explained above provides a valuable insight into the research process at the various stages in its lifecycle. Such activities help to gain the trust of the researchers facilitating the process of data curation within data repositories at a point early on in the research lifecycle, a fundamental key to the success of these information systems. In addition to gaining the trust of the researcher such engagement offers the opportunity to acquire the researcher's own thoughts, feelings and expectations as to how information services, policies and technologies may shape the future. Issues, such as who the technology is for, how it fits in with researchers' practices or what the purpose of the technology is, require prior consultation with those with a vested interest in the technology.

5 Discussion and Conclusion

In this paper we attributed the lack of researcher engagement with OA publication repositories to the fact that the main drivers behind their development were somehow distant from current research needs. This, we argue may be due to the lack of an appropriate iterative requirements analysis involving the main user community.

Research data repositories pose similar challenges. Our experience has shown that using open data as a message to engage researchers in curation activities makes it easy to become detached from current research needs in many disciplines. The heterogeneity of research practices and their datasets, some of which cannot be openly shared provides further evidence of the importance in understanding and appreciating the requirements from the different research communities. Moreover, we believe that the curation of research data requires trusted relationships achieved by working and conversing with researchers early on in their research process. This paper presents approaches from both Edinburgh and Oxford which try to articulate and understand

---

[1] Project led by HATII, University of Glasgow - http://www.data-audit.eu/

how researchers work with data and information, the barriers they find and their priorities for services required to assist them. We argue that failure to engage with the specific needs of researchers through these initiatives, may lead to the development of data repository services that are under-exploited or indeed may not even be used.

Further work on user engagement in data curation should be pursued to explore connections with other areas such as data citation and the academic reward system, data management tools, business models as well as institutional and funders' policies.

References

1. Harnard, S.: For whom the gate tolls? How and why to free the referred research literature online through author/institution self-archiving now (2001), http://cogprints.org/1639/
2. Raym, C.: The case for institutional repositories: a SPARC position paper (2002)
3. Salo, D.: Innkeeper at the Roach Motel. Library Trends 57(2) (2008)
4. Murray-Rust, P.: Open Data in science. Nature Proceedings (2008)
5. OECD: OECD Principles and guidelines for access to research data from public funding. Paris (2007), http://www.oecd.org/dataoecd/9/61/38500813.pdf
6. Gibbs, H.: DISC-UK DataShare: State of the art review (2007), http://www.disc-uk.org/docs/state-of-the-art-review.pdf
7. Lyon, L., Coles, S., Duke, M., Koch, T.: Scaling up: towards a federation of crystallography data repositories (2008), http://eprints.soton.ac.uk/51263/
8. Berman, H.M.: The Protein Data Bank: a historical perspective. Acta Crystallographica, 88–95 (2008)
9. Benton, D.: Recent changes in the GenBank online service. Nucleic. Acid. Research 18(6) (1990)
10. Steinhart, G.: DataStaR, a data staging repository for digital research data (2008), http://www.dcc.ac.uk/events/dcc-2008/programme/posters/DataStaR.pdf
11. Martinez-Uribe, L.: Findings of the scoping study and research data management workshop (2008), http://tinyurl.com/55fxgw
12. Ekmekcioglu, Ç., Rice, R.: Edinburgh data audit implementation project: Final report (2009), http://ie-repository.jisc.ac.uk/283/

Just One Bit in a Million:
On the Effects of Data Corruption in Files

Volker Heydegger

Universität zu Köln, Historisch-Kulturwissenschaftliche Informationsverarbeitung (HKI),
Albertus-Magnus-Platz, 50968 Köln, Germany
volker.heydegger@uni-koeln.de

Abstract. So far little attention has been paid to file format robustness, i.e., a file formats capability for keeping its information as safe as possible in spite of data corruption. The paper on hand reports on the first comprehensive research on this topic. The research work is based on a study on the status quo of file format robustness for various file formats from the image domain. A controlled test corpus was built which comprises files with different format characteristics. The files are the basis for data corruption experiments which are reported on and discussed.

Keywords: digital preservation, file format, file format robustness, data integrity, data corruption, bit error, error resilience.

1 Introduction

Long-term preservation of digital information is by now and will be even more so in the future one of the most important challenges for digital libraries. It is a task for which a multitude of factors have to be taken into account. These factors are hardly predictable, simply due to the fact that digital preservation is something targeting at the unknown future: Are we still able to maintain the preservation infrastructure in the future? Is there still enough money to do so? Is there still enough proper skilled manpower available? Can we rely on the current legal foundation in the future as well? How long is the technology we use at the moment sufficient for the preservation task? Are there major changes in technologies which affect the access to our digital assets? If so, do we have adequate strategies and means to cope with possible changes? and so on. These are just a few of the general questions decision-makers have to carefully weigh.

Although many of the factors are fairly unpredictable, there are still some constants involved in the process of digital preservation which are likely to persist in time: these are the fundamental concepts which draw the link from information to data.

On the lowest level this is the storage of information as *binary-coded data*. No matter if information was or is or will be stored on a punched card, floppy disk, hard drive or holographic memory, this concept is likely to remain. In contrast, the favourite storage technology for digital data has changed many times in the past.

M. Agosti et al. (Eds.): ECDL 2009, LNCS 5714, pp. 315–326, 2009.

Moreover, binary data must be set into a specific context, it needs to be formatted. The meaning of data is primarily determined through a specific model of information. For example, a raster image is typically described as a set of pixels with some other additional elements like the arrangement of the pixel values in a two-dimensional array or a statement on its chromaticity. Besides these descriptive characteristics, which make up the appearance of the specific model 'raster image', digital data is meant to be processed by software. Therefore formatting of data also includes several technical characteristics which have a strong relation to the technical environment.

Such a *data format* for a specific model of information is another essential concept that links between information and data. Hence, it will always be a core element in the digital preservation task.

Given this, it is obvious that file formats[1] are spotlighted when it comes to discussing the strategies for long-term preservation of digital information. The focus to date was namely on their implications for reducing storage space on storage systems, openness of specification for the public, level of adoption in the academic and commercial communities and several functional issues (e.g., searchable text, metadata support) [4, 5].

So far little attention has been paid to the factor *robustness*, i.e., *a file formats capability to keeping its information as safe as possible in spite of data corruption*. A reason for that may be the perception of data corruption as a phenomenon that is mainly related to the physical storage media since bit errors actually happen there and can be detected and corrected on this level. Another reason could be the circumstance that we usually talk about 'damaged files' rather than 'damaged file formats' although a file is basically nothing but an individual physical realisation of the file format in question. Last but not least, there is a lack of systematic surveys on the impact of physical data corruption to file integrity both under lab conditions and in the real world. However, recently published studies on the reliability of disks and storage systems in the real world [2, 10, 11] which verified the existence of disk failures over time are an indication for the latent risk of file corruption as well since files are manifest to storage media.[2]

The paper on hand reports on the first comprehensive research on the topic of file format robustness. It is based on the assumption that there is indeed a correlation between 1) the way information is encoded in a file with respect to the definitions of the underlying file format and 2) information consistency. If this is so, it should be possible to sustain the robustness of a file format. This would result in a significant reduction of information loss for the case of corrupted data in files.

Theoretical groundwork including preliminary results of experimental tests have already been undertaken and reported by the author [6]. Both have been extended to an in-depth study on the effects of data corruption in files of various file formats from the raster image domain.

---

[1] In the following, the term 'file format' is synonymous with 'data format'. Both terms refer to the same basic concept, whereas 'file format' and its instance, a file, additionally relate to the technical environment.

[2] In contrary, there are a considerable number of studies on the reliability of storage media in case of physical degradation which were examined under lab conditions, e.g. [7].

2 Approach

This research work is based on a study on the status quo of file format robustness. A controlled test corpus was built which comprises files of various file formats with different format characteristics. The files are the basis for data corruption experiments. A software tool which is able to damage files by flipping single bits (or bytes) was developed for that purpose. The damaged files are compared against the undamaged reference file, measuring the quantitative difference (i.e., the differences based on the pixel data) between both with metrics from statistical data analysis.

Conclusions are finally drawn on the measured values. A summary report on the study results is given in chapter 2.2.

2.1 Study Outline

2.1.1 Selection of File Formats and Test Corpus

The decision on which file formats to choose for the study was influenced by several criteria. On the one hand, those formats which have a factual relevance as an archival format for digital libraries and archives were considered, as well as including current trends (e.g., JPEG2000 as archival format). Second, the decision was also influenced by the (estimated) active usage of file formats in the real world, also including the commercial communities. Third, technical considerations were taken into account. There had to be sufficient support of the file format by software tools to successfully run the experiments for the study. Openness of the file format specification is also desirable in order to allow for an in-depth analysis of format features and structures. And fourth, the choice of file format was subject to theoretical considerations which are very strongly linked to content-related aspects of the research topic (e.g., strategies for improvement of robustness).

In consideration of all these criteria, the following file formats were selected: [3]
Tagged Image File Format, version 6.0 (TIFF), JPEG File Interchange Format (JPG), JPEG 2000 File Format (JP2), Portable Network Graphics, version 1.2 (PNG), Graphics Interchange Format, version 89a (GIF), Windows Bitmap (BMP).

Based on these six formats, a set of test files was arranged in the form of a 'controlled' test corpus. It offers the advantage to precisely choose such format characteristics which are of special interest for robustness and hence are the best for analytical research such as this. [4] This controlled test corpus is also assumed to be representative, since the files' characteristics are in most cases common ones. [5]

An evaluation of the file formats shows that there are many aspects which are basically important for robustness, yet their actual significance is often marginal if looked at in isolation. For example: TIFF defines the characteristic 'Photometric Interpretation'. For monochrome and grey-scaled images it enables an application to interpret the pixel data in either a white-is-zero (i.e., bit value 0 has to be interpreted as white) or black-is-zero format. If the data which carries this information is corrupted, an

---

[3] The abbreviations in parentheses will be used in the following sections of this paper.

[4] In contrast, a random sample of files would be targeted to descriptive research but this is not the primary objective here.

[5] Compare Table 1 in which the files are described.

application may either not be able to process the file at all or may process the data in the contrary sense (given that the application has not implemented a default behaviour for such cases). In both cases the effect of data corruption concerning this characteristic is fatal. However, viewed from a strict probabilistic point, the probability that such an error occurs at all is (for low corruption rates) fairly small since this information is encoded using only 12 bytes which is infinitesimal in comparison to possible file sizes of a million bytes or more.

One of the factors which is suspected to have a great influence on robustness is compression of data. The danger of losing data during transmission over noisy channels is a well-known fact[6]. Especially in conjunction with data being compressed the consequences of bad data can potentially become extremely drastic. Therefore it is also assumed that this is similar for data serialized on some storage medium. Thus, file formats which heavily build on data compression are suspected of being less robust than those that do not. Although this may not really be a new or even surprising insight, it is not always beyond all question: JPEG2000 codec not only supports simple error detection but also, in a quite advanced way, correction of errors via several features defined in the JPEG2000 compression algorithm.[7] There should be at least gradual variations in the robustness for files with different compression algorithms. The compression feature was included in those files of the test corpus which underlying file format allows for usage of a specific compression algorithm.

Table 1 in the appendix gives an overview of those corpus test files which are discussed in more detail in chapter 2.2 (report on the results).

2.1.2 Experimental Design and Technical Environment

The study is based on experiments in which the files of the test corpus are processed along a chain of different software components. First they are damaged ('Corrupter'), then converted to an analysis format ('Converter') and finally analysed ('Analyser') by applying metrics which basically compare the pixel data from the undamaged and corrupted file. 'Damaged' means: Bits (or bytes) are not totally dropped but changed, i.e. the bits are flipped.[8]

The Corrupter is equipped with different operation modes. The main one is a general corruption procedure on the whole file where randomly[9] chosen bits are flipped according to an arbitrary percentage.[10] By doing so, the introduced errors are fairly equally distributed. This pattern of error distribution is most suitable for this research task, since it is the most neutral pattern that allows for universally valid conclusions regarding robustness. In reality, there is no error pattern that can be verified as the

---

[6] Some file formats, especially those which are designed for application in the World Wide Web, provide control mechanisms. PNG for example has defined checksums for each of its chunks to enable error control for the processing software.

[7] A description would be beyond the scope of this paper; please see [8] for details on the JPEG 2000 compression.

[8] Dropping data was not considered for the experiments since pre-tests revealed strong evidence that this is a knock-out criterion in most cases even for very small corruption rates.

[9] This is done by a random number generator which implements the Mersenne Twister Algorithm [9].

[10] This is the operation mode for the results reported here.

general pattern. The error pattern depends on a large number of factors like the nature of storage media, logic of storage, source responsible for damage and many others.[11]

The Converter is mainly a wrapper for third-party file format converting tools. The Windows bitmap format (BMP) was chosen as the analysis format for it fits the requirements of such an analysis format best. It is a simple structured format, easy to process, widely supported by conversion tools and still capable of describing the characteristics (basically the pixel data) we need for analysis of file format robustness.

Conversion to an analysis format is not the only option one could apply at this stage of the experiment. Either way, we need some form of re-processing of the corrupted files since the analysis builds on a comparison of the pixel data. Thus, e.g., in case of compressed data, decompressing the data is absolutely essential in order to be able to compare the pixel data properly.[12] So in this second stage of the experiment, every corrupted file is converted to uncompressed BMP.

In the last stage, the files are compared against the reference file which was also converted to BMP. The Analyser has implemented a number of metrics (see below) which work on the pixel data. The results of measurement are finally written to a file for further analysis.

Several pre-tests were conducted in order to optimize the experimental design. One of the critical steps is the conversion of the corrupted files to the chosen analysis format. It was supposed by experience that the third party software used for that may perform differently. Three open source conversion tools were chosen for conversion: ImageMagick and IrfanView for all file formats and additionally Kakadu (JP2 only).[13] A complete test cycle (applying two different corruption rates with a smaller trial) was run on every test file using the three tools successively. Based on the analysis of the metric results, the tool with the best performance was finally chosen for the experiment.

Another decision that had to be made was related to the choice of corruption rates. This choice had to be made individually since the file sizes of the test files differ and therefore some very low rates could not be performed on small files. Moreover, the rates were 'calibrated' according to the results of pre-tests since it would not be sensible to run large tests on successive corruption rates where the differences in the results are marginal (this was true for some of the files in either the very low or very high rates). As a result, the span of corruption rates applied on the files varies from exactly one bit[14] corrupted (this was done for all files) up to 0.1% (of the individual file size). Furthermore, in some cases the introduced errors caused serious problems (e.g., sudden crashes, buffer overruns, refused termination), especially for very high

---

[11] Just because of this broad variation in the factors possibly determining an error pattern, the random pattern introduced here is most suitable for general conclusions on robustness. It is the intention of this work to draw general conclusions on robustness, independently from determinants which may vary by and by.

[12] During the process of format conversion, data is read into memory and re-encoded to the chosen target format. As a consequence, the data is also completely reinterpreted by the application, including the corrupted data as well.

[13] See http://www.imagemagick.org,
http://www.irfanview.com/, and http://www.kakadusoftware.com/.

[14] This is equivalent to a 1-byte corruption in most cases, since most file formats use at least one byte for storage of information units.

corruption rates. In such cases the experiments for the given rate could not performed to its end as well.

Finally, the size of the samples had to be determined. This is a rule of thumb estimate which can be figured out by pre-tests. These led to the decision to set the size of the samples to 1000 per run, i.e. for each corruption rate the test files were corrupted, converted, and compared to the undamaged reference file a thousand times.. All files, the corrupted ones as well as the converted analysis files, were saved on hard-disk to enable secondary analysis.

2.1.3 Metrics

The Robustness Indicator (RI) is a simple metric that counts the number of different pixels of two image data sets and relates it to the total number of pixels:

$$RI = D / n \qquad (1)$$

where

D   is the number of different pixels,
n   is the total number of pixels.

Interpretation: The smaller the RI, the better the robustness. Range is from 0 to 1. If multiplied by 100, it shows the percentage of different pixel values.

For the overall comparison of file format robustness, the RI values for each single file-to-file comparison are summed up and finally averaged over the sample size:

$$RImean = \sum RI / m \qquad (2)$$

where

$\sum RI$ is the sum of all RI,

m   is the sample size (usually 1000).

A very familiar image quality metric is the root mean squared error (RMSE). This was implemented as well. It is among a group of related metrics (e.g., peak-signal-to-noise ratio, mean squared error) which basically tell us the same facts in slightly different form.[15] RMSE is defined as:

$$RMSE = \sqrt{\frac{\sum (Xi - Yi)^2}{n}} \qquad (3)$$

where

Xi   is a pixel-constituting value of the reference image data
Yi   is a pixel-constituting value of the corrupted image data
n   is the total number of pixel constituting values

Interpretation: The higher the RMSE, the worse the robustness. Range is from 0 to RMSEmax.[16]

---

[15] Therefore these metrics were not computed.
[16] RMSEmax is one of the auxiliary metrics we defined. It is the highest (worst) possible value for RMSE for the given image data set of the reference file.

Just as for RI, the mean over all single RMSE (RMSEmean) of a sample was computed.[17]

Besides these two core metrics for measurement of robustness, some further metrics which aim to provide cross-checks on the validity of the measured values for RImean and RMSEmean were included. These are: RI and RMSE based on the actual successful comparisons (called RImean- and RMSEmean-), median, standard deviation, skewness and confidence intervals (based on either the median or the mean, depending on the skewness of the distribution of the single values for RImean and RMSEmean).[18]

There are many different metrics which could be applied to measure robustness. For reasons of paper space, they can not be discussed in detail.[19] Nevertheless, there is a motivation for the choice of the primary metrics for this study. Besides the fact that the implementation of these metrics is easy and effective[20], it was mainly influenced by two further points:

First, it was intended to make a statement of 'hard facts': How many of the pixels have changed after data corruption, based on the data that encode its information? RI is perfectly suited for this. It actually tells us if the data that keeps the pixel's information is still undistorted or not.

RMSE has been introduced since it is also aimed at drawing conclusions on the quality of what is left after data corruption. In contrast to RI, which is based on boolean comparison, it is a metric of gradual changes within the pixels. This turns it into a metric more suitable for quality measurement. Nevertheless, this is only true by tendency. RMSE is, as all the other measurements of distortion, more or less suited for quality measurement [1] since its significance highly depends on the type of distortion produced by the corrupted data.

2.2 Results

Tables 2 and 3 in the appendix contain the results for RMSEmean and RImean. Results for RMSEmean- and RImean- as well as the other auxiliary metrics are not included in here since they did not have an impact on the validity of the core observations we make in the next section.

2.2.1 Observations

In general, the results for the robustness metrics differ for various file formats in a wide range, depending on the specific format characteristic introduced to the test file. Many of our test files responded to data corruption in a very sensitive way, especially those with compressed data in it. A single bit can be extremely significant for robustness: the corruption of just one single bit out of the entirety of a million bits proves to be destructive for information consistency.

---

[17] The formula is very similar to the RI modification (additionally RMSE is divided by m), therefore a representation of it is not included at this place.

[18] Since these metrics are auxiliary metrics for RI and RMSE and also very common in statistics, they are not described in detail.

[19] See [1] for a compilation of measures.

[20] This is especially important for a study like this one, where we deal with a huge amount of data.

Observation 1: Three of the files we described in table 1 do not make use of any data compression (bmp_A, tiff_A1, tiff_A2). Among these three, the robustness for bmp_1 is the best for both RImean and RMSEmean. Most notably for the levels of higher data corruption (starting from 0.05 onwards, not shown in the table), bmp_A outperforms the two TIFF files.

Observation 2: The two TIFF files we described in table 1 are almost identical in all of their characteristics, except in the way they arrange the pixel data within the file. File tiff_A1 groups the image data in one continuous byte sequence. The position of this sequence within the file is stored in exactly one small sequence of bytes (the off-set) in the so-called TIFF header. If this essential sequence is corrupted, the entire file can not processed any more, since the application is not able to find the pixel data anymore. In contrary to tiff_A1, tiff_A2 contains 144 of these image data stripes, thus also 144 offsets which point to the stripes. This means at the same time, that file tiff_B contains 143 essential sequences of bytes more than file tiff_A1. Following the rules of probability, the lower robustness of tiff_A2 shown by our metrics is fairly evident.

Observation 3: All of the ten files containing compressed data show higher values for RImean (indicating less robustness) than those which do not comprise compressed data, over all levels of data corruption. For RMSEmean, these are eight out of ten. Only the JP2 files with fully enabled error resilience functionality (JP2_B, JP2_C) could achieve roughly the same results as the worst performing file (tiff_A2) out of the three files with uncompressed data.

Observation 4: The evaluation of the metrics results over all files which make use of compression must be regarded differently for RImean and RMSEmean. Regarding RMSEmean, the JP2 files are by far the best performing files whereas the data for RImean can not verify this. However, a secondary (visual) analysis of the corrupted files clearly show much better image quality levels than for those files with the other compression methods. Therefore the reflections on the applicability of distortion metrics in section 2.1.3 prove true.

Observation 5: Regarding the TIFF files: Uncompressed TIFFs (tiff_A1, tiff_A2) always perform better than TIFFs which make use of compression (tiff_B, tiff_C). For both RImean and RMSEmean, usage of JPEG compression appears to be better than ZIP compression (followed by LZW compression; this file is not included in table 1).

Observation 6: Regarding the JPEG files: JPEG compression appears to react quite differently to data corruption. Although jpeg_B is of higher compression (0.081) than jpeg_A (0.327) it shows much better robustness (based on the data for both RImean and RMSEmean). Robustness can additionally be improved if the JPEG compression feature 'progressive' is used.

Observation 7: Regarding the JP2 files: The file with lossless data compression (jp2_B) does not prove to be more robust than jp2_C which contains lossy com-pressed data. For RImean it performs even worse. Concerning error resilience: Espe-cially the data for RMSEmean show the advantages of the error resilience features introduced in JPEG 2000 part 2. The files in which we included error resilience fea-tures (jp2_B, jp2_C) performed significantly better than the ones that do not include these features, especially for higher corruption levels.

Further observations: We also included JP2 test files[21] to the corpus in which the different resilience options are varied with each other and did several tests on them. One of the main findings is that the usage of SOP and EPH[22] markers alone showed only moderate improvement of robustness compared to files not using any of the error resilience features. Best robustness improvements can be achieved if all of the error resilience options are enabled. One further finding is that additional usage of smaller so called precincts, which in [3] is also considered to improve robustness, only results in marginal improvement of robustness.

Pre-tests related to the tools we used in the Converter module brought to light that there is perhaps an obstacle which needs to be overcome for acceptance of JPEG 2000 compression in the preservation context for the future: From the three different tools we tested for handling the JP2 files, only the one with which the files were also created (Kakadu) was able to deal with the error resilience features in such a way that robustness was actually improved.

2.2.2 Conclusions

Our assumption concerning the negative relation of data compression and robustness could be verified. The result data reveal better performance of files with uncompressed data for almost all tested files. Nevertheless, this is not always as obvious as it is still assumed. The findings in observation 2 and observation 3 show that JP2 can be quite competitive to file formats not using data compression feature if, for the latter, other features which potentially have an influence on robustness are not considered.[23]

Data compression is a crucial feature regarding robustness. However, it is by far not the only one. Observation 1 and 2 clearly revealed this fact. In the first case, BMP uncompressed showed better robustness than TIFF uncompressed. The reason for this lies in the fact that a standard BMP file is fairly near to raw data. It contains only few essential data[24], i.e., data the application necessarily needs in order to process the file. Thus the probability of corruption of such essential data compared to files of other file formats is quite low. Observation 2 is a good example for the effects of structural determinations in file formats. In this case, the image data is split in many parts (strips), whereas each part is referenced by an individual offset. Unfortunately, every single offset to the image data is essential in order to process the referenced data sequence in a regular way. If an offset is corrupted, the effect of this corruption is not restricted to the offset's bytes alone, all of the bytes which carry the information of the referenced data sequence are affected as well. Given this, a single corruption of an offset is momentous as such - but even more if there is not only one single offset but many (as it is for tiff_A2), since the rules of probability have an increasing effect.

This study demonstrates that there are indeed factors, originating in file format design and functionality, which have an influence on the robustness of files. Taking

---

[21] Not included in table 1.

[22] For information on error resilience and other JPEG 2000 features, see [8].

[23] Given this, the findings of tests in [3] where JP2 is supposed to be even more robust than TIFF uncompressed could not be verified. The image of the distorted TIFF shown in this report (Figure 6b) strongly let us suppose that the TIFF files used for these tests are multiple striped TIFFs (corresponding to tiff_A2, see observation 2) which prove to be significantly less robust than 1-striped TIFFs.

[24] This refers to the categorization of file format data in [6].

these factors more carefully in consideration should lead to improvement of saving digital information beyond the existing strategies (such as error correction on hardware level or distributed and redundant storage of files).

3 Outlook

Although it is believed that research in this area is very fundamental for the overall goal of the preservation task of digital libraries and the maintenance of our (digital) cultural heritage, very little research has been undertaken on this topic so far [3, 6]. Further research by the author is currently in progress, concentrating on file formats which mainly deal with text information. This will be a fruitful contribution especially with regard to the latest trends in the discussion on the appropriate file format for text and hybrid content. It will also be the groundwork for suggestions on the improvement of robustness on file format level, which will be made in a final step.[25]

References

1. Avcıbas, I., Sankur, B., Sayood, K.: Statistical evaluation of image quality measures. Journal of Electronic Imaging 11(2), 206–223 (2002)
2. Bairavasundaram, L.N., et al.: An Analysis of Data Corruption in the Storage Stack. ACM Transactions on Storage 4(3) (2008)
3. Buonora, P., Liberati, F.: A Format for Digital Preservation of Images: A Study on JPEG 2000 File Robustness. D-Lib Magazine 7/8 (2008),
   http://www.dlib.org/dlib/july08/buonora/07buonora.html
   (accessed May 2009)
4. Chapman, S., et al.: Page Image Compression for Mass Digitization. In: Archiving 2007. Final program and proceedings, pp. 37–42 (2007)
5. Gilesse, R., Rog, J., Verheusen, A.: Life Beyond uncompressed TIFF: Alternative File Formats for the Storage of Master Image Files. In: Archiving 2008. Final program and proceedings, pp. 41–46 (2008)
6. Heydegger, V.: Analysing the Impact of File Formats on Data Integrity. In: Archiving 2008. Final program and proceedings, pp. 50–55 (2008)
7. Iraci, J.: The Relative Stabilities of Optical Disk Formats. Restaurator 26(2) (2005)
8. ISO/IEC 15444-5:2003. JPEG 2000 image coding system (2003)
9. Matsumoto, M., Nishimura, T.: Mersenne Twister: A 623-dimensionally equidistributed uniform pseudorandom number generator. ACM Trans. on Modeling and Computer Simulation 8(1), 3–30 (1998)
10. Panzer-Steindel, B.: Data Integrity, internal CERN/IT study (2007),
    http://indico.cern.ch/getFile.py/access?contribId=3&sessionId=0&resId=1&materialId=paper&confId=13797 (accessed May 2009)
11. Schroeder, B., Gibson, G.A.: Disk failures in the real world: What does an mttf of 1,000,000 hours mean to you? In: Proceedings of the 5th USENIX Conference on File and Storage Technologies, FAST (2007)

---

[25] This study reported in here is part of a larger project. All existing and future results will be published in the end of it. Please contact the author for any further questions.

Appendix

Table 1. Selection[26] of test corpus files with brief description of core characteristics. All images are of the same dimensions (498 x 719). The compression rate indicates the ratio of: file size / file size of uncompressed file.

format	name	characteristics
TIFF	tiff_A1	file size (byte): 1075526; image type: coloured; bits per pixel: 24 (8,8,8); compression: none; number of stripes: 1;
	tiff_A2	file size (byte): 1075682; image type: coloured; bits per pixel: 24 (8,8,8); compression: none; number of stripes: 144;
	tiff_B	file size (byte): 107697; image type: coloured; bits per pixel: 24 (8,8,8); compression: jpeg; compression rate: 0.101; number of stripes: 1;
	tiff_C	file size (byte): 253494; image type: greyscale; bits per pixel: 8; compression: ZIP; compression rate: 0.856; number of stripes: 1;
JPEG	jpeg_A	file size (byte): 352162; image type: coloured; bits per pixel: 24 (8,8,8); compression: jpeg; compression rate: 0.327;
	jpeg_B	file size (byte): 87010; image type: coloured; bits per pixel: 24 (8,8,8); compression: jpeg; compression rate: 0.081
	jpeg_C	file size (byte): 83133; image type: coloured; bits per pixel: 24 (8,8,8); compression: jpeg; mode: progressive; compression rate: 0.078;
JP2	jp2_A	file size (byte): 103048; image type: coloured; bits per pixel: 24 (8,8,8); compression: jpeg 2000; compression rate: 0.096; error resilience: no error resilience features included;
	jp2_B	file size (byte): 689227; image type: coloured; bits per pixel: 24 (8,8,8); compression: jpeg 2000; compression rate: 0.641; features: all error resilience features included[27], lossless compression;
	jp2_C	file size (byte): 103048; image type: coloured; bits per pixel: 24 (8,8,8); compression: jpeg 2000; compression rate: 0.096; features: all error resilience features included, lossy compression
PNG	png_A	file size (byte): 700847; image type: coloured; bits per pixel: 24 (8,8,8); compression: ZIP; compression rate: 0.652
GIF	gif_A	file size (byte): 207.637; image type: coloured; bits per pixel: 8; compression: LZW; compression rate: 0.575
BMP	bmp_A	file size (byte): 1.075.678; dimensions: 498 x 719; image type: coloured; bits per pixel: 24 (8,8,8); compression: none

---

[26] The test corpus contains 43 files in total, for reason of space, the table only contains those files which are discussed in more detail.

[28] Please see [8] for a description of the error resilience features. We used all of the features which are provided by the creation software Kakadu.

Table 2. Result data for metric RImean

	1 bit	0.0005	0.0010	0.0025	0.0050	0.0075	0.0100
tiff_A1	0.0000	0.0021	0.0072	0.0148	0.0222	0.0397	0.0463
tiff_A2	0.0017	0.0134	0.0296	0.0734	0.1340	0.1741	0.2413
tiff_B	0.2123	0.5307	0.7170	0.8507	0.9127	0.9379	0.9533
tiff_C	0.1598	0.7127	0.8434	0.9384	0.9695	0.9784	0.9882
jpeg_A	0.1886	0.8106	0.8927	0.9567	0.9790	0.9861	0.9903
jpeg_B	0.1658	0.4147	0.6335	0.8079	0.8929	0.9280	0.9466
jpeg_C	0.1978	0.4422	0.6667	0.8135	0.8857	0.9185	0.9362
jp2_A	0.2245	0.6435	0.8529	0.9838	0.9973	0.9987	0.9989
jp2_B	0.0401	0.7242	0.8910	0.9778	0.9898	0.9932	0.9944
jp2_C	0.1595	0.4593	0.6884	0.9034	0.9608	0.9761	0.9823
png_A	0.1680	0.9033	0.9539	0.9779	0.9915	0.9949	0.9962
gif_A	0.1700	0.5367	0.6500	0.7549	0.8180	0.8407	0.8582
bmp_A	0.0000	0.0011	0.0022	0.0041	0.0062	0.0109	0.0164

Table 3. Result data for RMSEmean. Please note: By tendency, result values from 100.0 onwards also go along with very poor visual image quality.

	1 bit	0.0005	0.001	0.0025	0.005	0.0075	0.01
tiff_A1	0.0308	0.6099	1.6920	3.2719	4.7602	8.0671	9.4261
tiff_A2	0.2790	2.5138	5.1467	12.4700	22.3199	29.4321	39.4085
tiff_B	14.1767	40.2629	59.7920	82.3622	96.8800	102.9344	107.0307
tiff_C	19.9272	74.3268	87.2172	101.3937	109.1695	113.6602	116.4384
jpeg_A	19.7448	94.3614	102.8009	106.8254	107.1547	109.3841	113.1354
jpeg_B	13.6913	35.8373	63.1653	89.7461	102.0462	108.1198	111.3786
jpeg_C	7.4957	17.8646	32.4221	54.0923	81.5385	98.3300	108.5839
jp2_A	1.9187	6.8776	10.4855	19.7557	31.4201	40.9756	48.0054
jp2_B	0.2605	4.8834	7.6121	15.7449	21.4953	32.9461	38.0008
jp2_C	1.0732	4.5185	8.7373	12.2292	20.8362	29.7704	35.1526
png_A	31.1893	132.2457	137.7804	139.0441	142.4394	143.4338	144.0673
gif_A	32.6047	83.0106	93.2021	102.1034	108.3719	112.9389	116.7700
bmp_A	0.0300	0.5241	0.6677	1.2994	1.9721	2.7480	3.9217

Formalising a Model for Digital Rights Clearance

Claudio Prandoni[1], Marlis Valentini[1], and Martin Doerr[2]

[1] Metaware S.p.A., Via Filippo Turati 43/45,
56125 Pisa, Italy
{c.prandoni,m.valentini}@metaware.it
[2] Institute of Computer Science, FORTH,
Vassilika Vouton, 71110 Heraklion, Crete, Greece
martin@ics.forth.gr

Abstract. Due to the increasing complexity and world-wide distribution of digital objects, identification and enforcement of digital rights have become too complex to be carried out manually. It is necessary to take into account the case-specific applicable laws, the complete creation history of a work and the existing licenses. However, no formal generic model has been presented so far integrating these aspects. This paper presents an innovative domain ontology of the Intellectual Property Rights. It distinguishes four levels of abstraction or control: (1) the legal framework, (2) the individual rights people hold, (3) the individual usage agreements right holders and others may issue, and (4) the particular actions that are restricted by IPR regulations or bring particular rights into existence. The ontology has the potential to enable wide semantic interoperability of digital repositories for identifying existing rights on digital objects and tracing the impact of particular actions on rights and regulations.

Keywords: Intellectual Property Rights, Digital Provenance, Digital Rights, Information Integration, Interoperability, Ontology, Domain Models, Ontology Engineering.

1 Introduction

Intellectual Property Rights (IPRs) are exclusive rights that the law grants to the authors of intellectual creations. They are divided in Industrial Property Rights and Copyright. Industrial Property Rights covers inventions (patents), industrial designs, trademarks and indications of source, while Copyright protects literary, artistic and scientific works [1]. Related Rights were introduced later to complement Copyright and to protect artistic performances, sound productions, databases, software and folklore.

This paper presents a domain ontology, hereinafter called the "Digital Rights Ontology" (DRO), that aims at providing a comprehensive conceptualisation of the basic relationships and entities in the Intellectual Property Rights domain, focusing in more detail on Copyright and Related Rights. This ontology has been developed in the framework of the European-funded IST project CASPAR (Cultural, Artistic and Scientific knowledge for Preservation, Access and Retrieval - IST-2005-2.5.10) in order to support the definition and the tracing of digital rights in the context of long term access and preservation of digital archive holdings.

M. Agosti et al. (Eds.): ECDL 2009, LNCS 5714, pp. 327–338, 2009.
© Springer-Verlag Berlin Heidelberg 2009

2 Application Scenarios

The Digital Rights Ontology presented in this paper aims at modelling within one single framework the most relevant aspects required in digital rights applications. These can be divided at a general level in two main categories:

- identification and reporting of rights;
- evaluation of rights and consequences of actions.

An archive must be able to determine Intellectual Property Rights because they impose legal limitations to the actions on the archived material. Final consumers are not allowed, by law, to freely use rights-protected creations unless they have the proper permissions. The same is true for the archival institution; unless it has obtained the complete rights ownership by means of a transfer of rights, it needs an explicit authorisation to hold any copy, to perform any kind of modification and to distribute it. If it does not possess these permissions, the right holder might have, in the worst case, the right to ask for compensation payments and for deleting all copies, which would mean not only the loss of an archive holding, but also the loss of all the effort spent to curate and preserve it through the years.

In order to identify all existing rights on digital objects, the *Digital Provenance* is the first aspect that must be analysed. Digital Provenance documents the actions that form the creation history of a digital work and its constituent parts; providing, among others, information on who participated in the creation and what their contribution was. A complication arises when there is a relationship with other existing creations; for instance, the incorporation of parts of expressions in other expressions [4], such as the use of images, sound tracks, etc., in multimedia productions. This requires the ability to trace inclusion and derivation chains in the production.

Further elements that must be considered are the country under which a creative work is protected and the happening of particular events, such as transfers of rights, death of an author and consequent copyright expiration after a given time, or voluntary release of a work under Public Domain.

The second important aspect in the digital rights clearance workflow is the evaluation of rights and the consequences of actions. In this case an analysis of existing licenses and applicable laws is required. This analysis then leads to the compilation of a set of rules, which can be modelled as patterns of constrained activities against which to check the specific individual actions performed by a particular user on a particular work. In this way, implementing a rights enforcing algorithm means to check if there is at least one pattern against which the single individual situation matches. If a match is found then the action is allowed, otherwise it is assumed to be prohibited. Detecting allowances rather than prohibitions guarantees that no illegal action is regarded to be permitted by the system, even if the information the decision is based on may be incomplete; however, an allowed case may be missed by the system. The precision of the result depends on how much information is available on the digital provenance and the licensed rights. In case of complicated licenses, the evaluation system may indicate potential allowances to the user that it cannot evaluate automatically.

3 Existing Rights Models

In Digital Rights Management (DRM) systems, technical enforcement measures are often used in conjunction with machine-understandable Rights Expression Languages (RELs). The latter are used to express license terms and conditions that can be checked automatically. Some standards already exist, such as MPEG-21REL and ODRL (Open Digital Rights Language). MPEG-21REL incorporates XrML and is probably the most widely adopted specification for commercial DRM; while ODRL was created as an international effort to provide an open standard for the DRM sector. Both come with their own rights model [5][6]. However, the types of rights that they conceptualise are only the permissions obtained by the consumers. Ownership rights, i.e. the exclusive rights held by the authors of the creative works, are not modelled.

Besides these standards, there are the Creative Commons licensing models [7] and the associated ccREL [8][9] language. CC licenses, likewise other Copyleft [10] licensing forms, aim at releasing most of the limitations that derive from Copyright. ccREL allows to specify permissions and related constraints, but not ownership rights.

In addition to RELs, other sources of information have also been consulted. The work carried out by R. García, in particular the analysis and conceptualisation reported in [11], and implemented in the ontologies IPROnto [1] and RightsOnto [12], has been used as a starting point. The driver of these ontologies was the management of IPR licenses in e-commerce multimedia applications. They model the core concepts for checking and retrieving usage permissions through DL reasoning and semantic queries, providing an alternative to the previous syntactic approaches to build interoperability between different RELs.

It is worthwhile to mention the <indecs> framework [13] as the precursor of many of the current existing rights metadata formats, both in the academic research and in the commercial field (MPEG21 RDD). This project has recognised that mechanisms to transform metadata into representations of events of Digital Provenance appear to provide the most powerful approach to interoperability. The event-based model in <indecs> does, however, only allow the user to trace rights, as was required in e-commerce systems, and to retrieve events, but it does not model how rights effect events.

The DRO was developed independently from any particular service implementation and, with respect to the aforementioned right models, it aims to provide a more comprehensive conceptualisation of rights. It includes ownership rights as well as permissions, and all the relevant entities related to provenance, content of legislation and content of licenses which influence the status of rights.

4 The Digital Rights Ontology

4.1 Modelling Approach

The development of the DRO builds on some existing standard or well-established core ontologies, such as CIDOC CRM [14][15][16] and FRBRoo [15][17][18][19].

The CIDOC (Documentation Committee of the International Council of Museums) Conceptual Reference Model provides definitions and a formal structure for modelling concepts and relationships used in cultural heritage documentation, and became an international standard, ISO 21127:2006. It is an event-centric model of the material history of people, their interactions and the objects they deal with. It has been in development since 1996 by abstracting over an empirical base of hundreds of database schemata and data structures from different cultural-historical domains. It already integrates the basic ideas of the <indecs> framework [14] and the subsequent ABC Harmony Model, which also describes the history of digital artefacts [20].

FRBRoo is an ontological interpretation created in 2003-2008 from FRBR (Functional Requirements for Bibliographic Records), the major expression of library conceptualisations created by IFLA (International Federation of Library Associations and Institutions) in the years 1992-1998. It identifies bibliographic entities (such as works, persons, events, places, etc.), their attributes and the relationships between them. It is formulated as a specialisation of ISO21127 and it is the result of careful harmonisation of the CIDOC and FRBR models. A major innovation of FRBRoo is the modelling of the mechanisms of *incorporation* of works, such as the creation of a song incorporating lyrics from an existing poem, or the use of a reproduction of a painting on a book cover.

The extension with certain concepts and terminology of the intellectual property rights domain relies on existing IPR-specific works. Indeed, several concepts have been adopted from the ontologies IPROnto and CopyrightOnto or reinterpreted in order to conform with the ontological rigor applied to the DRO. However, taxonomy of the adaptation is substantially different. In fact, the IPR-relevant actions were modelled in IPROnto as subclasses of the rights that govern that class of action. For instance, the *CommunicationRight* has *Communicate, Broadcast, Retransmit* and *MakeAvailable* as subclasses. As explained in [11], this ontologically wrong connection between rights and actions allowed the reduction of checking if an action is authorised, to just check if the right class subsumes the action class. However, it fails to represent correctly any other relevant events and to integrate other models. Instead, in the approach followed for the definition of the DRO, the checking of permissions does not consist in performing class subsumption, but matching a set of ontology instances (the circumstance of an intended action) against another set of ontology instances (the pattern that describes a certain permission).

In addition, some standard Rights Expression Languages, like MPEG21-REL [5], ODRL [6] and Creative Commons [8] have been used as a guide in the definition of the license-specific concepts and terminology. Finally, the documentation from WIPO (World Intellectual Property Organisation) has been used for consultation.

The extension and harmonisation of existing ontologies outlined above makes it possible not only to focus on the modelling of rights, but also to be aligned with the view of the curators (mainly library and museum archivists) and the view of the legislators (for the notion of ontological adequacy see [21]). This will allow, for instance, to simplify the mediation between different information sources, as long as they commit to a common ontology.

4.2 Rights as a Pattern of Constrained Activities

The core idea that characterises the Digital Rights Ontology is the distinction of 4 levels of abstraction or control, as indicated in Fig. 1:

- *legal framework*, consisting in the type of rights that are valid in a given country at a given time;
- *individual rights* owned by the right holders (also called *ownership rights*), e.g. Peter, creator of a digital work, owns the right to make a copy of it;
- *individual usage agreements* between right holders and other people, e.g. Peter issues to Martin a license which allows him to make copies of Peter's work under certain conditions;
- *particular actions*, which may fall under IPR-related regulations, e.g. Martin makes a copy of Peter's work. They may be allowed, disallowed, or allowed under particular conditions.

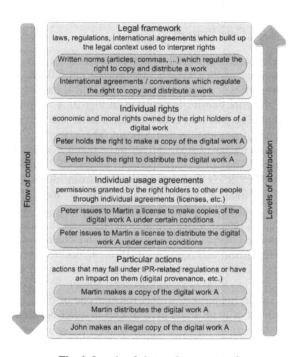

Fig. 1. Levels of abstraction or control

These levels can be also interpreted in a top-down manner as a flow of control: the legal framework documents and regulates the individual rights owned by the right holders; the latter regulate the individual permissions that right holders are allowed to grant to other people; and finally the individual permissions granted by the right holders, if any, regulate the particular actions that a user can legally perform on a digital work, overriding general prohibitions.

The underlying idea at the base of the DRO is that all the three upper levels: laws, rights ownership and licenses, are modelled as patterns of constraints for activities that determine if an activity is permitted or not, or if subsequent activities of particular patterns are foreseen. This means that each specific individual action is related to the agreements, rights and laws that regulate it. Martin can make a copy of a digital work because he either signed an individual agreement with the creator of the work, he is the right holder of the work, or there is a particular law which grants him the right to make such a copy. This is a pattern matching process that should be applied at three different levels and can be implemented by suitable procedural code.

This way to make the nature of rights explicit allows for formalising one of the key processes needed in the domain of rights, namely the checking of permissions to perform a certain activity. This process can be described as a matching procedure, where the permissions (patterns) are matched against a given situation. This is in fact what happens when permissions are verified, either if they are given by the right holders through licenses or if they are given by the law in form of "user rights". A concrete example of how this mechanism works is described in paragraph 5.2.

4.3   Description of the Core Entities

The last version (release 0.5) of the DRO is represented in RDFS, publicly available at [22], and is composed of 96 classes (9 inherited from CIDOC CRM and 5 from FRBRoo) and 44 properties (7 inherited from CIDOC CRM and 2 from FRBRoo).

An overview of the core entities is depicted in Fig. 2. A unique prefix is used to identify each concept. The prefix is composed of a letter that serves as a namespace identifier, and a number that identifies the concept. The letters are "E" and "C" for the concepts inherited from CIDOC CRM, "F" for those taken from FRBRoo and "D" for DRO specific concepts. The same for properties, "P" and "S" identify the ones inherited from CIDOC-CRM, "R" is used for those taken from FRBRoo and "I" for DRO specific properties. The final "F" means forward, while "B" means backward.

The main classes used to formalise the legal framework are *Regulation* and *NationalRightType*. Regulations describe patterns of situations that are permitted. This is what Rights Expression Languages aim to express and control. Regulation splits in two subclasses: *WrittenNorm* and *Agreement*. *WrittenNorm* represents all laws which are valid in a certain country at a certain time; *Agreement* describes both international agreements which override local laws and bilateral agreements between right holders and other people. A right holder may in fact transfer one or more rights that he owns through an *IPRContract* or grant some permissions to other people to act on his digital objects issuing them an *IPRLicense*. *NationalRightType* is used to represent the types of rights that have a validity in a given country at a given time.

As already pointed out, the DRO distinguishes between two kinds of rights: the ownership rights, which are the exclusive rights typically held by the authors of the creative works, and the permissions that are granted by the right holders to other persons to use such works. The class *OwnershipRight* models the first type of rights, while the class *Permission* includes all types of authorisations to make use of content, including the authorisations given by the law and those given by the right holders through licenses. In both cases it is a consequence of a *Regulation*, respectively a *WrittenNorm* and an *IPRLicense*.

The other core entities which are part of the DRO have been adopted from CIDOC CRM and FRBRoo, in particular the Creation and the Action models, i.e. concepts like *Actor*, *LegalObject* and *Activity* with their subclasses and relationships. These concepts have been linked to the concepts which are specific of the copyright domain through the use of suitable relationships. For instance an *Actor* owns a *Right* which *isOn* a *LegalObject* and a *Right* allows (or *disallows*) an *Activity*.

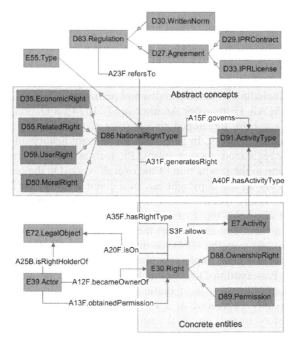

Fig. 2. Overview of DRO core entities

Another important characteristic inherited from the CIDOC CRM is the distinction between individual entities on one side, like persons, objects, licenses, etc., and general concepts on the other side, like types of rights, types of activities, constraints, and others, which are used to categorise individual entities. So, in the DRO there are more abstraction levels of rights entities: *Right* represents the instances of rights held by individual legal and physical persons on a precise object, while *NationalRightType* models types of rights. The same for *Activity* and *ActivityType*. This approach reflects looking also at the properties, so we have that a *RightType governs* an *ActivityType* and, at the corresponding individual level, a particular *Right allows* a specific *Activity*.

Therefore *OwnershipRight* and *Permission* represent rights of a given person on a precise object, while the pattern of situation that is allowed is represented by apposite entities, such as *ActivityType*, *PermissionPattern*, *Constraint*, *Condition*, *Validity* and others, together with properties such as *hasDuration*, *hasPurpose*, *hasExerciseLimit*, *hasAuthorisedPrincipals*, *hasFee*. Individual situations, expressed in terms of ontology instances, should then be matched against the general patterns, still expressed in terms of instances of the ontology.

Another important notion that was introduced is that of *Validity*, in order to address the problems due to the existence of different national legislations. Each *NationalRightType*, for instance the *DistributionRight* or the *AttributionRight*, is modelled as having a validity in space and time, and is linked to a set of applicable *Regulations*. Also the rights-generating activities, for instance *Derive*, *ProduceFixation* or *Perform*, have the dimensions of space and time as one of their attributes. This makes it possible to derive the precise instance of *NationalRightType* to which the individual ownership right corresponds, thus linking it to regulating Laws.

5   Examples

5.1   Identification and Reporting of Rights

The first application of the Digital Rights Ontology is to use it as a formal dictionary to express rights, including ownership rights and permissions.

Within the CASPAR Framework Architecture [23][24], a Digital Rights Management component has been developed that derives all the existing ownership rights given the provenance of a work. In particular, all relevant history information about a given work can be registered via a specific API, such as who performed which activity, thus contributing to the production of the overall work. On the basis of this provenance information and of the copyright law of a given country, the DRM component determines who holds any specific ownership right on any creative contribution to the work, for instance who holds the distribution on the lyrics of a musical piece.

The DRM component can also export the information about the ownership rights that it has derived, and uses the DRO for that purpose. In the CASPAR project, this rights export service has been implemented to generate provenance metadata, namely a subset of PDI (Preservation Description Information), to be associated to the archived content data objects. It is of particular benefit for preservation purposes to have PDIs analytically expressed in a knowledge representation language, as this simplifies the preservation of the PDI itself. In fact, the ontology provides both a dictionary and a formal semantics of its entities.

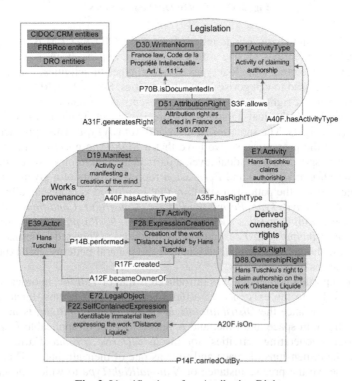

Fig. 3. Identification of an Attribution Right

Fig. 3 describes the pattern that is evaluated by the DRM component to derive a specific ownership right, namely the right to claim authorship on a given intellectual creation. The diagram highlights that there are two groups of entities that contribute in bringing an ownership right into existence. On one side there are the provenance events and related entities, which are specific to the work, and on the other side there are some more general applicable entities, defined by the legal framework of a given country. The latter express the rules by which certain types of activities generate rights-protected products, and the activity types on which the product creators have the exclusive right.

Looking at the figure, Hans Tutschku holds the exclusive right to claim authorship on the work "Distance Liquide", as a consequence that he composed this acousmatic work and that the copyright law assigns the attribution right to the creators of intellectual works.

5.2   Evaluation of Rights and Consequences of Actions

An innovative characteristic of the Digital Rights Ontology is that it makes the nature of rights explicit, namely that it expresses the permissions as patterns of constrained

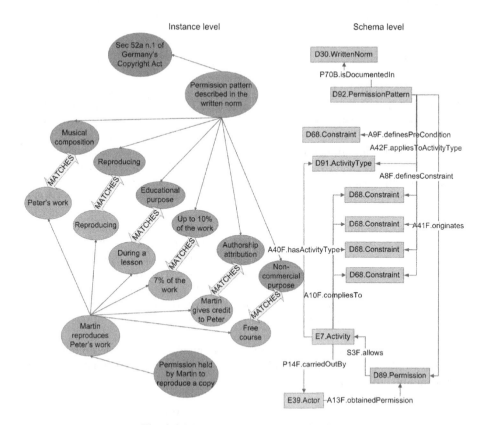

Fig. 4. Right evaluation as pattern matching

activities. Therefore the evaluation of permissions corresponds to the matching of real case situations against the patterns described in licenses or in written norms. This means that it is required from one side to be able to capture the details of the consumers' intended actions and on the other side to formulate the constituent parts of the permissions, be they given through licenses or by the law.

For instance, the educational user right can be modelled, likewise a license, as a set of *PermissionPatterns*, namely one for each type of activity. Fig. 4 depicts an example where the activity is the reproduction of a musical piece. The intended action is highlighted on the bottom half and the top half is the permission pattern. The individual right that the teacher Martin holds, allowing him to reproduce an excerpt of the musical piece during a lesson, is derived from the successful matching between the instances describing the intended action and the instances describing the pattern. Finally, this *PermissionPattern* is bound to a set of *WrittenNorms* that regulate it, which also declare the consequences of violating it.

Yet, in many cases it is difficult to implement the automatic execution of such a pattern matching process. The idea is that certain conditions would be cut out from the evaluation process, and would just be presented to the consumer. Furthermore, the norms related to the particular use case could be referenced.

So far, only the concepts necessary as parameters for this rights evaluation process have been identified. A related implementation is foreseen by procedural code.

6   Conclusions

The main purpose of the Digital Rights Ontology is to provide a standard model that allows digital repositories to identify existing rights on digital objects and to evaluate the consequences of potential actions which may fall under IPR-related regulations. This is achieved by making explicit all the fundamental aspects and events in the history of a work that influence the status of rights (i.e. digital provenance, licenses and legal framework) and by interpreting the rights as patterns of future activities.

This innovative idea also represents a key aspect for the purpose of determining, preserving and enforcing digital rights in the long term. In this perspective the DRO is being used, within the CASPAR Framework, as the reference conceptual model for digital rights. The implementation of the Digital Rights Management component relies upon it as the formal language for exporting information about ownership rights, once they are derived by applying the copyright norms of a particular country to the digital provenance information of a given work. The logics of the respective evaluation has been implemented in procedural code operating on a database, which addresses the scalability requirements of the CASPAR system better than a rule-based system, but this is not the subject of this paper. The ongoing status of development can be followed on [24].

The compatibility of the DRO with important ontologies for the description of intellectual work - CIDOC CRM, FRBRoo - makes it a promising candidate to support interoperability of potential global digital rights clearance services across on-line digital repositories.

Future activities related to the DRO include the prototype implementation of a rights enforcing algorithm, so far foreseen by procedural code, that would be based on the pattern matching model described in this paper. In principle, the pattern matching could also be done by KR reasoning systems, but for reasons of platform restrictions and performance concerns, we have not yet considered this approach.

In addition, the digital provenance sub-model and the derivation of rights will be further validated, in particular with respect to the challenging cases of incorporation and derivation of works, where determining the rights is particularly complex.

Acknowledgments. We would like to express special thanks to the European Commission for funding our work in the fields of Digital Rights and Ontology and to the useful cooperation of the research team of the CASPAR project, in particular David Giaretta, project coordinator. Thanks to Fiore Basile who conceived the work and to David Lamb and Francesca Fenili for their feedback for improving the presentation.

References

1. Delgado, J., Gallego, I., Llorente, S., García, R.: IPROnto: An Ontology for Digital Rights Management. In: Frontiers in Artificial Intelligence and Applications. 16th Annual Conference on Legal Knowledge and Information Systems. JURIX 2003, vol. 106, pp. 111–120. IOS Press, Amsterdam (2003), http://www.jurix.nl/pdf/j03-12.pdf
2. Nadah, N., Dulong de Rosnay, M., Bachimont, B.: Licensing digital content with a generic ontology: escaping from the jungle of rights expression languages. In: International Association for Artificial Intelligence and Law, pp. 65–69 (June 2007), http://doi.acm.org/10.1145/1276318.1276330
3. Coyle, K.: Rights expression languages - a report for the library of congress (February 2004), http://www.loc.gov/standards/relreport.pdf
4. Doerr M., Bekiari C.: FRBRoo, a Conceptual Model for Performing Arts. In: 2008 Annual Conference of CIDOC, Athens (September 2008), http://www.cidoc2008.gr/cidoc/Documents/papers/drfile.2008-06-18.1838425657
5. Multimedia Description Schemes (MDS) Group: Introducing MPEG-21 REL - an Overview. International Organization for Standardization (July 2005), http://www.chiariglione.org/MPEG/technologies/mp21-rel/index.htm
6. Iannella, R.: Open Digital Rights Language (ODRL) Specification. Version 1.1 (August 2002), http://odrl.net/1.1/ODRL-11.pdf
7. Logos and codes for the CC core licenses, http://creativecommons.org/about/licenses/meet-the-licenses
8. Description of ccREL, http://wiki.creativecommons.org/images/d/d6/Ccrel-1.0.pdf
9. Overview of ccREL vocabulary (RDF Schema), http://creativecommons.org/schema.rdf
10. Definition of Copyleft licensing, http://en.wikipedia.org/wiki/Copyleft
11. García, R.: A Semantic Web Approach to Digital Rights Management. PhD Thesis, Universitat Pompeu Fabra, Barcelona (2005), http://rhizomik.net/~roberto/thesis/
12. García, R., Gil, R., Delgado, J.: A web ontologies framework for digital rights management. In: Artificial Intelligence and Law, vol. 15(2), pp. 137–154. Kluwer Academic Publishers, Dordrecht (2007)

13. Bide, M., Rust, G.: The <indecs> metadata framework: Principles, model and data diction-
    ary. Godfrey Rust, MUZE Inc., Mark Bide, EDItEUR (June 2000),
    http://www.doi.org/topics/indecs/indecs_framework_2000.pdf
14. CIDOC CRM v0.4.2,
    http://cidoc.ics.forth.gr/official_release_cidoc.html
15. Doerr, M.: The CIDOC CRM – An Ontological Approach to Semantic Interoperability of
    Metadata. AI Magazine 24(3) (2003),
    http://cidoc.ics.forth.gr/docs/ontological_approach.pdf
16. CIDOC CRM Web Site, http://www.cidoc-crm.org
17. Doerr, M., Le Bœuf, P., Bekiari, C. (eds.): International Working Group on FRBR and
    CIDOC CRM Harmonisation: FRBR object-oriented definition and mapping to FRBR ER
    (version 0.9 draft) (January 2008),
    http://cidoc.ics.forth.gr/docs/frbr_oo/frbr_docs/
    FRBR_oo_V0.9.pdf
18. FRBRoo v1.0, http://cidoc.ics.forth.gr/frbr_drafts.html
19. Le Bœuf, P.: Functional Requirements for Bibliographic Records (FRBR): Hype or Cure-
    All? Haworth Information Press Inc. (2005) ISBN: 0789027984
20. Doerr, M., Hunter, J., Lagoze, C.: Towards a Core Ontology for Information Integration.
    Journal of Digital Information 4(1) Article No. 169 (April 2003)
21. Smith, B.: Ontology: An Introduction. In: Floridi, L. (ed.) The Blackwell Guide to the Phi-
    losophy of Computing and Information, pp. 155–166. Blackwell, Oxford (2003)
22. Digital Rights Ontology v0.5,
    http://www.casparpreserves.eu/publications/
    ontologies/RightsOntology
23. CASPAR project Web Site, http://www.casparpreserves.eu
24. CASPAR software, documentation and code,
    http://developers.casparpreserves.eu:8080/

Evaluation in Context[*]

Jaap Kamps[1,2], Mounia Lalmas[3], and Birger Larsen[4]

[1] Archives and Information Studies, University of Amsterdam
[2] ISLA, University of Amsterdam
[3] Department of Computing Science, University of Glasgow
[4] Information Studies, Royal School of Library and Information Science

Abstract. All search happens in a particular context—such as the particular collection of a digital library, its associated search tasks, and its associated users. Information retrieval researchers usually agree on the importance of context, but they rarely address the issue. In particular, evaluation in the Cranfield tradition requires abstracting away from individual differences between users. This paper investigates if we can bring some of this context into the Cranfield paradigm. Our approach is the following: we will attempt to record the "context" of the humans already in the loop—the topic authors/assessors—by designing targeted questionnaires. The questionnaire data becomes part of the evaluation test-suite as valuable data on the context of the search requests. We have experimented with this questionnaire approach during the evaluation campaign of the INitiative for the Evaluation of XML Retrieval (INEX). The results of this case study demonstrate the viability of the questionnaire approach as a means to capture context in evaluation. This can help explain and control some of the user or topic variation in the test collection. Moreover, it allows to break down the set of topics in various meaningful categories, e.g. those that suit a particular task scenario, and zoom in on the relative performance for such a group of topics.

1 Introduction

The history of information retrieval (IR) is a showcase of theoretical progress going hand-in-hand with experimental evaluation. The scientific evaluation of IR systems is rooted on the Cranfield experiments [4], and the main thrust in recent years has been the Text REtrieval Conference (TREC) and its various regional and task-specific counterparts such as CLEF, NTCIR, and INEX. The Cranfield tradition of test collection development tries to abstract away from individual differences between users [15]. Yet at the same time, it has been known for a long that individual differences are one of the greatest sources of variation in relevance judgments and system failures [3, 13]. In fact, even within the test collections built in the Cranfield tradition, the "topic effect" or "user effect" is the largest source of variation [2]. Nonetheless, the overwhelming success of experimental IR can be interpreted as a clear signal that the test collection abstraction is effective for evaluating document retrieval.

Digital libraries researchers are addressing search tasks of increasing complexity that require finer-grained judgments than standard document retrieval. Examples are

---

[*] This research was partly funded by DELOS (an EU network of excellence in Digital Libraries) through the INEX initiative for the evaluation of XML retrieval.

information pin-pointing tasks like Question Answering and XML Retrieval. For these tasks, it is more than plausible that individual differences have a much greater impact. As [14], p.15 puts it:

> TREC needs to engage, more positively and fully, with context and the nature of the whole setup information-seeking task T rather than just the experimental task X. ... I believe that it is important for TREC's system-building partici-pants to be encouraged to work forward from a fuller knowledge of the context, rather than limiting their attention to the attenuated form of the context that the D * Q * R environment normally embodies, and recovering what may be dis-tinctive about this – and hence somewhat indicative of significant features of the larger context – by working backwards from system results.

The "user effect" is keeping test collection builders in a double-bind: On the one hand, building a stable and reusable test collection requires abstracting from task and user differences. On the other hand, more realistic search tasks such as those occurring in digital libraries requires bringing some of the user and task context into consideration.

The main research question of this paper is to investigate if we can bring some of the user's context into the Cranfield paradigm. Our approach is the following: rather than fitting a retrieval task and its evaluation to a particular context, we will attempt to record the "context" of the humans already in the loop during the construction of a test collection: the topic authors/assessors. Recall that, as mentioned above, individual difference are by no means outlawed in the Cranfield tradition—they are the greatest source of variation in standard search tasks investigated at TREC and other evaluation forums. By designing targeted questionnaires, we can record salient features of the topic authors and their topics of request, and of the assessors and their judging behavior. The resulting questionnaire data becomes part of the evaluation test-suite as valuable contextual data. This can help explain and control some of the user or topic variation in the test collection. Moreover, it will allow to break down the set of topics in various meaningful categories, and zoom in on the relative performance for such a group of topics. This allows for the testing of a larger variety of research questions as the tests can be restricted to the appropriate subset of topics.

Our aims are closely related to those of the TREC HARD track (2003–2005) and its continuation in the ciQA task at the TREC QA track (2006–2007). The HARD tracks [1] investigated whether retrieval systems could improve by i) query metadata that better described the information need, ii) interaction with the searcher through clar-ification forms, and iii) passage level retrieval and relevance judgments. Here the query metadata—consisting of fields like familiarity, genre, geography, subject, and related text—is most closely related. The main focus of the HARD track (and its continuation) was on user system interaction focusing on additional information that can be directly used by retrieval systems, whereas our main focus is on recording the broader context of the topics and assessments to provide insight in the constructed test collection, and to aid further analysis.

We have conducted an exploratory experiment with our questionnaire approach during the 2007 evaluation campaign of the INitiative for the Evaluation of XML Re-trieval. Research in XML retrieval attempts to take advantage of the structure of explic-itly marked up documents to provide more focused retrieval. This is believed to be of benefit when users are searching large documents, such as those often contained in digital libraries. The task of XML retrieval is a much more complicated one than standard document retrieval. Not only must XML element retrieval systems be able to identify relevant content; in addition a suitable granularity of the returned elements must be decided on. As a consequence the creation of test collections for XML retrieval is a notable challenge in itself. During INEX 2007, topic authors completed a questionnaire immediately after submitting the final version of their topics, and assessors completed a questionnaire after finishing the judging of their topics. The topic author questionnaire consisted of 19 questions about the topic familiarity, the type of information requested and expected, results presentation, and the use of structured queries. The assessor questionnaire consisted of 13 questions about the topic of request, the meaning of their relevance judgments. We will focus here on the assessor questionnaire, and investigate the value of the context recorded by the questionnaires to help answer some important questions underlying focused retrieval in structured documents.

The rest of this paper is structured as follows: Section 2 presents some background on the INEX initiative, its search tasks, and some of the main underlying questions. Section 3 presents the questionnaire and presents an analysis of the main results. Section 4 discusses the test collection in context, where the questionnaire is related to the rankings of retrieval systems. In Section 5, we end by discussing our results and by drawing some conclusions.

2   Ad Hoc Retrieval at INEX

In 2002 until 2004, assessors judged pools of retrieved elements, although presented in their article context, using complex two-dimensional judgments [10]. These judgments where based on exhaustivity (basically topical relevance, whether the element contained enough relevant information) and specificity (whether the element contained no excess non-relevant information), both judged on a 4-point scale. The complexity of the assessments also led to complex measures, having to deal with a range of problems [9]. In 2005, this was substituted for by a much simpler assessment system, in which assessors are asked to highlight all, and only, the relevant text [10]. This greatly simplified the assessment process and obviated the need for complex rules to make assessments consistent over partly overlapping XML elements. This made the assessors tasks an intuitive one, leading to eliciting more natural judgments. However, from the start there was a lively discussion on the meaning of the highlighted passages: Are assessors highlighting relevant text according to some global criterion based on the narrative? Or are assessors highlighting the most relevant text according to the local article context?

This led to the introduction of two search tasks at INEX 2006: Relevant in Context and Best in Context, and the elicitation of a separate Best-entry-point judgment. The first task corresponds to an end-user task where focused retrieval answers are grouped per document, in their original document order, providing access through further navigational means. This assumes that users consider documents as the most natural units of retrieval, and prefer an overview of relevance in their original context.

Relevant in Context (RiC).   This task asks systems to return non-overlapping relevant document parts clustered by the unit of the document that they are contained within.

An alternative way to phrase the task is to return documents with the most focused, relevant parts highlighted within.

The second task is similar to Relevant in Context, but asks for only a single best point to start reading the relevant content in an article.

Best in Context (BiC). This task asks systems to return a single document part per document. The start of the single document part corresponds to the best entry point for starting to read the relevant text in the document.

The Relevant in Context Task is evaluated against the text highlighted by the assessors, whereas the Best in Context Task is evaluated against the best-entry-points. For both tasks, mean average generalized precision (MAgP) is used [8]. This is a MAP-like measure where the score per document varies between 0 and 1. Specifically, for Relevant in Context the score per document is determined by how well the retrieved text corresponds to the highlighted text, and for Best in Context the score per document depends on the distance between the retrieved entry point and the assessor's best entry point. In the paper, we will focus on these two tasks.

3   Questionnaires at INEX 2007

We designed and used topic creator and assessor questionnaires in the INEX 2007 evaluation campaign [5]. An IR test collection consists of a collection of documents, a set of search topics, and relevance judgments. For INEX 2007, the document collection is an XML'ified version of the English Wikipedia. Search requests or topics are authored (and also judged) by the INEX participants. At the INEX 2007 ad hoc track, a total of 130 topics was used and a total of 107 topics was assessed. Directly after submitting a candidate topic, the topic author was presented with a new page containing a questionnaire consisting of 19 questions and an open space for comments on the questionnaire. For 107 of the 130 ad hoc topic, this topic questionnaire is available.[1] Directly after assessing a topic, the judge was presented with a questionnaire consisting of 13 questions and an open space for comments. These 13 questions dealt with various issues on the topic of request, the meaning of the highlighted passages, and the meaning of the best-entry-point (BEP). In the rest of this paper, we will discuss the responses to the assessor questionnaire. We restrict the analysis to the 91 topics for which we have both a topic creator and assessor questionnaire (and judgments).

3.1   Topic of Request

We first discuss the responses to questions about the topic of request. Table 1 shows question C1. Almost 60% of the topics have been assessed by the original topic author. Table 2 shows question C2. The majority of judges is familiar with the subject matter of the topic at hand, although there are still 5% of the topics where assessors venture into unfamiliar terrain. Table 3 shows question C3. It is reassuring that the majority of the topics was easy to judge. Table 4 shows question C4. For over 80% of the topics, the Wikipedia is an obvious resource to look for information.

---

[1] Some topics were derived from other sources, such as the topics used at the INEX 2006 Interactive track, and hence we do not have a topic creator questionnaire for these topics.

Table 1. (C1) *Did you submit this topic to INEX?* **Table 2.** (C2) *How familiar were you with the subject matter of the topic?*

Answer		Freq.	Perc.	Answer		Freq.	Perc.
no		37	41%	Not familiar	1	5	5%
yes		54	59%		2	13	14%
					3	32	35%
					4	24	26%
				Very familiar	5	17	19%

Table 3. (C3) *How hard was it to decide whether* **Table 4.** (C4) *Is Wikipedia an obvious source to information was relevant?* *look for information on the topic?*

Answer		Freq.	Perc.	Answer		Freq.	Perc.
Very easy	1	16	18%	no		16	18%
	2	40	44%	yes		75	82%
	3	21	23%				
	4	14	15%				
Very difficult	5	0	0%				

Table 5. (C5) *Can a highlighted passage be* **Table 6.** (C6) *Is a single highlighted passage (check all that apply):* *enough to answer the topic?*

Answer	Freq.	Perc.	Answer		Freq.	Perc.
a single sentence	69	76%	None of them is	1	11	12%
a single paragraph	86	95%		2	17	19%
a single (sub)section	77	85%		3	30	33%
a whole article	62	68%		4	28	31%
			All of them are	5	5	5%

Table 7. (C7) *Are highlighted passages still in-* **Table 8.** (C8) *How often does relevant informa-* *formative when presented out of context?* *tion occur in an article about something else?*

Answer		Freq.	Perc.	Answer		Freq.	Perc.
None of them is	1	2	2%	Never	1	9	10%
	2	16	18%		2	35	38%
	3	22	24%		3	27	30%
	4	41	45%		4	20	22%
All of them are	5	10	11%	Always	5	0	0%

3.2   Highlighted Passages

We now show the responses to questions about the meaning of highlighted passages: Table 5 shows question C5. It is clear that there is no fixed unit of retrieval, for almost 1/4 of the topics a relevant passage cannot be a sentence, and for almost 1/3 of the topics a relevant passage cannot be a whole article. Table 6 shows question C6. There is also no consensus on whether a single passage could suffice as an answer. Table 7 shows question C7. Again no consensus on whether an isolated passage is still informative. Table 8 shows question C8. Relevant passages frequently occur in articles about

Table 9. (C9) *How well does the total length of* **Table 10.** (C10) *Which of the following two highlighted text correspond to the usefulness of strategies is closer to your actual highlighting: an article?*   *(I) the best passages, (II) all relevant text?*

Answer		Freq.	Perc.	Answer		Freq.	Perc.
Never	1	3	3%	I: best	1	12	13%
	2	20	22%		2	22	24%
	3	36	40%		3	5	5%
	4	25	27%		4	30	33%
Always	5	7	8%	II: relevant	5	22	24%

Table 11. (C11) *Can a best entry point be (check all that apply)*

Answer	Freq.	Perc.
the start of a highlighted passage	84	92%
the sectioning structure containing the highlighted text	55	60%
the start of the article	51	56%

Table 12. (C12) *Does the best entry point corre-* **Table 13.** (C13) *Does the best entry point corre-spond to the best passage?*   *spond to the first passage?*

Answer		Freq.	Perc.	Answer		Freq.	Perc.
Never	1	1	1%	Never	1	2	2%
	2	15	16%		2	21	23%
	3	26	29%		3	21	23%
	4	31	34%		4	29	32%
Always	5	18	20%	Always	5	18	20%

something else, which support the motivation behind focused retrieval. Table 9 shows question C9. There is no evidence for assumption that the length of a highlighted passage corresponds to its usefulness, an assumption that has been repeatedly proposed for evaluation measures at INEX. Table 10 shows question C10. There is a remarkable division over the two assessment strategies: the strategy I highlighting the "best passages" is chosen almost as frequently as the strategy II highlighting "relevant passages." Since the particular strategy will have an impact on the resulting assessments, where a judge using strategy I will regard less text as relevant than a judge using strategy II, this can have a large potential impact on the ranking of systems.

3.3 Best Entry Points

We now show the responses to questions about the meaning of the best entry point. Table 11 shows question C11. For almost all topics, the best-entry point can be the start of the highlighted passage (C11), but other types of BEPs occur [11]. Table 12 shows question C12. In the majority of cases, the BEP corresponds to the best passage. Table 13 shows question C13. Again, in the majority of cases the BEP corresponds to the first passage. The responses to C11-C13 may be in part explained by vast majority of relevant articles (4,581 out of 6,491) having only a single highlighted passage [5].

Table 14. Relationship between answers for pairs of questions (chi-square test at percentiles 0.95 and 0.99)

	C1	C2	C3	C4	C5				C6	C7	C8	C9	C10	C11			C12
					sen	par	sec	art						pas	sec	art	
C2	0.99																
C3	-	0.95															
C4	0.95	-	-														
C5 sen	-	-	-	-													
C5 par	-	-	-	-	-												
C5 sec	-	-	0.95	-	-	-											
C5 art	-	-	-	-	-	0.95	0.99										
C6	-	-	-	-	-	-	-	-									
C7	-	-	-	-	-	-	-	-	0.99								
C8	-	-	-	-	-	-	-	-	-	-							
C9	-	-	-	-	-	-	-	-	-	-	-						
C10	-	-	-	-	-	0.99	0.99	0.99	-	-	-	0.99					
C11 pas	-	-	-	0.95	-	-	-	-	-	-	-	-	-				
C11 sec	-	-	0.95	-	-	0.99	0.95	-	-	-	-	-	0.95	-			
C11 art	-	-	-	-	-	-	0.99	0.99	-	-	-	-	0.99	0.95	0.99		
C12	-	-	0.99	-	-	0.95	0.99	0.95	-	-	-	-	-	-	-	-	-
C13	-	0.95	-	-	-	0.99	-	0.95	0.95	0.99	-	-	-	-	0.99	-	-

3.4 Relations

We now analyze the relation between responses to different questions in the question-naire. Table 14 show the relations between pairs of questions in the questionnaire. Since most questions give nominal answers (e.g., yes/no) we use a chi-square test. In particular, we have found that there are a number of relations worth considering. There is an interesting, but not unexpected, relation between being a topic author (C1) and being more familiar with the topic at hand (C2), and also a relation between familiarity (C2) and ease of judging (C3). This confirms the importance of having topics assessed by the original topic author. The judging strategy, judging the best or all relevant passages (C10), is indeed clearly affecting judging behavior: it is related to the granularity of highlighted passages (C5) and to the choice of BEP (C11). There is also a relation with the correspondence between the amount of highlighted text and its usefulness (C9), which holds for assessors highlighting all relevant text, but not for assessors highlighting only the best passages. Finally, the choice of BEPs as best passages (C12) is related to relevant passages being smaller units of the document structure (sentences, paragraphs, (sub)sections), and the choice of BEPs as first passages (C13) is related to relevant passages being complete answers (a single passage is an answer, and still informative in isolation).

Perhaps the most striking observation is that there is such great variety in the responses of the topic authors. This suggests that there are distinct search tasks underlying XML retrieval. This also gives support to the decision at INEX to define a number of distinct search tasks, thus allowing the study of alternative search scenarios for digital libraries. There is rich contextual information in XML retrieval, and the questionnaires provide a means to extract it. But what is the relative importance of the contextual information? In the following section, we will investigate this in terms of system effectiveness.

Table 15. (C5 article): Can a highlighted passage be: a whole article?

Answer	Freq.	%	Relevant in Context task		Best in Context task	
			Overall rank of top 10	Tau	Overall rank of top 10	Tau
yes	62	68.13	1 4 2 8 5 6 3 **10** 7 12	0.9301	1 2 3 4 8 9 6 5 14 15	0.8938
no	29	31.87	6 7 1 2 4 5 18 3 22 14	0.7725	5 *25 22 18* 12 11 *10 21* 16 7	0.7175

4   Test Collection in Context

In Section 3, we reported the responses to the questionnaire as a survey amongst assessors. The outcomes exemplify the wide range of what we can consider contextual information regarding the topics and their assessments. In this way, the questionnaire data also becomes part of the evaluation test-suite constructed during INEX 2007. Our conjecture is that the questionnaire data can provide valuable contextual data on the topics of request and their topic authors. In this section, we start to explore how this additional data can be used.

We investigate how context affects the system ranking—what systems turn out to be effective?—by looking at the relative ranking of systems for a subset of the topics corresponding to a particular context. We analyze the official submissions to the INEX 2007 ad hoc retrieval track's Relevant in Context and Best in Context tasks. There were in total 66 valid runs submitted for the Relevant in Context task, and 71 valid runs for the Best in Context task. For each question in the questionnaire, we can break down the set of topics over the answer categories allowing us to investigate the performance of systems for a particular context. That is, for each of the subsets of the topics we can calculate how the 66 or 71 retrieval systems are ranked, and compare the resulting ranking to the ranking over all topics (i.e., the official outcomes over 104 assessed topics). Below, we will discuss some of the questions in detail. For each answer category, we calculate the rank-correlation (we use Kendall's tau) between the score over all topics, and the score over the selected topics corresponding to a particular answer category.[2] We will show the ten best performing systems for both Relevant in Context (RiC) and Best in Context (BiC) for topics corresponding to each of the answer categories.

Table 15 shows the results of question C5 (only the "article" part) over the 91 topics for which we have both the questionnaires and assessments. Each answer category (column 1) selects a number of topics (columns 2 and 3). In columns 4–13 (Relevant in Context) and columns 15–24 (Best in Context) we show the ten best performing systems for this subset of the topics. The systems are labeled with their system rank over the whole topic set (i.e., based on the official scores). That is, in columns 4 and 15 we find the best scoring system for the subsets, which in case of the subset of 62 topics with response "yes" is also labeled "1" (for both tasks) and hence the best overall system. Over the 29 topics with response "no", the best RiC system was ranked 6th over all topics, and the best BiC system was ranked 5th over all topics. Columns 14 and 25 show

---

[2] Since all topics in the subset are necessarily included also in the whole topic set, the subset scores will automatically approximate the overall scores depending on the size of the subset. This makes it difficult to compare the rank correlations for different subsets of the topics, especially if they contain different numbers of topics.

Table 16. (C7): Are highlighted passages still informative when presented out of context?

Answer	Freq.	%	Relevant in Context task		Best in Context task	
			Overall rank of top 10	Tau	Overall rank of top 10	Tau
1+2 (none)	18	19.78	6 9 8 14 4 1 22 26 3 5	0.8079	5 25 2 3 4 6 22 9 15 14	0.7328
3	22	24.18	1 2 5 7 4 3 6 10 9 8	0.8424	1 2 5 11 8 21 9 13 6 10	0.7859
4+5 (all)	51	56.04	10 1 4 2 6 7 5 8 12 3	0.8937	1 3 4 8 2 17 15 14 9 10	0.9155

Table 17. (C10): Which of the following two strategies is closer to your actual highlighting: (I) the best passages, (II) all relevant text?

Answer	Freq.	%	Relevant in Context task		Best in Context task	
			Overall rank of top 10	Tau	Overall rank of top 10	Tau
1+2 (I)	34	37.36	6 1 2 7 4 8 3 11 5 14	0.8844	5 2 13 9 1 22 4 3 10 12	0.8334
3	5	5.49	8 13 6 10 12 21 5 11 14 16	0.6942	9 1 8 5 20 27 13 6 31 37	0.5670
4+5 (II)	52	57.14	1 4 2 5 3 7 6 8 10 9	0.8918	1 3 4 2 8 15 14 11 6 10	0.8777

the rank correlation between the ranking over all topics and the ranking based on the subset of the topics. We see that the ranking over the "yes" topics corresponds well with the overall ranking (rank correlations around 90%). The ranking over the "no" topics shows remarkable upsets. To aid the analysis we have highlighted two types of runs. We show runs retrieving only whole articles in **bold**, since they were remarkably effective for BiC. We show runs based on the structured (or CAS, content and structure) query in *italics*, since they seemed not to improve over standard keyword query runs. A structured query contains references to the document structure and generally suggests the retrieval of particular XML elements, and participants could use either the structured query or a standard keyword query. We see, especially for BiC, that many of the runs that perform well on the "no" topics use such structured queries. Again for BiC, the run performing best over all topics, and over the "yes" topics, is always retrieving the start of the article as BEP—a strategy not particularly effective for the "no" topics. The rank correlation between the two subsets of the topics selected by responses "yes" and "no" is 0.7268 for RiC, and only 0.6483 for BiC.

Table 16 discusses C7, whether highlighted passages are still informative out of context (1=None of them is, 5=All of them are). We have collapsed the 5-point scale to a three point scale, to have a sufficient number of topics per answer category. The topics where passages are self-contained, response "4+5 (all)," correlate the best with the overall ranking. On these topics, runs retrieving whole articles (indicated in bold) are effective for BiC, but also for RiC. The system ranking for "1+2 (none)" shows remarkable upsets, especially from runs using the structured query (indicated in italics). The effectiveness of article retrieval seems counter-intuitive, and is is partly due to the Wikipedia's structure, where individual entries are exclusively focused on a single topic, and are often relatively short [7].

Table 17 discusses C10, about the two assessor strategies highlighting the best information, or all relevant information (1=strategy I, 5=strategy II). Strategy II is leading to fully highlighted articles, and this is favorable for article retrieval strategies (indicated in bold). Despite the radical differences between the strategies, the effect on the

Table 18. (C11 article): Can a best entry point be: the start of the article?

Answer	Freq.	%	Relevant in Context task		Best in Context task	
			Overall rank of top 10	Tau	Overall rank of top 10	Tau
yes	51	56.04	5 6 1 8 4 10 2 3 7 12	0.8993	1 2 3 4 8 6 5 9 14 15	0.8801
no	40	43.96	1 2 7 4 6 3 11 5 18 8	0.8751	18 5 10 12 11 9 1 3 4 7	0.8358

Table 19. (C12): Does the best entry point correspond to the best passage?

Answer	Freq.	%	Relevant in Context task		Best in Context task	
			Overall rank of top 10	Tau	Overall rank of top 10	Tau
1+2 (never)	16	17.58	4 5 1 3 2 7 8 13 6 9	0.7837	1 9 8 2 17 20 11 28 5 7	0.7787
3	26	28.57	8 4 10 1 2 6 7 5 3 12	0.8294	1 8 3 4 17 9 14 15 6 11	0.8117
4+5 (always)	49	53.85	1 2 6 4 7 5 3 8 14 11	0.9152	5 2 13 3 4 6 10 1 9 12	0.8922

Table 20. (C13): Does the best entry point correspond to the first passage?

Answer	Freq.	%	Relevant in Context task		Best in Context task	
			Overall rank of top 10	Tau	Overall rank of top 10	Tau
1+2 (never)	23	25.27	2 1 5 7 3 4 6 11 9 8	0.9105	21 10 18 12 7 11 13 3 4 1	0.8036
3	21	23.08	6 8 14 9 4 3 7 15 5 23	0.8396	2 5 4 3 14 15 1 23 16 6	0.8455
4+5 (always)	47	51.65	4 1 10 2 8 5 6 12 7 13	0.8443	1 8 17 9 6 2 5 20 3 4	0.8406

system rankings is not dramatic—the system-rank correlation between "1+2 (I)" topic and "4+5 (II)" topics is 0.7855 (RiC) and 0.7481 (BiC). This also suggests that the retrieval techniques effective for finding the best information are also effective for finding all relevant information, and the other way around.

Table 18 discusses C11 (article part), whether the best entry point can be the start of the article. For BiC, the "yes" topics resemble the overall ranking closely for the top runs, and the "no" topics favors systems using the structured query (similar to the breakdown over C5 in Table 15). For RiC the ranking of the top runs is more similar for the ranking over the "no" topics. This suggests that the RiC and BiC tasks do have a different nature. Note also that the article run labeled "1" for BiC is exactly identical to the article run labeled "10" for RiC. The system rank correlation between "yes" and "no" topics is 0.7893 for RiC, and 0.7465 for BiC.

Table 19 shows C12, whether the BEP corresponds to the best passage (1=Never, 5=Always). For BiC, article retrieval (indicated in bold) seems particularly effective for the "1+2 (never)" and "3" topics. For RiC, the overall ranking is more resembling the ranking over "4+5 (always)" topics, again suggesting differences between the two tasks.

Table 20 shows C13, whether the BEP corresponds to the first passage (1=Never, 5=Always). For both tasks, article retrieval is effective for "4+5 (always)," which makes sense since the first passage will be closer to the start of the article. Also, for both tasks the structure query is effective for "1+2 (never)," which makes sense since elements matching the structural constraints of the query need not occur early in the article. Again, there is a divergence between the overall ranking of RiC resembling the "1+2 (never)" topics, and the overall ranking of BiC resembling the "4+5 (always)" topics.

The main general conclusion is that context matters for the relative ranking of systems: we see varying levels of agreement between the ranking over all topics, and the ranking on subsets of the topics sharing particular context.

5   Discussion and Conclusions

One of the greatest achievements of the field of IR is the development of a rigorous methodology to evaluate retrieval effectiveness [4]. The Cranfield approach as continued by TREC, CLEF, NCTIR and INEX has served us very well: virtually all progress in IR owes directly or indirectly to test collections built within the Cranfield paradigm. The Cranfield tradition of test collection development tries to abstract away from individual differences between topic authors and assessors [15]. However, more complex search tasks that are a closer approximation of real-world information seeking in action, such as those prevalent in digital libraries, seem to require, in contrast, that some of the user's context is taken into account [6, 14].

This paper experimented with a new approach: rather than fitting a search task and its evaluation to a particular context, we record the "context" of the humans already in the loop: the topic authors/assessors. In particular, we investigated the use of a dedicated questionnaires to elicit and record salient aspects of the topic author's and assessor's context. The questionnaire data helps control the construction of a test collection. Moreover, the questionnaire data becomes part of the evaluation test-suite as contextual data of the search topics and their topic authors and assessors. There is a risk that extensions to Cranfield will limit the reusability of the resulting test collection. The extension proposed in this paper keeps the original test collection (documents, topics, relevance judgments) completely intact. In fact, the extension seems more likely to increase reusability, e.g., by allowing researchers to analyse their system's performance over topics that best match their intended search context.

How does the questionnaire data improve the evaluation in comparison with the traditional "bag of topics"? On the one hand, the questionnaires can be used to control the test collection building. The responses also give an overview of the composition of the topic set, and highlights sources of divergence. This can be crucial during the selection of candidate topics to be assessed, or when combining test collections from different years. On the other hand, the questionnaires can be used directly by participants to investigate which contextual aspects impacted their system's performance (and how). We have shown this in Section 4. It is important to note that we do not envision the main system comparison to change, such comparison requires the common ground provided by the entire topic set, and the ultimate aim is to have a systems that performs well on all topic types. But in addition to this, the contextual data gathered is facilitating deeper analysis of what worked and what not and why it worked and why not—especially in the case of failure analysis, this can be insightful.

It is also clear that the questionnaires are no a panacea. Interpreting the data is hard: What does it mean precisely when for n% of the topics a certain response is given? What does it mean precisely if my system performs well for a particular subset of topics? This may not be immediately clear, although the contextual data in the questionnaire will help focus on "interesting" contextual aspects and subsets of the topics, and will at least give a good hint about the interpretation. This is partly because we are venturing in new terrain, and cannot compare to data from earlier years or to other well-understood data.

Evaluating a comprehensive search systems, such as a digital library, is a complex and difficult undertaking [12]. All searches happen in a particular context—such as the particular collection of the digital library, its associated search tasks, and its associated users. Evaluation should take relevant parts of this context into account. The most striking observation overall is that there is such great variety in the responses of the assessors—much greater than we expected beforehand. Perhaps equally surprising is that the system rankings are nonetheless reasonably robust—much more robust than we expected beforehand. This can also be interpreted as strong support that Cranfield is working: despite the differences in context (or noise in terms of Cranfield) the system rankings are remarkably stable with significant system rank correlations above 0.6 for all reasonably sized subsets of topics (also between different subsets of topics). That is, to a large extent we find ourselves in the same position as [16], despite all the upsets detailed in Section 4, there is also broad agreement on separating the good systems from the bad systems.

Acknowledgements. Jaap Kamps was supported by the Netherlands Organization for Scientific Research (NWO, grants # 612.066.513, 639.072.601, and 640.001.501). Mounia Lalmas position is funded by Microsoft Research/Royal Academy of Engineering.

References

[1] Allan, J.: HARD track overview in TREC 2005: High accuracy retrieval from documents. In: The Fourteenth Text REtrieval Conference (TREC 2003). National Institute of Standards and Technology, pp. 500–255. NIST Special Publication (2004)

[2] Banks, D., Over, P., Zhang, N.-F.: Blind men and elephants: Six approaches to TREC tasks. Information Retrieval 1, 7–34 (1999)

[3] Buckley, C.: Why current IR engines fail. In: Proceedings of the 27th Annual International ACM SIGIR Conference, pp. 584–585. ACM Press, New York (2004)

[4] Cleverdon, C.W.: The Cranfield tests on index language devices. Aslib 19, 173–192 (1967)

[5] Fuhr, N., Kamps, J., Lalmas, M., Malik, S., Trotman, A.: Overview of the INEX 2007 ad hoc track. In: Fuhr, N., Kamps, J., Lalmas, M., Trotman, A. (eds.) INEX 2007. LNCS, vol. 4862, pp. 1–23. Springer, Heidelberg (2008)

[6] Ingwersen, P., Järvelin, K.: The Turn: Integration of Information Seeking and Retrieval in Context. Springer, Heidelberg (2005)

[7] Kamps, J., Koolen, M., Lalmas, M.: Locating relevant text within XML documents. In: Proceedings of the 31th Annual International ACM SIGIR Conference on Research and Development in Information Retrieval, pp. 847–849. ACM Press, New York (2008)

[8] Kamps, J., Pehcevski, J., Kazai, G., Lalmas, M., Robertson, S.: INEX 2007 evaluation measures. In: Fuhr, N., Kamps, J., Lalmas, M., Trotman, A. (eds.) INEX 2007. LNCS, vol. 4862, pp. 24–33. Springer, Heidelberg (2008)

[9] Kazai, G., Lalmas, M., de Vries, A.P.: The overlap problem in content-oriented XML retrieval evaluation. In: Proceedings of the 27th Annual International ACM SIGIR Conference, pp. 72–79. ACM Press, New York (2004)

[10] Piwowarski, B., Trotman, A., Lalmas, M.: Sound and complete relevance assessments for XML retrieval. ACM Transactions in Information Systems 27(1) (2008)

[11] Reid, J., Lalmas, M., Finesilver, K., Hertzum, M.: Best entry points for structured document retrieval: Parts I & II. Information Processing and Management 42, 74–105 (2006)

[12] Saracevic, T.: Digital library evaluation: Toward evolution of concepts. Library Trends – Special issue on Evaluation of Digital Libraries 49(2), 350–369 (2000)

[13] Saracevic, T.: Relevance: A review of and a framework for the thinking on the notion in information science. JASIS 26, 321–343 (1975)

[14] Sparck Jones, K.: What's the value of TREC – is there a gap to jump or a chasm to bridge? SIGIR Forum 40, 10–20 (2006)

[15] Voorhees, E.M.: The philosophy of information retrieval evaluation. In: Peters, C., Braschler, M., Gonzalo, J., Kluck, M. (eds.) CLEF 2001. LNCS, vol. 2406, pp. 355–370. Springer, Heidelberg (2002)

[16] Zobel, J.: How reliable are the results of large-scale information retrieval experiments? In: Proceedings of the 21st Annual International ACM SIGIR Conference on Research and Development in Information Retrieval, pp. 307–314. ACM Press, New York (1998)

Comparing Google to Ask-a-Librarian Service for Answering Factual and Topical Questions

Pertti Vakkari and Mari Taneli

University of Tampere, FIN-33014, Finland
Pertti.Vakkari@uta.fi

Abstract. This paper evaluates to which extent Google retrieved correct answers as responses to queries inferred from factual and topical requests in a digital Ask-a-Librarian service. 100 factual and 100 topical questions were picked from a digital reference service run by public libraries. The queries inferred simulated average Web queries. The top 10 retrieval results were observed for the answer. The inspection was stopped when the first correct answer was identified. Google retrieved correct answers to 42 % of the topical questions and 29 % of factual questions. Results concerning the characteristics of queries and retrieval effectiveness are also presented. Evaluations indicate that public libraries' reference services answer correctly 55 % of the questions. Thus, Google is not outperforming Ask-a-Librarian service, although it seems to perform relatively satisfactory in retrieving answers to topical questions.

1 Introduction

The Internet has brought vast amounts of information and search tools within the reach of people. Search engines have enhanced the accessibility of those resources by providing easy means to search, identify, locate and download information items needed. Finding information and answers to requests is only a few clicks away.

The number of Internet users is growing steadily especially in industrialized countries. The use of search engines has gained popularity world-wide. They are considered easy to use. Easy access to information resources on the Internet from one's own desk with the help of search engines is challenging the functions of libraries, public libraries in particular (D'Elia & al. 2002, Huysmans & Hillebrink 2008). Material and services provided by libraries are also available on the Internet. This may result in a potential change in consumer demand for the types of services provided by libraries. People can compare the suitability and performance of the services of the library and the Internet and make their choice based on that evaluation.

There is not much empirical evidence on this change. D'Elia & al. (2002) showed that in the year 2000 the use of public libraries and the Internet in the US were complementary. Most of those who used the library used the Internet and vice versa. Using the Internet for various purposes was associated neither with the reasons why people use the library or their frequency of use.

It is evident that the use of the Internet influences differently the use of various library services. Borrowing novels or reference services are probably influenced

M. Agosti et al. (Eds.): ECDL 2009, LNCS 5714, pp. 352–363, 2009.

differently by the Internet (Huysmans & Hillebrink 2008). It is likely that the demand of the reference services is changing due to use of search engines. Numminen & Vakkari (2009) have shown that of the questions sent to an Ask-a-Librarian digital reference service the proportion of topical questions has decreased and that of factual questions increased significantly between the years 1999 and 2006. They suggest that this change may be due to the way users are searching on the Internet.

Although there have been several evaluations of search engines, they are compared with each others (Spink & Jansen 2004, Lewandowski 2008). Similarly, digital Ask-an-Expert-services - including librarian - are evaluated by comparing these services with each other (Janes & al. 2001). It is very rare to do across search space evaluations, e.g. how search engines function compared to digital library services like question answering or even retrieving information from a digital library (cf. McCown & al. 2005). However, it is necessary to evaluate search engines with appropriate services provided by traditional or digital libraries, because people may increasingly use search engines as surrogates for library services. The aim of this study is to explore to what extent Google retrieves correct answers to questions addressed to a digital Ask-a-Librarian reference service. Specifically, we analyze what is the proportion of correct answers Google retrieves as responses to queries inferred from factual and topical requests in an Ask-a-Librarian service.

2   Related Research

2.1 Evaluation of Search Engines

Search engines can be evaluated on several dimensions, but retrieval effectiveness has been the major theme (Lewandowski 2008). By effectiveness is meant systems' ability to retrieve relevant information items. It is measured typically by precision and recall. Due to the differences between search engines and digital Ask-a-Librarian services, the reasonable feature to compare their functioning is their ability to provide correct answers to questions. In our study we analyze how effectively Google retrieves answers to queries inferred from questions to an Ask-a-Librarian service. Therefore, retrieved correct items are the indicator on which we base the evaluation.

There are several studies analyzing the effectiveness of search engines. The major studies in the 1990s (Chu & Rosenthal 1996, Gordon & Pathak 1999, Leighton & Srivastava 1999) have used queries drawn from real reference questions. The number of queries used is small varying in those studies from 5 to 15. The number of queries in later studies has varied from 25 to 70 (Lewandowski 2008).

The number of retrieval results analyzed for the evaluation has typically been either 10 or 20 first information items (Lewandowski 2008). This is consistent with the findings of empirical Web search studies, which show that the majority of searchers view only the first result page. A typical user views on average 8 Web documents per session (Spink & Jansen 2004). The studies also show that the number of words in queries is typically 2.5-3 (Spink & Jansen 2004), and that topical queries are shorter compared to factual ones. Kellar & al. (2007) have shown that on average topical queries consist of 3.3 words, whereas factual queries consist of 4.7 words.

The precision of search engines in studies varies somewhat. Precision is the proportion of relevant items among those retrieved. In Gordon & Pathak (1999) precision within the top ten varied from 40.6% to17.6%, whereas in Chu & Rosenthal (1996) the range was from 78 % to 55 %. In a more recent study precision within the top 20 varied from 49 % to 34 % (Lewandowski 2008).

There is one study, which has compared a search engine and a digital library service in their ability to produce useful results. McCown & al. (2005) measured the pedagogical usefulness of results returned by the National Science Digital Library (NSDL) and Google. Teachers evaluated search terms and search results based on the Standards of Learning (SOL) for schools. They ranked on a five-point scale the links returned by Google as more relevant to the SOL than the links returned by NSDL. On average neither of the search engines generated search results that would satisfy the educational requirements in the SOL. However, the proportion of Web pages rated as adequate of all results (precision) was in Google 38 % and in NSDL 17 %. Thus, Google seemed to outperform NSDL in retrieving educationally useful results.

2.2 Web Queries

Web queries can be divided into informational, navigational and transactional (Broder 2002). The aim of an informational query is to obtain information about the topic of the query (Rose & Levinson 2004). Navigational queries aim at reaching a particular site. They can be described as known item searches (Broder 2002, Rose & Levinson 2004). Transactional queries seek to obtain something else than information, typically resources from the Web (Rose & Levinson 2004).

Most of the queries on the Web are informational. The proportion of informational queries was about 60 % in Rose & Levinson (2004) and 40 % in Broder (2002). Topical queries were distinguished only in Rose & Levinson by the name "undirected queries". Their share varied between 23-31 %. Fact-finding i.e. directed queries comprised only 3-7 % of all queries in Rose & Levinson (2004). Navigational queries are equal to known item questions. Their share was about a quarter in Broder (2002) and about one seventh in Rose & Levinson (2004).

Thus, topical queries seem to be much more common than factual queries and somewhat more common than known item queries among Internet searches. Based on the proportions of these query types, we may expect that the use of search engines challenges reference services in public libraries most in answering topical questions, but not so much in answering factual or known item questions.

2.3 Evaluation of Digital Reference Services

There is a long tradition of evaluating reference services in libraries. Unobtrusive observer studies are used to assess the quality of question responses. In unobtrusive observation evaluations, trained questioners act as real clients to pose reference questions in the actual reference setting (Katz 2002). The accumulated results show that about 55 % of the questions are answered accurately (Katz 2002). This approach has also been used in the evaluations of digital reference services. Kaske & Arnold (2002) found that in 36 libraries, chat reference services answered correctly 55 % and email services 60 % of the 12 questions asked. Kwon (2007) showed that public libraries' chat services answered completely 78 % of factual questions and 70 % of topical questions. Arnold & Kaske (2005) found an even greater proportion (92 %) of correctly answered questions among 3000 questions to a public library's chat service.

It is difficult to assess validly the trends emerging from these results due to the varying nature of the data used. The 55 % rule seems to apply to those studies using unobtrusive testing, whereas higher figures are derived from the answers to users' real questions as assessed by researchers. Although the findings hint that a growing proportion of users' requests are answered correctly especially through digital reference services, this hypothesis requires further empirical evidence.

3 Research Design

The aim of this study is to explore to which extent Google retrieves correct answers to questions addressed to a digital Ask-a-Librarian reference service. The specific research questions are as follows:

- What is the proportion of correct answers Google retrieves as responses to queries inferred from factual and topical requests in an Ask-a-Librarian service?
- How are query characteristics related to search effectiveness in question types?

3.1 Questions

We focus on factual and topical questions. A factual question has typically one definite answer and it can be answered by consulting one or two reference tools (Katz 2002). We categorized the factual questions into sub-categories based on our data. The following kind of factual question types emerged from the data: definition of a concept (What is the cost of living index?), origin or meaning of a name or saying (What is the meaning of the name Maximus?), instruction how to make a product (Instruction to make a broom?), translation of a word (What is depression (in Finnish) in Swedish?), and other facts (How many state officials are there in Finland?).

A topical question requests materials or information on a certain subject (Katz 2002). We divided topical questions into questions concerning individual persons, and other topical questions. The former mostly included questions concerning e.g. artists or writers (I need information about the novelist Anni Swan). The latter typically focused on a subject matter (I need materials about the early history of globalization).

We selected questions from an Ask-a-Librarian digital reference service (http://www.libraries.fi/en_GB/ask_librarian). This service is run by about 50 public libraries. Questions are sent via a Web form and answered by email. The staff running the service typically directs questions to the local public library of the requesters. The question answer pairs are stored in the public archive of the service.

Our data consists of 100 factual and 100 topical questions. They were selected from the archive of the service from September 2008 backwards so long that the required number of both question types was achieved. Both researchers classified the questions. The inter-rater reliability among factual questions was 94 % and among topical questions 90 %, which can be considered as high (Krippendorf 2004).

3.2 Queries

We formulated queries of the factual and topical questions. We sought to simulate typical users of search engines, who formulate short queries of 2.5-3 words and view only the first result page (Spink & Jansen 2004). In queries we used only words, which were used in the questions. For selecting the expressions for the queries, we made conceptual query plans for each question (Lancaster & Warner 1993). A facet is a concept identified from, and defining one exclusive aspect of a request (Sormunen 2000). We identified the major facts of the requests. The maximum number of facets per query was three. Each facet was expressed typically by one word in the basic form picked from a question. There were some exceptions when a facet was expressed by more than one word. Those exceptions consisted mainly of phrases from requests concerning the origin of sayings like the first line of a poem or other necessary phrases for expressing a facet. Synonyms were not used. On average factual queries consisted of 2.2 facets and 3.5 words and topical queries of 1.8 facets and 2.5 words. Thus, on average the length of both factual and topical queries seem to resemble typical Web queries (Spink & Jansen 2004). Our factual queries contained more words than topical ones, which corresponds with the findings in Kellar & al. (2007).

We calculated between both researchers the inter-rater reliability of the queries formulated. The reliability rate among factual queries was 92 % and among topical queries 90 %, which can be considered as high (Krippendorf 2004).

3.3 Answers and Evaluation

The answers given by the Ask-a-Librarian service formed the basis for assessing the correctness of the answers retrieved. The information provided in the pages and links retrieved should correspond to the answer given to a request by the service but not necessarily be exactly the same. For instance, links and document lists or information in items retrieved to topical questions should not correspond totally with the information in the original answer, but to cover its main content. The point was that the content of the information item provided a satisfactory answer to the question in line of the answer by the reference service.

We categorized the correctness of the information item as an answer as follows: It was considered as correct if it contained the main content of the answer. It was partially correct if it contained some aspects of the answer. It was categorized as "mention" if the information item provided only a mention of the answer.

To identify the answer we observed the first ten information items on the result page. First we assessed whether the answer could be found in the results descriptions on the result page containing the first ten items. If it was not identified on the result page, the first link from each of the results was inspected from the first to the tenth. The inspection was stopped when the first correct answer was found.

We calculated for all queries the proportion of answers found, and for those queries retrieving an answer the average rank of the answer and its average degree of correctness.

We supposed that the answers of the reference service were correct, but this is not necessary the case. For assessing the quality of the answers we inspected them. Although we did not check the sources of the answers, they however, seemed reliable to us. This is the major limitation of our study, which may bias the results.

4 Results

4.1 The Answers Retrieved

By using Google answers were found to 47 % of the questions. Significantly more answers were retrieved for topical than factual questions (t=3.3, df=198, p=.000). The proportion of answers for the former was 65 % and for the latter 29 %. Thus, Google produced over twice as many correct replies to topical than factual queries.

Table 1. The average rank of answers and the correctness of answers by question type

	Factual Q (n=29)	Topical Q (n=65)	p
Rank of answers	2.1	2.0	.820
Degree of correctness	3.0	2.5	.000

If we observe only answered questions there were no significant differences in the rank of references providing the correct answer to factual or topical questions (Table 1). In both groups on average the second item on the result list contained the correct answer. The first item included the correct answer in 62 % of factual questions and in 65 % of topical questions. The three first items contained the correct answer in 83 % of factual questions and 86 % of topical questions. Thus, in most of the cases in both groups the correct answers were ranked in the very beginning of the result list.

We scored the correctness of the answers as follows: totally correct=3, partially correct=2, and mention=1. The correctness of the answers differed significantly between factual and topical questions (t=.3.7, df=92, p=.000) (Table 2). All the answers to the factual questions totally covered the answer, whereas 65 % of answers to the topical questions were totally correct. In 35 % of topical requests the answers did not include all the necessary information. Thus, it seems that if Google was able to retrieve an answer to a factual query, it was always totally correct, whereas in about one third of the cases the answer to a topical query was incomplete.

Table 2. The degree of correctness in answers by question type (%)

Correctness	Factual Q (n=29)	Topical Q (n=65)	Total (n=94)
Totally	100	65	76
Partially	0	15	11
Mention	0	20	14
Total	100	100	101

4.2 Characteristics of Queries

The average number of facets in factual queries was 2.2. and in topical queries 1.8 (Table 3). A two-way ANOVA indicated that this difference was marginally significant (F=3.1, p=.08).

Table 3. The average number of facets in queries

Answer	Factual Q (n=100)	Topical Q (n=100)	Total (n=200)
Found (n=94)	1.9	1.7	1.7
Not found (n=106)	2.3	2.2	2.3
Total (n=200)	2.2	1.8	2.0

The queries retrieving correct answers consisted of significantly fewer facets than queries producing no answers (F=18.3, p=.000). The former consisted of 1.7 facets and the latter of 2.3 facets. It seems that queries representing requests by fewer facets retrieve more correct answers. The significant correlation between the number of facets and whether the answer was found (-.35, p=.000) confirms that.

On average queries contained three words (Table 4). Factual queries consisted of 3.5 words and topical queries of 2.5 words. This one word difference in the length of query was significant (F=8.4, p=.004). Results also indicate that queries that retrieve correct answers were significantly shorter (2.5 vs. 3.5 words) than those that did not retrieve correct answers (F=9.7, p=.002). This was regardless of the question type (factual: t=2.1, df=98, p=.04; topical: t=2.3, df=98, p=.009).

Table 4. The average number of words in queries

Answer	Factual Q (n=100)	Topical Q (n=100)	Total (n=200)
Found (n=94)	2.9	2.3	2.5
Not found (n=106)	3.8	2.9	3.5
Total (n=200)	3.5	2.5	3.0

It seems that queries representing requests by fewer words retrieve more correct answers. The significant correlation between the number of words and whether the answer was found was -.30 (p=.000) confirming that. It is likely that those queries containing fewer words are simpler in the sense that they consist of fewer facets. Simple requests containing fewer facets are likely to produce more correct answers as response to queries than complex requests with several facets.

The number of facets was significantly associated with the correctness but not with the rank of the answers found. The correlation between the number of facets and the rank of correct answers was -.04 (p=.74) and between facets and the degree of correctness -.28 (p=.006). The corresponding coefficients between the number of words and rank order was .09 (p=.40) and the level of correctness -.15 (p=.14). Thus, the fewer facets the question contained the higher the correctness of the answer re-trieved. The number of words in queries was associated neither with the rank or the correctness of the items retrieved.

4.3 Categorizing Factual and Topical Questions

Based on data we categorized both factual and topical questions to more specific groups. Factual questions were divided into questions concerning definition of concepts, origin of words, or phrases and sayings, instructions to make a product, translation of a word or phrase, and typical factual questions. Topical questions were divided into questions concerning persons, and other topical questions. We first analyze factual question types and then topical question types.

Although the number of cases in groups other than typical facts is small, the figures hint marginally at certain trends (Chi2=2.8, df=5, p=.091) (Table 5). It seems that Google retrieves a high proportion of definitions of concepts (71 %), whereas it is not successful in locating origins of sayings or phrases, or providing searchers with translation of words. These categories produced zero hits. About one third of typical facts, origin of words, and instructions were found by Google.

Table 5. Answers found and the number of words in the queries by factual question type (%)

Type of factual question (n)	% found	# words
Definition of a concept (n=7)	71	1.7
Origin of a word (n=6)	33	2.3
Origin of a saying or a phrase (n=10)	0	6.2
Instructions to a product (n=6)	33	3.7
Translation (n=6)	0	4.5
Typical fact (n=65)	31	3.3
Total (n=100)	29	3.5

As we showed earlier there was a significant inverse correlation (-.30, p=.000) between answers found and the number of query words. An ANOVA confirmed that there are significant differences in the length of queries between the groups (F=7.4, p=.000) (Table 5). The number of words in queries concerning concepts was significantly smaller than concerning typical facts, origins of sayings or instructions (Dunnett C: p<.05). The proportion of correct answers in this question type was higher than in other factual questions, and its query length was the shortest of all, on average 1.7 words. The length of queries in those question types with zero hits was the greatest, from 4 to 6.2 words. There seems to be a inverse rank order between the proportion of correct answers and the length of the query across question types.

The proportion of correct answers in queries concerning persons was 72 %, and in other topical queries 60 %. The difference is not significant (t=1.5, df = 63 p=.14). Neither did the rank of answers differ between these groups (t=0.5, df=63, p=.61). Thus, Google tends to produce to the same degree correct answers to queries in both groups and rank them about equally.

The average degree of correctness in answers to questions concerning persons was significantly greater than in other topical question (t=2.2, df=63, p=.032) (Table 6). In the former the answers tend to be correct whereas in the latter they were on the partially correct side.

The number of facets was significantly greater in other topical queries compared to queries concerning persons (t=4.8, df=63, p=000). The number of words was about the same (2.2 vs. 2.4) in both groups (t=1.4, df=63, p=.16). The name of a person comprises one facet, which can be expressed by two words (i.e. first name and family name). This explains in part the difference between the groups in the number of facets and not in the number of words.

Table 6. Means of correctness and query variables by topical question type

	Persons (26)	Topics (39)	p
Rank of answers	1.9	2.1	.610
Degree of correctness	2.7	2.3	.032
Number of facets	1.2	2.0	.000
Number of words	2.2	2.4	.160

It seems that it is easier to represent questions concerning persons as queries compared to other topical questions, and consequently the proportion of correct answers and the correctness of the answers is higher in the former group.

In all, a more specific categorization of question types revealed that the positive associations between short queries and the higher proportion of answers found, and the higher correctness of answers were in part produced by certain question types (definitions of concepts, and persons), which required fewer facets and words.

5 Discussion and Conclusions

Digital libraries are challenged by the increasing provision of information and tools for accessing information on the Internet. It is important to analyze to which degree services provided by libraries are able to compete with the Internet. Topical queries are popular on the Web compared to factual queries (Rose & Levinson 2004). There are signs that the proportion of topical questions in the digital reference services has decreased, and the proportion of factual questions increased (Numminen & Vakkari 2009). It is proposed that this is produced by the way people are searching on the Net.

We have compared the functioning of one particular digital library service to one typical service provided by the Internet. We have analyzed to which degree Google retrieves correct answers to queries inferred from topical and factual questions answered by an Ask-a-Librarian digital service. We formulated queries from requests by simulating average Web searchers. It is known that they use short queries of about three words and look only at the first result page (Spink & Jansen 2004). We analyzed whether the queries produced a correct answer within the top ten retrieval results.

Our study extends our knowledge of how an Ask-a-Librarian service performs compared to Google in answering topical and factual questions. We have used 200 queries, which is much more than in typical studies assessing search engines (Lewandowski 2008) and not less than on average in evaluations of public libraries' reference services. Also the high inter-rater reliability of classifying question types and formulating queries enhance the validity of our findings.

5.1 Search Results

On average Google retrieved correct answers to 47 % of the questions. The correct answer was retrieved to 65 % of topical questions and to 29 % of factual questions. Thus, Google retrieved twice as many correct answers to topical than factual requests. This result supports our hypothesis, that the use of search engines challenges reference services in public libraries most in answering topical questions, but not so much in answering factual questions (cf. Numminen & Vakkari 2009).

On average the second item on the result list provided the correct answer regardless of the question type. The correct answers were typically ranked in the very beginning of the retrieval results, in most of the cases among the three first items.

The level of answers' correctness differed between the question types. All correct answers to factual questions covered the answer totally, whereas 35 % of the answers to topical questions did not include all the necessary information.

It is not possible to compare our findings with the earlier studies on the effectiveness of search engines due to the differences in the measurement of effectiveness. We calculated the proportion of correct answers retrieved, whereas the typical measure of effectiveness has been precision (eg. Gordon & Pathak 1999, Lewandowski 2008).

Evaluation of question answering in public libraries' reference services has produced the rule that 55 % of the questions are answered accurately (Katz 2002). In digital reference services this regularity varies to some extent. Kaske & Arnold (2002) found that 60 % of questions asked by e-mail from 36 public libraries were answered correctly. Kwon (2007) showed that public libraries chat services answered correctly78 % of the factual questions and 70 % of the topical questions.

Compared to these evaluation results it seems that Google performs relatively well when it retrieves correct answers to 65 % of topical questions. However, if we take into account only totally correct answers retrieved, then the proportion of accurate answers drops to 42 %. This is a somewhat poorer performance than average in answering topical questions in digital reference services.

The digital reference services in public libraries seem to outperform Google as provider of answers to factual questions. Google retrieved totally correct answers to 29 % of factual questions. This is clearly less compared to 55 % rule or to the finding in Kwon (2007), which indicated that 78 % of factual questions in public libraries' chat service were answered correctly.

If we suppose that searchers are willing to invest more effort in inspecting the results, the picture changes somewhat in favor of Google. By allowing that each information item on the result list can be inspected three clicks away instead of the one click we used as criterium, the proportion of totally correct answers retrieved to topical questions was 52 % and to factual questions 40 %.

In all, it seems that Google is not severely challenging public libraries' reference services. It does not reach the 55 % proportion of correctly answered questions, which is regular in public libraries. However, if searchers are willing to invest more effort on inspecting the retrieval results, then Google may approach the regular performance level of public libraries' reference services in answering topical questions.

5.2 Characteristics of Queries

Our factual queries contained more facets and words than topical ones. This resembles the finding in Kellar & al. (2007), that factual Web queries contain more words than topical ones. Our results also indicated that queries with fewer facets and words retrieved significantly more correct answers than longer queries regardless of question type. Thus, the fewer facets and words queries contained, the more correct answers they retrieved.

An elaboration of question types showed that factual questions concerning definition of concepts and topical questions concerning persons retrieved more correct answers compared to other question types within the major question categories. The former retrieved 71 % correct answers, and the latter 72 % correct answers. Both types of queries also contained the lowest number of words and facets compared to other question types in the respective major question categories.

The association between the low number of facets and words, and success in retrieval seem to hint that those questions, which are relative unambiguous and can be expressed by few words, also perform best as queries. Questions concerning the names of persons and definition of concepts are such clear questions. This explains why shorter queries tend to perform best in our study.

5.3 Conclusions

We have shown that Google challenges public libraries' digital reference services most in answering topical questions. It retrieved twice as many correct answers to topical than factual questions. This finding supports the hypothesis that people increasingly make topical searches on the Web (Rose & Levinson 2004), and this is likely to lead to the decrease in the proportion of topical questions addressed to digital reference services (Numminen & Vakkari 2009). In particular, Google was successful in retrieving answers to unambiguous questions, which can be expressed by few facets and words. Although search engines are challenging the role of public libraries' reference services, on average they still outperform search engines. It seems that the engines are currently not able to reach the 55 % level of correct answers, which public libraries regularly provide to their customers according to evaluations.

Google's performance figures improved, when results were inspected following further links they provided. If people are willing to invest more effort on inspecting search results, this increases the likelihood of finding correct answers. Thus, also in this respect the future of libraries' reference service is in the hands of people looking for information.

References

Arnold, J., Kaske, N.: Evaluating the quality of a chat service. Portal 5(2), 177–193 (2005)

Broder, A.: A taxonomy of web search. SIGIR Forum 36(2), 3–10 (2002)

Chu, H., Rosenthal, M.: Search engines for the world wide web: A comparative study and evaluation methodology. In: ASIS 1996 Annual Conference Proceedings (1996)

D'Elia, G., Jörgensen, C., Woelfel, J.: The impact of the Internet on public library use. JASIST 53(10), 802–820 (2002)

Gordon, M., Pathak, P.: Finding information on the world wide web: The retrieval effectiveness of search engines. Information Processing & Management 35(2), 141–180 (1999)

Huysmans, F., Hillebrink, C.: The future of the Dutch public library: Ten years on. SCP, The Hague (2008)

Janes, J., Hill, C., Rolfe, A.: Ask-an-Expert services analysis. JASIST 52(13), 1106–1121 (2001)

Kaske, N., Arnold, J.: An unobtrusive evaluation of online real time library reference services. Paper presented at the Library Research Round Table, ALA Annual Conference, Atlanta, GA. (2002), http://www.lib.umd.edu/groups/digref/LRRT.html

Katz, W.A.: Introduction to reference work, vol. 1. McGraw-Hill, Boston (2002)

Kellar, M., Watters, K., Shepherd, M.: A field study characterizing web-based information seeking tasks. JASIST 58(7), 999–1018 (2007)

Krippendorf, K.: Content analysis. Sage, Thousand Oaks (2004)

Kwon, N.: Public library patrons' use of collaborative chat reference service: The effectiveness of question answering by question type. Library & Information Science Research 29, 70–91 (2007)

Lancaster, W., Warner, A.: Information retrieval today. Information Resources Press, Arlington (1993)

Leighton, W., Srivastava, J.: First 20 precision among world wide web search services. JASIST 50(10), 870–881 (1999)

Lewandowski, D.: The retrieval effectiveness of web search engines: considering results descriptions. Journal of Documentation 64(6), 915–937 (2008)

McCown, F., Bollen, J., Nelson, M.: Evaluation of the NSDL and Google for obtaining pedagogocal resources. In: Rauber, A., Christodoulakis, S., Tjoa, A.M. (eds.) ECDL 2005. LNCS, vol. 3652, pp. 344–355. Springer, Heidelberg (2005)

Numminen, P., Vakkari, P.: Question types in public libraries' digital reference service in Finland: Comparing 1999 and 2006. JASIST 60 (2009),
http://www3.interscience.wiley.com/journal/122219062/abstract
DOI: 10.1002/asi.21047

Rose, D., Levinson, D.: Understanding user goals in web search. In: WWW 2004, New York, May 17-22, 2004, pp. 13–19 (2004)

Sormunen, E.: A method of measuring wide range performance of Boolean queries in full-text databases. Acta Universitatis Tamperensis 748. Tampere University Press (2000)

Spink, A., Jansen, B.: Web search: Public searching of the Web. Kluwer, Dordrecht (2004)

How People Read Books Online:
Mining and Visualizing Web Logs for Use Information

Rong Chen[1], Anne Rose[2], and Benjamin B. Bederson[2]

[1] Department of Computer Science and Technique
College of Computer Science, Sichuan University
Chengdu, 610065, China
[2] Human-Computer Interaction Lab
Department of Computer Science
University of Maryland
College Park, MD 20770, USA
chen-rong@cs.scu.edu.cn, {rose,bederson}@cs.umd.edu

Abstract. This paper explores how people read books online using the International Children's Digital Library (ICDL). We analyzed usage of the ICDL in an attempt to understand how people read books from websites. We propose a definition of reading a book (in contrast to others who visit the website), and report a number of observations about the use of the library in question.

Keywords: Web Log Analysis, Information Visualization, Web Usage Mining, ICDL, Reading Online.

1 Introduction

There is now a wide range of online books from sources such as Google Book Search[1], Amazon[2], our own International Children's Digital Library (ICDL)[3], and others. While there is significant effort to understand how people use websites through services such as Google Analytics[4] and various tools to process web logs, these services fall short when trying to understand how people read books online.

The issue is that the existing approaches aggregate data and combine individuals. They support understanding e-commerce activities such as understanding "conversions", and knowing whether certain goals have been reached – such as if a product has been purchased, or whether a book has been downloaded. You can even find out how many pages of some content area have been accessed – so it is possible to discover how many pages of a certain book have been read. But it is impossible using traditional techniques to discover how many individuals have read a book. Or how many pages of a book are typically read by individuals. Or how many books an individual reads. In sum, we want to know how people read books online.

---

[1] http://books.google.com
[2] http://www.amazon.com
[3] http://www.childrenslibrary.org
[4] http://www.google.com/analytics

M. Agosti et al. (Eds.): ECDL 2009, LNCS 5714, pp. 364–369, 2009.
© Springer-Verlag Berlin Heidelberg 2009

In this paper, we analyze and visualize web log data. While it would be ideal to actually observe individual reading online, that is not scalable. So, instead we focus on book-centered reading behavior with the actual logs from the ICDL.

This analysis was done on the public usage of the ICDL from one week (20 October 2008 through 25 October 2008), which represents just over 23,000 unique visitors, 26,000 visits, and 336,000 page views.

2 Review of Related Literature

People's online reading behavior has increasingly become an area of empirical and theoretical exploration by researchers from a wide range of disciplines, such as psychology, education, literacy studies, information science and computer science. Different disciplines have diverse ways of probing these questions.

Many researchers use active observation: Some researchers have performed experiments on understanding changes in reading behavior with paper-reading [1][2].

Web page centered research is used by some web usage mining tools [3]. Google analytics gives the average time on page and average reading count of pages in general, but it focuses on each webpage other than each book. So Google analytics can't describe the progress of book reading and how never reports what individuals do.

A number of web log analytics tools exist such as Webalizer, Web Log Expert, Web Log Suite and WUM [4]. However, they are limited in their support for site-wide analysis of the kind we are pursuing to understand how people read books.

3 Visualizing Book Reading Sessions

Because the ICDL is free and allows anonymous usage, relatively few people register with the site. Thus, it is difficult to track an individual's reading progress. But with a bit of effort, we can analyze and track what we call *book reading sessions* (BRS) with reasonable accuracy. We extract BRS as follows.

Step 1: Clean Data: We filtered out records from the Apache web server logs with any error status code as well as records which reference embedded image files.

Step 2: Parse URL: The fields of the web log that we use are IPaddress, Agent, Begintime, Referrer, and URL (which contains fields separated by "&".) For example: /icdl/BookPage?bookid=husblsk_00040002&pnum1=10&pnum2=11&twoPage=true &route=simple_0_0_blue%20sky_English_0&size=0&fullscreen=false&lang=Englis h&ilang=English

Step 3: User Identification & Session Identification: It is a complex process to analyze web logs, but many papers have discussed it [5][6][7]. We follow such identification methods, and define BRS as a time series data set, which includes one "Book Reader" web page (Figure 1) and many "BookPage" web pages (Figure 2).

We can then observe how many pages people read and how much time they spent in each reading session by using visualization software such as LifeLines. LifeLines supports visual exploration of multiple records of categorical temporal data and allows alignment of data on sentinel events, showing intervals of validity [8][9].

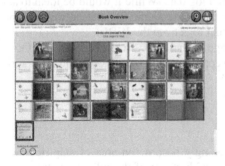
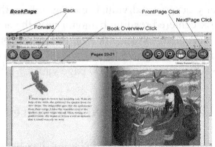

Fig. 1. A sample *BookReader* page   **Fig. 2.** A sample *BookPage* page

We use LifeLines to visualize BRS (Figures 3, 4). Each *BRS* called a record is vertically stacked on alternating background colors. It is identified by its ID on the left, and its page number ("*Page No*") in this reading session is listed under the session ID in order. Each *BookPage* (called an "event") appears as colored triangle icons on the timeline in the middle of the main display area. The beginning time of the first event (*Page 001*) are aligned vertically.

Seeing this different kind of reading behavior brings us to a key question – what do we mean to "read a book" online? Clearly there are many different styles. Some sessions clearly represent reading and some clearly do not. So, what do we do?

4   Definition of RBRS – Real Book Reading Session

By looking at the BRS data, we now define what we call a Real Book Reading Session (RBRS). It is reasonable to consider a book "read" if every page is looked at for a reasonable time. However, it is difficult to draw a clear boundary between reading and non-reading. For example, if someone reads ¾'s of the pages of a book, while skipping the introductory and ending matter, most people would probably consider that to also be reading the book. What if they skipped two chapters in the middle? Since this is a subjective decision, and our primary purpose was to distinguish people that were doing some reading compared to those that weren't, we decided on a simple and unambiguous definition.

We define a book to be considered read if an individual has looked at more than half of the pages of the book. Therefore, a real book reading session (RBRS) is defined as a book reading session where the percentage of distinct pages read is greater than 50%.

We collected 21,060 sessions, in which 900 books were visited by users (Figure 5). We chose the turning point (50%) in this curve to be the threshold in the definition of RBRS. Based on this definition, 1,197 sessions out of the 21,060 were RBRS, which includes 331 distinct books.

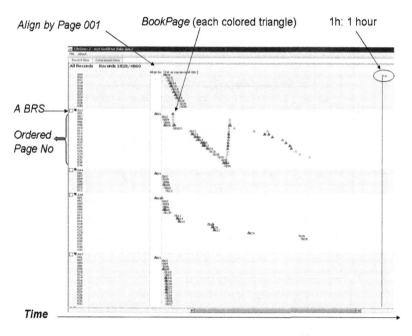

Fig. 3. This is part of one book's BRSs, includes five book reading sessions .The first BRS on the top spends about 5 minutes on a whole book. The second BRS reads the whole book, then goes back over each page, and reads the entire book a second time more slowly. The third BRS only looks at the first six pages quickly, and then leaves. The fourth BRS looks at the entire book, but there are significant pauses after every few pages. Over one hour is spent on this book. The fifth one looks at every page in the book, but does this so quickly that the entire book is scanned in just one minute.

Fig. 4. There are 12 BRSs, each of which includes only a few book pages. They each start at the beginning of the book, and then leave relatively quickly.

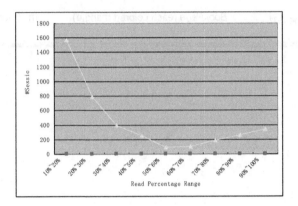

Fig. 5. *# Session* vs. *Read Percentage.* The number of reading sessions (BRS) that represents a user reading the indicated percentage of pages. There is a low value (near 50%) in the figure which motivated us to pick 50% as the number to specify the RBRS cut-off.

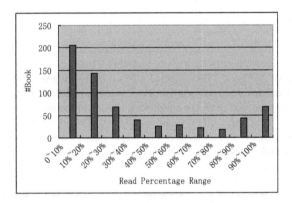

Fig. 6. *# Books* vs. *Read Percentage.* Each bar shows the number of books that had the indicated percentage of pages looked at in a reading session.

5 Web Usage Analysis

We now look at how different books are read. Figure 6 show, as expected, that there are many books which remain mostly unread. And the number of books that have more of them read decreases with the amount of the book that is read – to a point. Surprisingly, the number of books that have 80% or more of the pages read increases. We have no evidence to support an explanation of this. However, we observe that it makes sense that when reading books, it is natural for highly engaged readers to read all of the book – even end paper, etc.

Another thing we looked at was how much time people spent on each page of a book. If people were reading the entire page, then we would expect that the time spent on each page would roughly correlate to the amount of text on that page. To examine this, we picked one book ("Three Little Pigs") that had a varying amount of text on each page that was read relatively often. The results are somewhat surprising in that while the time spent on each page is clearly lower when there are very few words on a page, it clearly does not increase directly with the amount of text on each page. Again, we don't have any data that explains this, but some possible explanations are that many readers may be simply looking at pictures. Another possibility is that our hypothesis is incorrect, and people naturally spend a roughly constant amount of time per page – perhaps looking at pictures more, or reading more slowly when there is less text.

In sum, this short paper just scratches the surface in looking at how people actually read books in the ICDL. The data indicates some expected, but some surprising results – which clearly indicate the need to study this in more detail.

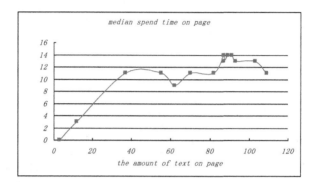

Fig. 7. This shows the *median spent time* (in seconds) correlated with *the amount of text on page* (in words), for the book "Three Little Pigs"

References

1. Liu, Z.: Reading behavior in the digital environment Changes in reading behavior over the past ten years. Journal of Documentation 61, 700–712 (2005)
2. O'Hara, K., Sellen, A.: A comparison of reading paper and on-line documents. In: Proceedings of Human Factors in Computing Systems(CHI 1997), pp. 335–342. ACM Press, New York (1997)
3. Srivastava, J., Cooley, R., Deshpande, M., Tan, P.N.: Web Usage Mining: Discovery and Applications of Usage Patterns from Web Data. In: ACM SIGKDD Explorations Newsletter. ACM Press, New York (2000)
4. http://hypknowsys.sourceforge.net/wiki/The_Web_Utilizat
5. Cooley, R., Mobasher, B., Srivastava, J.: Data Preparation for Mining World Wide Web Browsing Patterns. J. Knowledge and Information Systems 1(1), 5–32 (1999)
6. Wu, K.-L., Yu, P.S., Ballman, A.: Speed tracer: a web usage mining and analysis tool. IBM Systems Journal 37(1), 89 (1998)
7. Yang, Z.L., Wang, Y.T., Kitsuregawa, M.: An Effective System for Mining Web Log. In: Zhou, X., Li, J., Shen, H.T., Kitsuregawa, M., Zhang, Y. (eds.) APWeb 2006. LNCS, vol. 3841, pp. 40–52. Springer, Heidelberg (2006)
8. Plaisant, C., Milash, B., Rose, A., Widoff, S., Shneiderman, B.: Lifelines: visualizing personal histories. In: Proc. CHI (1996)
9. Wang, T.D., Murphy, S., Plaisant, C.: Aligning Temporal Data by Sentinel Events: Discovering Patterns in Electronic Health Records. In: Proc.CHI 2008. ACM Press, New York (2008)

Usability Evaluation of a Multimedia Archive: B@bele

Roberta Caccialupi[1], Licia Calvi[2], Maria Cassella[3], and Georgia Conte[1]

[1] Multimedia Production Centre, University of Milano-Bicocca,
viale dell'Innovazione, 10 20125 Milano, Italy
{roberta.caccialupi,georgia.conte}@unimib.it
[2] Lessius College, K.U.Leuven, Sint-Andriesstraat, 2
2000 Antwerp, Belgium
licia.calvi@lessius.eu
[3] University of Torino, Via Po, 17
10124 Torino, Italy
maria.cassella@unito.it

Abstract. In institutional repositories, simple discovery and submission interfaces help increase documents deposit as scholars have very little time to self-archive. So far, however, usability evaluation of such interfaces has been limited. In this paper, we present the usability evaluation of a repository interface, i.e., the interface of B@bele, the DSpace installation of the Multimedia Production Centre (CPM) of the University Milano-Bicocca. The results of this evaluation point out the most important shortcomings of the present DSpace interface: difficulties with browsing within communities and collections; problems with the submission interface due to scarcely familiar terminology (metadata) or terms that are not relevant in the specific academic context (community); problems in the submission process due to some ambiguous buttons, to the lack of authority files, and to the lack of clearly marked compulsory fields. In this way, this study will help improve not only B@bele, but also all other installations of DSpace currently available.

Keywords: User interfaces, institutional repositories, DSpace, usability evaluation.

1 Introduction

Institutional repositories (IRs) are one of the most innovative and creative components of digital libraries nowadays. Their function is to "manage, preserve and promote access to the knowledge base produced within an institution" [5].

Institutional repositories are a powerful and complex mean to disseminate academic knowledge. Their complexity is reflected in the wide spectrum of aspects that institutional repositories cover: technological developments, management issues, the need to analyze their adoption and use by the different scientific communities, case studies and best practices in the use of repositories, studies on the state of the art in local contexts, the marketing of repositories [1], [2], [4], [5], [10], [11], [13], [16], [17]. There is however still an aspect in the literature on institutional repositories that

M. Agosti et al. (Eds.): ECDL 2009, LNCS 5714, pp. 370–376, 2009.

has remained rather unexplored: it is the usability of their applications which may prevent an increased use of institutional repositories.

With the release of DSpace 1.5, based on the new Manakin interface, the need to evaluate DSpace usability has become more urgent. In this paper, we present the usability evaluation both of the discovery and of the submission interfaces of B@bele, the DSpace-based repository of the Multimedia Production Centre (CPM) of the University Milano-Bicocca.

2 Literature Review

The literature on usability in repositories [3], [6], [8], [9], [12], [14], [15] comprises a couple of comparative studies on the usability of EPrints and DSpace [8], [12] and a few studies on the usability of DSpace alone [3], [6], [15]. Kim Jihyun [12], for instance, reports the results of a usability evaluation on the interface of both the EPrints and DSpace applications in use at the Australian National University (ANU). His tests were performed by18 students, equally divided between EPrints and DSpace. The tasks they had to perform mainly covered simple and advanced search activities. The submission interface was not evaluated in this study.

A study by Cunningham et al. [8] involved librarians and university researchers testing both the search and the submission interfaces.

Micheal Boock's [6] evaluated the submission interface of the DSpace repository for electronic theses/dissertations with 6 users from the Oregon State University (OSU). He reported some difficulties from the users to register in the repository and emphasizes the importance to follow documents submission with clear messages and an adequate feedback from the system itself.

Finally, Ottaviani [15] carried on a much deeper analysis of Deep Blue, the DSpace application at the University of Michigan. Its submission interface was tested by a group of lecturers, while graduate and undergraduate students evaluated the usability of the search interface. His conclusions on the search interface are rather different from those reported by Jihyun [12]: Ottaviani's experiments point out some problems relative to the advanced search interface, to the reporting in DSpace as well as problems in the submission interface, for what concerns the terminology adopted, the license agreement and the absence of any feedback from the system.

Another aspect that has been evaluated with DSpace was its information retrieval capabilities [3], both in its simple and advanced search functionality and the reporting to the end users. Both functionalities are experienced as inadequate for an electronic archive of Master's and Ph.D. dissertations.

3 A Case Study: B@bele

With the release of DSpace 1.5, based on the new Manakin interface, the need to evaluate its usability has become urgent. B@bele is an application of DSpace 1.5 in use at the Multimedia Production Centre (CPM) of the University of Milano-Bicocca. B@bele is not an institutional repository in the strict sense of the term, but it is employed within CPM as a digital archive for mainly multimedia material by individual researchers and administrative personnel. At present, it hosts six communities, each with a personalized layout.

A number of modifications had to be introduced on the original Manakin interface. These include the full translation of the terminology into Italian along with the name of the repository itself which refers to the multimedia character of B@bele collections. The translation tried to adhere to the original English terminology as much as possible. Significant differences between the Italian and the English versions will be signaled throughout this paper when they apply.

It has been reported [7], [9] that the success or failure of digital repositories is proportional to the simplicity, the easiness and the velocity of the operations that are necessary to submit documents into the archive. Since these systems rely on self-archiving, the documents submission process and their connected metadata become a critical element in determining the usability of such platforms. To evaluate them, we analyzed:

1. The functionality of the search interface both in its simple and advanced version and navigation within each community and collection.
2. The functionality of the submission interface, i.e., the process of inserting items and documents with reference to the main options during workflow.
3. Metadata management, both in their submission phase and in their modification phase.

For all the examined activities, we intended to evaluate users' familiarity with the terminology adopted (for instance, metadata, Open Access), the degree of difficulty in the execution of the predefined tasks and the level of users' satisfaction while interacting with B@bele. In order to test B@bele on different user categories, we selected 15 end users, subdivided into three homogeneous typologies. They were: administrative personnel, faculty and Ph.D. students. They were all not familiar with the use of B@bele.

The methodology we followed was a participant observation combined with a thinking aloud protocol, consisting in a series of predefined tasks that each user had to perform separately and sequentially. The complete test has been recorded using CAMTASIA. A pre-test questionnaire and a post-test one were also submitted to the selected users.

4   Discussion of the Results

In this section, we highlight the most important results drawn from this study. We group our findings into four categories and indicate for each of them the problems that were experienced by the test users. Some problems are very much B@bele-centered (for instance terminology), the others actually pertain the DSpace 1.5 version and can therefore be generalized to all other instances of it. In the last section, we will present the guidelines to improve the usability of DSpace 1.5.

Terminology. Problems with the terminology were related to the inadequate adherence of the DSpace phrasing into Italian, which was causing some inconsistencies due to bad translations. The button "save and exit" at the bottom of the submission page, for instance, has been translated as "salva e annulla", literally from the previous Manakin interface version, causing a lot of confusion in many users who did not know what to choose between the button "next", to continue with the submission, and the button "save and exit", that would have caused their submission to be ended. The use of the term "save" did indeed induce many users to think that the saved item was published as well and submission was over. What was also missing was a clear and unambiguous nomenclature for each navigation button that could reflect its real function (for instance, a "continue submission" button instead of "continue" alone) and an exhaustive explanation of the more complex processes, like the workflow process, for example (see further).

Browsing in communities and collections. Our subjects had difficulties in distinguishing the concepts of "community" and "collection", in identifying the dependencies between them (i.e., from community to sub-community and collection) and in understanding the exact meaning ascribed to these two terms in this context since they are normally associated with social network research. An additional problem was the difficulty most test subjects showed in finding the link to the communities and the collections on the Home page.

Finally, once a community is selected, it is not intuitive for users to understand how many documents are included in a community and which, among those, are really downloadable in full-text.[1]

Upload. The problem with submitting a new document depends on the layout of the upload page. Finding the upload link on that page is not simple since it is not visually recognizable as the link is inserted in the middle of the page. This task becomes especially difficult if other processes are not yet concluded and are therefore still active.

Workflow. Our test pointed out the need to systematically revise the workflow process which is now fragmented, redundant and not very intuitive.

5 Guidelines to Improve the Usability of DSpace 1.5

On the basis of the empirical analysis discussed in the previous section, the following guidelines to improve the usability of DSpace 1.5 can be drawn. Although some refer directly to the specificity of B@ele (see points 1 and 2), the majority of the following guidelines are meant to improve the usability of the DSpace 1.5 interface itself.

1. Adapt the DSpace/B@bele terminology to the context of use, so that users with familiarity within a specific domain can recognize the terms adopted and assign a meaning to them.

---

[1] Some documents as a matter of fact may be embargoed and therefore are not downloadable.

2. Although the need for a help function is normally considered a sign of bad design, some form of help and user's support (that is missing in the present version of B@bele) should be included, for instance in the form of a glossary accessible with a hyperlink to clarify the meaning of specific DSpace terms.

Terminology

3. Replace the phrase "save and exit" with a more appropriate and intuitive phrase, for instance, "save and continue later", since the former induced many users to think that saving would automatically determine the document publication and terminate the task.
4. Rename the ADD button that is currently used to add multiple elements, for instance authors' names, as "add co-authors" or make its function more explicit in the text. The ADD button was indeed found mainly ambiguous: many subjects thought it had to be used to confirm the submission of the inserted information, with the results that they continued to add the same information.
5. Specify the meaning of technical or librarian-like jargon terms like "series" or "reports".

Navigation

6. Include back and forth buttons to make navigation more user-friendly.
7. Make the community page more comprehensible by:
    a. Compressing the list of communities into a drop down menu, so as to visualize each sub-level only when the higher level is selected. This would reduce the list length and give a clearer hierarchical structure although at the loss of some visibility and of the general overview.
    b. Indicating for each collection the number of documents there included, and their status, as public or not, using, for instance, a lock icon. In this way, users could get an immediate and intuitive idea over the archive consistency and over the accessibility of its content.
8. Distinguish visually between active and disabled breadcrumbs.
9. Include an author authority system to speed up the process of inserting personal data and to prevent typing mistakes.

Submission

10. Mark clearly all compulsory fields in order to speed up data submission and to allow users to insert only the really necessary metadata.
11. Make the link to a new submission more visible, in a position where it can be easily found by the users.

Workflow

12. Insert a separate section for the tasks the reviewer should perform, since this function is at the moment by far intuitive and clear.
13. Restructure the insertion page in order to reflect a more logical sequence from the point of view of a user executing a task.

6 Conclusion

In this paper, we have presented the usability evaluation of B@bele, an archive for the publication of multimedia material. This evaluation was not only intended to improve B@bele, but was also intended to provide the missing link in the usability evaluation of discovery and of submission interfaces, whose analysis has so far has been rather limited.

Our next step is to improve its usability in the way indicated in this paper. This will surely help the diffusion of digital archives in general and of open archives like DSpace in particular.

References

1. A DRIVERS guide to European repositories. Edited by Kasia Weenink, Leo Waaijers, Karen van Godtsenhoven. Amsterdam University Press, Amsterdam (2008)
2. Allinson, J., François, S., Lewis, S.: SWORD, Simple Web-service Offering Repository Deposit. ARIADNE 54 (2008),
   http://www.ariadne.ac.uk/issue54/allinson-et-al/
3. Atkinson, L.: The Rejection of D-Space: selecting Theses Database Software at the University of Calgary Archives. In: Proceedings 9th International Symposium on Electronic Theses and Dissertations, Quebec City, QC, Canada (2006)
4. Barton, M.R., Waters, M.M.: Creating an institutional repository: LEADIRS workbook. MIT Libraries (2004-2005)
5. Bevilacqua, F.: L'organizzazione dei depositi istituzionali DSpace in Italia. Biblioteche oggi 26(6), 17–25 (2008)
6. Boock, M.: Improving DSpace@OSU with a usability study of the ET/D submission process. ARIADNE 45 (2005), http://www.ariadne.ac.uk/issue45/boock/
7. Carr, L., Harnad, S.: Keystroke economy: a study of the time and effort involved in self-archiving. Technical report, University of Southampton (2005)
8. Cunningham, S.J., et al.: An Ethnographic Study of Institutional Repository Librarians: Their Experiences of Usability. In: Proceedings Open Repositories, San Antonio, TX, USA (2007)
9. Davis, P.M., Connolly, M.J.L.: Institutional Repositories: Evaluating the Reasons for Non-use of Cornell University's Installation of DSpace. D-Lib Magazine 13(3/4) (2007),
   http://www.dlib.org/dlib/march07/davis/03davis.html
10. Fried Foster, N.F., Gibbons, S.: Understanding faculty to improve content recruitment for institutional repositories. D-Lib Magazine 11(1) (2005)
11. Gierveld, H.: Considering a marketing and communications approach for an institutional repository. ARIADNE 49 (2006),
    http://www.ariadne.ac.uk/issue49/gierveld/
12. Jihyun, K.: Studies of searching behaviour: finding documents in digital institutional repositoryies, DSpace and EPrints. In: Proceedings of the American Society for Information Science and Technology 42(1) (2006) DOI: 10.1002/meet.1450420173
13. Markey, K., et al.: Census of Institutional Repositories in the United States MIRACLE Project Research Findings. In: Council on Library and Information Resources, Washington, D.C (2007),
    http://www.clir.org/pubs/reports/pub140/pub140.pdf

14. McKay, D.: Institutional repositories and their 'other' users: usability beyond authors, ARIADNE 52 (2007), http://www.ariadne.ac.uk/issue52/mckay/
15. Ottaviani, J.: University of Michigan DSpace (a.k.a. Deep Blue) Usability Studies: Summary Findings (2006)
16. Phillips, S., et al.: Manakin: a new face for DSpace. D-Lib Magazine 13(11/12) (2007)
17. Van de Sompel, H., Lagoze, C.: Interoperability for the discovery use and re-use of units of scholarly communicaton. CTWatch Quaterly 3(3)

Digital Libraries, Personalisation, and Network Effects - Unpicking the Paradoxes

Joy Palmer[1], Caroline Williams[1], Paul Walk[2], David Kay[3], Dean Rehberger[4],
and Bill Hart-Davidson[5]

[1] JISC National Data Centre, UK
{joy.palmer,caroline.williams}@manchester.ac.uk
[2] UKOLN, University of Bath, UK
p.walk@ukoln.ac.uk
[3] Sero Consulting Ltd., UK
david.kay@sero.co.uk
[4] Digital Humanities Research Centre, Michigan State University, USA
mfegan@mail.matrix.msu.edu
[5] Writing in Digital Environments Research Center (WIDE),
Michigan State University, USA
hartdav2@msu.edu

Panel Description

The focus of this panel presentation is on personalisation (including adaptive personalisation) and the constructions of 'digital societies' around digital libraries and collections. Panelists will represent a variety of perspectives - NEH (USA) JISC (UK) & EU - ranging from developers of highly specialised academic digital libraries, to directors of national digital libraries that aim to achieve system-wide aims.

The 'Library 2.0' and emerging 'Archives 2.0' movements are understandably asking what library and archives can do to exploit 2.0 and adaptive technologies to potentially enhance catalogues, finding aids and repositories. User-generated content, especially from subject experts, has the potential to enrich catalogue entries and digital objects and aid the processes of learning, teaching and research. Collections thus become potential 'architectures of participation'; the bases for robust digital communities that support education and research. In addition, questions are being asked over the role of attention data in personalisation, which can be used to widen participation and strengthen those communities, as well as providing new pathways through resources, yielding 'long tails' of digital library content. Webscale systems such as Google and Amazon can leverage network effects in astoundingly efficient ways, aggregating individual and group data to enhance search and adaptive personalisation functions. How might national digital libraries adapt such a model to support learning and research?

This panel will discuss the technical and cultural challenges that emerge when the Amazon-like model for supporting adaptive personalisation is transferred to the digital library domain. Technical challenges concern the significant shift in approaches to application development and 'system-wide' initiatives, approaches which increasingly privilege a dispersed and fragmented data-model. The drive for centralised digital library systems is being questioned, as a growing trend emerges whereby web technologies are used to knit or mash data and systems together at the presentation level.

M. Agosti et al. (Eds.): ECDL 2009, LNCS 5714, p. 377, 2009.
© Springer-Verlag Berlin Heidelberg 2009

DL Education in the EU and in the US: Where Are We?, Where Are We Going?

Vittore Casarosa[1], Anna Maria Tammaro[2], Tatjana Aparac Jelušic[3],
Stefan Gradman[4], Tefko Saracevic[5], and Ron Larsen[6]

[1] ISTI-CNR and DELOS Association
casarosa@isti.cnr.it
[2] University of Parma and DILL International Master
annamaria.tammaro@unipr.it
[3] University of Zadar, Croatia
taparac@ffzg.hr
[4] Humboldt University, Germany
stefan.gradmann@ibi.hu-berlin.de
[5] Rutgers University, USA
tefko@scils.rutgers.edu
[6] University of Pittsburgh, USA
rlarsen@sis.pitt.edu

Panel Description

The EU i2010 policy framework to build the European Information Society has positioned digital libraries as a critical component for its realization. The i2010 Digital Libraries initiative sets out to make all Europe's cultural resources and scientific records – books, journals, films, maps, photographs, music, etc. – accessible to all, and preserve it for future generations. In the US, digital libraries have been indicated as a critical component in the long-term realization of the promise of cyberinfrastructure, as tools that will change how science and humanities research is organized, stored, disseminated, and curated (Blue Ribbon Advisory Panel on Cyberinfrastructure). Since the advent of digital libraries in the early nineties, a consistent focus of DL research and development, beyond Computer Science, has been the application of DLs in educational contexts. Many projects have explored how best to teach WITH digital libraries, but only more recently has research been conducted on the best way to teach ABOUT digital libraries.

Some LIS schools have focused on specialized education programs, providing advanced certificates in "digital librarianship" beyond the first professional degree. Others have focused on integrating education for digital librarianship into the general LIS professional degree education program. A few have pursued programs of education that are independent of the first professional LIS education degree. North America and Europe present patterns of education that represent all three approaches. Due to different disciplinary and cultural influences, DL education programs have developed along different paths, and as a result the focus of DL education is on different approaches to DL development and different applications and implementations of DLs in practice.

M. Agosti et al. (Eds.): ECDL 2009, LNCS 5714, pp. 378–379, 2009.
© Springer-Verlag Berlin Heidelberg 2009

DL education in the US has traditionally (in the Web age, 15 years are enough to make a tradition) emphasized two aspects of DL development: collection development (digitization and metadata) and information architecture (applications and network infrastructure). In short, DL education practices in the US have been heavily influenced by both traditional library education and elements of computer science. DL education in the EU has evolved from a different tradition. On one side we have the extension of LIS curricula towards digitization issues, formats, digital access and delivery. On another side it has been growing out of what could be called humanities computing and museum informatics.

The main purpose of this panel is to debate the strengths and the weaknesses of those different approaches, and to identify commonalities and complementarities, which could guide further development of DL education. It is understood that there is a variety of approaches to DL education, and certainly the panelists will not be able to represent all of them. Participation from the audience will be encouraged and solicited in order to get a picture of the current situation as wide as possible (not necessarily restricted to US and EU), and in order to find the greatest common denominator that could be the base for a "common" shared view of DL education.

Conceptual Discovery of Educational Resources through Learning Objectives

Stuart A. Sutton and Diny Golder

University of Washington and JES & Co.
sasutton@u.washington.edu, dinyg@jesandco.org

Abstract. This poster reports on current work with the NSF-funded Achievement Standards Network (ASN) to support discovery of educational resources in digital libraries using conceptual graphs of officially promulgated achievement standards statements. Conceptual graphs or knowledge maps of achievement standards reveal the macrostructure of the learning domain modeled by those standards and support higher-level understanding by teachers and students. The work builds on the conceptual framework of the AAAS knowledge maps by providing the means to flexibly define and deploy new relationship schemas to fit the disparate modeling needs of the nearly 740 learning standards documents in the ASN repository. Using an RDF-based, node-link representation of learning goals and the relationships among them, the ASN Knowledge Map Service will provide the framework to correlate educational resources to nodes in conceptual models in order to augment more conventional mechanisms of discovery and retrieval in digital libraries.

Keywords: digital libraries; content and curriculum standards; knowledge maps; conceptual discovery.

1 Introduction

UNESCO identifies the various stages or levels in pre-college or pre-university education as Level 0 (pre-primary education) through Level 3 (upper secondary).[1] As part of this framework, jurisdictional authorities in most nations have formally promulgated achievements standards that specify what students studying at these levels should know and be able to do as a result of their education. Some such standards are promulgated at the national level and others at the level of states, provinces and other governmental subdivisions. Some nations have a single controlling set of standards and others like the United States have as many as 51 complete sets at the level of the states and the District of Columbia and even more at the local level within states.

The names provided these formally promulgated standards vary substantially across nations with little rationale for the variety in nomenclature. Some nations call them curriculum standards, others simply call them content standards, and still others identify them as frameworks. The component parts of these controlling standards are also identified by means of a rich array of names—standards, benchmarks, assessments and indicators. The list of names goes on *ad infinitum*. For the purposes of our

M. Agosti et al. (Eds.): ECDL 2009, LNCS 5714, pp. 380–383, 2009.

research, we call all standards of this class, *achievement standards* since they all relate broadly to learning goals and student achievement.

Even given this rich nomenclature for achievement standards, there is a fairly universal framing of their intent and content. All achievement standards more or less indicate the knowledge and skills, the ways of thinking, working, communicating, reasoning, and investigating expected of students studying at UNESCO education Level 0 through Level 3. Achievement standards also enumerate the most important and enduring ideas, concepts, issues, dilemmas, and knowledge considered essential to the domain of study that should be taught and learned in schools under the jurisdiction of the standard's promulgating authority.

So defined, achievement standards are ontological (and frequently political) in nature—modeling the learning expectations of a people in their children and youth. Achievement standards reveal the macrostructure of the domain they model and, therefore, provide an additional mechanism for digital library access to educational resources correlated to those standards. Thus, metadata describing lesson plans, learning objects and other educational resources useful in meeting specific learning objectives may be assigned the identifiers for the achievement standards in which those objectives are embodied. Such assignments support searching by teachers, parents, students, curriculum developers and school administrators for appropriate resources to meet jurisdictional needs. In the United States, these achievements standards are beginning to be used as assessment categories for student learning through standards-based report cards.

This poster reports on the preliminary research involving advanced uses of a national repository of U.S. achievement standards called the Achievement Standards Network (ASN) that supports both research in standards-based education and the correlation of educational resources to achievement standards for various purposes ranging from enhanced information retrieval through standards compliance in teaching and learning.[2] In this research, we build on the information modeling of our earlier ASN research by exploring the power inherent in the standards data and the implicit and explicit relationships they embody in revealing the macrostructure of standards domains with the goal of enhancing the use of the standards by teachers and students in domain comprehension, exploration and resource discovery.

The goal of the original ASN work was to develop a conceptual schema and a networked repository of machine-addressable achievement standards that would serve immediate needs for true-to-source representations of the standards while being fully amenable to the Semantic Web. Using the Dublin Core Abstract Model as the framework in the original work, two entities were defined—the standard document and the statement. Statements represent the component achievement assertions contained in the standards documents. The standards documents are atomized into their component statements and represented in RDF with each standards document and each statement being assigned a dereferencable URI. To date, the ASN contains over 740 current and historical state and national U.S. standards documents with initial forays underway into including standards from non-U.S. jurisdictions. These documents are atomized into over 350,000 individual achievement statements.

The ASN data model also defines a set of structural relationships between individual statements creating hierarchical taxon paths comprised of RDF-triples that reconstruct the inter-statement context of the standards document. The work reported here is exploring additional, non-structural paths through the standards data based on semantic relationships deemed useful in interpreting and using the standards. These additional paths take the form of knowledge maps. One of the goals of the research is to provide a means for creating the most useful maps and the definition of the new properties necessary to the generation of those maps. The new properties defined will be used by ASN in the refinement of its standards authoring tool to support creation of new knowledge maps by authors of standards in the ASN repository.

"Knowledge maps are node-link representations in which ideas are located in nodes and connected to other related ideas through a series of labeled links."[3] The final form of the knowledge map is a directed acyclic graph. In work pre-dating the ASN, the node-link representation of achievement standards data was used by the NSF-funded National Science Digital Library (NSDL) to create a navigable visual representation of an achievement standards knowledge map.[4] The NSDL work demonstrated the utility of such maps in supporting development of higher-order cognitive skills necessary to knowledge acquisition and more successful retrieval of educational resources when those resources are mapped to nodes in the knowledge map. [5, 6].

The NSDL knowledge map work was based on the learning goals in the *Benchmarks for Science Literacy* [7] as visualized in the *Atlas for Science Literacy* [8] and provides visual representations that emphasize the interconnectedness of science concepts and the connections between learning goals and digital resources in the NSDL. Working closely with the authors of the Atlas, the NSDL researchers defined the set of explicit semantic relationships set out in Table 1 to serve as map edges.[9] In sum, the NSDL researchers derived a single schema of properties to represent the relationships inherent in a single standard—*Benchmarks for Science Literacy*.

Table 1. Supported relationships in the NSDL Concept Space Interchange Protocol (CSIP)

Prerequisite	Is similar to
Post-requisite	References
Contributes to achieving	Is associated with
Contains	Is referenced by
Is aligned to	Supports
Is closely related to	Contributes to and relies upon
Is part of	Needs or requires

While the research reported here builds on the NSDL knowledge map conceptualization, it does not assume a fixed set of explicit semantic relations. Instead, we are exploring the nature of the explicit knowledge map relationships across standards documents from ten U.S. states to determine: (1) whether there is a common set of relationships inherent in standards knowledge maps regardless of the varying characteristics of the standards modeled; and (2) whether providing the capability of a core set of relationships with the ability to extend that set at the time of knowledge map creation results in more useful, expressive maps.

In addition to the investigation of explicit knowledge map relations—i.e., those explicit relationships defined in the authoring tool's configuration for a specific map, we are also exploring whether the automatic visual mapping of the structural relationships inherent in the existing ASN data produces mapping results more or less as useful as the human-authored knowledge maps. The goal of this second thread of inquiry is to determine whether the return-on-investment for the knowledge map authoring by humans exceeds that of the automatically generated mappings based solely on inherent document structure. Using the NSDL relationship schema noted in Table 1, preliminary results indicate that the *prerequisite*, *post-requisite*, *contributes to achieving*, *contains* and *is part* might prove useful and amenable to automatic identification. We are also exploring whether a hybrid system that relies first on the automatically generated map using the standards inherent structural properties as a base for human mapping augmentation serves user cognitive needs for visualizations of the macrostructure of the standards.

References

1. United Nations Education, Scientific and Cultural Organization (UNESCO): International Standard Classification of Education (1997),
   `http://www.uis.unesco.org/TEMPLATE/pdf/isced/ISCED_A.pdf`
2. Sutton, S.A., Golder, D.: Achievement Standards Network (ASN): An Application Profile for Mapping K-12 Educational Resources to Achievement Standards. In: Proceedings of the International Conference on Dublin Core and Metadata Applications, pp. 69–79 (2008)
3. O'Donnell, A.M., Dansereau, D.F., Hall, R.H.: Knowledge Maps as Scaffolds for Cognitive Processing. Educational Psychology Review 14, 71–86 (2002)
4. NSDL Strand Maps, `http://knowledgemaps.nsdl.org/`
5. Butcher, K.R., Bhusham, S., Sumner, T.: Multimedia Displays for Conceptual Discovery: Information Seeking with Strand Maps. Multimedia Systems 11(30), 236–248 (2006)
6. Sumner, T., Ahmad, F., Bhushan, S., Gu, Q., Molina, F., Willard, S., Wright, M., Davis, L., Janée, G.: Linking Learning Goals and Educational Resources through Interactive Concept Map Visualizations. International Journal of Digital Libraries 5, 18–24 (2005)
7. American Association for the Advancement of Science: Benchmarks for Science Literacy. Project 2061. Oxford University Press, Oxford (1993)
8. American Association for the Advancement of Science, National Science Teachers Association: Atlas of Science Literacy. Project 2061. Washington, D.C (2001)
9. Schema – CSIP.xsd, `http://knowledgemaps.nsdl.org/cms1-2/docs/cms/cms1-2/CSIPSchema/csip_doc.jsp`

Organizing Learning Objects for Personalized eLearning Services

Naimdjon Takhirov, Ingeborg T. Sølvberg, and Trond Aalberg

Department of Computer and Information Science
Norwegian University of Science and Technology
NO-7491 Trondheim, Norway
{takhirov,ingeborg,trondaal}@idi.ntnu.no

Abstract. In this paper we present a way to organize Learning Objects to achieve personalized eLearning. PEDAL-NG is a system that supports personalization based on the user's prior knowledge and the learning style in an existing and operational eLearning environment. The prior knowledge assessment and the learning style questionnaire proved to be simple and useful tools to gather necessary information about the user in order to deliver personalized eLearning experience.

1 Introduction

The popularity of the Internet and recent developments in educational digital libraries have created new possibilities for online learning [1]. However, the problem of the "one-size-fits-all" may result in high dropout rates and low levels of motivation and satisfaction [2]. Human beings are different, and as such they learn and process information in different ways. Past experience, prior knowledge, skills, learning style and interests are among the factors that may affect the individuals' requirements for effective learning [3]. Instead of delivering the same content to all users, eLearning systems should be able to adapt to each individual's characteristics in order to increase the relevance and appropriateness of the learning material. Although the range of eLearning applications has increased substantially both in academia and industry, in reality, users are often represented with the same learning style that students have been presented with for 50 years [4]. Learning is a personal and adaptive process, and thus individual student's needs and preferences should be considered from start to finish [5].

2 Background

The objective of personalization in web-based systems generally is to provide users with the information they want or need without asking for it explicitly [6]. In the context of eLearning, personalization may be achieved in several dimensions [7]. These dimensions include, but not limited to, personalized selection of Learning Objects (LOs), dynamic and adaptive sequencing of LOs, the output format (output on a computer screen, PDF or output tailored to mobile devices).

M. Agosti et al. (Eds.): ECDL 2009, LNCS 5714, pp. 384–387, 2009.

In this project, the learning style and prior knowledge have been taken into account as personalization parameters. The prior knowledge may be adjusted to adapt learning material as the user completes activities. Adapting learning material implies adapting content, in runtime, by observing learner behavior during the progression of the course and matching metadata associated with LOs and the metadata stored in learner profile. As this process occurs during runtime, any activity may affect the structure and sequence of the LOs.

The process of personalized course construction starts with collecting information about the user's existing knowledge. This information is then used to match the level of LOs. For simplicity, three levels have been employed: easy, intermediate, and advanced. The second step is to map the learning style of the user. To do that, a simple learning style questionnaire consisting of sixteen questions about daily life situations was used. This questionnaire is based on VAK (V-isual, A-uditory, K-inesthetic) learning style model [8]. Each possible value in the questionnaire reflects a specific learning style. Having gathered information about the user's prior knowledge and learning style, PEDAL-NG constructs a preliminary learning path which is subsequently adjusted during course progression. The architecture of PEDAL-NG is presented in [9] and the detailed description of the project can be found in [10].

3   Learning Objects in PEDAL-NG

PEDAL-NG is developed as a plugin module on top of Inspera LCMS[1]. This eLearning environment is widely used by elementary schools and high-schools throughout Norway. The Inspera LCMS is basically a digital library of LOs organized as courses.

PEDAL-NG maintains a dynamic index of the course in the memory. This index is rebuilt upon authentication of the user. Since both LOs and users include the *learning_style* and *level* metadata elements in their metadata schema, it is straightforward to execute search for relevant LOs within a specific subset of nodes. The structure of the sample course which was used to evaluate the PEDAL-NG prototype is illustrated in Figure 1.

Each node in the tree (except for assessment nodes) represent a topic within the course. For example, a course on Databases normally has a topic on SQL or Entity Relationship modeling. Each topic contains different types of LOs. These LOs may be of type theory items, examples, exercises and assessment. Each item is stored with specific learning style and level. As an illustration, in the course module SQL, example of different item types are:

– figures and pictures for visual users. These visual artifacts illustrate how an SQL query is executed against the database and the results of the query being fetched from the database;
– sound items for auditory users. Sound items are textual items converted on-the-fly using text-to-speech technology;

---

[1] Inspera is a software company based in Oslo, Norway, specializing in delivering software services to educational sector. For more info visit www.inspera.com

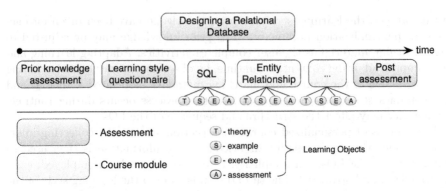

Fig. 1. The structure of the sample course *"Designing a Relational Database"* showing some of the course modules

- a live interactive database was made available for kinesthetic learners. Kinesthetic users had an opportunity to execute a query and obtain results of the query applying the "learning-by-doing" paradigm.

LOs are described using internal metadata schema of Inspera LCMS. The schema was extended with elements such as *learning_style* and *level*. For instance, visual users want content to be presented with texts and pictures, auditory learners prefer the content to be read by the system (using text-to-speech conversion technology or pre-recorded sound items), and kinesthetic learners would learn best when material is accompanied by exercises so that learners can practice themselves [11]. Recognized metadata standards such as IEEE LOM has elements named InteractivityType, InteractivityLevel, and Difficulty. These elements cannot be used explicitly to describe the learning style and the level of a learning object.

The selection of LOs is performed at runtime by the adaptive engine executing simple queries. The adaptive engine is the core of PEDAL-NG. It provides services to achieve adaptivity and delivers personalized course. The adaptive engine selects LOs based on the metadata associated with LOs, metadata associated with learners, and usage logs. Infrastructure for usage logs is provided by the built-in tracking mechanism in the Inspera LCMS. For example, if SQL represents a topic on a course on "Designing Relational Databases", assuming this topic object has id 565479, the user has a visual learning style and was found to have some prior knowledge in the subject (intermediate), the query string would be *+parentId:565479 +learning_style:visual +level:intermediate*. This query will return all candidate LOs that are mapped to SQL topic. The crucial part of selecting matching LOs is to ensure that course designers have followed the convention of what learning style model the system uses.

4   Evaluation and Further Work

The working prototype PEDAL-NG integrates prior knowledge assessment and learning style questionnaire to map prior knowledge and learning style of users before commencing a course. Organization and selection of LOs in a way described in this paper does not require complex algorithm and programming effort. The prior knowledge assessment and the learning style questionnaire proved to be simple but useful tools to gather necessary information about the user in order to deliver personalized eLearning experience. The working prototype was evaluated by a group of fourteen users on a sample course on *Designing a Relational Database*. Upon course completion the participants were asked to complete a survey. This survey results indicate that users favored personalization and the results from this experiment constitute a solid ground for further work.

References

1. Gu, Q., Chica, S., Ahmad1, F., Khan, H., Sumner, T., Martin, J.H., Butcher, K.: Personalizing the Selection of Digital Library Resources to Support Intentional Learning. In: Christensen-Dalsgaard, B., Castelli, D., Ammitzbøll Jurik, B., Lippincott, J. (eds.) ECDL 2008. LNCS, vol. 5173, pp. 244–255. Springer, Heidelberg (2008)
2. Tyler-Smith, K.: Early attrition among first time elearners: A review of factors that contribute to drop-out, withdrawal and non-completion rates of adult learners undertaking elearning programmes. Journal of Online Learning and Teaching 2(2) (2006)
3. Kyung-Sun, K., Moore, J.L.: Web-based learning: Factors affecting students' satisfaction and learning experience. First Monday 10(11) (2005)
4. Bchner, A., Patterson, D.: Personalised e-learning opportunities - call for a pedagogical domain knowledge model. In: Proc. of 15th International Workshop on Database and Expert Systems Applications, Zaragoza, Spain (2004)
5. Boticario, J.G., Santos, O., Rosmalen, P.: Issues developing standard-based adaptive learning management systems. Paper for the EADTU 2005 Working Conference: Towards Lisbon 2010: Collaboration for Innovative Content in Lifelong Open and Learning (2005)
6. Maurice, D.M., Anand, S.S., Bchner, A.G.: Personalization on the net using web mining: introduction. Communications of the ACM 43(8) (2000)
7. Conlan, O., Wade, V., Bruen, C., Gargan, M.: Multi-model, Metadata Driven Approach to Adaptive Hypermedia Services for Personalized eLearning. In: Proc. of AH and Adaptive Web-Based Systems (2002)
8. Fleming, N.: VARK. A guide to learning styles (2008),
   http://www.vark-learn.com
9. Takhirov, N., Sølvberg, I.: Adaptive personalized eLearning on top of existing LCMS. In: Proc. of Joint Conference on Digital Libraries, JCDL 2009, Austin, TX, USA (2009)
10. Takhirov, N.: Adaptive personalized eLearning. Master thesis, Norwegian University of Science and Technology, Trondheim, Norway (2008),
    http://ask.bibsys.no/ask/action/show?pid=082894205&kid=biblio
11. Felder, R.M., Silverman, L.K.: Learning and teaching styles. Engineering Education, 78(7) (1988)

Gaining Access to Decentralised Library Resources Using Location-Aware Services

Bjarne Sletten Olsen and Ingeborg T. Sølvberg

Norwegian University of Science and Technology
Department of Computer and Information Science
Sem Sælands vei 7-9
NO-7491 Trondheim
Norway
bjarnesl@stud.ntnu.no
http://www.idi.ntnu.no

Abstract. The paper describes a prototype that enables library users to use their mobile devices to find the physical location of specific services or objects in a branch of a distributed library. It guides the users to this location using external map services, location-awareness and navigational tools. The architecture of the system is briefly described together with the integrated services.

Keywords: Location-aware applications, mobile applications, library resources.

1 Introduction

As the number of mobile phones sold reaches new heights every day, ubiquitous computing is becoming a natural part of the modern life, and new applications emerge, providing functionalities that not long ago would have been unthinkable. Examples of this are WhosHere that enables users to broadcast their profile and to detect friends or people with common interests nearby[1], and Sekai Camera that enables users to tag real-life objects with comments using the camera on the mobile phone along with GPS[2]. Many of these functionalities take advantage of the location of the user. Such location-aware applications are able to provide contextual information, reconfigure, trigger actions or select nearby objects based on this location. In a geographical area there normally are several libraries or library branches, each holding documents on different subjects. Usually the library catalogues are available using a web-based search. The system described in this paper enables library users to use their mobile devices to find the physical location of specific services or objects in a branch of the library, and to guide the users to this location using maps and navigational tools. In this paper we describe a prototype that demonstrates the possibilities at the library of the Norwegian University of Science and Technology (NTNU). An overview of the functionality is presented along with the design of the prototype as well as plans

M. Agosti et al. (Eds.): ECDL 2009, LNCS 5714, pp. 388–391, 2009.

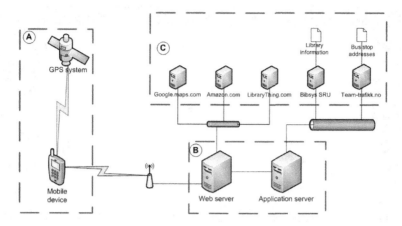

Fig. 1. Overall design with three modules. The Mobile application (A), the server-side application (B) and adapters and services (C).

for further development. The goal of this project is to demonstrate the possibilities offered by location-aware applications, as well as to show how already existing services can be integrated with this system.

2   The Prototype

The system consists of three modules as shown in Figure 1. Module A is the application running on the mobile device, handling communication with the GPS system. Module B is the part of the application running on servers and consists of the web server, the application server and the local database. Module C consists of adapters that handles communication with the external services, ensuring an identical interface for the application server, independent of what service the information is to be retrieved from. The server-side application communicates with the following services: BIBSYS grants access to the Norwegian library catalogue, containing more than 4.3 million titles, including the holdings of NTNU Library's 11 branches[3]. Google maps[4] provide a service that enables the construction of JavaScript-based maps. These constructors accept parameter's representing information that can be included in the map, such as location of Points of Interest (POI). The system is able to retrieve bus schedules from the local bus company Team Trafikk[5]. They provide a natural text search engine that responds to queries on bus schedules. The queries have to be on the form "When is the next bus for point A leaving from point B?". The response is in natural text, and has to be parsed. Amazon.com and Librarything.com are used as sources for book cover images.

2.1   A Walk-Through Example

A library user is in the city and needs a specific book right away. A search in BIBSYS, using the prototype running on his or hers mobile device, informs the

Fig. 2. Running prototype with the user, the departure bus stop, the arrival bus stop and the library indicated

user that the desired book resides in several of the university libraries branches. Information on the status of the document(chekd out, available and so on) is part of the BIBSYS system, and access to information is not provided as of now. The user selects the nearest library with the given document. The system generates a map, indicating the last registered position of the user as well as the position of the selected branch library. When the user wants to go to the branch library by bus, the system queries the local bus company on how to get there. A route to the nearest bus stop with connections to the library is then displayed on the map, along with the time of the next departure. The example is shown in Figure 2. When she or he arrives at the library, a map of the library is made available and guides the user to the shelf in which the document resides.

3   The System and Technologies

The mobile applicaiton was written using C#[1]. As newer browsers are already experimenting with the direct access of user coordinates through JavaScript, a stand alone application will not be necessary in the future, and the present mobile-side application is used only for testing. As platform independence is not required for the server-side application, this is written using C#. One of the goals with the development of the prototype is to demonstrate that integration with other existing services is feasible. An adapter is created for each of the services the prototype needs to communicate with. In order to add new services a new adapter must be created for each service, and the map genereation must be updated to handle the new information available.

---

[1] http://msdn.microsoft.com/en-us/vcsharp/aa336789.aspx

4   Future Work

As the system described in this paper is a prototype, there are some services that are not yet included, although planned for. One of these is the addition of a news-service, presenting the user with the latest news for the nearby area in map. This can be displayed either on city-, district- or street-level depending on the quality of the location-information associated with the news. As not all mobile devices support the use of GPS, some WiFi based location services have been considered. In the city of Trondheim, where NTNU resides, a localisation service named GeoPos[7] is available. This enables localisation services on all devices with a WiFi card installed. This would add the possibility of indoor-localisation to the application, facilitating the creation of additional services.

References

1. WhosHere, http://myrete.com/WhosHere.html
2. Gizmodo, http://www.gizmodo.com.au/2008/09/sekai_camera_turns_on_worlds_balloon_help-2.html
3. BISBSYS Ask, http://www.bibsys.no/wps/wcm/connect/BIBSYS+Eng/
4. Google Maps API, http://code.google.com/apis/maps/
5. Team Trafikk home page, http://team-trafikk.no/
6. JSR 179: Location API for J2ME, http://www.jcp.org/en/jsr/detail?id=179
7. GeoPos, http://www.geopos.no/geopos/

Thematic Digital Libraries at the University of Porto: Metadata Integration over a Repository Infrastructure

Isabel Barroso[1], Marta Azevedo[2], and Cristina Ribeiro[3]

[1] Faculty of Fine Arts, University of Porto
[2] Faculty of Nutrition and Food Sciences,
University of Porto
[3] Faculty of Engineering,
University of Porto/INESC-Porto

Abstract. The University of Porto has a well-established set of specialized libraries serving the research and student population of its 14 schools. Thematic digital libraries can be valuable for organizing specific collections and for supporting emergent communities. This work focuses on two case studies, one in the area of the Fine Arts and the other in the area of Food and Nutrition. For building both digital libraries we propose to use the existing university repository infrastructure and to establish a metadata workflow that makes use of available descriptions in the library catalogues and in the university information system. We expect that such an approach, which takes into account the institutional context and resources, can be used in other collections at our university and inspire similar initiatives elsewhere.

Keywords: metadata integration, institutional repositories, DSpace, digital libraries, description and cataloguing.

1 Thematic Digital Libraries at the University of Porto

The University of Porto is the largest Portuguese university, serving a population of over 28,000 students and covering research areas that range from the humanities, social sciences, fine arts, architecture, economics and business studies to science, engineering, medicine, sports and law. The university currently has three information systems that handle bibliographic records. The online library catalogues, managed with the Aleph system, are used for cataloguing, acquisition, loan and administration. The university information system developed on the in-house SIGARRA technology [1] includes the publications module used by the research community to deposit scientific publications, which are then validated by the library services. The university repository, built on the DSpace platform [2], is intended to manage and give open access to the scientific production of the university.

Our goal is to create thematic collections of digital objects, in the form of digital libraries, taking into account the information management context in the university and reusing existing bibliographic descriptions. This will guarantee quality descriptions in the libraries, avoid duplicate description, and use the technological support of an existing repository service. We build on earlier work concerning integration platforms and the extension of institutional repositories [3,4] and create specific procedures for metadata integration.

M. Agosti et al. (Eds.): ECDL 2009, LNCS 5714, pp. 392–395, 2009.

2   The Fine Arts and the Food and Nutrition Digital Libraries

The Library of Fine Arts holds an important collection with almost 230 years of history. The selected collection includes a variety of documents from the historical archive and the library: ancient books and magazines, engravings from 16th to 19th century, as well as minutes from ordinary and general conferences, letters, personal files of artists as students and as teachers and manuscripts. Researchers, students and teachers within and outside the University often request the collection. Significantly, none of these documents is accessible on the online catalogue. By providing the integration of collections from both the historical archive and the library, we will enable users to have full access to the collection and contribute to its preservation. As an example, a search for Silva Porto, a well known Portuguese painter, will retrieve all kinds of documents, including exhibition catalogues, photographs, his personal file, books written by and of him or even minutes where his name is mentioned, letters received and delivered all over his academic life.

The Food and Nutrition Digital Library aims to preserve and ensure access and reuse of the materials developed by the faculty community for scientific, academic, didactic and dissemination purposes. The prototype digital library will be based on the collection of digitized documents authored by one of the mentors of the institution, Dr. Emílio Peres. Dr. Peres, considered the father of Food and Nutrition Sciences in Portugal, is recognized for his seminal and unique work, for his brilliant communication skills and for his fresh scientific vision. The collection includes valuable manuscripts concerning the Food Education National Campaigns in the 80's, sets of slides used in his first lectures, the myriad articles published in local and national media, as well as radio conferences and TV interviews and programs.

Regarding both case studies, the main issue is that some of the objects are already described in the online bibliographic catalogue or in the university information system, but others will have to be described. The need to reuse existing quality metadata for describing digital objects presented a challenge to integrate metadata from different systems into the digital libraries.

3   Integrating Metadata in DSpace

The DSpace software [2] was conceived mainly to accommodate institutional repositories of scientific production and uses the Dublin Core format to register metadata that can be imported using an XML format. Using DSpace for our thematic digital libraries has required attention to three issues: the integration of the digital libraries in the university infrastructures; the reuse of cataloguing materials and the adaptation of DSpace features to the thematic digital libraries purposes.

The first problem required a clear view of the existing resources and some agreement with the university services. The digital libraries are sharing existing library and IT staff and this has led to the choice of DSpace, which is already supported by the university.

The second problem has led to the development of two metadata workflows to gather existing object descriptions in the catalogues and in the university information system. The diagram in Figure 1 illustrates the metadata creation for the digital library based on the new and existing descriptions.

Fig. 1. Metadata sources for the digital libraries

After analyzing the existing resources within the University we decided to establish methods for exporting record metadata, and the associated digital objects when available, from the Aleph catalogue to DSpace and from the SIGARRA system to DSpace, in order to reuse bibliographic records.

To guarantee that all required record data were exported from Aleph to DSpace we have designed a coding schema for the collections and used it on the UNIMARC records in Aleph, where CCL commands then allow us to select records for export. The resulting XML document with chosen UNIMARC fields is mapped to Dublin Core format [5] using a custom-designed XSL. The export process is controlled by publication date on the bibliographic catalogue.

The export procedure from SIGARRA, the university information system, to DSpace, currently under development, will be supported on procedures designed for the university repository.

The third problem mentioned above, regarding the design of a digital library interface in DSpace, is still being addressed and will require several development and test steps using the intended target audience.

Since the digital libraries will accommodate mostly reserved and ancient collections we decided to publish documents with image resolution of 200 dpi in JPEG format, keeping restricted access to files using 600 dpi TIFF format for preservation purposes. Depending on their condition, the documents will be either digitized or photographed. For text documents, a TIFF file will be used for preservation and full-text will be extracted for indexing if available.

4   Prototype Digital Libraries

DSpace organizes repositories by communities that can be used to represent the units in an organization- departments, libraries, museums. Collections are organized for our communities according to themes or document types. The diagram in Figure 2 illustrates the community and collection structure of the two case studies in DSpace.

Since the Fine Arts collection has documents of different sources we had to consider organizing the communities by services, and collections by document type. The archival cataloguing rules typically use a hierarchical structure capturing document provenance [6]. We created two sub-communities, archive and library, and three sub-communities within the archive community (Portuense Academy of Fine Arts, School of the Arts and Portuense Museum). For the library community there was no need to create sub-communities.

The nature of the prototype collection for the Food and Nutrition Digital Library, which is composed by different types of work, from scientific articles and books to newspaper and magazine articles, lecture presentations and didactic material, justifies a functional organization of the digital library in different areas featured as communities of action for faculty members: Research; Teaching and Learning; Extension and

Training; Media Communication. These main action fields are functionally viewed as dynamic because some documents types and materials can be of interest to different audiences and used in different contexts by the professionals. Generically, the collection organization tries to respond to the uneven goal of disseminating science for the public interest: from original scientific materials to the local newspaper.

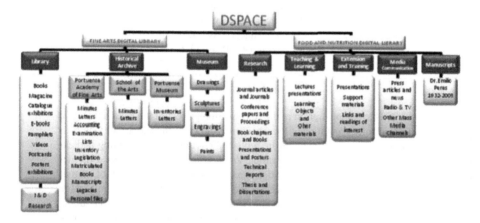

Fig. 2. Communities and collections for the two case studies

References

1. Ribeiro, L.M., David, Gabriel, Azevedo, Ana, Santos, dos Marques, J.C.: Developing an information system at the Engineering Faculty of Porto University. FEUP, Porto (1997), http://repositorio.up.pt/aberto/handle/10216/606
2. Smith, M., Bass, M., McClellan, G., Tansley, R., Barton, M., Branschofsky, M., Stuve, D., Walker, J.H.: DSpace: An Open Source Dynamic Digital Repository. D-Lib Magazine 9(1) (2003)
3. Pyrounakis, G., Saidis, K., Nikolaidou, M., Lourdi, I.: Designing an integrated digital library framework to support multiple heterogeneous collections. In: Heery, R., Lyon, L. (eds.) ECDL 2004. LNCS, vol. 3232, pp. 26–37. Springer, Heidelberg (2004)
4. Wise, M., Spiro, L., Henry, G., Byrd, S.: Expanding roles for the institutional repository. OCLC Systems & Services 23(2), 216–223 (2007)
5. Permanent UNIMARC Committee. UNIMARC/Dublin Core Mapping, http://www.unimarc.net/dubin-core-map.html
6. ISAD(G): General International Standard Archival Description, 2nd edn, http://www.ica.org/en/node/30000

Recollection: Integrating Data through Access

Laura E. Campbell

The Library of Congress
Office of Strategic Initiatives
101 Independence Ave SE
Washington, DC 20540-1310
lcam@loc.gov

Abstract. This demonstration of the Recollection project of the National Digital Information Infrastructure and Preservation Program at the Library of Congress will showcase a prototype platform, tools and environment for sharing and access to diverse born-digital collections. As the Program has addressed the development of distributed preservation through a national community of partner institutions, the challenge of access and interoperability has become more urgent. The network needs to be able to strategically bring collections under stewardship and keep an inventory without excessive burden on the collecting organizations. The data under stewardship is very diverse and follows standards acceptable within each content domain. These circumstances require an infrastructure that enables the community of NDIIPP Partners to share their collections and data on an ongoing basis. This allows NDIIPP to maintain the benefits of a distributed network of partners and also take advantage of the collections speaking to one another.

Keywords: User Interfaces, Interoperability and Data Integration, Digital Archiving and Preservation, Collection Development and Management and Policies, Knowledge Organization Systems, Semantic Web Issues in Digital Libraries.

1 Description

The National Digital Information Infrastructure and Preservation Program at the Library of Congress is an initiative to develop a national strategy to collect, archive and preserve digital content for current and future generations. It is based on an understanding that digital stewardship on a national scale depends on active cooperation between communities. The NDIIPP network of partners have collected a diverse array of digital content, including social science datasets; geospatial information; websites and blogs; e-journals; audiovisual materials; and digital government records [1].

These diverse collections are held in the dispersed repositories and archival systems of over 130 partner institutions where each organization collects, manages, and stores at-risk digital content according to what is most suitable for the industry or domain that it serves. This practice is necessary in a federated network of heterogeneous infrastructures but creates challenges in providing meaningful access across collections. However, it is clear that digital content grows in value exponentially as it is

M. Agosti et al. (Eds.): ECDL 2009, LNCS 5714, pp. 396–397, 2009.

integrated and interconnected. As the Library of Congress and its partners develop a framework for a national digital collection, they have recognized a requirement to share and integrate partner collections in the interest of a coherent strategy.

The Recollection platform uses semantic technologies to enhance discoverable access for NDIIPP collections, making them easier to find, access, and share, and especially to integrate with other digital information sources. The demonstration data comes from a variety of sources and formats including Excel spreadsheets, databases, and XML datasets. Data guidelines focusing on key access points are shared and an open interface enables third parties to plug services and applications into the Recollection framework, encouraging community participation [2].

2 Demonstration Outline

The demonstration will include a live website demonstration of the Recollection tool platform. The demonstration will show how data is loaded and managed in the system, how data managers can select data attributes and create tag clouds, timelines, maps and graphs to provide access to complex digital content such as geospatial data, social science datasets, and web archives. It will demonstrate how data can be augmented using web services.   Technical requirements are: a computer connected to a projector, PowerPoint software, and an Internet connection. Browser requirement is the latest version of Mozilla Firefox.

3 Target Audience

This demonstration of Recollection, an interactive tool platform to enhance access to preserved digital materials, is targeted toward digital collection curators, metadata managers, web service designers, digital archivists, and anyone interested in collaborative approaches to information services.

References

1. National Digital Information Infrastructure and Preservation Program,
   http://www.digitalpreservation.gov
2. Zepheira, http://zepheira.com/

A Gateway to the Knowledge in Taiwan's Digital Archives

Tzu-Yen Hsu, Ting-Hua Chen, Chung-Hsi Hung, and Keh-Jiann Chen

Research Center for Information Technology Innovation, Academia Sinica,
Taipei 115, Taiwan
{ciyan,james,johan,kchen}@iis.sinica.edu.tw

Abstract. Taiwan's digital archives cover a broad range of cultural and natural assets. More than 2 million objects have been accumulated since the project was launched in 2002. As the number and diversity of digitised objects have increased rapidly, it has become increasingly difficult for people to gain a clear picture of the contents of the archives. To disseminate the abundant and diverse resources to the public, we are building a knowledge structure that consists of categorized keywords extracted from objects' metadata, and developing a function called "Tagging Tool" to facilitate fast and efficient mining of resources. For example, users who want to read an article enriched with archive collections can utilize the tool to identify archive specific keywords in the text automatically and annotate them with references to relevant resources. As a result, users can save a great deal of time on keyword searching, and contextualize various entities, such as historical events and people's names.

Keywords: keyword extraction, text annotation, digital library.

1 Introduction

The goal of the Taiwan e-Learning & Digital Archives Program[1] (TELDAP), which was launched in 2002, is to permanently preserve assets, such as natural resources, historical artifacts, intellectual property, and spatial data, in digitized forms. To help users retrieve archive resources through an integrated interface, a portal website called Union Catalog[2] (UC) was established in 2005. Its main functions are full-text searching, category browsing, and keyword browsing.

When archive resources grow rapidly and diversely, users tend to favor hot shared links, keywords, or social tags; however, we estimate that an increasing number of resources are never, or seldom, viewed. Motivated by Wikipedia, the Semantic Web, and ontology sharing, we propose the concept of knowledge building to enhance archive utilization. On the basis of knowledge construction, objects or documents can be classified into appropriate categories and their relationships can be mapped. Once loosely organized resources become tightly

---

[1] Taiwan e-Learning & Digital Archives Program, http://teldap.tw/en/

[2] Union Catalog, Taiwan, http://catalog.digitalarchives.tw/

M. Agosti et al. (Eds.): ECDL 2009, LNCS 5714, pp. 398–401, 2009.

connected, it is possible to present users with more useful and meaningful resources and engage them in exploring other resources that are seldom viewed. We believe that through services, such as keyword detection and suggestions and recommendations about related resources based on knowledge construction, TELDAP's resources could be disseminated more widely and trigger more innovative add-on value. Initially, resource dissemination or knowledge sharing is performed by the Tagging Tool, which we describe in the following.

2   Knowledge Building

Our knowledge representation is inspired by the notion of ontology, which is frequently defined as an explicit specification of a conceptualization[1]. Basically, ontology is composed of "vocabulary for representing and communicating knowledge about some topic and a set of relationships that hold among the terms in that vocabulary[2]." For this reason, our knowledge model consists of categorized/hierarchical keywords, which are derived from given keywords and also extracted from the titles and descriptions of resource metadata. The steps of the knowledge building process are as follows.

- **Preprocessing:** Our data source consists of resource metadata, which describes resources based on Dublin Core's 15 elements[3] . Usually, data inside metadata is expressed as pure text; however, in our case some redundant information exists, for example, punctuation, HTML markups, and other information decoration. Therefore, we need to filter out such information in order to obtain complete sequences of Chinese characters.
- **Tokenization:** This step converts a sequence of Chinese characters into a group of tokens, which are considered meaningful units or lexicons, so as to facilitate further part-of-speech analysis and lexicon statistics. However, tokenization in Chinese does not have clear separation rules like the spaces and periods in English, and consecutive features make lexicon recognition difficult. For example, a four-character segment could be analyzed as a lexicon or recognized as two separate two-character lexicons. It is almost impossible to implement the Chinese tokenization process solely by pure logical computing. Hence, we developed a word segmentation system[4] to complete this task by using a rich lexicon database.
- **Keyword selection:** This step decides if a token should become a TELDAP keyword or if it should be abandoned. We have tried several algorithms to generate a better result at an acceptable cost. Currently, a token needs to satisfy three criteria to become a valid keyword. First, the token must be a noun type. Second, it must have a high frequency in TELDAP texts, but a low frequency in a general corpus. This means that it is not a general term, but it is relevant to TELDAP. Third, a semi-manual review decides whether the token becomes a valid keyword based on its original metadata.

---

[3] Dublin Core Metadata Element Set, http://dublincore.org/documents/dces/
[4] Word segmentation system, http://ckipsvr.iis.sinica.edu.tw/

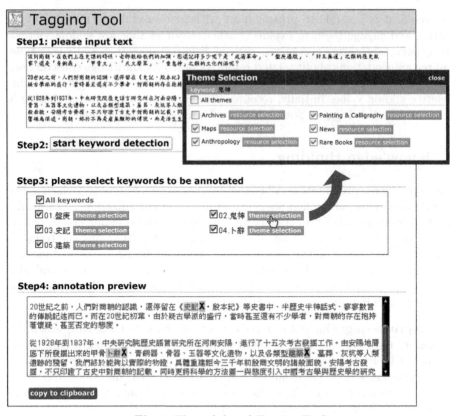

Fig. 1. The web-based Tagging Tool

There is an important task at the end of this phase. Specifically, all qualified keywords need to be added to the lexicon database to enable the subsequent tokenization process to recognize those keywords.

– **Keyword classification:** A keyword might easily relate to several themes; however, such keyword ambiguity could reduce the quality of the search results and force users to sort search lists themselves. The metadata of each resource contains at least one record of theme classification assigned by the academic institute or organization that owns the actual resource; therefore, we can categorize valid keywords into main themes based on the corresponding resources. A keyword displayed by classified resources can help users retrieve information rapidly from related categories.

3   Tagging Tool

Based on the knowledge structure, the Tagging Tool's key functions are TELDAP-specific keyword detection and selective keyword and resource annotation. The friendly step-by-step user interface is shown in Fig.1. Initially, users can input or paste any text that they want to enrich with links to TELDAP's resources, and start the keyword detection process (Fig.1, step2), which completes text tokenization and keyword matching between text tokens and the TELDAP keyword database. Second, users can select keywords, themes, and resources relevant to the input text (Fig.1, step3) or just leave them as default settings. Finally, users can preview the annotated text and decide if each keyword is appropriate for their needs (Fig.1, step4).

The current Tagging Tool, shown in Fig.1, requires active users to decorate their articles with links to TELDAP's resources step by step. However, the user interface may not be straightforward for some users, especially those who want to get background information immediately. Therefore, it would be better if readers can have an alternative means of using Tagging Tool in a form like Google widgets or Internet browser add-ons. Obviously, the way users interact with the Tagging Tool can be developed further.

4 Future Work

The Tagging Tool is just a beginning to examine and demonstrate the potential of the knowledge structure, which still needs a more definite and transparent building process to avoid wrong calculations and incomplete thinking. After we gain more user feedback to refine keywords and the connections between keywords and resources, the TELDAP keyword repository can be shared with intended applications and institutes. It is an effective means of disseminating TELDAP's resources more widely. Moreover, a valuable keyword source linked to the rich archive collections could quite possibly inspire further innovation.

Acknowledgement

This work was supported by the National Science Council of Taiwan under Grant No. NSC98-2631-H-001-008.

References

1. Gruber, T.: A translation approach to portable ontologies. Knowledge Acquisition 5(2), 199–220 (1993)
2. Gruber, T.: What is an ontology, http://www-ksl-svc.stanford.edu:5915/doc/frame-editor/what-is-an-ontology.html

Developing a Digital Libraries Master's Programme

Elena Macevičiūtė, Tom Wilson, and Helena Francke

Swedish School of Library and Information Science, University of Borås,
Allégatan 1, Borås, 50190, Sweden
{elena.maceviciute,wilsontd}@gmail.com, helena.francke@hb.se
http://www.hb.se/

Abstract. The changes in Swedish education following the Bologna requirements resulted in the first Master's programmes in Library and Information
Science. Two of them target information professionals working with digital
resources and services and seeking to develop and update their knowledge. One
programme is oriented to foreign students from all over the world, another to
Swedish students. A Venn diagram illustrating the relationships among the elements of LIS was used to develop the curriculum for the international Master's
programmes in Digital Libraries and Information Services. As this programme
is delivered in distance learning mode there was a need to find ways of organizing the study process and deliver the study materials so that it suited this mode
of education. The poster describes the content and design of the programme, as
well as student reactions.

Keywords: Digital libraries, curriculum, Master's programme, Sweden, international students.

1 Background

The 'Bologna process', whereby higher education in Europe is coordinated within a
common structure of degrees, is opening up opportunities for the development of new
programmes. In Sweden, the existing structure of the Magister degree (a four-year
undergraduate programme) was such that the development of specialisations was
rather difficult. When the opportunity arose, the Swedish School of Library and
Information Science (SSLIS - a joint institute of the University of Borås and Göteborg
University), started developing both undergraduate and Master's programmes.

The School had offered distance learning versions of its Magister degree for many
years and it was felt that a programme aimed at the further development of practising
librarians would be most likely to attract students if offered in the distance mode.
There was also a desire on the part of the University of Borås to attract more
participation from overseas and the distance mode, particularly using e-learning,
offered the possibility of attracting students from almost anywhere in the world.

One of the first programmes to be proposed under this framework is the Master's
in Library and Information Science: direction Digital Library and Information
Services. It is taught in English and offered through the University's e-learning
platform, Ping Pong.

M. Agosti et al. (Eds.): ECDL 2009, LNCS 5714, pp. 402–407, 2009.
© Springer-Verlag Berlin Heidelberg 2009

The programme was under development from 2004. The work started with a broad investigation of the competence requirements for this new area in Sweden [1] and with identifying the similar programmes developed elsewhere. As soon as the Digital Library Curriculum Project (http://curric.dlib.vt.edu/) was initiated by the University of North Carolina, Chapel Hill and Virginia Tech, SSLIS could also draw on their expertise, including evaluation of the programme presented here. SSLIS also followed the development of the ERASMUS MUNDUS project on International Digital Library Learning (DILL) that is run by Oslo University College, University of Parma and Tallinn University (http://dill.hio.no/).

In addition to the English language programme directed to a broad international community, SSLIS started working on a master's programme in Digital Services – Culture, Information & Communication for Swedish students. What specifically characterises this work is the close collaboration between SSLIS and a number of partner organizations in the library field, including the National Library of Sweden, the Gothenburg municipal library, BTJ AB (a leading supplier of media services), the University of Borås library, and the Swedish Library Association. Collaboration will continue, and be broadened to additional organisations, when the programme is up and running. The strong element of collaboration with external organizations is also the reason why the planning of the programme receives generous support from the Swedish Knowledge Foundation.

This text introduces the elements of the SSLIS Master's programmes in Digital Libraries and Services.

2   Digital Library Elements: Programme Content

Libraries today are predominantly 'hybrid' libraries, with traditional storage of print materials and, increasingly, access to digital resources. These resources may be held by the library itself or may be freely accessible resources from elsewhere, or access provided by subscription to, for example, encyclopaedias and e-journals. Academic libraries are also involved in the development of institutional repositories for the storage and retrieval of the research outputs of academic members of staff. Some other institutions concentrate of digitization of cultural heritage.

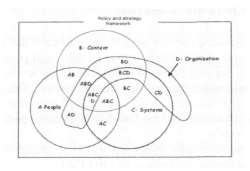

Fig. 1. Venn diagram for library and information work [2]

The management of this diverse array of resources, therefore, calls for a variety of skills, ranging from the technical skills required for the digitization of the library's own resources (very often special collections of rare materials); to the negotiation skills required in managing subscriptions, or the financial management skills needed to run a digitization unit.

In breaking down the broad area, we drew upon Wilson's [2] Venn diagram of the scope of librarianship (see Figure 1). Within the English language Master's programme the managerial orientation is the main focus because the programme is directed to practicing librarians and information specialists. The programme seeks to educate a broad competence required for management of digital resources and services in a variety of institutions. Consequently, the irregular area labelled D in the diagram constitutes the core, since it identifies the organizational framework within which *people* are both served and (in the case of staff) managed, where *content* is organized, digitized and made available for use, and where *systems* are developed and employed to manage the digitized information resources. Space does not allow a detailed interpretation of the diagram beyond this, but the relationship of the curriculum to the diagram should become clear in the following sections. The Swedish programme goes beyond the D area and is directed towards more general areas marked as content (mainly, cultural heritage), systems, and framework.

The knowledge and competence developed by these Master's programmes include not only managerial capabilities, but also socio-cultural and institutional issues, theoretical and practical issues of digitization processes and information retrieval for digital libraries (including hands-on experience of project planning, mark-up languages, image editing, database management, etc.) [3]. A special effort is made to provide strong understanding of information behaviour and seeking in digital environments, human-computer interaction, digital service economics and evaluation approaches. These were competences named by the initial survey participants [1].

Each teacher involved in the project is building his or her course on the latest research on digital libraries (e.g., DELOS reference model for DLMS), and as a rule conducts related research within SSLIS (e.g., for a doctoral dissertation in digitization, open access, etc. or in a project like Sustaining Heritage Access through Multivalent ArchiviNg SHAMAN, etc.).

3 Programme Elements: Modules

The international programme consists of six themes, whereas the Swedish programme allows for more elective courses. Both include an individual project to be undertaken by a student in his or her own library or in one for which access is available, which then can become an intrinsic part of the Master's thesis. The content of the two programmes is slightly different, with the international programme focusing more on managerial issues and libraries, and the Swedish programme putting the emphasis on culture and communication.

The list of courses is shown below in Table 1, including the ECTS. Both programmes meet the European 'standard' of 120 ECTS.

The relations of the courses with broad areas within the Venn diagram (Fig. 1) are shown by the capital letters in the brackets for each item.

Table 1. Courses in SSLIS Master's programmes for Digital Libraries

International Master's programme LIS: direction Digital Libraries and Information Services	Master's programme LIS: direction Digital Services – Culture, Information and Communication
1. Digital library management (15 ECTS) (BCD in Venn diagram)	1. Digital media in the culture and information sectors (7,5 ECTS) (BC)
2. Users and information practices in digital environments (15 ECTS) (ABC)	2. Technology of digital libraries I (7,5 ECTS) (C)
3. Information retrieval for digital libraries, I and II (15 ECTS) (BC)	3. Users and information practices in digital environments (15 ECTS) (ABC)
4. Technology of digital libraries I and II (15 ECTS) (C)	4. Project management (7,5 ECTS) (BCD)
5. Digitisation and digital preservation (15 ECTS) (BCD)	5. Elective courses (22,5 ECTS)
6. Digital library research methods (15 ECTS)(Outside the diagram)	6. R & D methods within digital services (15 ECTS)(Outside the diagram)
7. Project work practicum (7.5 ECTS) (ABCD)	7. Work place-related project (15 ECTS) (ABCD)
8. Master's thesis (22.5 ECTS) (ABCD)	8. Master's thesis (30 ECTS) (ABCD)

Both programmes are starting in autumn 2009, however, the international Master's programme has already piloted several courses in the distance mode in 2008 and in spring of 2009 for international students from Europe, Asia, Australia and Africa.

4   Content Delivery

This part includes examples of the course pages developed for the international Master's programme.

As noted earlier, the programme uses the Ping Pong virtual learning environment. The screen-shot below shows the initial page for one of the courses - *Users and information practices in digital environments*.

Fig. 2. Initial page in a course of the Programme (e-learning platform Ping Pong)

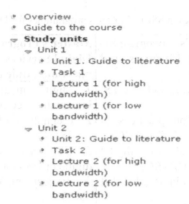

Fig. 3. A side bar showing the types of teaching materials

The 'study units' are the individual units of the course and each study unit consists of a number of elements, as shown in the next screen-shot.

The programme consists not only of traditional documentary learning materials, but also includes videotaped lectures. Ping Pong also includes facilities for discussion among students and with teachers and a variety of other virtual learning functions. The teachers are using a combination of word-processed files and Web pages. The following screen-shot shows one of the Web pages written for the unit on managing digital libraries.

Web pages include a guide to the literature, materials for discussions, links to relevant Web sites and lists of literature with an indication of where to find it (the links to the texts that are in the public domain are embedded).

The programme also includes two residential periods each year. This provides an opportunity for the students to meet their classmates and teachers, to attend lectures, to engage in face-to-face discussions and to experience hands-on learning in the classroom. The latter is an essential part in the courses related to technology and information retrieval.

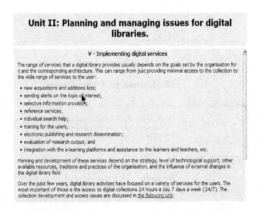

Fig. 4. A Web page in a course

5   Student Reactions

Most of the students who have studied two courses in the international programme have succeeded in completing the first one and, at the time of writing, are studying the third unit of the second course. They have noted the logical link between the courses. Different ways in which the course materials are delivered in both courses help them to stay motivated and stimulate their interest in the process of study.

Students have deemed the schedule for the delivery of the tasks to be demanding but realistic in terms of planned reading and working. It has also been appreciated that the deadlines, though set, are not cast in iron and students have felt that the schedule has been reasonably flexible.

The students have appreciated the service of the University Library and Learning Resources as the free access to the databases is very important to them. The provision of the lecture texts on the platform in Word and PowerPoint formats has been regarded as a feature of usability, especially by those who have to pay for the network connection time. The use of WebCite [4] for the online material is useful as it helps to reduce the problem of broken links for the students (and teachers as well).

References

1. Maceviciute, E.: Edukacja dla potrzeb bibliotek cyfrowych i informacji. In: M. Kocójowa (ed.), Profesjonalna informacja w internecie, pp. 26--34. Wydawnictwo universitetu Jagiellońskiego, Kraków (2005)
2. Wilson, T.D.: Mapping the curriculum in information studies. New Library World 102(11/12), 436–442 (2001)
3. Dahlström, M., Doracic, A.: Digitization education: courses taken and lessons learned. D-Lib Magazine 15(¾) (2009),
   http://www.dlib.org/dlib/march09/dahlstrom/03dahlstrom.html
4. WebCite, http://www.webcitation.org/

The DiSCmap Project: Overview and First Results

Duncan Birrell[1], Milena Dobreva[1,3], Gordon Dunsire[1],
Jillian Griffiths[2], Richard Hartley[2], and Kathleen Menzies[1]

[1] Centre for Digital Library Research (CDLR), Information Resources Directorate (IRD),
University of Strathclyde, Livingstone Tower, 26 Richmond Street Glasgow,
G1 1XH United Kingdom
{duncan.birrell,milena.dobreva,g.dunsire}strath.ac.uk,
klmenzies@cis.srtath.ac.uk
[2] Centre for Research in Library and Information Management (CERLIM),
Department of Information and Communications, Manchester Metropolitan University,
Geoffrey Manton Building, Manchester M15 6LL, United Kingdom
{R.J.Hartley,J.R.Griffiths}@mmu.ac.uk
[3] Institute of Mathematics and Informatics, bl. 8, Acad. G. Bonchev St.,
1113 Sofia, Bulgaria

Abstract. Traditionally, digitisation of cultural and scientific heritage material
for use by the scholarly community has been led by supply rather than demand.
The DiSCmap project commissioned by JISC in 2008, aimed to study what re-
focussing of digitisation efforts will suit best the users of digitised materials,
especially in the context of the research and teaching in the higher education
institutions in the UK. The paper presents some of its initial outcomes based on
quantitative and qualitative analysis of 945 special collections nominated for
digitisation by intermediary users (librarians, archivist and museum curators),
as well as end users' study involving a combination of online survey, focus
groups and in-depth interviews. The criteria for prioritising digitisation
advanced by intermediaries and end users were analysed and cross-mapped
to a range of existing digitisation frameworks. A user-driven prioritisation
framework which synthesises the findings of the project is presented.

Keywords: selection, appraisal, user-defined criteria, digitization, special
collections.

1 Introduction

The DiSCmap project (Digitisation of Special Collections: mapping, assessment,
prioritisation) was commissioned by JISC[1] in 2008 to CDLR. The work on the project
has been completed between September 2008 and May 2009 by CDLR and CERLIM
(The Centre for Research in Library and Information Management) at the Manchester
Metropolitan University. The project had as its primary goal to study the user needs in
digitised special collections in the higher education institutions in the UK.

---

[1] Joint Information Systems Committee (JISC), http://www.jisc.ac.uk/

M. Agosti et al. (Eds.): ECDL 2009, LNCS 5714, pp. 408–411, 2009.

Traditionally, digitisation of cultural and scientific heritage material for use by the scholarly community has been led by supply rather than demand. JISC's recent Digitisation Strategy [3], however, makes clear their commitment to re-focussing digitisation efforts to make them most valuable to direct users of digitised materials, including researchers, teachers and students.

The project was constructed as a set of inter-connected tasks aimed at assessing the current landscape of digitisation of special collections from the point of view of the needs of the researcher and teachers within UK higher education institutions. It included the following basic components:

- Mapping and assessing existing digitised special collections in UK higher education institutions;
- Identifying and summarising best practice in digitisation within the public and private sector and both nationally and internationally;
- Preparing an inventory, via interaction with librarians, archivists and museum curators, of collections as-yet not digitised;
- Constructing a framework of criteria for the assessment of the potential value and impact in the digitisation of individual collections;
- Using innovative methods, including social networking and participatory workshops, to engage and talk directly to direct users in order to discern their views on current and future digitisation needs.

2   First Outcomes from DiSCmap

DiSCmap collected data on 945 collections nominated for digitisation by intermediary users (librarian, archivists and museum curators) from over half of the 196 higher education institutions in the UK. Over half of the collections (54%) came from the libraries in higher education institutions, 38% are archival collections, 7% are museum collections and 1% - departmental collections.

The implementation of DiSCmap was organised seeking to make its results:

- Representative (by a fair UK wide regional distribution).
- Non-hierarchical (includes Ancient, Old, Redbrick, Post-Robbins and New/Post-1992 higher education institutions).
- Granular (by surveying both intermediaries and end users).
- Functional (deliver the resources users want and need).

The online survey helped to collect evidence on the reasons for nomination of the various collections and this allows understanding better how intermediary users perceive the impact of digitised collections on research and teaching. In parallel, the project studied the direct users' views on anticipated impact of digitised special collections through a combination of web survey, focus groups and in-depth interviews. The interviews with end users showed a different set of criteria for advancing special collections for digitisation. The nominated criteria by both groups of users were cross-mapped and compared with other existing frameworks suggested by a number of projects and previous research publications [1,2, 4 – 8].

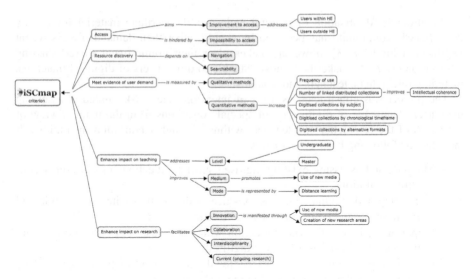

Fig. 1. A concept map of DiSCmap user-driven prioritisation framework

The synthesis of digitisation cases suggested by intermediaries and of the reasons for digitisation by advanced by end users is presented on the concept map below (see Fig. 1).

The distribution of collections across subject domains, regions and types of materials, languages of written materials combined with nominated criteria allows constructing short lists of collections matching different sets of prioritisation criteria. Although the project addressed the needs of higher education institutions in the UK, the comparison of criteria for digitisation with other existing frameworks is of interest to the digital library community. In addition DiSCmap gathered rich evidence on criteria advanced by intermediary and end-users and the differences in their points of view on priorities in digitisation.

3   Conclusions

Based on the analysis of the data gathered, the project produced the following recommendations:

1. The long list of collections should be harmonized and sustained into the future.
2. The user-driven framework developed by DiSCmap can be seen as a tool to support a flexible approach to the prioritising digitisation of special collections.
3. A comprehensive collection description and finding utility is needed in the UK.
4. Granularity issues of collection description facilities need to be revisited.
5. Metadata issues for collection level description need to be better addressed.
6. A stronger connection should be established with the actual use of digitised resources in the wider context of research/learning/entertainment.

7. Information literacy related to resources presenting collections can be further enhanced.
8. Further work can be done on the impact of "to-be" digitised resources (qualitative and quantitative methods).

DiSCmap has analysed a comprehensive range of end user digitisation priorities that are directly related to teaching and research. In doing so it has made considerable advances in identifying and understanding *the actual digitisation needs of the scholarly community*. It has done so with the aim of removing the element of guesswork and assumption hitherto inherent in our understanding of user requirements in this area. Additionally, its combination of intermediary and end user studies provides a richness of view points which highlight the many important and differing aspects related to the end user dimension in digitisation.

References

1. Ayris, P.: Guidance for selecting materials for digitisation. In: Joint RLG and NPO Preservation Conference: Guidelines for Digital Imaging, September 28-30, 1998, Warwick, UK, (unpublished) (1998), http://eprints.ucl.ac.uk/492/1/paul_ayris3.pdf
2. Cornell University Library. Selecting traditional library materials for digitization. Report of the CUL task force on digitization (2005), http://www.library.cornell.edu/colldev/digitalselection.html
3. JISC digitisation strategy (2008), http://www.jisc.ac.uk/media/documents/programmes/digitisation/jisc_digitisation_strategy_2008.doc/
4. MINERVA Working Group 6 (2004). Good practices handbook, p.21. DIGIT STAG report (2002), http://www.minervaeurope.org/structure/workinggroups/goodpract/document/goodpractices1_3.pdf/
5. National Digital Forum. Digitisation selection work: Position Paper (2007), http://ndf.natlib.govt.nz/downloads/NDF%20Digitisation%20Selection%20Work%20edited.pdf
6. National Library of Australia Collection digitisation policy (2008), http://www.nla.gov.au/policy/digitisation.html
7. Report of the meeting of the digitalization of natural history collections STAG of GBIF (2002), http://circa.gbif.net/Public/irc/gbif/digit/library?l=/meetings/digit_stag_meeting&vm=detailed&sb=Title
8. Ross, S.: Strategies for selecting resources for digitization: source-orientated, user-driven, asset-aware model (SOUDAAM). In: Coppock, T. (ed.) Making information available in digital format: perspectives from practitioners, pp. 5–27. The Stationery Office Ltd., Edinburgh (1999)

Securing the iRODS Metadata Catalog for Digital Preservation

Gonçalo Antunes[1] and José Barateiro[2]

[1] INESC-ID, Information Systems Group, Lisbon, Portugal
[2] LNEC - Laboratório Nacional de Engenharia Civil, Lisbon, Portugal
goncalo.antunes@tagus.ist.utl.pt, jbarateiro@lnec.pt

Abstract. Digital preservation is the ability to retrieve, access, and use digital objects through time, while ensuring the authenticity and integrity properties of these objects. Data grids represent a model of storage systems designed for data management and sharing, which concept also has been proposed for digital preservation. However, since data grids are not specifically designed for this purpose, they present weaknesses that have to be handled. This poster will present a set of services to address a problem in the metadata catalogue of the iRODS data grid, strengthening that platform for digital preservation purposes.

Keywords: Digital Libraries, Digital Preservation, Data Grids.

1 Introduction

The creation of OAIS by the CCSDS [1] acknowledges the importance of the preservation of scientific data to scientific research (*e-Science*), where research may span several decades and produce large quantities of data that is crucial to preserve. Moreover, preservation of scientific data may be mandatory by law, as it is the case of the Portuguese Dam Safety Legislation [2], which establishes that the National Laboratory of Civil Engineering of Portugal[1] is responsible of keeping an updated archive of data concerning dam safety.

Data grids offer distributed computation and storage of massive data sets, and allow sharing and collaboration, which are crucial for *e-Science*. iRODS[2], is an open-source data grid system that has been proposed for digital preservation purposes [3]. An important component of the iRODS data grid system is the metadata catalogue, which maintains crucial information like the physical location of files and nodes, state of operations and processes, etc. This metadata catalog is supported by a centralized database system, which represents a point of failure when used in the context of a preservation scenario. If a failure, disaster or intentional attack affects the metadata catalog, it can cause a total data loss, even if the data stored on the other nodes remain intact. Previous work summarizes the threats that can affect the metadata catalogue [4].

In order to reduce the risk of threats to the iRODS metadata catalog, we propose the following set of extensions as new services: (*i*) a Backup Service, which exports

---

[1] http:///www.lnec.pt
[2] http://www.irods.org

M. Agosti et al. (Eds.): ECDL 2009, LNCS 5714, pp. 412–415, 2009.

the contents of the metadata catalogue database to a structure of XML files and repli-cates them to other nodes of the data grid; and (*ii*) a Recovery Service, a service to recover the metadata catalogue database in case of failure.

This work has been done in the scope of two projects: the international project SHAMAN[3], and the national project GRITO[4].

2   The Backup Service

The backup service is intended to be sensitive to aspects such as the workload of the system, in order to not interfere with the normal functioning of other regular proc-esses that might be running on the system.

Accordingly, the measure of the current workload conditions has to take place early in the process. We consider the CPU and memory used at a determined time by the node running the backup service as a workload measure. To make the backup policy less intrusive we took the decision of implementing an incremental backup policy. Figure 1 presents the UML activity diagram of replicating the metadata catalogue to other nodes in the data grid.

The process starts with the evaluation of the conditions for copy which assesses the workload of the node running the backup process (the central node) by measuring the CPU and memory in use. If the conditions are met, the process proceeds with the evaluation of the type of backup that must be done. If the conditions are not met, a log is registered (to notify the administrator) and a new backup process is scheduled. If the process is running for the first time, a full backup of the catalogue is made, includ-ing the schema of the database. The actual policy determines that the replicas

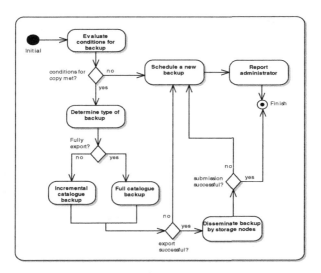

Fig. 1. High-level activity diagram of the Backup service

---

[3] http://www.shaman-ip.eu
[4] http://grito.intraneia.com

are disseminated through all of the nodes in the system using built-in replication features. Along with those replicas, a list containing the physical path and hostname of each of the replicas is created in a folder in the host file system, therefore outside the grid virtual system. The reason for this is that, in case a disaster affects the metadata catalogue, the metadata elements with the location of the recovery files will not be lost. So, by creating this list with the recovery files, we can bypass the catalogue and recover from the disaster.

We could also replicate the recovery files outside the grid virtual system, but using the grid to store the files we can use the replication mechanisms it provides and so ensure the preservation of the backups by applying to them to the same preservation policies as for the other preserved objects (including the audit processes).

3 The Recovery Service

The recovery service is illustrated in Figure 2. After setting up a new central node, the administrator can initiate the recovery process by specifying the location (full path) of the backup lists and the hostname of one of the nodes containing the backup files.

If the backup files do not exist (they might had been affected by the failure or disaster), the process fails and the administrator has the chance to give another location path or host. If the backup files exist, the service accesses the list and retrieves the most recent full backup replica, parsing the document from the beginning to the end. When it finds the most recent backup file, it tries to access its first listed replica. If none of the replicas of the full backup is accessible, the process fails. Otherwise, the process continues with the recovery until it concludes.

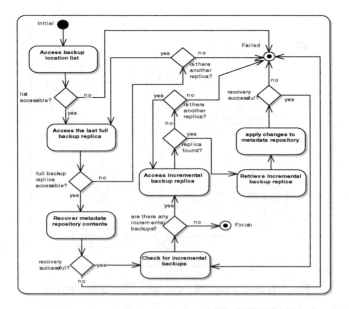

Fig. 2. High-level activity diagram of the Recovery service

After this full recovery, the process checks for incremental backups. As the incremental backups stand for the history of the changes applied to the records in the catalogue, the recovery process applies the changes to the catalogue in the correct order until it reaches the state prior to the failure.

4 Conclusion

In this poster, we present a set of services with the objective of better adapting the iRODS data grid system to preservation purposes. The developed services may present a conceptual solution to some of the problems that may arise from the use of a data grid for preservation, independently of its implementation details, as long as the grid makes use of a metadata catalogue.

We implemented the proposed services using the extensibility properties of the iRODS system, as it allows the addition of new functionalities through the development of micro-services, which are small and well-defined functions/procedures (written in the C language), to execute a determined micro-level task. Users and administrators can chain those micro-services in order to create Actions (macro-level functionalities), also named Rules. The two services were implemented through the definition of Rules.

Future work will consist in testing and refining the developed services so that the gap that separates data grids from digital preservation is filled.

References

1. CCSDS: Reference Model for an Open Archival Information System (OAIS) - Blue Book. National Aeronautics and Space Administration (2002)
2. RSB. Dam safety regulation, decreto-lei (344), Diário da República, Lisbon (October 15, 2007) (in Portuguese)
3. Hedges, M., Hasan, A., Blanke, T.: Management and preservation of research data with iRODS. In: Proceedings of the ACM First Workshop on Cyberinfrastructure: information Management in eScience, CIMS 2007, Lisbon, Portugal, November 09, 2007, pp. 17–22. ACM, New York (2007)
4. Barateiro, J., Antunes, G., Cabral, M., Borbinha, J., Rodrigues, R.: Using a Grid for Digital Preservation. In: Buchanan, G., Masoodian, M., Cunningham, S.J. (eds.) ICADL 2008. LNCS, vol. 5362, pp. 225–235. Springer, Heidelberg (2008)

JSTOR - Data for Research

John Burns, Alan Brenner, Keith Kiser, Michael Krot, Clare Llewellyn,
and Ronald Snyder

301 E. Liberty, Ste. 330,
Ann Arbor, MI 48104
USA
{john.burns,alan.brenner,keith.kiser,michael.krot,
clare.llewllyn,ronald.snyder}@ithaka.org

Abstract. JSTOR is a not for profit organization dedicated to helping the scholarly community discover, use and build upon a large range of intellectual content in a trusted digital archive. JSTOR has created a new tool called "Data for Research" that allows users to interact with the corpus in new ways. Using DfR researchers can now explore the content visually, analyze the text and the references, and download complex datasets for offline analysis.

Keywords: JSTOR, text analysis, data, research, users, dataset, corpus.

1 Introduction

In pursuit of an overarching goal of technological innovation [1, 2], JSTOR is working in close collaboration with other researchers throughout the academic community to build new tools in order to enhance, explore, and allow more effective use of the content in the archive. This will bring significant additional value to the communities that JSTOR serves. Data for Research (DfR) [3] is a service that provides a web based visualization tool to explore and analyze the data within JSTOR and the ability to create, refine and download datasets of metadata, word frequency counts, references, n-grams and key words.

2 Visualizing JSTOR

A web interface allows users to explore content in the archive, discover features of interest, and ultimately generate downloadable datasets. Along with full-text searching of all content, the tool also provides faceting by discipline, journal, publisher, author, language, reviewed work/author, article type, the presence of references, time span or specific date. To assist in analysis of the data, graphs are employed to highlight key aspects about the slice of the JSTOR corpus the user has selected. These graphs include; frequency of articles returned by year, relative weight of articles returned adjusted to reflect total number of articles held in JSTOR by year, and distribution of articles across disciplines (Fig. 1). Furthermore, users are presented with total number of articles available in each facet for subsequent dataset refinement.

M. Agosti et al. (Eds.): ECDL 2009, LNCS 5714, pp. 416–419, 2009.

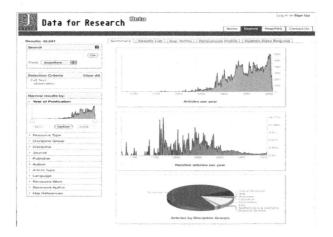

Fig. 1. Data for Research Summary view with a full text search for the word "observatory"

In addition to graphing tools, users have access to individual article metadata, key terms, references and n-grams. Key terms for the entire dataset (generated using TF/IDF [4]) are displayed as tag cloud where word size is in relative proportion to term weight. The tool also provides a dataset level view of references with a textual summary of references in and charts that show average number of references and average age of references by year (Fig. 2).

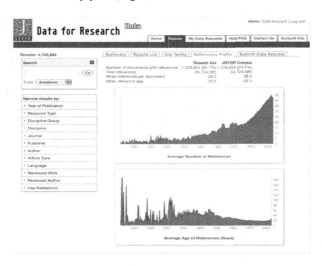

Fig. 2. Data for Research Summary Reference View for the entire archive

3   Creating a Dataset

Once users have defined a dataset they can choose to download it one of several data types (Table 1) in either CSV or XML formats. After the dataset is generated, users receive an email with a link that allows them to download the files and perform offline analysis.

Table 1. Available Data Types

Type	Description
Word Counts	Word occurrence frequency counts
Citations	Basic citation metadata
Bigrams	Two-word sequences that occur consecutively
Trigrams	Three-word sequences that occur consecutively
Quadgrams	Four-word sequences that occur consecutively
Key Terms	Auto extracted key terms (TF-IDF)
References	A list of works cited by articles in the data set

Optionally, users interested in programmatic access can use a combination of SRU (Search and Retrieve via URL) [5] and CQL (Context Query Language) [6] to download data directly. The DfR SRU service is found at http://dfr.jstor.org/sru. Pointing a browser at this address will produce a web form for defining CQL queries. The data returned by an SRU *searchRetrieve* query contains basic bibliographic data such as a unique document identifier, the document title, author names, publisher, and date of publication. The document identifier can then be used to download all data types from Table 1 in an XML format.

4 System Architecture

Various pieces of readily available, open-source software have been integrated atop the JSTOR corpus to form the Data for Research analysis framework. At the time of writing, JSTOR is home to nearly 5 million articles from over 1300 journals in over 50 languages spread across nearly 350 years. Given the very large amount of data, a chief requirement of any software was that it continues to perform at a large scale.

DfR employs the web framework Django for user account and permission management as well as the front-end interfaces [7]. Searching and faceting work by storing article data in a Lucene [8] index in conjunction with the search server Solr [9]. In addition to these, for speed and simplicity, a Berkeley Database [10] is used to store word counts and n-grams used in dataset production. Custom Java and Python code are also used throughout the data the application.

5 User Need

Throughout the years, JSTOR has received many requests for data from researchers from a wide variety of scholarly disciplines and with varying degrees of technical ability. Before the Data for Research tool, each individual request required a response to be hand crafted. Needless to say, this was inefficient, time consuming, and often resulted in long waiting periods for the end user. This experience informed development of the Data for Research tool.

Initial observations about users of the Data for Research site indeed confirm a large variety in our user base, many of whom have helped inform feature development. Since its inception, interest has been strong from all segments of the scholarly community. Further functionalities have been added when they were requested via feedback from the site. Linguists requested the n-gram functionality, various groups requested access to reference information, and the API was developed through close consultation with users from the University of California Digital Library who expressed an interest in having such a tool.

6 Example Analysis

Initial analysis of the JSTOR corpus using DfR has highlighted some interesting case studies that may provide a simple illustration of how the system can be used:

- The long s – the character found extensively in pre-1800 documents that looks like an "f" – rapidly dropped out of use after 1800.
- Use of the word "hath" dies out of general use around 1900. Hath is only used in language and literature and historical journals after 1900.
- Searching for the word *chymistry*, gives a line graph that demonstrates how usage of the word fading as time increases, until suddenly at the turn of this century there is a spike in usage. Analysis shows the occurrence of this word is in citations – modern scholars are using digitized works on the Internet in their research.

References

1. JSTOR, http://www.jstor.org
2. Schonfeld, R.: JSTOR a History. Princeton University Press, Princeton (2003)
3. Data for Research, http://dfr.jstor.org
4. Salton, G., Buckley, C.: Term-weighted approaches in automatic text retrieval. Information Processing & Management (1988)
5. SRU: Search/Retrieval via URL, http://www.loc.gov/standards/sru/
6. CQL: The Context Query Language, http://www.loc.gov/standards/sru/specs/cql.html
7. Django, http://djangoproject.com
8. Lucene, http://lucene.apache.org/
9. Solr, http://lucene.apache.org/solr/
10. Yadava, A.: The Berkeley DB Book. Apress (2007)

Improving Information Retrieval Effectiveness in Peer-to-Peer Networks through Query Piggybacking

Emanuele Di Buccio, Ivano Masiero, and Massimo Melucci

Department of Information Engineering, University of Padua, Italy
{dibuccio,masieroi,melo}@dei.unipd.it

Abstract. This work describes an algorithm which aims at increasing the quantity of relevant documents retrieved from a Peer-To-Peer (P2P) network. The algorithm is based on a statistical model used for ranking documents, peers and ultra-peers, and on a "piggybacking" technique performed when the query is routed across the network. The algorithm "amplifies" the statistical information about the neighborhood stored in each ultra-peer. The preliminary experiments provided encouraging results as the quantity of relevant documents retrieved through the network almost doubles once query piggybacking is exploited.

1 Introduction

One of the potentials of P2P networks when adopted as "a federated search layer for digital libraries" [1] is to allow each Digital Library (DL) node, namely each peer, to contribute to the document collections by pushing their own documents, yet without delivering the full content. Although P2P is a promising paradigm, it poses new daunting challenges: the main one is that the content to be searched is anarchically distributed across large networks of not necessarily cooperative peers. Such a little cooperation implies that each peer has got a little knowledge about the content of the other peers and, more specifically, it has got no knowledge about most peers. A suitable solution is unstructured networks built on top of different indexes which summarize groups of peers and not only the collections of the peers [1,3]. In particular, the P2P networks considered in this work are unstructured (i.e. no DHT-like data structures), hybrid (i.e. the simultaneous presence of peers and ultra-peers) and hierarchical (i.e. each peer refers to one and only one ultra-peer which serves a group of peers acting as a hub/router for queries sent by a user in its group). Figure 1 depicts an instance of this kind of P2P networks, where rounds represent ultra-peers and rhombi represent peers.

At searching time, ranking ultra-peers is necessary for selecting those groups to which peers storing most relevant documents belong. The specific algorithm adopted in this paper is the one proposed in [3] and implemented in the SPINA software architecture [2]; such algorithm is based on a statistical model used for ranking documents, peers and ultra-peers. In SPINA each ultra-peer maintains the information to rank its neighbors in its *ultra-peer index* – in this index, the

M. Agosti et al. (Eds.): ECDL 2009, LNCS 5714, pp. 420–424, 2009.
© Springer-Verlag Berlin Heidelberg 2009

elements in the posting list associated to each feature are the identifier of the neighbor and the weight of the feature in that neighbor. Because of its content, the ultra-peer index has "ultra-peer granularity". The set of *neighboring ultra-peers* or *neighbors* of an ultra-peer X are the ultra-peers whose aggregate statistics are maintained in the ultra-peer index of X. In Figure 1, an arc connecting X and Y denotes that Y is a neighbor of X. Let suppose a query is sent to ultra-peer E. At each hop, the query is forwarded to the m top-ranked neighbors. After two hops, ultra-peers H, I, L are possibly contacted (gray colored area) during the query routing. In this scenario, starting from E, if the ultra-peer A stores relevant documents, these will not be retrieved and recall decreases.

This paper describes an algorithm which aims at improving recall in similar scenarios. The algorithm is based on the framework proposed in [3] together with a "piggybacking" technique performed at query routing time. Although the use of piggybacking is a well known technique in computer networks for managing the messages carried at the lowest levels of the network protocols, to our knowledge, no research work reported the use of piggyback-

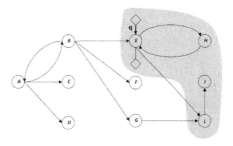

Fig. 1. Instance of unstructured hierarchical hybrid P2P network

ing for Information Retrieval (IR) across P2P networks at *retrieval level* in the way reported in this paper.

2   Improving Recall through Query Piggybacking

An ultra-peer Y is a *known ultra-peer* with respect to the ultra-peer X if X has just a partial knowledge of Y, that is, it stores a subset of the aggregate statistics of Y. This notion differs from the one of *neighboring ultra-peer* where a complete knowledge of the aggregate statistics about the ultra-peer is required. The aggregate statistics are stored in the Dynamic History Index (DHI) and refer to ultra-peers which are *not* in the neighborhood of X. Let suppose the user expresses his information need by submitting a query as a bag of features f_i's, e.g. keywords. At the beginning, the DHI of X is empty. When X receives the query, it selects some top-ranked neighbors. Then, the weight of each feature in the neighbors of X are *piggybacked* onto the message used to forward the query. The message is received by the neighbors which (i) forward the query to the top-ranked peers of the group governed by the ultra-peer and (ii) extract the piggybacked ranking information for each feature of the query. The features and the weights which do not refer to the neighbors are used to increment the DHI. Following this, each ultra-peer ranks the set of its *neighboring* and *known* ultra-peers searching across the ultra-peer index and the DHI as they were a single index. The process is then iterated for each contacted ultra-peer.

Table 1. Indexes and ranking steps in the ultra-peer A (Tab. 1a) and B (Tab. 1b); k_i is a feature in the indexes, and f_i is a feature of the submitted query

(a)

k_i	A	B	C	D		f_i	A	B	C	D
a	2	1	0	2		b	4	1	1	1
b	4	1	1	1		c	2	2	1	2
c	2	2	1	2		e	5	2	1	5
d	0	5	0	0		$\sum f_i$	11	5	3	8
e	5	2	1	5						

(b)

k_i	A	B	E	F	G		k_i	C	D		f_i	A	B	C	D	E	F	G
a	2	1	0	1	5		b	1	1		b	4	1	1	1	2	3	4
b	4	1	2	3	4		c	1	2		c	2	2	1	2	7	2	1
c	2	2	7	2	1		e	1	5		e	5	2	1	5	1	0	2
d	0	5	0	1	1		$\sum f_i$				$\sum f_i$	11	5	3	8	10	5	7
e	5	2	1	0	2													

Consider, for example, the network depicted in Fig. 1 where $q' = \{b, c, e\}$ is the query submitted by a user (peer) in the group led by ultra-peer A. Let $m = 3$ be the number of top-ranked ultra-peers to which the query is routed by another ultra-peer. Let assume that the DHIs of A and B are empty at the beginning and A, B, C, D are the neighbors of A and A, B, E, F, G are the neighbors of B.

The left part of Table 1a represents the ultra-peer index of A listing the weights for each feature k_i in the index. When the query arrives at A, the ultra-peers in its neighborhood are ranked and the m top-ranked ones are selected to forward the query q'. The right part of Table 1a shows the selected top-scores (shaded in grey). Furthermore, A sends the query to the top-ranked ultra-peers, adding the ranking information about each feature, including its own scores. Ultra-peer B receives the query, forwards q' to the top-ranked peers of its group, extracts the weights for each feature of q' and increments its DHI using the statistics about known ultra-peers (not in the set of neighbors of B). Afterward, it iterates the ranking process by accessing its ultra-peer index and DHI, which act as a single index. As for B, Table 1b represents (from left): the index terms k_i's and their weights in the ultra-peer granularity index, the updated contents of the DHI and, finally, the results of the ranking step. B sends the query to the top-ranked ultra-peers (in this case E, F, G), adding to query message all the ranking information about each feature of q'. Note that ultra-peers A, B, D are not selected because they have already processed the query. A mechanism to avoid a re-forwarding of the query to the same receivers can be easily implemented.

Fig. 2 depicts the changes in the overlay network after two hops w.r.t. ultra-peer E. The dashed lines point out the increment of the DHI's size due to the propagation of aggregate statistics for each feature of the query. For example, during the last query's hop to ultra-peer E, high granularity information on five new ultra-peers – namely A, B, C, D, F, G – are added to the DHI of E, with regard to the features b, c and e. This example shows how the "horizon" of

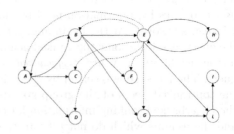

Fig. 2. Changes in the topology after two hops w.r.t. ultra-peer E

the ultra-peer E can be widened by the query piggybacking technique. If a new query with the features b and e starts from the ultra-peer E, the most promising ultra-peer A can be reached in one hop by exploiting the information stored in the DHI.

3   Experiments

The experiments were based on a realistic P2P network: the 1,421,088 docu-ments distributed across 2,500 collections of the DLLC (Digital Libraries Lu and Callan) collection were distributed on an actual, real peer network. The functionalities for indexing, retrieval and communication among peers were pro-vided by SPINA [2]. The connections between the ultra-peers were generated randomly. Each ultra-peer had no less than 3 and no more than 5 neighbors.

Some preliminary experiments run on the TREC topics numbered from 451 to 500 are reported in this paper. The starting ultra-peer was drawn at random and the draw was repeated ten times. Time-To-Live (TTL) was the number of times the query was routed to an ultra-peer, m was the number of top-ranked neighbors to which a query was routed by an ultra-peer, whereas k was the number of top-ranked peers to which a query was routed by the ultra-peer of the group, and n was the number of top-ranked documents retrieved from each selected peer. Table 2 reports the average number of relevant retrieved documents per starting ultra-peer for two runs when TTL $= 2$, $m = 5$, $k = 1$, $n = 50$. Run 1 did not make use of the query piggybacking technique starting from an empty DHI — a DHI for each ultra-peer was created —, Run 2 exploited the data stored in the DHI during Run 1. The obtained results suggest that the query piggybacking technique has a positive effect over the resource selection step.

Table 2. Number of relevant retrieved documents per starting ultra-peer for two runs

Starting ultra-peer	A	B	C	D	E	F	G	H	I	L
Run (1)	60	57	55	66	74	62	54	72	59	64
Run (2)	135	111	84	94	154	133	103	125	128	112

4   Concluding Remarks

This paper has illustrated an algorithm based on a query piggybacking technique and some preliminary experimental results using a reference test collection. The obtained results are encouraging and suggest an extensive evaluation of the pro-posed technique. The impact of the query distribution and small connected net-work topologies on the effectiveness of the proposed algorithm will be matter of future investigation. To our knowledge, this infrastructure will be the first realistic peer network for large scale P2P-IR experiments.

References

1. Lu, J., Callan, J.: Full-text federated search of text-based digital libraries in peer-to-peer networks. Information Retrieval 9(4), 477–498 (2006)
2. Di Buccio, E., Ferro, N., Melucci, M.: Content-based information retrieval in SPINA. In: Proceedings of IRCDL 2008, Padova, Italy, January 2008, pp. 89–92 (2008)
3. Melucci, M., Poggiani, A.: A study of a weighting scheme for information retrieval in hierarchical peer-to-peer networks. In: Amati, G., Carpineto, C., Romano, G. (eds.) ECiR 2007. LNCS, vol. 4425, pp. 136–147. Springer, Heidelberg (2007)

The Planets Interoperability Framework*
An Infrastructure for Digital Preservation Actions

Ross King[1], Rainer Schmidt[1],
Andrew N. Jackson[2], Carl Wilson[2], and Fabian Steeg[3]

[1] Austrian Research Centers GmbH - ARC
Donau-City-Strasse 1
1220 Vienna, Austria
firstname.lastname@arcs.ac.at
[2] The British Library
Boston Spa, Wetherby, West Yorkshire
LS23 7BQ, United Kingdom
firstname.lastname@bl.uk
[3] Universität zu Köln
Albertus-Magnus-Platz
50923 Cologne, Germany
fabian.steeg@uni-koeln.de

Abstract. We report on the implementation of a software infrastructure for preservation actions, carried out in the context of the European Integrated Project *Planets* – the Planets *Interoperability Framework* (IF). The design of the framework was driven by the requirements of logical preservation in the domain of libraries and archives. The IF is a Java-based software suite built on a number of open source components and Java standards. Specific features of interest are a web service architecture including specified preservation service interfaces for the integration of new and existing preservation tools and a workflow engine for the execution of service-based preservation plans.

1   Introduction

We describe a software infrastructure developed in the context of the EU Integrated Project Planets known as the Planets Interoperability Framework (IF). A more detailed overview of the Planets project can be found elsewhere [1].

2   State of the Art

Significant effort in the preservation community has been dedicated towards producing OAIS-compliant archiving infrastructures, although OAIS itself is primarily a conceptual model and does not provide guidelines for implementation [2].

---

* Work presented in this paper is partially supported by the European Community under the Information Society Technologies (IST) Programme of the 6th FP for RTD - Project IST-033789.

M. Agosti et al. (Eds.): ECDL 2009, LNCS 5714, pp. 425–428, 2009.
© Springer-Verlag Berlin Heidelberg 2009

While Planets has concentrated on the specific OAIS component "Preservation Planning", the CASPAR project [3] places an emphasis on representation information and a complete OAIS-compliant archive implementation.

One component of the overall CASPAR architecture that is comparable to the Planets Interoperability Framework is the Preservation Data Store (PDS) [4]. The PDS approach delegates preservation-related functionality to the storage component. As in Planets, the PDS architecture supports integration with existing archives. However, PDS lacks the flexibility of the Planets IF plug-in approach to preservation tools and workflows.

The closest parallels to the Planets approach are the Australian project PANIC (Preservation services Architecture for New media and Interactive Collections) [5] and The United Kingdom's National Archives Seamless Flow framework [6]. Both are based on service-oriented architectures (SOA) in which preservation actions are invoked through Web Services.

Planets Interoperability Framework: Architecture

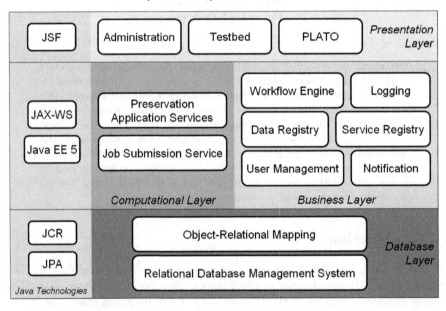

Fig. 1. Planets Interoperability Framework Architecture

3   The Planets Interoperability Framework

The Interoperability Framework provides an infrastructure to carry out digital preservation actions in the form of flexible, service-based workflows. Planets is driven by the requirements of memory institutions, primarily national libraries and archives. These institutions generally already have archiving systems in place, and replacing such systems is neither feasible nor desirable. Therefore the IF was designed to run in parallel with existing archive systems; it is not intended to replace these or to provide archiving functionality.

For this short description of a selection of IF components and their functionality, refer to the *Business Layer* of figure 1. The *Data Registry* provides storage and persistence services to IF users, components and services. In particular the storage API provides methods for storing complex Digital Objects and Preservation Events. The *Service Registry* enables users and service providers to look up and publish information about preservation services, and enables Planets system administrators to manage information about these services. In order to support service discovery, we provide extensible service categorization mechanisms that allow the Service Registry to be queried using these categories. The Planets IF *Workflow Engine* implements a component-oriented enactor that governs the orchestration of the preservation components, including functionalities like session-management, communication, and preservation metadata handling. Distributed preservation workflows are conducted from high-level components that abstract the underlying protocol layers.

3.1   Implementation Details

The Interoperability Framework provides a Java-based infrastructure that leverages a number of standards and open source tools. Referring to the *Java Technologies* indicated in figure 1, the core of the IF implementation is the Java Platform, Enterprise Edition (Java EE 5) standard, which among other things provides a framework for the efficient implementation of Web Services and Web applications. In particular we make use of Sun Microsystems' Web Services Interoperability Technology (WSIT) suite and the underlying JAX-WS (Java API for XML Web Services) standard. The IF provides a pre-configured JBoss[1] application server as its default deployment environment. We chose JBoss as a stable, well-supported open source implementation of the Java EE 5 standard. Data persistence is provided through the Java Persistence API (JPA), supported by the underlying Apache Derby[2] relational database management system (RDBMS). Derby was chosen because its small footprint and pure Java implementation allow it to be easily packaged and installed with minimum user expertise. However, as we use only the standard Java Database Connectivity (JDBC) API and no RDBMS-specific features, it is possible to configure the IF for operation with other open-source or commercial databases in a production environment.

3.2   Performance and Scalability

A crucial aspect of our preservation system is the establishment of a distributed, reliable, and scalable computational tier. Advances in virtualization allow the deployment of entire preservation environments, including operating systems and applications, to distributed computational nodes. This allows one to instantiate sets of transient system images on demand, which can be federated as a

---

[1] http://www.jboss.org/
[2] http://db.apache.org/derby/

virtualized cluster. We have implemented a prototype Job Submission Service (JSS) that can manage such infrastructures and execute Planets preservation workflows on a virtual computing cluster or *Cloud*. Initial experiments with the Amazon Elastic Compute Cloud (EC2) reported elsewhere [7] have demonstrated the feasibility of this approach for the Planets service architecture.

4 Conclusions and Future Work

After the third project year, the Planets Interoperability Framework provides a stable preservation infrastructure, which is available for download[3] in a platform-independent installation package. In the fourth and final year of the project, we will continue research on scalable services for preservation actions based on virtual computational clusters. In addition, field tests of the Planets Software Suite will be carried out at partner institutions, demonstrating how the Interoperability Framework and the associated Planets applications and services can act as an added-value preservation action system for existing digital repositories at national libraries and archives.

References

1. Farquhar, A., Hockx-Yu, H.: Planets: Integrated Services for Digital Preservation. International Journal of Digital Curation 2(2) (2007)
2. ISO Standard 14721:2003: Space Data and Information Transfer Systems – A Reference Model for an Open Archival Information System (OAIS). International Organization for Standardization (2003)
3. Giaretta, D.: The CASPAR Approach to Digital Preservation. The International Journal of Digital Curation 2(1) (2007),
   http://www.ijdc.net/ijdc/article/view/29/32
4. Factor, M., Naor, D., Rabinovici-Cohen, S., Ramati, L., Reshef, P., Ronen, S., Satran, J., Giaretta, D.: Preservation DataStores: New storage paradigm for preservation environments. IBM Journal of Research and Development on Storage Technologies and Systems 52(4/5) (2008)
5. Hunter, J., Choudhury, S.: A semi-automated digital preservation system based on semantic web services. In: JCDL 2004: Proceedings of the 4th ACM/IEEE-CS joint conference on Digital libraries, pp. 269–278 (2004)
6. Brown, A.: Developing Practical Approaches to Active Preservation. The International Journal of Digital Curation 2(1) (2007),
   http://www.ijdc.net/ijdc/article/view/37/42
7. Schmidt, R., Sadilek, C., King, R.: A Service for Data-Intensive Computations on Virtual Clusters. In: INTENSIVE 2009: Proceedings of The First International Conference on Intensive Applications and Services (to appear)

---

[3] http://gforge.planets-project.eu/gf/project/if_sp/

Improving Annotations in Digital Documents

Jennifer Pearson, George Buchanan, and Harold Thimbleby

Future Interaction Technology Laboratory
Swansea University & City University, UK
{csjen,h.w.thimbleby}@swansea.ac.uk, george.buchanan.1@city.ac.uk

Abstract. Annotation plays a major role in a user's reading of a document: from elementary school students making notes on text books to professors marking up their latest research papers. A common place for annotations to appear is in the margin of a document. Surprisingly, there is little systematic knowledge of how, why and when annotations are written in margins or over the main text. This paper investigates how margin size impacts the ease with which documents can be annotated, and user annotation behavior. The research comprises of a two part investigation: first, a paper study that examines margins and their use in physical documents; secondly, we evaluate document reader software that supports an extended margin for annotation in digital documents.

Keywords: Annotation, digital documents, document triage.

1 Introduction

Annotations on documents have appeared in many forms for centuries: making notes on a document, marking assignments or even professionally annotating books: annotation allows users the freedom to make their own mark on pre-existing literature.

In this paper, we report research that demonstrates the significance of margin space in annotating both digital and physical documents. Though the issue of margin space may seem trivial, there is a lack of concrete research across much of the field of annotation. We provide detailed evidence on our chosen topic and demonstrate how digital document reader software can be significantly improved by changing their interaction design, informed by observation of actual user behavior.

1.1 Research Motivation

The topic of annotations has been repeatedly been studied by researchers over recent years. In 1998 Alder et al [1] reported that users spend nearly half their time working with documents either annotating or note taking. Cathy Marshall [2,3], Abigail Sellen [1,4] and others have also explored various areas of this topic. The cumulative impact of this research is a clear understanding that annotation plays a critical role in how users process, examine, and manage information. However, though annotation is important, Marshall, Sellen, etc. concur that it is poorly supported in digital documents.

This paper studies the significance of the position of an annotation within a document: e.g. annotating over the document content itself, or writing in the margins. Where

M. Agosti et al. (Eds.): ECDL 2009, LNCS 5714, pp. 429–432, 2009.

space is insufficient for a user's notes, or the original material cannot be marked, supplementary notes can be taken on a separate medium such as "Post-it" notes or a notebook. However, whilst Marshall and Sellen noted the importance of position in relating an annotation to the content it illuminates, we lack detailed knowledge of this connection.

We report a two–phase investigation that probes the role of location in annotation. An initial study of annotation on paper is followed by a second comparative experiment with electronic annotation.

2   Paper Study

There have been a number of studies of how users annotate paper printed documents (e.g. [2,5]). However, the issue of how users exploit and manage space when annotating has received little attention. Superficially, this issue is straightforward: annotations will appear near to the material that they relate to. In printed documents, space is limited, and using separate materials (e.g. notebooks) requires the user to co–ordinate more objects and demands more of the working environment (e.g., simply, somewhere for these to be kept). Cathy Marshall [2] noted three major locations for annotations: on the document content itself, in the margins of the document, or on a separate medium (e.g. post–its, scrap paper). However, she did not observe the actual creation of the annotations, so how materials were used or created was unclear.

We undertook a paper–based comparative user study, to observe the issues raised by Marshall in a "live" context. This tested each of the placement choices listed above to determine the most common methods, and the factors affecting users' choice of position and method of marking. Throughout, our primary interest was to comprehend users' decisions to either annotate over the document content itself or write in the margins.

Each of our 10 participants was provided with two varieties of printed PDF: some with a minimal margin; and a contrasting version with an expanded and uniformly wide margin. In both formats, the size of text in any one document was the same. The expanded margin documents were a full A4 size, with the original document presented in the middle. This typically resulted in the original taking 50% of the total surface area of the page (the size of the original content varied). In contrast, the minimal margin texts were always trimmed to give small margins of approximately 5-7mm. This extreme difference was used to ensure we observed the strategies users deployed when margins fail to provide sufficient space for notetaking.

We created a set of seven tasks (one document per task) to complete for each of the two document formats (marginless and extended margins). Each participant undertook 14 tasks assigned on a latin–square design to balance ordering, pairwise and other effects. Each task included five open and two closed sub–tasks for the assigned document. We wished to observe annotation in as natural and unconstrained an environment as possible, and thus the two closed sub–tasks provided a minimally intrusive baseline against which other activity could be compared.

A visual inspection of the materials produced by our participants revealed that the margin was the preferred location for written annotation. Whilst highlighting was typically, on the textual content, most written material appeared at the edges of the documents.

Annotating on top of the document is cumbersome and restrictive. Seven participants reported that obscuring the original text with annotations makes both the text and notes more difficult to read, with two even suggesting that it would discourage them from writing more notes.

The small space between lines of text no doubt contributed to the very small number of comments made there. Occasionally, we observed annotations being written at an angle or vertically to provide large unbroken space. Our participants were choosing location and space strategies to make either writing or reading of comments easier.

Participants answered a set of 8 questions after a completing each document, to identify the subjective impact on the margin size on the user's ability to create annotations. Most ratings resulted in a higher score for the extended margin presentation: e.g. legibility of notes rated 8.8 vs 7.1 ($p=0.007$, $t=3.04$).

3   Comparison Study

We recruited 16 participants from the research sector to participate in Comparative study, testing a PDF reader with a large "virtual" margin area, against a traditional interaction where annotation must be on the logical page. As with the paper study (Section 2) participants with a research–based background were chosen intentionally due to the high likelihood that they regularly annotate material in their working life.

3.1   Results

The outcome of this study was in line with the expectations raised by our paper study. Participant's subjective feedback gave many significant benefits to the enlarged margin area. For example, Margin notes had better legibility ($p=0.01$, $t=-2.36$, 7.56 vs 8.81/10), obscured the text less ($p<0.001$, $t=8.01$), and was subjectively faster ($p=0.099$).

There were some advantages to annotating directly on the PDF: e.g. when highlighting particular words, or connecting parts of the content together or to notes, and indicating spans of content ($p=0.01$, 8.81 vs 7.56), etc. A good running solution would contain both approaches.

These results strongly imply that using the margin of the document as an annotation area is not an alternative to marking the PDF, but rather a useful supplement to it. Some tools are invaluable for use on the margins, but unhelpful on the PDF content, and vice versa for other tools. The question that we now address is whether this pattern is explained by the higher–level tasks that are associated with each tool.

3.2   Summary

To answer the outstanding research questions surrounding the implemented system, a systematic comparison study was undertaken focusing on four sections of research.

Our investigations confirm that there are some tools that are best suited to the margin while others are best suited to the PDF. This information strongly suggests that the margins are a useful *addition* to marking up the PDF as opposed to a straightforward *alternative*. This concurs with our findings from the paper study (Sec. 2).

The study also confirms that together the tools in the system make up a complete set. Each excels at a particular type of task (highlighting specific points, highlighting specific areas, making connections between points, illustrating, making notes).

4   Discussion

Marshall [2] and O'Hara and Sellen [4] concur that at present it is easier to read and annotate on paper than on digital media. However, the scope of this earlier work has been to identify limitations of digital texts, not to remedy them.

Our work here has attempted to understand how users exploit print media, and to replicate some of those advantages in digital documents. We believe that much of the gap emerges from subtle but critical affordances in the physical world, that disappear in electronic texts. Space for annotation is an issue for users of both printed and digital literature. Providing larger margins is of benefit in both domains, and the previous arbitrary limitation of marginal space in the digital world is an unnecessary constraint.

The study investigating traditional paper based annotation methods has proved that margins form an integral part of the physical annotation process. Thus far however, no attempt has been made to utilise this system in the digital document sector. Fast advancing desktop screen technology now affords us more space than ever to extend applications making additional margins on digital documents a reality. Digital document readers are far from being a replacement for paper; the margin annotation system however endeavors to bridge the gap between the physical and digital domains in order to make the digital annotation process significantly less cumbersome. The systematic comparison study performed upon the completed digital systems clearly confirms the popularity of the margins as well as the additionally implemented tools.

Acknowledgements

Jennifer Pearson is sponsored by Microsoft Research Limited; Harold Thimbleby is a Royal Society-Leverhulme Trust Senior Research Fellow, and both gratefully acknowledges this support. This research is supported by EPSRC Grant GR/S84798.

References

1. Adler, A., Gujar, A., Harrison, B.L., O'Hara, K., Sellen, A.: A diary study of work-related reading: design implications for digital reading devices. In: Procs ACM SIGCHI, pp. 241–248. ACM Press/Addison-Wesley Publishing Co. (1998)
2. Marshall, C.C.: Annotation: from paper books to the digital library. In: DL 1997: Procs. ACM International conference on Digital libraries, pp. 131–140. ACM Press, New York (1997)
3. Marshall, C.C.: Toward an ecology of hypertext annotation. In: Procs. ACM Hypertext 1998, pp. 40–49. ACM Press, New York (1998)
4. O'Hara, K., Sellen, A.: A comparison of reading paper and on-line documents. In: CHI 1997: Procs. ACM SIGCHI conference, pp. 335–342. ACM Press, New York (1997)
5. Wu, C.-S.A., Robinson, S.J., Mazalek, A.: Turning a page on the digital annotation of physical books. In: TEI 2008: Proceedings of the 2nd international conference on Tangible and embedded interaction, pp. 109–116. ACM Press, New York (2008)

Workspace Narrative Exploration: Overcoming Interruption-Caused Context Loss in Information Seeking Tasks[*]

Youngjoo Park and Richard Furuta

Center for the Study of Digital Libraries
and Department of Computer Science and Engineering
Texas A&M University
College Station, TX 77843, USA
{yjoo9317,furuta}@cse.tamu.edu

Abstract. As digital libraries become more prevalent and as the amount of accessible online information grows, users often must consult diverse information collections in carrying out tasks. Simultaneously, the impact of frequent interruptions on task performance gets more severe. To manage the negative effects of interruptions on work performance, workers often engage in task management activities to ensure they are better prepared to resume suspended tasks. However, managing tasks causes additional cognitive burden and incurs a time cost to users who already are experiencing demands on their attention and time. We describe a system that allows users to browse their previous workspace intuitively, and enhances continuity of their tasks by supporting them to retrieve desired work context more quickly and easily.

Keywords: Context browser, narrative, task, contextual cue, discontinuity, user interfaces.

1 Introduction

The diversity and volume of online information are growing daily. As a result, tasks performed by information workers often include various information pieces carefully selected from many different collections on the World-Wide Web or on huge local hard disk drives. For information workers, performing tasks typically coincides with frequent interruptions and resumptions, which leads to the need for an efficient mitigating interface for minimizing negative effects of interruptions. In particular, a significant task, such as a problem solving task, tends to be complex and to last a longer period of time, making the task and the users vulnerable to interruption [1]. When an interruption occurs, the relationship between the task and a set of associated information resources easily can be broken as the task's environment—i.e., desktop status—starts being adjusted to the new task's requirements; e.g., as users close windows and open new ones. Unfortunately, the current computing environment lacks a service by which users may seamlessly 1) restore the context when the task was suspended and 2) resume the unfinished task with rich contextual artifacts.

---

[*] This material is based upon work supported by the National Science Foundation under Grant No. IIS-0534314.

M. Agosti et al. (Eds.): ECDL 2009, LNCS 5714, pp. 433–436, 2009.

Search

Time

Input density
graph

Tag & bookmark

File activity

Date

Restored desktop

Window object

Time track bar

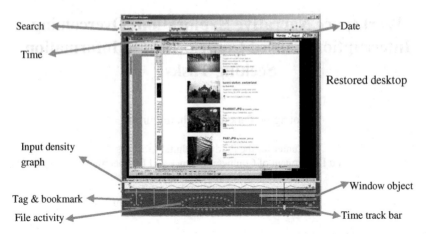

Fig. 1. Timeline interface of a context browser

In this paper, we briefly present an interface for users' workspace narrative exploration, *Context Browser*, and describe a study in which subjects performed multiple tasks and encountered various interruptions in a more realistic working environment. The work reported here extends our previous study [2], which showed that the subjects with the context browser were able to prepare for task resumption more quickly and retrieve more associated information than the subjects without the context browser.

2   Context Browser: Keeping Workspace Narrative Close

A user's desktop fluidly accommodates various work settings; a workspace narrative is a stream of a user's activities spread across various work settings, i.e., a contextual archive. We hypothesized that digitizing workspace narratives would enable the novel opportunity to index, restore, reuse, manipulate, explore and, most importantly, search users' cognitive and task contexts in which associated windows were visible. This is easier for users than is requiring explicit user intervention to save discrete information at a specific moment, which is typical in existing task management tools.

The *Context Browser* implements the idea of placing a user into an appropriate context to carry out a task with less cognitive and time costs. Its timeline viewer is the primary interface that users use to explore narratives of a desktop to find their desired context or any necessary information resources; see Figure 1. A reconstructed desktop in the timeline view interface supports typical desktop interactions with a mouse such as window selection and dragging, and even double-clicking to launch a window. The interface also supports launching the whole desktop status to retrieve a desired work context. The system is equipped with 1) a tag and bookmark mechanism to allow future revisiting and reuse and 2) a search engine as well that utilizes a vector space model to measure cosine similarities between a query and a metadata element in the contextual archive, such as window titles, URLs, Web page titles and process names.

3   User Study and Findings

24 subjects (5 female), with ages ranging from 20 to 39, participated in the study. They were not compensated. 6 out of the 14 married subjects had a child(ren). To perform between subject measures, we divided subjects evenly into two separate groups. Subjects in **group 1** used the context browser and those in **group 2** did not.

To make a typical office environment with frequent interruptions and multiple tasks, we prepared a set of three tasks and assigned them to subjects prior to the study. The study was conducted in two separate sessions; each taking approximately one hour. The period between the two sessions was about 24 ~ 48 hours.

Task #1 asked the subjects to find pictures of given 9 cities (5 pictures a city). From now on, we refer to this task as the *City pictures task*.

Task #2, the *Van Gogh task*, asked them to search for Van Gogh's famous paintings with their metadata, i.e., location, title, year, and save them in the document.

In **Task #3**, subjects planned a move to San Jose, CA, due to a job offer with a monthly salary of $5,000 after taxes. They needed to seek information on housing or apartments and to find at least 3 candidates to move in. We refer to this task as the *San Jose task*.

3.1   Recall and Precision

The City pictures task echoes such tasks that are associated with a considerable amount of information resources, such as graphs and images. In reality, there might be the situation in which users need to retrieve and review all resources they had already seen previously either to verify that they have a proper result or modify the current result. To simulate this situation, in the 2nd session we additionally asked subjects to do the following—"Given 10 minutes, find the images that you believe that you saved while performing this task."

Group 1 (avg=0.643, sd=0.03) showed better recall rate than group 2 (avg=0.4, sd=0.14), t=5.45, df=22, p<0.0001. Particularly, the difference in precision is more obvious than the one in recall. By having the subjects in group 1 exposed to narratives of prior desktop status, group 1 (avg=0.933, sd=0.46) collected more images with much higher precision within a limited period than group 2 (avg=0.59, sd=0.15) did, t=7.23, df=22, p<0.0001.

3.2   Time Lag to Recover Search Contexts

The San Jose task forced subjects to execute cognitively loaded activities, i.e., repeated search activities, to narrow down their search list to a manageable size. They had to resolve various parameters, such as a choice between house and apartment, a number of bedrooms, prices, regions and so forth. In the 2nd session, we asked them to look for two more houses or apartments sharing similar specification to the one they liked most previously. In terms of a time cost to restore a desired search list, group 1 (avg=245 sec., sd=70) successfully retrieved the context faster than group 2 (avg=521 sec., sd=225), t=-4.06, df=22, p<0.001. The subjects in group 1 found the context browser useful to get them back to the state where they were evaluating houses or apartments on the list, which was the fruit of various search activities. However, the subjects in group 2 again had to go through all the steps that they took previously to find them, which sometimes painfully cost a long period to execute the given task.

3.3 Performance After a Lengthy Period

We wondered if the Context Browser would continue to be useful after some time had passed, especially for the subjects in group 2, who had not used it previously. We thus conducted one additional experiment with re-invited 6 subjects from group 2 six weeks after the previous study. We let them use the context browser at this time to see if there was any improvement in handling the problems that they did previously after more than a month delay. During this additional session, we again assigned the same protocols, i.e., finding the images that you saved previously and a similar house or apartment, and measured as we did 6 weeks earlier. We performed a paired-samples T test comparing these results with their previous performance and discovered that with the context browser they were able to retrieve more images, i.e., a better recall, (mean difference=-0.17, t=-6.7, df=5, p<0.001) and the images more correctly, i.e., a better precision, (mean difference=-0.218, t=-8.26, df=5, p<0.001) even after the 6 week delay. They also easily found the house or apartment they liked and successfully searched similar places less stressfully and more quickly (mean difference=177.16 sec., t=3.47, df=5, p=0.018). This finding also indicates that by only browsing the timeline, subjects were able to fetch their desired information and contexts since the archives for group 2 were not arranged except by time.

4   Conclusion

With the Context Browser, the study showed that users are better able to remember and recognize semantic information presented by the system than with a typical desktop environment. This is because the workspace narratives successfully captured semantically meaningful contextual cues and the context browser provided users with services, such as timeline browsing, tag and bookmark, search, and so forth, by which they can easily browse and retrieve appropriate work settings. In addition, the subjects expressed that they found the context browser very useful since it handled a tedious information organization task and kept their desired information within a close distance asking less both time and cognitive costs to find it.

References

1. Czerwinski, M., Horvitz, E., Wilhite, S.: A diary study of task switching and interruptions. In: Proceedings of CHI, pp. 175–182 (2004)
2. Park, Y., Furuta, R.: Keeping narratives of a desktop to enhance continuity of on-going tasks. In: Proceedings of JCDL, pp. 393–396 (2008)

SyGAR – A Synthetic Data Generator for Evaluating Name Disambiguation Methods

Anderson A. Ferreira, Marcos André Gonçalves, Jussara M. Almeida,
Alberto H.F. Laender, and Adriano Veloso

Department of Computer Science
Federal University of Minas Gerais
31270-901 Belo Horizonte-MG Brazil
{ferreira,mgoncalv,jussara,laender,adrianov}@dcc.ufmg.br

Abstract. Name ambiguity in the context of bibliographic citations is one of the hardest problems currently faced by the digital library community. Several methods have been proposed in the literature, but none of them provides the perfect solution for the problem. More importantly, basically all of these methods were tested in limited and restricted scenarios, which raises concerns about their practical applicability. In this work, we deal with these limitations by proposing a synthetic generator of ambiguous authorship records called SyGAR. The generator was validated against a gold standard collection of disambiguated records, and applied to evaluate three disambiguation methods in a relevant scenario.

1 Introduction

It is practically a consensus that *author name disambiguation* in the context of bibliographic citations is one of the hardest problems currently faced by the digital library community. To solve this problem, a disambiguator is applied to correctly and unambiguously assign a citation record to one or more authors, already or not present in the digital library, despite the existence of multiple authors with the same name (or very similar names – polysems), or different name variations (synonyms) for the same author in the data repository.

The complexity of dealing with ambiguities in digital libraries has led to a myriad of methods for name disambiguation [1,2,3,4,5,6]. Most of these methods demonstrated to be effective in specific scenarios with limited, restricted, and static snapshot collections. This leads to the question: *Would any of these methods effectively work on a dynamic and evolving scenario of a living digital library?*

In this paper, we propose a Synthetic generator of ambiguous Groups of Authorship Records (SyGAR) that is capable of generating synthetic authorship records of ambiguous groups, and thus can be used to simulate the evolution of a digital library over time. The use of a synthetic generator to evaluate disambiguation methods makes it possible to generate and simulate several controlled, yet realistic, long term scenarios to assess how distinct methods would behave under a number of different conditions.

M. Agosti et al. (Eds.): ECDL 2009, LNCS 5714, pp. 437–441, 2009.
© Springer-Verlag Berlin Heidelberg 2009

2   Generating Synthetic Ambiguous Authorship Records

2.1   SyGAR Design

SyGAR takes as input a real collection of ambiguous groups previously disambiguated. Each such authorship record is composed of the author name (*author*), a list of her coauthors' names (*coauthors*), a list of terms present in the work title, and the publication venue title. For each ambiguous group in the input collection (*input group*), the number of unique authors N_A and the total number of authorship records N_R to be generated are also inputs to SyGAR.

As output, SyGAR produces a representative list of synthetically generated authorship records (*output group*) using a set of attribute distributions that characterize the publication profiles of each group and of its individual authors.

Building Author and Group Publication Profiles from Input Groups. Each publication profile of an author a is extracted from the corresponding input group by summarizing her record list into: (1) the distribution of the number of coauthors per a's record - $P^a_{nCoauthors}$; (2) a's coauthor popularity distribution - $P^a_{Coauthor}$; (3) the distribution of the number of terms in a work title by a - P^a_{nTerms}; (4) a's term popularity distribution - P^a_{Term}; and (5) a's venue popularity distribution - P^a_{Venue} (i.e., the distribution of the number of a's records with the same venue title). We assume that these attribute distributions are statistically independent, the terms appearing in the work title are independent from each other, and so are the work coauthors. Finally, we build a group profile with the distribution of the number of records per author - $P^g_{nRecordsPerAuthor}$.

Generating Records for Existing Authors. Each synthetic authorship record is created by following the steps: (1) select one of the authors a of the group according to $P^g_{nRecordsPerAuthor}$; (2) select the number a_c of coauthors according to $P^a_{nCoauthors}$; (3) repeat a_c times: select one coauthor according to $P^a_{Coauthor}$; (4) select the number a_t of terms in the title according to P^a_{nTerms}; (5) repeat a_t times: select one term for the work title according to P^a_{Term}; and (6) select the publication venue according to P^a_{Venue}.

Adding New Authors. SyGAR may be used to create records for new authors. Currently, SyGAR uses a knowledge base with the distribution of the number of records with the same coauthor - $P_{Coauthor}$, and the attribute distributions of the publication profile of each coauthor in the input collection. A new author is created by selecting one of its coauthors a, using $P_{Coauthor}$. The new author inherits a's profile. All generated records will have a as one of its coauthors. This strategy mimics the case of an author who follows the areas of one that will be a frequent coauthor.

2.2   Validation

We validate SyGAR by comparing real ambiguous groups against corresponding synthetically generated groups, assessing whether the synthetic groups capture

Table 1. SyGAR Validation across State-of-the-Art Name Disambiguation Methods

Method	Ambiguous Group	Real MicroF$_1$	Real MacroF$_1$	Synthetic MicroF$_1$	Synthetic MacroF$_1$
SVM	A. Gupta	0.879±0.009	0.650±0.027	0.894±0.009	0.651±0.027
	C. Chen	0.761±0.015	0.611±0.025	0.779±0.012	0.580±0.018
	D. Johnson	0.809±0.027	0.623±0.026	0.817±0.018	0.615±0.029
SLAND	A. Gupta	0.916±0.008	0.809±0.025	0.947±0.006	0.807±0.028
	C. Chen	0.866±0.007	0.781±0.013	0.903±0.007	0.795±0.016
	D. Johnson	0.896±0.028	0.731±0.041	0.905±0.013	0.747±0.023

the aspects that are relevant to disambiguation methods. The real groups used, "C. Chen", "D. Johnson" and "A. Gupta", are selected from the collection of groups extracted from DBLP by Han *et al* [2].

For each real group, ten synthetic groups were generated. The number of authors and records per author in the synthetic group are set to be the same as in the input group. Table 1 shows average results of the disambiguation with 95% confidence intervals, with two supervised methods, an SVM-based method [1] and SLAND [6], under micro and macro F$_1$ measures. For all metrics, methods and groups, the results obtained for the real group are very close to those for the corresponding synthetic group, with a maximum error under 6%. In fact, six out of the twelve pairs of results are statistically indistinguishable with 95% of confidence. Thus, SyGAR is able to accurately capture aspects of real groups that are key to evaluate state-of-the-art name disambiguation methods.

3 Evaluating Disambiguation Methods with SyGAR

To illustrate a use of SyGAR, we evaluate three state-of-the-art disambiguation methods, namely an SVM-based method, a K-way spectral clustering method (KWAY), and SLAND, in realistic scenarios that encompass a live digital library (DL) evolving over a period of ten years. We perform experiments on ambiguous group "A. Gupta". The DL, at its initial state s_0, consists of records from the real group. At the end of each year, a load is performed into the DL with synthetic records generated by SyGAR, parameterized with the real group as source of author profiles. The distribution of the number of records generated to each author in the group is built based on the distribution of the average number of publications per year per (existing and new) author. These distributions were extracted from the DBLP for the analyzed group during the period of 1984-2008.

Starting at state s_i, for each new load, SyGAR generates records to authors already in the DL as well as to new authors (it is specified as a fraction f equals to 0%, 2%, 5%, and 10% of the total number of authors in the DL at the state s_i). If either SVM or SLAND is used, all the records making up state s_i are used as training data, and the data in the new load are used as test data for the disambiguators. If KWAY is used, the generated records are first incorporated into the current state of the DL and the disambiguation is done with all records using the correct number of authors in the DL. The DL evolves then into a new state s_{i+1}, and the micro-F$_1$ values are calculated for the whole DL in state s_{i+1}.

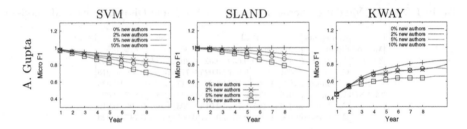

Fig. 1. Evolving DL and Addition of New Authors

The results reported next are averages of five runs, with a standard deviation typically under 5% (and at most 15%) of the mean.

Figure 1 shows the results in each state of the digital library over the ten-year period. There is an increase in the ambiguity for both SVM and SLAND and all values of f (but $f=0$ for SLAND) with sucessive data loads. Moreover, in any state of the DL, the increase in the ambiguity is higher for larger values of f, as expected. In comparison with SVM, SLAND makes fewer erroneous predictions during its application, dealing better with new authors.

Interestingly, KWAY tends to improve over time, as there is incrementally more information about each author, helping it to better characterize them. However, we also see a trend for performance stabilization typically after the 5th or 6th data load. Nevertheless, KWAY slightly outperforms SVM after ten years for $f > 2\%$, although the improvement does not exceed 10%. In comparison with SLAND, KWAY is inferior in all cases.

4   Conclusions and Future Work

In this paper, we presented SyGAR, a synthetic generator of ambiguous groups of authorship records that is capable of generating synthetic records, and used it to evaluate three state-of-the-art disambiguation methods in scenarios that capture relevant aspects of real-world bibliographic digital libraries.

As future work, we intend to further experiment with other disambiguators and scenarios, enhance SyGAR with more sophisticated mechanisms to add new authors and to dynamically change existing author profiles, and investigate the robustness of several disambiguators to errors in the original input collection.

Acknowledgments. This research is partially funded by projects INCTWeb (grant number 573871/2008-6) and InfoWeb (grant number 55.0874/2007-0), and by the authors' individual grants from CAPES and CNPq.

References

1. Han, H., Giles, C.L., Zha, H., Li, C., Tsioutsiouliklis, K.: Two supervised learning approaches for name disambiguation in author citations. In: JCDL, pp. 296–305 (2004)

2. Han, H., Zha, H., Giles, C.L.: Name disambiguation in author citations using a k-way spectral clustering method. In: JCDL, pp. 334–343 (2005)
3. Huang, J., Ertekin, S., Giles, C.L.: Efficient name disambiguation for large-scale databases. In: Fürnkranz, J., Scheffer, T., Spiliopoulou, M. (eds.) ECML/PKDD 2006. LNCS (LNAI), vol. 4213, pp. 536–544. Springer, Heidelberg (2006)
4. On, B.W., Lee, D., Kang, J., Mitra, P.: Comparative study of name disambiguation problem using a scalable blocking-based framework. In: JCDL, pp. 344–353 (2005)
5. Song, Y., Huang, J., Councill, I.G., Li, J., Giles, C.L.: Efficient topic-based unsupervised name disambiguation. In: JCDL, pp. 342–351 (2007)
6. Veloso, A., Ferreira, A.A., Gonçalves, M.A., Laender, A.H.F., Meira Jr., W., Belém, R.: Cost-effective on-demand associative name disambiguation in bibliographic citations. Technical Report RT DCC.001/2009, DCC-UFMG (under review) (2009)

Searching in a Book

Veronica Liesaputra, Ian H. Witten, and David Bainbridge

Department of Computer Science, University of Waikato
Hamilton, New Zealand
{vl6,ihw,davidb}@cs.waikato.ac.nz

Abstract. Information has no value unless it is accessible. With physical books, most people rely on the table of contents and subject index to find what they want. But what if they are reading a book in a digital library and have access to a full-text search tool?.

The paper describes a search interface to Realistic Books, and investigates the influence of document format and search result presentation on information finding. We compare searching in Realistic Books with searching in HTML and PDF files, and with physical books.

Keywords: Within-Document search, Electronic book, Flash application.

1 Introduction

Libraries and others are digitizing books and enabling people around the world to access them online. These projects enable full-text retrieval over vast collections of books. But once a book has been reached, how do users find information inside and navigate within it?

Subject indexes are an important access tool for physical books, whose utility frequently depends on the quality of the index. As large volumes of text migrate to electronic formats, it is easy to assume that such indexes become superfluous.

Most electronic document readers have a text search function. Readers type what they seek into the search box and are taken to the next closest occurrence, with matching strings highlighted and buttons to move to the next and previous occurrences. However, literal string matching has limitations. Readers who do not know exactly what they seek or how it might be expressed may issue search queries that match many irrelevant passages—or no passages at all.

This paper examines how people retrieve relevant and appropriate passages from within a document. We describe a user study comparing search performance with the Realistic Book format [1] against HTML, PDF and physical books, based on outcome and process measures [3].

2 Experimental Procedure

We conducted a user study that measured performance with each format—HTML, PDF, Realistic Books, and physical books—in terms of both outcome and process measures [3]. Outcome measures concern what readers get from the text. They evaluate *efficiency*,

M. Agosti et al. (Eds.): ECDL 2009, LNCS 5714, pp. 442–446, 2009.
© Springer-Verlag Berlin Heidelberg 2009

the time spend by readers in answering each question, *effectiveness*, the number of questions answered correctly, and *searching experience*, a subjective assessment of usefulness, likeability, and ease of use. Process measures investigate what search and navigation functions participants use to complete the task. The experiment is designed to test four hypotheses:

H1: Realistic Book users will find answers faster than with other document formats.
H2: Realistic Book users will make fewer task errors than with other formats.
H3: Realistic Book users will report higher satisfaction than with other formats.
H4: Regardless of format, users will use the most effective search tools to answer questions, i.e. BoB for index questions and search tools for full-text search questions.

To help mitigate variability, we recruited 32 high school and university students aged 15–40 from a variety of disciplines. In order to bias the experiment *against* the Realistic Book format, participants were already familiar with the HTML and PDF formats.

For our experiments we used a book that was available in electronic form and had a professional-quality subject index whose terms accurately represented the contents. We chose a university-level text called *Data Mining: Practical Machine Learning Tools and Techniques* [4]. We used only the first three chapters, and modified the table of contents and subject index to remove all references to other parts of the book, retaining the original design and layout.

The book was presented in three electronic formats: HTML, PDF and Realistic Books. Each contained the title, a hyperlinked table of contents (ToC), the main text, and a hyperlinked back-of-book subject index (BoB). To eliminate layout effects all three formats were paginated in exactly the same way as the physical book.

There were four sets of tasks, which participants undertook in the same order. Each task asked four questions that involved seeking information in the book. A different format was used for each task, and participants were exposed to the formats in different orders. The 24 possible orderings were allocated evenly: eight participants performed each task using the same document representation. Because participants did not know they were being timed, they felt no pressure and worked at their own pace. The functions they used were recorded.

To address H4, each task posed two full-text search questions and two BoB questions, in an order that changed from one task to another. Thus participants would use search tools or the BoB at least once while searching with each document format. To provide a balanced assessment of the utility of search, the questions have varying degrees of difficulty: one easy and one hard question for each question type.

3   Results

3.1   Efficiency

A one-way analysis of variance shows that differences due to format are statistically significant at the 1% level for all tasks except the first. For the last 3 tasks, the improvement from Realistic to Physical books is statistically significant at the 1% level (*t*-value ranges from 3.8 to 4.3), and its improvement over PDF and HTML is statistically significant at the 5% level (t-value from 2.1 to 2.6). The differences between the Physical book and the PDF and HTML formats are significant at the 10% level (t-value from 1.7 to 2.0). There is no significant difference between PDF and HTML for any task.

H1 (speed): Participants consistently produce answers quicker with the Realistic Book.

3.2 Effectiveness

Participants relied on full-text search to find answers in the electronic document formats. Because the subject index questions were designed to be difficult to answer without using the BoB, participants generally got them wrong. With the physical book, most participants failed on the full-text search questions.

H2 (accuracy): A one-way analysis of variance found no significant differences between document formats in the number of task errors made. Thus H2 is not upheld.

3.3 Search Experience

Having completed all tasks, participants were asked which formats were most useful, easy to navigate and locate information, pleasant and engaging, and preferred overall.

H3 (satisfaction): Participants found Realistic Books the most useful, engaging and easy to use format, combining a good reading environment with a good searching experience.

The formats were judged useful for different types of documents and activities. HTML is preferred for short documents. Physical books are best for reading activities, but not for information seeking. For searching, PDF or Realistic Books are preferred.

Participants found it hard to understand the structure of the document in HTML and PDF. They easily became disoriented, not knowing where they were in the document and finding it difficult to return to a specific location. Some participants (20%) found it difficult to step through the search results in PDF. Participants felt that they always knew where they were in Realistic Books, and navigated around more freely than with other formats. Bookmark tabs and page edges make it easier to move between search results without losing orientation.

3.4 Participants' Strategies

Before beginning each task, most participants briefly overviewed the document and went to the ToC page.

In physical book, they then sought terms they thought relevant in the subject index. If they failed to find the answer through the ToC and BoB, they carefully read sections of the book they thought might contain the relevant information.

In electronic formats, all participants typed into the Find box a word or a phrase that they thought would lead to the relevant information, to see whether it returned any results.

In HTML and PDF, the only time most participants used the BoB was when a search term appeared in it and nowhere else. Nearly all of them neglected the BoB, even though they had been told about it before beginning their tasks and it is listed in the book's ToC, relying instead on search for navigation.

In the Realistic Book, they also used BoB to seek more appropriate search terms when full-text search failed to find the answer, or when they could not think of any more suitable search terms. Once participants were familiar with the document structure, most used the spatial reference of passages in the book to help them remember where information was.

Having exhausted all search terms they could think of, participants would go to the ToC and guess which section contains the information they seek.

H4 (appropriate choice of search tool): Full-text search is the principal information-finding strategy for all formats; participants consulted the BoB as a secondary strategy only for the Realistic Book format. Thus H4 is partially upheld.

4 Conclusions

Back-of-book indexes are considered by readers to be important access mechanisms for physical books. Index terms are carefully chosen to represent the key ideas in the book, and to bridge the author's perspective of the topic to keywords that readers might use. A comparative evaluation of a subject index and full-text search showed that participants found information more effectively and efficiently with the former [2]. However, a post-task questionnaire revealed that participants still preferred search to the subject index.

This paper investigates whether document format influences search behavior and performance. We observed participants performing information-finding tasks in HTML, PDF, and Realistic Book formats, and in physical books. The tasks were designed to test four specific hypotheses.

The results upheld Hypothesis 1: users of Realistic Book found answers significantly faster (99% confidence level) than with conventional document displays.

Hypothesis 2, that Realistic Book users make fewer task errors than with other formats, was not upheld. There were no significant differences between Realistic Book and other formats in terms of the number of correctly answered questions.

Hypothesis 3, that Realistic Book users report higher satisfaction than with other formats, was strongly upheld (although no statistical analysis was performed).

As for Hypothesis 4, that regardless of format users choose the most effective search tools to answer questions, we found that all participants chose full-text search first, whether it was appropriate or not. Only with Realistic Books did they consult the back-of-book index as a secondary choice; for the HTML and PDF formats nearly all participants neglected it. Thus Hypothesis 4 is partially upheld; in addition, we found that whether the back-of-book index is used at all depends on the document display format.

The only time at participants used the back-of-book index for HTML and PDF documents was when one of their search results matched a term in it. Such terms were usually synonyms or bridging words that point users to the actual word used in the text. Without them, readers would not have been able to find the answers. In contrast, readers of Realistic Books also used the back-of-book index to suggest more appropriate search terms when their initial full-text search failed to find an answer, or when they could not think of any more suitable search terms. Thus it is helpful for electronic documents to have well-constructed subject indexes.

Overall, the results reported here strongly uphold the idea that Realistic Books are an effective document format for finding information compared with other formats.

Acknowledgments. We acknowledge the entire New Zealand Digital Library Project team for their unstinting work in providing an environment that makes this kind of research meaningful—and fun. We also acknowledge the support of the European Media Lab where part of the work was done. This research is funded in part by Google.

References

1. Liesaputra, V., Witten, I.H., Bainbridge, D.: Creating and reading Realistic Electronic Books. Computer 42(2), 46–55 (2009)
2. Ryan, C., Henselmeier, S.: Usability testing at Macmillan. Keywords 8, 188–202 (2000)
3. Schumacher, G., Waller, R.: Testing design alternatives: A comparison of procedures. In: Designing usable texts. Orlando, FL (1985)
4. Witten, I.H., Frank, E.: Data mining, San Francisco, CA (2005)

Searching Archival Finding Aids:
Retrieval in Original Order?

Junte Zhang[1] and Jaap Kamps[1,2]

[1] Archives and Information Studies, Faculty of Humanities, University of Amsterdam
[2] ISLA, Faculty of Science, University of Amsterdam

Abstract. Archival principles as Provenance (keeping material from the same creator together) and its corollary Original Order (keeping the order of creation intact) could help improve access to the archival materials. We investigate the importance of relevance ranking and 'Original Order' when searching finding aids in EAD using XML Retrieval. Our experiment shows that relevance ranking is of paramount importance, although Original Order may help the retrieval of the first few results because these tend to cluster within the original order.

1 Introduction

Information in digital libraries and on the Web often has a rich internal structure – think of the document structure of books in a digital library. Such structure could be exploited to give direct access to relevant parts in these documents. A particular example of long and richly structured documents are archival finding aids created in Encoded Archival Description (EAD, [5]), which are structured in exactly the same way as the material they describe. First, by the principle of Provenance, all material created or received by the same individual, family, or organization is kept together. Second, by the principle of Original Order, all material of the same creator is stored in its original organization and sequence.

Archivists consider these principles crucial for archival access, though questioned [1], these have never been tested empirically [4]. The archival principle of Original Order corresponds to the preservation of the document hierarchy in an archival description. Physical re-arrangement (such as re-ordering by topic, time or geography), which could enhance user access to the archives, is rejected [3, 4]. Specifically, as stated in [4], even the re-arrangement of archives to suit the needs of historians is disallowed. In recent years, however, archival finding aids have been put online, giving it a new function as an information retrieval and discovery tool for users. Therefore, the actual impact of sticking to the original order of an archive when retrieving and presenting information needs to be examined.

Archives may span 100s or 1000s of meters of material, and the main purpose of the archival description is to help searchers identify the exact parts of the archive to consult. There is a direct and natural parallel between locating parts of the archival finding aids in EAD, and the focused access of other XML documents: XML Retrieval (XML IR) can be used to exploit the internal structure

M. Agosti et al. (Eds.): ECDL 2009, LNCS 5714, pp. 447–450, 2009.

of an EAD file. This structure could consist of elements that represent lengthy biographies, nested components, all the way down to the single item.

However, each and any of these elements can be returned in any order, either by respecting the Original Order, or returning them according to relevance only, or any other criterion. The retrieval effects of Original Order are not known, so *given the retrieval of any and arbitrary EAD/XML elements according to the relevance with a query, what are the retrieval effects of returning it in Original Order?* On the one hand, Original Order could enhance information access as relevant items may appear close to each other due to the intellectual organization by the archival creator. The would correspond with the Cluster Hypothesis, which deposits that 'closely associated documents tend to be relevant to the same requests' [2]. On the other hand, it may not be a useful feature to improve retrieval because the Cluster Hypothesis may not hold on archival finding aids.

2   Experimental Setup

The search requests were formulated as reference questions. The used queries consisted from 3 upto 13 keywords. Each of these reference questions was judged against a narrative in which the information need is clearly stated, including what is considered relevant. The relevance is determined by locating particular units of archival materials that will likely contain the sought answer. Descriptions of individual files and records tend to be very succinct – seldomly more than a single sentence. Additionally, a finding aid also contains contextual background descriptions of the archive, which may directly contain relevant information.

An assessment tool was created to facilitate the relevance assessments within an archive. For each search request, the most relevant archive was located – which may range up to a 1,000 pages and many thousands of XML elements, and all and only material in this archive was judged. This resulted in a relatively modest test collection (*qrels*) of in total 73 relevant elements in 5 archival finding aids in EAD, which we collected from the National Archives of the Netherlands.

The system used in our experimentation has been described in [7]. We indexed the collection without stopword removal, used the Dutch snowball stemmer, and standard parameters. For the retrieval of any arbitrary elements, we employ statistical language models (LM) [6], i.e. the probability distribution of all possible term sequences is estimated by applying statistical estimation techniques.

3   Results

3.1   Relevance versus Original Order

We first look at the whole run with 1,000 results in Table 1. In the top 1,000 results, both approaches obtain a reasonable recall of 62 out of 73 relevant elements ($R = 0.8493$). Considering that we look for very short descriptions (often a single sentence), the relevance ranking is performing quite well with a MAP of 0.1454 and a P@10 of 0.1600. What happens if we rank these 1,000 results in

Table 1. Retrieval Performance for first top N results for each topic

Run	Top 1,000		Top 500		Top 100		Top 10		Top 5	
	P@10	MAP	P@10	MAP	P@10	MAP	P@10	MAP	P@10	MAP
Relevance	0.1600	0.1454	0.1600	0.1398	0.1600	0.1321	0.1600	0.0917	0.1400	0.0822
Original Order	0.0000	0.0296	0.0000	0.0260	0.0400	0.0517	0.1600	0.0660	0.1400	0.1031

Table 2. DOM Tree distances in the qrels.

Topic ID	Count	Mean Depth	Mean Distance	Total count $<c_n>$
1.04.02	11	10.000	3.545	17,184
2.19.123	37	10.811	1.492	5,661
2.19.124	7	6.000	1.286	1,491
2.03.01	7	10.143	1.429	14,017
2.21.286	4	9.750	0.750	2,035

their original order? The score plummets to almost zero; the relevance ranking is crucial. It should be noted that we are reranking the top N of results, and usually there are just a handful of relevant results (also see Table 2).

Given that set of results must contain many non-relevant ones, reranking the top 1,000 on original order may not fairly reflect the utility of the original order. What will happen if we rerank a smaller set of results? The remaining columns of Table 1 show the results for different sets. The MAP of the relevance ranking drops as expected for the shorter runs. As the cut-off level is decreased, we see that the precision of the original order ranking increases. Interestingly, we see that the MAP for Original Order is higher than the standard element ranking when the cut-off level for each topic is set to 5. This signals that although the relevance ranking is of paramount importance, there is also still potential value in the original order, because relevant results have a tendency to cluster.

3.2   Cluster Hypothesis Effects

We want to further investigate the Cluster Hypothesis – how near are the relevant results in the original order of the document? We do this by measuring the distance between the relevant elements in the DOM tree. We restrict our attention to the relevant elements in the hierarchical descriptions of the archive (i.e., the component $<c_n>$ elements) in EAD. The components $<c_n>$ are nested within each other in <ARCHDESC> given the n, where $n \in \{01, ..., 12\}$. A component can also be unnumbered. The results are shown in Table 2. The first topic has 11 results with a mean depth of 10 nodes in the DOM tree. For each pair of results, we look at the distance to a common ancestor, which could be at most the depth itself (i.e., 10). For the first topic the mean distance over all pairs is 3.5 – which signal that the results are somewhat scattered through the archive. However, for the other topics the mean distance is between 0.75 and 1.5, which shows that relevant results occur in close proximity within the archive, especially given the large quantity of $<C_n>$ elements per topic (see Table 2). For example, in topic *1.04.02* only 11 out of 17,184 (or 0.06%) $<C_n>$ elements were seen as relevant.

3.3 Sparse Data on the Item Level

A challenge for effective XML IR is the sparse data, especially in the unit titles, which is a distinct property of the archival descriptions. When we analyze the selected relevant elements, we see that there are very short phrases, sometimes without the occurence of a keyword, e.g. *"Diverse stukken,* (Unit ID: 824)" (in English: *"Several pieces,"*). The sparse data on the item level can be attributed to idea of *inheritance of description* – each lower level inherits the description of the container [5], where this context is a crucial cue for assessing relevancy as related relevant items tend to be located in short distance from each other.

4   Conclusions and Future Work

We empirically examined the impact of the archival principle of Original Order on the ranking of search results by comparing it with a metadata retrieval system using XML IR techniques. Our results show that the relevance ranking is of paramount importance, but that the results have a (weak) tendency to cluster.

The Principle of Original Order is useful, because physical materials can only be ordered in a single way, and here the traditional archival practices make much sense. The question arises whether it will continue to be as useful in this digital age. With the advent of digital archives, we are no longer bound to the physical and practical limitation of before and we could construct multiple ordering of the same material inclusing those based on a search request or search profile at hand. This opens up a wealth of possibilities to improve archival access, and unleash the valuable treasures now hidden deeply inside the archive.

Acknowledgments. We gratefully acknowledge the National Archives of the Netherlands, and Henny van Schie, for their support. This research is supported by the Netherlands Organisation for Scientific Research (NWO) under project #639.072.601.

References

[1] Boles, F.: Disrespecting Original Order. American Archivist 45, 26–33 (1982)
[2] Jardine, N., van Rijsbergen, C.: The use of hierarchical clustering in information retrieval. Information Storage and Retrieval 7, 217–240 (1971)
[3] Jenkinson, H.: Reflections of an Archivist. Contemp. Review 165, 355–361 (1944)
[4] Lytle, R.H.: Intellectual Access to Archives: I. Provenance and Content Indexing Methods of Subject Retrieval. American Archivist 43(Winter), 64–75 (Winter 1980)
[5] Pitti, D.V.: Encoded archival description: The development of an encoding standard for archival finding aids. American Archivist 60(Summer), 268–283 (Summer 1997)
[6] Ponte, J.M., Croft, W.B.: A language modeling approach to information retrieval. In: SIGIR 1998, pp. 275–281. ACM Press, New York (1998)
[7] Zhang, J., Fachry, K.N., Kamps, J.: Access to Archival Finding Aids: Context Matters. In: Christensen-Dalsgaard, B., Castelli, D., Ammitzbøll Jurik, B., Lippincott, J. (eds.) ECDL 2008. LNCS, vol. 5173, pp. 455–457. Springer, Heidelberg (2008)

Digital Preservation and Access of Audio Heritage: A Case Study for Phonographic Discs

Sergio Canazza[1] and Nicola Orio[2]

[1] Lab. AVIRES, Dept. of Historical and Documentary Science
University of Udine, Via Petracco, 8, 33100, Udine (Italy)
sergio.canazza@uniud.it
[2] Information Management Systems Research Group, Dept. of Information Engineering
University of Padova, Via Gradenigo 6/a, 35100, Padova (Italy)
orio@dei.unipd.it

Abstract. We investigate differences among the approaches to the digitization of phonographic discs, using two novel methods developed by the authors: a system for synthesizing audio signals from still images of phonographic discs and a tool for the automatic alignment of audio signals. Results point out that this combined approach can be used as an effective tool for the preservation of and access to the audio documents.

Keywords: Audio cultural heritage, Audio documents access, Audio alignment.

1 Introduction

The importance of transferring *audio analogue* documents into the digital domain, known as active preservation, has been recognized by the international archive community, which provided a number of guidelines in particular for carriers in risk of disappearing. Within this group of documents the sound recordings on discs have been the most spread in the period from 1898 until about 1990. This wide time span in which these formats have been developed makes it difficult to select the correct playing format for the carriers, which may differ in speed, number of tracks, and chemistry characteristics. In order to provided the most complete information about the carrier, the auditory information can be augmented with cross-modal cues. We believe that it is interesting to deal with other information regarding the carrier corruption and imperfection occurred during the A/D conversion. This work describes different approaches for the A/D transfer of phonographic discs and provides two case studies where the similarities and differences of the approaches to re-recording are highlighted.

2 Phonographic Discs Re-recording Systems

Four typologies of playing equipment exist.

Mechanic. It uses a mechanical phonograph, the most common device for playing recorded sound from the 1870s through the 1950s, where the stylus is used to vibrate a diaphragm radiating through a horn.

M. Agosti et al. (Eds.): ECDL 2009, LNCS 5714, pp. 451–454, 2009.

Electro-Mechanical. Turntable drive systems with pickup (piezo-electric crystal or magnetic cartridges), operating on the physics principle of electromagnetic induction.

Opto-Mechanical. A laser turntable is a phonograph that plays gramophone records using a laser device as the pickup. Laser pickup reduces many problems associated with physical contact of the stylus. Unfortunately, the it is susceptible to damage and debris and very sensitive to surface reflectivity.

Opto-Digital. Digital image processing techniques can be applied to the problem of extracting audio data from recorded grooves, acquired using an electronic camera or other imaging system. The images can be processed to extract the audio data also in the case of a broken disc. In literature there are several approaches to this problem [1].

The authors have developed an HW/SW system (*Photos of GHOSTS: PoG* [2]) which: a) is able to recognize different rpm and to perform tracks separation; b) works with both low-cost hardware (starting from a 2D reconstruction of the grooves) and not-trained personnel; c) is robust with respect to dust and scratches; d) outputs de-noised and pitch corrected sound by means of novel restoration algorithm [3]; e) can apply hundreds of different equalization curves. The software automatically finds the disc center and radius from the scanned data, then performs groove rectification and track separation. Starting from the light intensity curve of the pixels in the image, the groove is modeled and the audio samples are obtained.

3   Case Study

As case study, we selected the double-sided 78 rpm 10" shellac disc recorded in New York on February the 23rd 1928:

- Rosina Gioiosa Trubia – *Sta terra nun fa pi mia (This land is not for me)* (R. Gioiosa, arr. R. Romani). Brunswick 58073B (E 26621/2);
- Rosina Gioiosa Trubia – *Mi vogghiu maritari (I want to get married)* (R. Gioiosa, arr. R. Romani). Brunswick 58073A (E 26617/8).

The audio signal was digitized using a sampling rate of 48 kHz with 24 bit resolution. The re-recording has been carried out in four ways:

a) mechanical: phonograph *His Master Voice, Monarch* model, equipped with a pickup *HMV Exhibition* and a horn *Columbia*. The re-recording was carried out in a 4 m^2 room, using a microphone Rhøde NT23, in parallel to the horn axis, 7.5 cm away and central respect to the horn. By means of a *Motu* 828 MkII, the signal is transferred in the digital domain.

b) Electro-mechanical: turntable *Diapason* model 14-A, arm *Diapason*, pickup *Shure* M44/7, Stylus *Expert*. The setting was: 78.26 rpm; 3.5 mil; 4g; truncated elliptical; FFRR equalization curve. *Prism* A/D Converter Dream AD-2.

c) Opto-mechanical: player laser *ELP* mod. LT-1XRC. The setting was: 78.26 rpm, FFRR equalization curve. *Prism* A/D Converter Dream AD-2.

d) Opto-digital: *PoG* system, 4800 dpi, 8 bit grayscale, no digital correction.

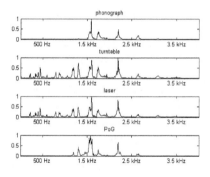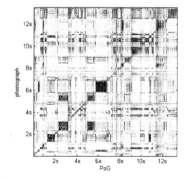

Fig. 1. *Left*: periodograms of about 14 seconds of *Sta terra nun fa pi mia*, taken with the four re-recording systems (the amplitudes have been normalized to ease the comparison). *Right*: Graphical representation of the local similarities between PoG and phonograph recordings.

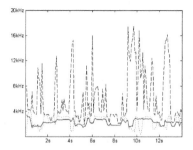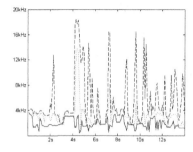

Fig. 2. Rolloff of three aligned signals for Sta terra nun fa pi mia (left) and Mi vogghiu maritari (right): PoG (solid), turntable (dashed), and laser (dotted)

The four audio signal spectral density estimations (periodograms) of the first case study, shown in Fig. 1 (left), highlight that, even if turntable and laser are more sensitive, both to the original signal and to the artifacts (scratch and dust) on the carrier, the four systems give comparable results. A closer analysis of the turntable spectrum, showed a peak at about 11 Hz, which is due to the mechanical vibrations of the arm and probably to a resonance between the arm and the engine. The same peak is not observed in the three other spectra. Similar considerations apply to the second case study.

An audio alignment method [4] was used to compare the similarities between these audio signals. At first we compared the playback speed of the mechanical equipment with *PoG*, which is a useful reference because it is not subject to mechanical variations. For the two case studies, we noticed that all the systems had constant speed, with no variation from the nominal value. Fig. 1 (right) highlights this regular trend for the phonograph, the dark areas on the main diagonal represent a high similarity between the two recordings, while occasional lighter areas represent differences that can be due to a different robustness to carrier corruptions.

Moreover, we focus on the differences between the electro- and opto-mechanical systems, using PoG as a reference, to investigate the possibility to automatically recognize the type of modern re-recording system. We computed and aligned the most relevant features proposed in the audio processing literature. In particular, the first four spectral moments (centroid, spread, skewness, and kurtosis), brightness, and spectral rolloff have been computed using an analysis window of 8192 points. As it can be noted from Fig. 2, spectral rolloff is a good indicator of the differences among the recording systems. Moreover, we computed mean and variance of the point to point differences of the spectral rolloffs, using *PoG* as a reference: (i) turntable versus *PoG*; (ii) laser versus *PoG*. Results are reported in Table. 1. In particular, the value of the variance allows us to identify the kind of re-recording system.

Table 1. Mean and variance of the difference between spectral rolloff using *PoG* as a reference

Case study	Re-recording systems	Mean	Variance
Sta terra nun fa pi mia	turntable	4287	4139
Sta terra nun fa pi mia	laser	10.32	768.4
Mi vogghiu maritari	turntable	3701	4473
Mi vogghiu maritari	laser	1320	1810

4 Conclusion

Audio archivists can take advantage from a variety of equipments for the re-recording of phonographic discs. We presented two case studies using the four different paradigms. We propose a combined approach: by applying the procedure described in this work, that is synthesizing an audio signal using *PoG* and aligning it to the existing digital documents, it is possible to identify the kind of A/D process, its eventual defects, and retrieve the original analogue recording.

References

1. Fedeyev, V., Haber, C.: Reconstruction of mechanically recorded sound by image processing. Journal of the Audio Engineering Society 51(12), 1172–1185 (2003)
2. Canazza, S., Ferrin, G., Snidaro, L.: Photos of GHOSTS: Photos of Grooves and HOles, Supporting Tracks Separation. In: Proceedings of XVII CIM Venezia, October 5-17, 2008, pp. 171–176 (2008)
3. Bari, A., Canazza, S., De Poli, G., Mian, G.A.: Toward a Methodology for the Restoration of Electro-Acoustic Music. Journal of New Music Research 30(4), 365–374 (2001)
4. Orio, N., Snidaro, L., Canazza, S.: Semi-automatic metadata extraction from shellac and vinyl discs. In: Proceedings of AXMEDIS 2008, Firenze, Italy, pp. 38–45 (2008)

Data Recovery from Distributed Personal Repositories

Rudolf Mayer[1], Robert Neumayer[2], and Andreas Rauber[1]

[1] Institute of Software Technology and Interactive Systems,
Vienna University of Technology, Austria
[2] Department of Computer and Information Science,
Norwegian University of Science and Technology, Trondheim, Norway

Abstract. We present an approach to personal disaster recovery, e.g. after a hard-disk crash, based not on an explicitly ex-ante defined recovery plan with a rigid backup regime, but rather on naturally accumulated and distributed sources of personal data, such as e-mails and their attachments. We aim to restore as much data as possible, and to provide means to organise it in a meaningful folder structure. Employing information retrieval techniques, we semi-automatically establish the context of and relations between the data objects along several different dimensions, to identify relations and groups. Different views at multiple levels of granularity then allow an interactive organisation into folders.

1 Introduction

Disaster recovery is commonly associated with database systems, professional archives, or large businesses. Even though private users or small and medium enterprises are often lacking either financial means or technical skills needed to apply large-scale rescue and backup strategies, they also have a strong need for personal archiving and disaster recovery. The importance of backup and preservation for SMEs and current shortcomings therein are pointed out in [4].

Commercial services for data recovery mostly focus on hardware failure and recovery of data from failed disks, which might not be possible in all cases, or an expensive process. However, the Internet has brought a certain shift in terms of data storage – a few years ago, one could assume users have most of their data stored on hard disks and other storage media, but this assumption does not hold any more. People increasingly use free, virtually unlimited e-mail accounts to store their documents as attachments in messages remotely [1]. Specific types of files might be stored in social media service sites such as Flickr, and in collaborative work sites such as Wikis. An approach to use newsgroup services and mail attachments as back-up strategy is described in [3]. However, it requires that the data is deliberately spread across external sites, rather than relying on the inherent replication in personal mail or groupware services.

In this work, we rather focus on the scenario that a user is suddenly left without his or her main file storage on his desktop computer and no backups exist or that they only comprise a very limited set of data. We emphasise methods

M. Agosti et al. (Eds.): ECDL 2009, LNCS 5714, pp. 455–458, 2009.

for analysing, mining and recovering from online, distributed data storage that do not require planning beforehand. Structuring huge amounts of attachments so that they resemble a usable structure is a challenge, requiring sophisticated analysis of the objects and especially their relations to each other. Many sources provide contextual information in terms of people involved in the communication, e-mail text, blog entries, or various comments. We thus aim at identifying the context of objects used in the same *time*, within the same *project*, with similar *content*, with a certain set a of *people*, and of a certain *type*. We present a prototype that offers visualisations of such context, and allows users to interactively organise their objects into meaningful folders.

2   Establishing Context of Digital Objects

Our application scenario is disaster recovery, i.e. to recover as many files as possible after a data loss. Rather than aiming at exact reconstruction, we organise the objects in a new, semantically meaningful way, allowing the user to adjust the automatically established structure. We focus on the following steps.

Object Recovery. We extract all digital objects from the various data sources, such as cooperative work platforms as BSCW and Wikis, e-mail accounts and other sources. This step is rather simple, and requires tools using the correct protocols to access remote information, and processing rules to identify objects worth harvesting.

Context Extraction. We then automatically extract semantic contextual information of objects. Context is present in several forms, ranging from low-level technical context in which the object was created, via its immediate context of use, such as people involved or the activity it is related to, up to a wider sociological or legal context. In this work, we consider only context that can be established in a semi-automatic manner, along several partially orthogonal dimensions. The principle of using various different dimensions is inspired by the concept of data warehouses and on-line analytical processing (OLAP). While the number of potential dimensions that digital objects can be organised by may be larger, we currently use the following in our first prototype: (1) the *time* of object creation, modification and use, (2) the content/file *type*, (3) the *people* involved, and (4) the *content*, across different sub-categories, such as *the topic, the genre, acronyms*, for example in project names (consult [2] for more details).

Combining Context Dimensions and Storage. These dimensions can well be used separately, but the true potential lies in combining and contrasting them with each other. Combining the temporal and content/project dimensions can for example identify sets of documents of various file formats that belong to one 'instance' of a yearly-repeating project reporting – slide-shows from the presentation, spreadsheets detailing financial aspects, and text-documents describing the outcomes. Using the time-dimension only, we could not figure out which documents belong to the same project, while the file type dimension would wrongly

Fig. 1. Pivot-table View on E-mail Attachments

separate the documents. Analysing data in a data warehouse is often performed using pivot-tables, which allow to e.g. see all the sales for a specific city for a certain product at various different levels of aggregation. We provide a similar tool to combine and contrast the relation of objects on those orthogonal dimensions, depicted in Figure 1. It allows users to organise the digital objects along certain dimensions deemed important, by grouping the objects along two different dimensions, and filter the documents along the same or other dimensions.

3 Experiments

We chose e-mails for initial evaluation, due to them being widespread, easily accessible, and safe from local data loss because stored on remote servers. Even though some e-mail corpora are available, we could not use any of those for the lack of attachments in most of the publicly available ones. We thus used two personal mailboxes of the authors of about 19.000 and 23.500 e-mails, from early 2005 and 2006 until early 2009, resp. 2.310 and 1.287 e-mails contained a total of 2.515 and 5.923 attachments. One important aspect in e-mail communication is that often a set of documents get sent packed in an archive format, e.g. as 'zip' files, while in a user's home directory, they are often in their original form. Thus, we also consider the files contained in archive files. The mailboxes exhibit similar structures: both were used primarily for work-related communication, mostly with people inside the same group, European and national project partners, students, and few private e-mails. Topics cover e.g. scientific publications and reviews, project management, and communication with regard to teaching. Differences between the mailboxes are mainly in the focus on different projects and conferences, but not in result performance, thus we focus on the first user.

An exemplary disaster recovery process related to scientific papers could be done as follows: (1) select the time and acronym dimensions for the grouping, (2) select only those acronyms that denote conferences, (3) combine cells on the date dimension to correctly identify the various time periods for each paper submission to each conference in each year, and optionally (4) mark only those cells that shall be exported, for example, de-select the cells 'ECDL / 2005' and 'ISMIR 2007', which rather denote materials for organising the conferences, and finally, (5) start the export by creating 'papers' as the parent folder; the system will assign each cell to a separate sub-folder, which is named by the acronym and date. After each step, we can choose to have the files just exported removed from, or kept in the list of available files. Thus, the set of documents to be stored can be reduced step by step, also facilitating the analysis process. Alternatively, objects can be recovered redundantly in several locations.

For a quantitative evaluation, the number of files recovered from the mailboxes were compared with the respective home directories of the users. To facilitate initial analysis, we opted for a sub-set of the most *relevant* folders, selecting those to data files, and omitting those that have separate back-up regimes, such as programming/development parts that are covered via version management repositories. The total file-count in this directories was 10,000, of which we further skipped temporary files. This resulted in 5,500 files, out of which we could match 1,730 documents (31%) with the files on the specified home directory folders. This folder still contained a lot of 'bulk' files, e.g. some data used for experiments. Therefore, we selected folders containing all the files of the user related to conference papers, peer reviews for conferences, teaching, and project proposals and reporting. After applying the same filtering, the total file-count in these directories was 3,000, of which we could recover 1,250 files (42.5%).

4   Conclusions and Future Work

We presented analysis tools to assist users in recovering their files after a data loss. We recover files from online repositories, and then semi-automatically establish the context of and relationship between those digital objects and allow for analysis based on pivot-tables. The experiments in this work showed promising results in the academic domain. Future work will thus focus on testing our approach in other settings, like small and medium sized enterprises, as well as evaluating the approach focusing on user satisfaction.

References

1. Marshall, C.C., Bly, S., Brun-Cottan, F.: The long term fate of our digital belongings: Toward a service model for personal archives. In: Proc. IS&T Archiving 2006 (2006)
2. Mayer, R., Rauber, A.: Establishing context of digital objects' creation, content and usage. In: Proc. Int. Workshop on Innovation in Digital Preservation (2009)
3. Smith, J.A., Klein, M., Nelson, M.L.: Repository replication using NNTP and SMTP. In: Gonzalo, J., Thanos, C., Verdejo, M.F., Carrasco, R.C. (eds.) ECDL 2006. LNCS, vol. 4172, pp. 51–62. Springer, Heidelberg (2006)
4. Strodl, S., Motlik, F., Stadler, K., Rauber, A.: Personal & sme archiving. In: Proc. 8th ACM IEEE Joint Conf. on Digital Libraries (2008)

A Web-Based Demo to Interactive Multimodal Transcription of Historic Text Images

Verónica Romero, Luis A. Leiva, Vicent Alabau,
Alejandro H. Toselli, and Enrique Vidal

ITI- Instituto Tecnológico de Informática,
Universidad Politécnica de Valencia, Spain
{vromero,luileito,valabau,ahector,evidal}@iti.upv.es

Abstract. Paleography experts spend many hours transcribing historic documents, and state-of-the-art handwritten text recognition systems are not suitable for performing this task automatically. In this paper we present the modifications on a previously developed interactive framework for transcription of handwritten text. This system, rather than full automation, aimed at assisting the user with the recognition-transcription process.

Key words: Handwritten recognition, Interactive framework, Web, HCI.

1 Introduction

Nowadays, there is an increasing number of on-line digital libraries publishing a large quantity of digitized legacy documents. These documents need to be transcribed in order to provide historians and other researchers new ways of indexing, consulting and querying them. Up-to-date Handwritten Text Recognition systems (HTR) cannot replace the experts on this task, since there are no perfect accuracy solutions. Therefore, once the full recognition process of one document has finished, the human expert revision is required to produce a quality transcription. Such post-edition solution is rather inefficient and uncomfortable for the user.

As an alternative to post-editing, a multimodal interactive approach is proposed in this work. The user feedback allows to improve the system accuracy [1,2], while multimodality increases system ergonomics and user acceptability.

2 Review of the MM-CATTI Framework

In the MM-CATTI framework, the user is involved in the transcription process since she is responsible of validating and/or correcting the HTR output [3]. The protocol that rules this process, is formulated in the following steps:

- The HTR system proposes a full transcription a handwritten text line image.
- The user validates the longest prefix and amends the first error in the suffix.
- A new extended prefix is produced based on the previous validated prefix.
- Using this new prefix, the system suggests a suitable suffix.
- These previous steps are iterated until a perfect transcription is obtained.

M. Agosti et al. (Eds.): ECDL 2009, LNCS 5714, pp. 459–460, 2009.

3   Demo Description

The demo presented in this paper is a web-based demo. First, a series of available documents is shown. After selecting the document, the user must choose a page of the document to work with. Finally, the user must transcribe the handwritten text images line by line, making corrections with pen strokes and also using the keyboard. If pen strokes were available, the MM-CATTI server uses an on-line HTR feedback subsystem to decode them. Then, taking into account the decoded word and the off-line models, the MM-CATTI server responds with a suitable continuation to the prefix validated by the user.

The web-based demo proposed in this paper differs from the demo presented in [4] mainly in the client-server communication. In the previous one, the communication was made asynchronously via Ajax and PHP. On the contrary, the new approach communicates much faster through sockets. Furthermore, it allows a more flexible architecture featuring multiple user connections and server load balancing.

4   Evaluation Results

Several experiments were carried out on a corpus corresponding to a historic handwriting document identified as "Cristo Salvador" [2,5]. This document was kindly provided by the Biblioteca Valenciana Digital (BIVALDI). The estimated human effort to produce error-free transcription using MM-CATTI is reduced by a 15% on average, with respect to the classical HTR system. Therefore, from every 100 words misrecognized by a conventional HTR system, a human post-editor will have to correct all the 100 erroneous words, while a MM-CATTI user would correct only 85 - the other 15 would be automatically corrected by the system.

Acknowledgment

This work has been supported by the EC (FEDER), the Spanish MEC under grant TIN2006-15694-C02-01 and the research programme Consolider Ingenio 2010 MIPRCV (CSD2007-00018) and by the UPV (FPI fellowship 2006-04).

References

1. Toselli, A.H., et al.: Computer assisted transcription of handwritten text. In: Proc. of ICDAR 2007, pp. 944–948. IEEE Computer Society, Los Alamitos (2007)
2. Romero, V., Toselli, A.H., Rodríguez, L., Vidal, E.: Computer assisted transcription for ancient text images. In: Kamel, M.S., Campilho, A. (eds.) ICIAR 2007. LNCS, vol. 4633, pp. 1182–1193. Springer, Heidelberg (2007)
3. Toselli, A.H., et al.: Computer assisted transcription of text images and multimodal interaction. In: Popescu-Belis, A., Stiefelhagen, R. (eds.) MLMI 2008. LNCS, vol. 5237, pp. 296–308. Springer, Heidelberg (2008)
4. Romero, V.: et al.: Interactive multimodal transcription of text images using a web-based demo system. In: Proc. of the IUI, Florida, pp. 477–478 (2009)
5. Romero, V., others: Improvements in the computer assisted transcription system of handwritten text images. In: Proc. of the PRIS 2008, pp. 103–112 (2008)

Geographic Information Retrieval and Digital Libraries

Ray R. Larson

School of Information
University of California, Berkeley
Berkeley, California, USA, 94720-4600
ray@sims.berkeley.edu

Abstract. In this demonstration we will examine the effectiveness of
Geographic Information Retrieval (GIR) methods in digital library interfaces. We will show how various types of information may benefit
from explicit geographic search, and where text-based place name search
may be sufficient. We will also show how implicit geographic search (or
geographic browsing) can be used to dynamically generate geographic
searches in geographic interfaces like Google Earth. In this demostration
we will show the algorithms used for Geographic search and how these
may be combined with text search. In addition we will show results from
the GeoCLEF IR evaluation for text-based search.

1 Geographic Information Retrieval

The goal of Geographic Information Retrieval (GIR) is to retrieve *relevant* information resources in response to queries with geographic constraints. GIR implies
that the indexing and retrieval of objects in a digital library collection takes into
account some form of georeferencing[2], and may use various forms of geographical proximity, containment, or other spatial relations in estimating or predicting
relevance. Systems that provide searches using GIR methods, including geographic digital libraries, and location-aware web search engines, are based on a
collection of georeferenced information resources and methods to spatially search
these resources with geographic location as part of their search specifications.

Information resources in digital library collections can be considered georeferenced if they are spatially indexed by one or more regions or points on the
surface of the Earth, where the specific locations of these regions are encoded
using spatial coordinates directly (*geometrically*), or indirectly by toponyms
(*place names*).

One common approach in digital libraries has been to use place names as
a geographical search surrogate. However, place names have well-documented
lexical and geographical problems [3]. Lexical problems include lack of unique-
ness, variant names or spellings, and name changes. Geographical problems include boundaries that change over time and geographic features or areas without
known place names. While geographic coordinates provide can an unambiguous
and persistent method for locating geographic areas or features, they also present

M. Agosti et al. (Eds.): ECDL 2009, LNCS 5714, pp. 461–464, 2009.
© Springer-Verlag Berlin Heidelberg 2009

Fig. 1. Geographic Searching in the Incunabula Short Title Catalog (ISTC)

their own set of challenges for efficient implementation. Among these challenges is the fact that the most popular interface for search systems (the simple search box), is extremely cumbersome for entering geographic searches based on co-ordinates. Users will seldom, if ever, know accurate coordinates for the places they are interested in. They can, however, often find them on a map. In this demonstration we will show how map-based interfaces (using Google Earth and Google Maps) can be used in conjunction with GIR search methods for retrieval of digital library information.

1.1 Probabilistic Spatial Ranking

A search method that employs the "Probability Ranking Principle", is one in which information objects are ranked and presented to the user in decreasing order of their estimated probability of relevance to the user's information need[6]. In previous work [5,1] we have described the development and testing of a probabilistic GIR retrieval model based on logistic regression. The form of that model used in this demonstration estimates the probability of relevance for a particular query and particular record in the database $P(R \mid Q, D)$, using the equivalent "log odds" of relevance expressed $\log O(R \mid Q, D)$ for a set of coefficients, c_i, associated with a set of S statistics, X_i, derived from the query and database, such that:

$$\log O(R \mid Q, D) = c_0 \sum_{i=1}^{S} c_i X_i \tag{1}$$

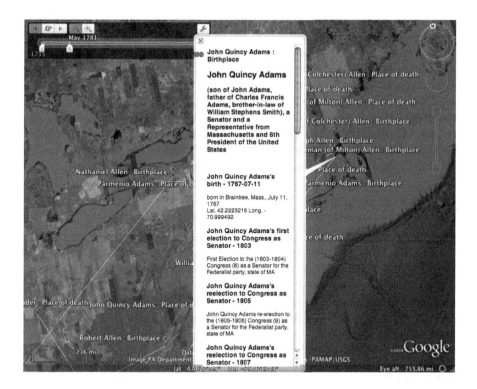

Fig. 2. Geographic Searching in the Congressional Biography Database

where c_0 is the intercept term of the regression. The spatial ranking, or probability of relevance, can then be simply determined from the log odds. For our retrieval approach, the explanatory statistics or feature variables of Geographic Information Objects (GIOs, i.e., the georeferenced items in the database being searched) included in the logistic regression model are fairly simple:

$X_1 =$ area of overlap(query region, candidate GIO) / area of query region
$X_2 =$ area of overlap(query region, candidate GIO) / area of candidate GIO

X_1 and X_2 are based on the extent of the area of overlap and non-overlap between the query and candidate regions. As described in [1] the c_i coefficients were estimated from a sample of geographic documents, and the resulting algorithm was tested on a different experimental set, showing significantly better performance than any previously described geographic ranking algorithm. In addition we will show how text search can be effectively used in mixed geographic and topical search context using another logistic regression-based algorithm based on our results from the GeoCLEF evaluation[4].

In the search system that will be demonstrated, we use the user's interaction with Google Earth to determine the query region, i.e., the query is based on the user's current view of the world as seen in Google Earth, specifically the bounding coordinates of the area currently visible. A new query is sent to the search system each time the user changes their view by moving, zooming in, or zooming out. The algorithm above is used to search for data in the database that overlap that search region using the algorithm described above (in the case of point data, the candidate GIO is assumed to be a small region surrounding the geographic point. Figures 1 and 2 show screen shots of this interface for data from the British Library's Incunabula Short Title Catalog (ISTC), and from an RDF database of events in the lives of U.S. Congressional representatives and Senators. The individual georeferenced items in these collections are automatically linked to other topically related databases including the official ISTC database site, Wikipedia, and the official US Congressional Biography site.

Acknowledgments

The work presented draws on two projects partially supported by Institute of Museum and Library Services National Leadership Grants: Support for the Learner: What, Where, When, and Who (http://ecai.org/imls2004/) and Bringing Lives to Light: Biography in Context (http://ecai.org/imls2006/). The system, and demo, would not have been possible without the work of PhD students Patricia Frontiera and Ryan Shaw and the efforts of Co-PIs Michael Buckland and Fredric C. Gey.

References

1. Frontiera, P., Larson, R.R., Radke, J.: A comparison of geometric approaches to assessing spatial similarity for GIR. International Journal of Geographical Information Science 22(3), 337–360 (2008)
2. Hill, L.L.: Georeferencing: The Geographic Associations of Information. MIT Press, Cambridge (2006)
3. Larson, R.R.: Geographic information retrieval and spatial browsing. In: Smith, L., Gluck, M. (eds.) GIS and Libraries: Patrons, Maps and Spatial Information, pp. 81–124. University of Illinois at Urbana-Champaign, GSLIS, Urbana-Champaign (1996)
4. Larson, R.R.: Cheshire at GeoCLEF 2008: Text and fusion approaches for GIR: CLEF working notes (2008),
   http://www.clef-campaign.org/2008/working_notes/larson_GeoCLEF.pdf
5. Larson, R.R., Frontiera, P.: Spatial ranking methods for geographic information retrieval (gir). In: Heery, R., Lyon, L. (eds.) ECDL 2004. LNCS, vol. 3232, pp. 45–56. Springer, Heidelberg (2004)
6. Robertson, S.E.: The probability ranking principle in ir. Journal of Documentation 33, 294–304 (1977)

Active Preservation

Robert Sharpe[1] and Adrian Brown[2]

[1] Tessella, 26 The Quadrant, Abingdon Science Park, Abingdon, Oxfordshire OX14 3YS,
United Kingdom
[2] Parliamentary Archives, Houses of Parliament, London, SW1A 0PW
(formerly The National Archives, Kew, Richmond, Surrey, TW9 4DU)

Abstract. In order to perform long-term digital preservation it is necessary to
be (i) understand the technology of the material being stored, (ii) be able to decide whether this technology is obsolete (and, if so, what to do about it) and (iii)
perform verifiable actions to remove the causes of this obsolescence (e.g., via
format migration). This demonstration will show a real-life solution for dealing
with these challenges. It is based off pioneering work performed mainly in conjunction with the UK National Archives' Seamless Flow programme and the
Planets project[1] and is now deployed in a variety of national libraries and
archives around the world.

Keywords: Digital Archiving and Preservation, Characterization, Preservation
Planning, Migration.

1 Introduction

Modern libraries receive a large quantity of digital material. This needs to be preserved and yet suffers from a very different form of degradation risk than traditional
material. This risk takes at least two forms:

- Storage media degradation. This problem can be mitigated by retrieving the content from volatile media to a central storage location and then applying an appropriate backup regime and policy of regular integrity checking.
- The ability to view digital objects requires a technical environment including information on the file format(s) involved, application software that can render
  these formats, an operating system that can support the application software and
  hardware on which to run the operating system. In other words, the ability to use
  digital objects depends on a stack of technical components with each component
  in this stack often rapidly becoming obsolete. There are three main approaches to
  dealing with this issue:
    - the "museum" approach which attempts to preserve all the compo-
      nents;
    - the emulation approach which accepts that some technical compo-
      nents need to change but attempts to preserve others (e.g., by

---

[1] The Planets project is co-financed by the European Union's Sixth Framework Programme for
Research and Technological Development (FP6).

M. Agosti et al. (Eds.): ECDL 2009, LNCS 5714, pp. 465–468, 2009.
© Springer-Verlag Berlin Heidelberg 2009

preserving the original format, application software and operating system and emulating this operating system on new hardware);

o   the migration approach which transforms the digital object to a new format and uses new application software running on an appropriate operating system / hardware platform.

The Open Archival Information Standard (OAIS) [1] has provided a conceptual framework for repository systems and a useful language to enable practical discussion of the problem. In particular, the standard makes a distinction between Information Objects (the conceptual entity that needs to be preserved) and Digital Objects (the physical entity that is initially created or is created as a result of some subsequent activity).

The Planets project, has built up a three-fold approach to addressing digital preservation [2]:

- Preservation Characterization. The need to characterize both Information Objects and Digital Objects to both determine the most appropriate actions to take and to provide a basis for validating those actions.
- Preservation Planning. The need to assess the preservation needs of digital material based on characterization information and plan any appropriate action.
- Preservation Action. The need to perform this action including verification of the resulting migration (or emulation).

We call this approach "Active Preservation" and distinguish it from "Passive Preservation" (the steps needed to preserve the original). In this demonstration a practical "Active Preservation" framework is introduced. This includes the ability to plug tools into this framework to deal with some formats and the ability to extend the supported toolset to both deal with other formats and improve existing tools.

2   Characterization

2.1   Technical Characterization

The first step towards preserving material that is ingested into the archive is to ensure that the files that constitute such material are technically characterized. This uses a framework created as part of the Planets project and involves four main steps:

- Identification of the format of every file using DROID [3, 4].
- Validation of that identification using a format-specific tool (e.g., Jhove [5]).
- Extraction of key properties about each file using a format-specific tool.
- The identification of embedded bytestreams within each file if it is a container format (e.g., bytestreams within a ZIP file). This, too, uses a format-specific tool. If a new bytestream is identified, it is then characterized in turn.

The framework allows for extension of new tools as they become available: they simply need to be wrapped. The PRONOM database service [3, 6] is used as a source of information for what to do at each step for each format after initial identification and is automatically queried by the framework using web services. For example, PRONOM holds a prioritized list of identification, validation and property extraction tools for each format and also, importantly, determines which properties to measure and assigns each such property a unique identifier for future comparison.

2.2 Conceptual Characterization

Material then goes through a second stage of characterization that divides Information Objects into atomic conceptual units of preservation called "components". For example, a web site might be divided into its constituent documents (e.g., web pages and PDF documents), images etc. that are too numerous to be catalogued and described by humans. The characteristics of each of these components are measured by aggregating the properties of its constituent files. This is an important step since these components form the atomic units of migration whilst allowing the number or structure of files that manifest them to vary depending on the technology of the day.

Now that characterization has been completed, the material can be ingested into an archive.

3 Preservation Planning

At some time in the future, material may have become obsolete. The system uses PRONOM to monitor this obsolescence using a risk-based system. This allows each format to be assigned a risk score based on configurable criteria. This can be queried (either via a user interface or an automated web service) to determine which formats are at risk. It is also possible to specify a risk associated with format properties (e.g., Word documents with track changes on or containing macros might be considered to be a bigger risk than those without). The output of this process is thus a list of all the formats (or formats and property combinations) that are currently at risk.

This list can then be used to automatically search an archive to find out all the Information Objects whose current technological manifestation is at risk. These can then be dealt with one by one (or in a batch process). For each such Information Object manifestation, the system can ask PRONOM to determine the optimum migration pathway (and the tool to use). PRONOM also holds a list of all the component properties that should be invariant under a migration. The framework can accept a configurable degree of tolerance (i.e. allow for controlled loss of significant characteristics if this is unavoidable). Hence, the framework can also be used to create presentation copies (e.g., lower resolution images for transport over the web) in a controlled manner (i.e. with a known and measurable degree of degradation).

4 Preservation Action: Migration

Once preservation planning has been completed, the next step is then to carry out the migration. This involves running the selected tool and then re-characterizing the output to both discover the technical characteristics of the new files created and to check that all the components are still present, the relationships between them are intact and that the list of properties described above have indeed remained invariant. In addition, there is the option to perform further tool-specific validation tests (e.g., for image migration compare the color distribution of the before and after images).

Once this process has been completed, the new manifestation can be ingested into the archive and, if appropriate, the old one marked as inactive so it does not seed further migrations (although the new one might at some time in the future).

5 Demonstration

The features described above will be demonstrated live showing the complete lifecycle of ingest, characterization, migration and object download using a range of Information Objects and Digital Objects that are relevant to the real data held by Libraries and Archives. This is based on software currently deployed at 7 libraries and archives.

6 Conclusion

Hence, the end result is a fully-automated digital preservation workflow provided in a framework that allows the addition of further characterization and migration tools as needed while allowing librarians and archivists to control detailed policy information and workflows.

References

1. Reference Model for an Open Archival Information System (OAIS) CCSDS 650.0-B-1 Blue Book (January 2002),
   http://public.ccsds.org/publications/archive/650x0b1.pdf
2. Farquhar, A., Hockx-Yu: Planets: Integrated Services for Digital Preservation. Int. Journal of Digital Curation 2(2), 88–99 (2007)
3. Brown, A.: Automating preservation: New developments in the PRONOM service. RLG DigiNews 9(2) (2005)
4. DROID home page, http://droid.sourceforge.net
5. Jhove home page, http://hul.harvard.edu/jhove
6. PRONOM home page, http://www.nationalarchives.gov.uk/pronom

Cultural Heritage Digital Libraries on Data Grids

Antonio Calanducci[1], Jorge Sevilla[1], Roberto Barbera[1], Giuseppe Andronico[1],
Monica Saso[2], Alessandro De Filippo[2], Stefania Iannizzotto[2], Domenico Vicinanza[3],
and Francesco De Mattia[4]

[1] INFN Catania
Via S.Sofia 64, 95128 Catania, Italy
{antonio.calanducci,jorge.sevilla,roberto.barbera,
giuseppe.andronico}@ct.infn.it
[2] Università di Catania, Facoltà di Lettere e Filosofia
Piazza Dante 32, Catania, Italy
lamusa@unict.it
[3] Conservatorio di Parma, ASTRA Project
Via del Conservatorio 27/A, 43100 Parma
francesco.demattia@conservatorio.pr.it
[4] Delivery Advanced Network Technology to Europe UK, ASTRA Project
City House 126-130 Hills Road, Cambridge CB2 1PQ
domenico.vicinanza@dante.co.uk

Abstract. Data Grids offer redundant and huge distributed storage capabilities,
providing an ideal and secure place for the long-term preservation of digitized
literary works and documents of artistic and historical relevance. In this demo,
we are going to show how we deployed some digital repositories of ancient
manuscripts making use of gLibrary, a grid-based system to host and manage
digital libraries

Keywords: Digital Libraries, Grid Computing, Data Grid, Cultural Heritage,
Digital Preservation.

1 Introduction

Data Grids offer redundant and huge distributed storage capabilities, providing an
ideal and secure place for the long-term preservation of digitized literary works and
documents of artistic and historical relevance.

In fact, digitization has been progressively used as a means for avoiding the loss of
literary heritage on paper, caused by physical ageing and the environmental conditions in which documents are kept. Document consultation is another problem that
leads to additional deterioration. Multiple copies of high resolutions scans stored in a
distributed environment and made available for consultation with an easy to use interface is a means to guarantee conservation of cultural heritage. Grid authentication and
authorization mechanisms allow a fine-grained access to archives by single users,
groups or entire communities. Moreover, metadata services permit a structured
organization of scanned files for quick searches.

M. Agosti et al. (Eds.): ECDL 2009, LNCS 5714, pp. 469–472, 2009.

Two use cases have been considered to demonstrate how grid digital libraries can guarantee enduring preservation of literary heritage: the archives of the work of Italian writer Federico De Roberto, made up of almost 8000 scans, and the musical and the musical archives of the "Civiltà Musicale Napoletana" project, made up of more than 250,000 digitizations.

A working prototype of the De Roberto digital repository has been implemented on the gLibrary platform, a grid-based system to host and manage digital libraries developed by INFN Catania, on the Sicilian e-infrastructure of the COMETA consortium.

Fig. 1. First page of notes for the De Roberto "I Viceré" work

2   The gLibrary Platform

gLibrary challenge is to offer an extensible, robust, secure and easy-to-use system to handle digital assets widespread stored on a distributed Grid infrastructure. This goal can be achieved exploiting a series of Grid services together with a proper business logic, and providing an intuitive front-end, accessible from everywhere and anytime. Users do not have to care about the complexity of the underlying systems and the geographical location of their data and they can consider the available Grid storage as a huge virtual disk.

gLibrary can be used to store, organize, search and retrieve any kind of digital assets represented as files in a Grid environment. Consequently, it can be useful for different users that need a secure way to save and share their assets. Assets are saved on the grid storage servers and can be encrypted and replicated on several storage servers, assuring maximum security and high availability to the users' data.

All entries in a gLibrary repository are organized according to their type: a list of specific attributes to describe each kind of asset to be handled by the system. These are the same attributes that can be queried by users.

Each type can have multiple subtypes with additional attributes and all types share a common attribute list (root type), that is fixed by design (in the next release it will be the Dublin Core set). Before users can start uploading assets, a hierarchy of types has to be defined by the repository administrator.

A filtering system, similar to the ones used by the Apple iTunes application to organize iPod/iPhone multimedia collections, is available to browse each deployed repository: some of the attributes of each types can be selected as filters, and their cascading application narrow the result set dynamically, allowing the user to find the interested asset with few mouse clicks.

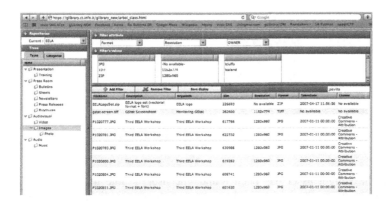

Fig. 2. gLibrary browsing front-end: type tree and filtering

gLibrary servers can host multiple libraries, which can have their own hierarchy of types and can be accessed by different users. Permissions are set up using a fine-grained authorization mechanism. Each asset and/or type has a set of ACLs (Access Control Lists) based on X.509 certificate that restricts its usage, allowing asset owners to grant access to a whole organization, selected groups of users or just a single user. Those entries and types, on which users do not have permissions, are also not visible from the browsing interface.

Both uploads and downloads to/from the grid are carried out over HTTPS or GSIFTP using respectively the user X.509 certificate on the browser or a local grid proxy.

3   Demonstration Content

Two repositories has been actually deployed with gLibrary on the Grid: the archives of the work of Italian writer Federico De Roberto (1861-1927), made up of almost 8000 scans, and the musical archives of the "Civiltà Musicale Napoletana" project, made up of more than 250,000 documents.

During the demonstration, the gLibrary front-end will be used to browse through the repositories and to look for items with some given properties, exploiting the filtering system of gLibrary. For example, a scholar may need to look for all the De Roberto drafts, printed in the 1919. He will first select the *De Roberto* repository, then select *Scansioni* (scans) from the type tree, then *testi a stampa* (printed drafts). Choosing the *PublicationYear* as filter, and selecting *1919* as value among the available years, he will get back a result set of all the assets satisfying his request. The search can be further refined choosing *Publisher* as second filter, to group drafts by publisher.

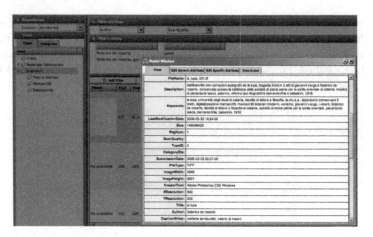

Fig. 3. Metadata full set of a chosen manuscript

Once the sought asset has been found, the user is able to inspect the complete
metadata set and finally is able to choose one of the replica links for downloading the
file from the proper Storage Element to his desktop/laptop.

GROBID: Combining Automatic Bibliographic Data Recognition and Term Extraction for Scholarship Publications

Patrice Lopez

European Patent Office, D-10969 Berlin, Germany
plopez@epo.org

Abstract. Based on state of the art machine learning techniques, GRO-BID (GeneRation Of BIbliographic Data) performs reliable bibliographic data extractions from scholar articles combined with multi-level term extractions. These two types of extraction present synergies and correspond to complementary descriptions of an article. This tool is viewed as a component for enhancing the existing and the future large repositories of technical and scientific publications.

1 Objectives

The purpose of this demonstration is to show to the digital library community a practical example of the accuracy of current state of the art machine learning techniques applied to information extraction in scholarship articles. The demonstration is based on the web application at the following addresse: http://grobid.no-ip.org.

2 Bibliographical Data Extraction

After the selection of a PDF document, GROBID extracts the bibliographical data corresponding to the header information (title, authors, abstract, etc.) and to each reference (title, authors, journal title, issue, number, etc.). The references are associated to their respective citation contexts. The result of the citation extraction can be exported as a whole or per reference following different formats (BibTeX and TEI) and as COInS[1].

The automatic extraction of bibliographical data is a challenging task because of the high variability of the bibliographical formats and presentations. We have applied Conditional Random Fields to this task following the approach of [1] implemented with the Mallet toolkit [2], based on approx. 1000 training examples for header information, and 1200 training examples for cited references. An evaluation with the reference CORA dataset showed a reliable level of accuracy of 98,6% per header field and 74.9% per complete header instance, 95,7% per citation field and 78.9% per citation instance.

---

[1] See http://ocoins.info

M. Agosti et al. (Eds.): ECDL 2009, LNCS 5714, pp. 473–474, 2009.

By selecting the *consolidation* option, GROBID sends a request to Crossref web service for each extracted citation. If a core of metadata (such as the main title and the first author) is correctly identified, the system will possibly retrieve the full publishers metadata. These metadata are then used for correcting the extracted fields and for enriching the results. Interestingly, the instance accuracy for citations goes up to 83.2% on the CORA dataset with this option.

3   Multi-level Term Extraction

If the option *term extraction* is selected, the header will be enriched with two lists of terms; a list of disciplinary domains for the purpose of a general categorization of the article and a list of the most significant terms extracted from the whole body of text. In addition, each citation is enriched with a third list of terms extracted from the different citation contexts in order to capture the important discriminant aspects for which the reference is used.

The usage of terms of domain-specific terminologies is admittedly viewed as one of the most distinguishing features of scientific and technical documents. Term extraction in GROBID follows the approach of [3]. A term is characterized by two scores; one representing its *phraseness* (or lexical cohesion), i.e. the degree to which a sequence of words can be considered a phrase, and one representing its *informativeness* , i.e. the degree to which the phrase is representative of a document given a collection of documents. A linguistic chain comprising language identification, sentence segmentation, word tokenization, POS tagging and lemmatization is first applied. Noun phrases are extracted as term candidates. The Dice coefficient is computed for evaluating the *phraseness* of a term. The *informativeness* is evaluated based on the estimation of the deviation between the document and a backgroung HMM language model based on a 18 millions corpus of English Wikipedia articles.

4   Application to Digital Libraries

We believe that the text processing modules will be a central component of the future digital libraries. The goal of GROBID is to support various user tasks in relation to large article repositories, in particular the assistance for self archiving of articles, the pre-processing of documents for information retrieval, the generation of reliable OpenURL links or automatic citation suggestions.

References

1. Peng, F., McCallum, A.: Accurate Information Extraction from Research Papers using Conditional Random Fields. In: Proceedings of HLT-NAACL (2004)
2. McCallum, A., Kachites, A.: MALLET: A Machine Learning for Language Toolkit (2002)
3. Tomokiyo, T., Hurst, M.: A language model approach to keyphrase extraction. In: Proceedings of ACL Workshop on Multiword Expressions (2003)

Building Standardized Digital Collections: ResCarta Tools, a Demo

John Sarnowski and Samuel Kessel

The ResCarta Foundation, Inc.
info@ResCarta.org
http://www.rescarta.org

Abstract. ResCarta Tools are a suite of open source software applications which can assist in the creation of standardized digital objects. ResCarta Tools are open and modular in their design. Modules for creating digital objects store the metadata in Library of Congress METS/ MODS/MIX XML formats. Collection and indexing modules create LUCENE indexes for high speed fielded and full text retrieval of objects. The tools have been used to create digital collections from a variety of analog and digital sources. Collections can be hosted on the web using Apache TOMCAT and the ResCarta WEB application, which provides inline metadata using COINS. Integrating the use of DLESE OAI is done using the Collection Manager METS XML data. The tools have been used by small public libraries to host a dozen pamphlets and aerospace manufactures to host tens of thousands of documents and millions of pages.

Key words: Search, open source, METS, MODS, MIX,.

1 Introduction

In its funding request for the Making of America 2, the member libraries stated, "The emerging national digital library has the potential to elevate resource sharing to a new level, as it will be possible for users anywhere to find and use entire books, journal articles, and primary source materials directly over the Internet."

They also warned, "However, this potential will be realized only if the library community agrees to new practices and standards that will allow digital library materials to be easily located and used. Without community standards, each library will store its electronic content in a proprietary format in proprietary computer systems.

They went on to further state "To create a national digital library, it will be necessary to define: a) community standards for the creation and use of digital library materials and; b) a national software architecture that allows digital materials to be shared easily over the network."

The ResCarta Foundation, a non-profit organization, was founded to encourage the development of a single set of open community standards and open source implementations of those standards.

M. Agosti et al. (Eds.): ECDL 2009, LNCS 5714, pp. 475–476, 2009.

Our goal is to create, through collaboration, digital content production standards and open source applications that allow users to access disparate digital collections in a simple, user friendly process, leading to interoperability.

2 Conclusion

ResCarta tools are professional grade, open source software applications that produce imagery which passes best practices testing, provides the end user with simple forms to create archive quality metadata in accepted Library of Congress XML formats, creates collection level metadata and provides web applications for the end user to host or have hosted their full resolution archive without the need for html coding. The collections produced can be validated from time to time with a checksum validation tool. Hosted collections using the ResCarta Web application can provide inline metadata for consumption by other open source tools like Zotero and collection level XML files to produce extract files for supporting open source OAI/PMH delivery of metadata.

Existing collections were made from city directories[1], newspaper archives[2], microfilmed journals[3], and born digital electronic documents[4].

3 The Demonstration

Starting with a scanner and a digital camera, we will produce raster images in TIFF and JPEG formats. The ResCarta Metadata Creation tool will be used to add metadata to the images. The ResCarta Data Conversion tool will convert these images along with their respective metadata elements into digital objects with unique identifiers, embedded metadata in MODS/MIX. Existing PDF multipage documents (Normal, Image and Text and Image Only) will also be converted to digital objects. The ResCarta Textural Metadata editor will be used to verify the full text data from the PDF sources. Then the ResCarta Collection Manager will be used to gather these objects and other preexisting digital objects into a collection and generate a METS XML file. An APACHE TOMCAT web server will be installed, the objects will be indexed and a fully functional website will be produced from the collection. An OAI/PMH service will be setup for distribution of the collections metadata. The collection will have each items checksum verified with the ResCarta Checksum Verification tool. Attendees will be given handouts with technical details and a disk with the software tools and sample data. The handouts will allow them to follow along, as the raster images become information on the web in less than twenty minutes. Questions will be taken from the audience.

---

[1] http://rescarta.lapl.org:8080/ResCarta-Web/jsp/RcWebBrowse.jsp
[2] http://murphylibrary.uwlax.edu/Racquet/jsp/RcWebBrowse.jsp
[3] http://www.lib.pu.ru/dcol/jsp/RcWebBrowse.jsp
[4] http://www.ama-assn.org/ama/pub/about-ama/our-history/
    ama-historical-archives.shtml

Digital Libraries - New Landscapes for Lifelong Learning? The "InfoLitGlobal"-Project

Heike vom Orde

Bayerischer Rundfunk -IZI-, Rundfunkplatz 1,
80335 Munich, Germany
Heike.vomOrde@brnet.de

Abstract. "InfoLitGlobal" is an international educational digital library which was created by the Information Literacy Section of the International Federation of Library Association and Institutions (IFLA) on behalf of UNESCO. "Info-LitGlobal" was conceived as a best practice and collaboration tool for information professionals who want to share their IL resources and materials with an international community and who are interested in learning from colleagues all over the world.

Keywords: educational digital libraries, information literacy, lifelong learning.

1 Introduction

In times of "digital societies", digital libraries must be more than mere repositories. Modern infrastructures and content management systems give us the chance to enhance the model of a stand-alone digital archive where objects are just deposited for subsequent access and download. Educators and learners in the field of lifelong learning are demanding for digital libraries that can effectively deliver quality materials in formats that are readily accessible. Furthermore these information systems should provide user interfaces similar to web portals, giving access to an actively managed collection of community-constructed educational resources and offering web 2.0 technologies which allow practitioners to connect.

This concept of a "learning digital library" helps the users to improve the quality of learning and education by promoting best practice solutions and by fostering an active community of learning and innovation where resources are developed and shared together. Research tells us that there is a positive correlation between learning and user interaction with a digital library – not only by using the content of a digital library but also by learning to effectively handle the information system itself. Thus digital libraries can have positive effects on learning and information search behaviour in many ways.

2 The "InfoLitGlobal" Project

The UNESCO / IFLA project "InfoLitGlobal" is based on a conceptual design of a "learning digital library". "InfoLitGlobal" is conceived as a best practice and

M. Agosti et al. (Eds.): ECDL 2009, LNCS 5714, pp. 477–478, 2009.

collaboration tool for information professionals who want to share their information literacy (IL) resources and materials with an international community and who are interested in learning from colleagues all over the world. In recent years information professionals worldwide have become aware of their role in the process of information literacy instruction. Currently, a fundamental change in institutional, faculty and library recognition of IL is taking place. There are different levels and various practices of IL worldwide. Sometimes the aims of IL programmes are restricted to conventional learning outcomes (following e.g. the ALA IL Competency Standards), in other cases librarians follow a wholistic vision of facilitating a conscious information behaviour. Anyway, the concept of information literacy offers the promise that people will be able to become critical thinkers and independent learners. At best, they are empowered to become lifelong learners.

UNESCO's IL policy consists of awareness-raising about the importance of information literacy at all levels of the education process. An essential element of this strategy is the integration of libraries into information literacy programmes. According to tradition they provide resources and services in an environment that fosters free and open inquiry and serve as a catalyst for the interpretation, integration and application of knowledge in all fields of learning. Corresponding to this policy, "InfoLitGlobal" aims mainly at putting successful IL concepts into practice.

The "InfoLitGlobal" digital library includes information on IL communication, such as blogs, conferences and websites, information literacy products, details of relevant organisations, publications and documentation on training the trainer. The library is browsable by category and country, searchable by keyword and has a listing of the most recent additions. Users who register will be able to upload their own resources to the directory. Recognizing the potential of Web 2.0 technologies the coordinators of "InfoLitGlobal" are planning to implement wikis and blogs to foster interactivity and personalization and to improve and enrich the access of the digital library.

The target audiences of my "InfoLitGlobal" digital library demonstration are information professionals and instructional librarians who are interested in developing new conceptual views of educational digital libraries.

Heike vom Orde,

Head of Documentation IZI

Standing Committee Member Information Literacy Section of the International Federation of Library Association and Institutions (IFLA)

Munich, 28 May 2009

REPOX – A Framework for Metadata Interchange

Diogo Reis[1,2], Nuno Freire[1,2], Hugo Manguinhas[1,2], and Gilberto Pedrosa[1,2]

[1] INESC-ID, Rua Alves Redol 9, Apartado 13069,
1000-029 Lisboa, Portugal
[2] IST – Department of Information Science and Engineering, Instituto Superior Técnico,
Lisbon Technical University, Portugal
{diogo.menareis,nuno.freire,hugo.manguinhas,
gilberto.pedrosa}@ist.utl.pt

Abstract. This demonstration presents an XML framework for metadata inter-change. REPOX has two goals: to be a means for libraries and other cultural institutions to provide OAI-PMH access to their metadata records, independently of their original format, with a tool that is easy to install, use and deploy; and to be used as an aggregator of OAI-PMH Data Sources. The records are stored internally in XML and there is a metadata transformation service that allows for translation to desired formats. This demonstration will show the usage scenarios, technologies and current results.

1 Introduction

Across libraries, we find many heterogeneous Library Management Systems (LMS) that use diverse metadata schemas to represent bibliographic data. Libraries face the need to make their bibliographic databases available by OAI-PMH to some European initiatives associated with digital libraries, like Europeana[1] and The European Library[2]. Because many vendors of LMS don't support OAI-PMH, libraries need to implement custom made solutions, often using open source software which requires some technical expertise, often not found in the libraries staff.

REPOX [1] is a tool that can be deployed to have OAI-PMH access to bibliographic databases, requiring little technical knowledge, having a fast start process (installation and configuration). REPOX is focused on having support for the specific schemas used in the libraries. It also contains a metadata transformation facility so they can provide their records through OAI-PMH in the desired schemas.

2 REPOX

The supported record ingesting processes are: a file system folder containing the records; and an OAI-PMH source. REPOX supports any XML schema, but it has built in support for schemas frequently used in libraries, such as MarcXchange, Dublin Core

---

[1] http://dev.europeana.eu/
[2] http://www.theeuropeanlibrary.org

M. Agosti et al. (Eds.): ECDL 2009, LNCS 5714, pp. 479–480, 2009.
© Springer-Verlag Berlin Heidelberg 2009

and the metadata profiles of Europeana and The European Library. ISO2709, a non XML encoding, is also supported since it is the main schema used by libraries to export/import data from the LMS. REPOX supports many product specific variations of this standard and several character encodings, including those that are library specific.

The record identifiers used in REPOX can be associated in two ways: generated by REPOX or extracted from each record using an XPath expression. The advantage of using extracted identifiers is that it is possible to update just the changes because the records can be recognized by the identifier.

The implementation is completely in JAVA. There is an installer for Windows and for Unix/Linux. The Web server used is Jetty, because it does not require a separate installation. For the same reason, the database is Derby, embedded in REPOX.

To expose the records by OAI-PMH, it is required at least the metadata schema oai_dc. REPOX has two approaches to solve this: sending an XSLT transformation from the local schema to oai_dc, or creating a transformation to oai_dc with a visual tool in a web page (using a JavaScript library).

4 Results and Future Work

REPOX is being deployed or tested in the national libraries of Portugal, Hungary, Slovenia, Poland, Russia and Spain in the context of TELplus[3]. In project FUMA-GABA[4], it has been tested by the national libraries of Albania, Bulgaria and successfully deployed at Ukraine. As an aggregator of metadata collections, REPOX was used in project DIGMAP to fulfill the requirements for submission of metadata records and their retrieval by the other DIGMAP services [2].

Current work focus on the use of REPOX as an aggregator, scalable to hundreds of data sources with several millions of metadata records. To accomplish that, the database API will be adapted for other available databases that can handle large scale reads and writes.

REPOX will be used in the infrastructure of Europeana and also in the project EuropeanaLocal[5] starting in 2009.

References

1. Freire, N., Manguinhas, H., Borbinha, J.L.: Metadata spaces: The concept and a case with REPOX. In: Sugimoto, S., Hunter, J., Rauber, A., Morishima, A. (eds.) ICADL 2006. LNCS, vol. 4312, pp. 293–302. Springer, Heidelberg (2006)
2. Martins, B., Manguinhas, H., Borbinha, J.L.: Extracting and Exploring Semantic Geographical Information from Textual Resources. In: Proceedings of the Second IEEE International Conference on Semantic Computing (ICSC) (August 2008)

---

[3] http://www.theeuropeanlibrary.org/telplus/
[4] http://www.theeuropeanlibrary.org/portal/organisation/cooperation/fumagaba/
[5] http://www.europeanalocal.eu/

A Visualization Tool of Probabilistic Models for Information Access Components

Lorenzo De Stefani, Giorgio Maria Di Nunzio, and Giorgio Vezzaro

Department of Information Engineering, University of Padua, Italy
{destefan,dinunzio,vezzarog}@dei.unipd.it

Abstract. Since massive collections of textual documents become more and more available in digital format, the organization and classification of these documents in Digital Library Management System (DLMS) becomes an important issue. Information access components of a DLMS, such as automatic categorization and retrieval components of digital objects, allow users to interact with the system in order to browse, explore, and retrieve resources from collections of objects. The demonstration presents a two-dimensional visualization tool of Naïve Bayes (NB) probabilistic models for Automated Text Categorization (ATC) and Information Retrieval (IR) useful to explore raw data and interpret results.

1 Introduction

DLMSs are an example of systems that manage collections of multi-media digitalized data and include components that perform the storage, access, retrieval, and analysis of the collections of data. The visualization of the results returned by these components may be a key point for: firstly, system designers during the process of raw data exploration; secondly, users to interpret results more clearly and possibly interact with them.

In this demonstration, we present a tool for the visualization of NB probabilistic models for information access components of a DLMS that represents digital objects on the two-dimensional space [2,1]. This tool demonstrates to be a valid visualization tool for understanding the relationships between categories of objects, and helps users to visually audit the classifier and identify suspicious training data. This model defines a direct relationship between the probability of an object given a category of interest and a point on a two-dimensional space. In this light, it is possible to graph entire collections of objects on a Cartesian plane, and to design algorithms that categorize and retrieve documents directly on this two-dimensional representation. The demonstration will applied to the task of automatic text classification and text retrieval.

2 The Two-Dimensional Representation of Probabilities

In the two-dimensional representation of documents, the equation of the ranking or the classification function has to be written in such a way that each coordinate

M. Agosti et al. (Eds.): ECDL 2009, LNCS 5714, pp. 481–482, 2009.

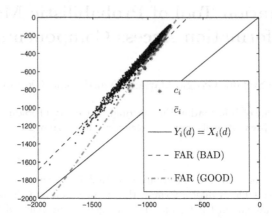

Fig. 1. An example of different separating lines for relevant and non relevant documents

of a document is the sum of two addends: a variable component $P(d|c_i)$, the probability of a document given a category of interest, and a constant component $P(c_i)$, the prior of the category of interest [1]. For example, in the case of NB models the equation becomes:

$$\underbrace{\log\left(P(d|c_i)\right) + \log\left(P(c_i)\right)}_{X_i(d)} > \underbrace{\log\left(P(d|\bar{c}_i)\right) + \log\left(P(\bar{c}_i)\right)}_{Y_i(d)}$$

In this demonstration we show the functionalities of the actual prototype on standard benchmark collections: how the ranking or classification functions are learned from the data as separating lines; how particular unbalanced distribution of documents can be corrected by means of parameter estimation; how the multivariate model and the multinomial model perform on different languages; how blind and/or explicit relevance feedback affect ranking list, and how the selection of relevant documents changes the shape of the clouds of relevant and non-relevant documents.

In Figure 1, a screen-shot of the main window of the visualization tool si shown. Relevant documents are plotted against non-relevant documents (respectively c_i and $\bar{c}_i$), three separating lines are shown to demonstrate how the same algorithm, the Focused Angular Algorithm (FAR), produces different separating lines according to different estimates of the parameters.

References

1. Di Nunzio, G.: Using Scatterplots to Understand and Improve Probabilistic Models for Text Categorization and Retrieval. International Journal of Approximate Reasoning (in press, 2009), http://dx.doi.org/10.1016/j.ijar.2009.01.002
2. Di Nunzio, G.M.: Visualization and Classification of Documents: A New Probabilistic Model to Automated Text Classification. Bulletin of the IEEE Technical Committee on Digital Libraries (IEEE-TCDL) 2(2) (2006)

User Interface for a Geo-Temporal Search Service Using DIGMAP Components

Jorge Machado[1,2], Gilberto Pedrosa[1], Nuno Freire[1], Bruno Martins[1],
and Hugo Manguinhas[1]

[1] INESC-ID / Instituto Superior Técnico
[2] Instituto Politécnico de Portalegre
{jorge.r.machado,gilberto.pedrosa,nuno.freire,
bruno.martins,hugo.manguinhas}@ist.utl.pt

Abstract. This demo presents a user interface for a Geo-Temporal search service built in the sequence of DIGMAP project. DIGMAP was a co-funded European Union project on old digitized maps and deals with resources rich in geographic and temporal information. This search interface followed a mashup approach using existing DIGMAP components: a metadata repository, a text mining tool, a Gazetteer, and a service to generate geographic contextual thumbnails. Google Maps API is used to provide a friendly and interactive user interface. This demo will present the resulting geo-temporal search engine functionalities, whose interface uses WEB 2.0 capabilities to provide contextualization in time and space and text clustering.

Keywords: User Interfaces, GeoTemporal Retrieval, Software Architectures, Metadata, Information Retrieval, Lucene.

1 Introduction

Nowadays the WEB is used not only by users but also by automatic services using published application programming interfaces (API). This demonstration explores this kind of development, known as mashup, to provide a search engine interface that incorporates geographic, temporal and textual dimensions. The search engine uses the infrastructure of the DIGMAP project[1]. This work follows previous studies presented in [3][4][5] where the authors refer to the relevance of Geo-Temporal user interfaces. Our approach aimed at achieving two goals: provide a friendly and interactive user interface exploring Geo-Temporal properties; separate the geographic and temporal dimensions from the textual one to let the user build the queries in these different dimensions.

To reach first goal are used five internal web services available from DIGMAP were: Nail Map[2] to generate contextual thumbnails; Metadata Repository[3] to manage

---

[1] http://www.digmap.eu
[2] http://nail.digmap.eu/
[3] http://repox.ist.utl.pt/

M. Agosti et al. (Eds.): ECDL 2009, LNCS 5714, pp. 483–486, 2009.
© Springer-Verlag Berlin Heidelberg 2009

and retrieve metadata records; Gazetteer[4] together with GeoParser[5] [1][2] to recognize and disambiguate the names of places and temporal expressions given in the text, also assigning documents to the encompassing geo-temporal scopes that they discuss as a whole. The Google Maps[6] API's is also used to provide a new interactive search interface. Sections 3 and 4 detail search and results interface respectively.

2   Search Interface

The search interface[7], illustrated in Fig. 1, is structured in four dimensions: text, space, time and collections. In text dimension we provide both a simple form, using only a single text box, and an advanced form combining different metadata elements, are made available. Users can provide restrictions on the textual dimension using one of these forms. The form corresponding to the temporal dimension is presented on top of the form that corresponds to the geographical dimension. In these dimensions we give the user the chance to select a time interval, specified in years, and a geographic box using a map.

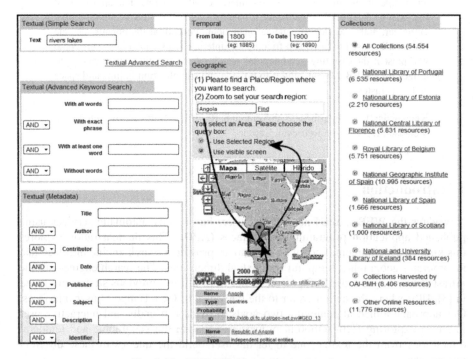

Fig. 1. The DIGMAP Search Service Interface

---

[4] http://gaz.digmap.eu/
[5] http://geoparser.digmap.eu/
[6] http://code.google.com/apis/maps/
[7] http://digmap2.ist.utl.pt:8080/mitra (Experimental Service).

The records were indexed in time, space, text and collections. To index geographic and time dimensions the Geoparser service was used to find geographic scopes and dates, in that sense almost every record, about 90%, have coordinates, latitudes and longitudes, and dates assigned, what turns possible to use this interface with success using a Geo-Temporal retrieval machine. The geographic panel provides a disambiguation service implemented using Geoparser and Gazetteer services from DIGMAP, and Google Maps API to provide an interactive map background. The users could zoom the map to choose the region where they want to search. Some records have geographic polygons attributed automatically by Geoparser or manually by cataloguers. In these cases when the user selects the region marker associated with a polygon the interface ask the user if the search box will be de polygon or the visible region.

3 Results Interface

The results page, illustrated in Fig. 2 (in last page), shows the records found for the search requested in Fig. 1. The component Nail Map is invoked in two different moments, first to get one thumbnail with the map miniature, second to get a thumbnail to contextualize the record in the searched region. The time period of the record is also contextualized in the search interval above the context map. When a user clicks on the map or on a record title the interface will invoke the metadata repository to return an HTML version of that record. During the presentation of results the client interface will invoke the search component to obtain clustered text fragments for each one of

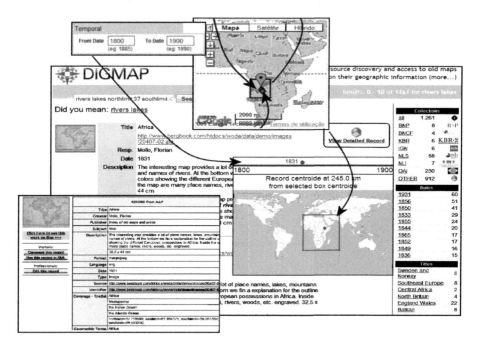

Fig. 2. Search Results in the new DIGMAP search interface

the metadata fields, dates and collections. All of these calls are transparent to the end user, who just sees as soon as possible a list of results, and after that, a constant evolution of the interface contextualizing him in text, space and time.

4 Conclusions

This user interface reveals very good possibilities along a small group of testers whose appreciate the new facilities in Geo-Temporal dimensions. We plan to continue this work adding more features both in Geo-Temporal and Text clustering domain. We plan also to perform usability tests to prove our interface effectiveness join the community.

References

[1] Manguinhas, H., Martins, B., Borbinha, J.: A Geo-Temporal Web Gazetteer Integrating Data From Multiple Source. In: Proceedings of the 2008 IEEE international Conference on Digital Information Management, November 13-16 (2008)
[2] Martins, B., Freire, N., Borbinha, J.: Complex Data Transformations in Digital Libraries with Spatio-Temporal Information. In: Proceedings of the 2008 International Conference on Asia-Pacific Digital Libraries, December 02-05 (2008)
[3] Buckland, M., Chen, A., Gey, F.C., Larson, R.R., Mostern, R.: Geographic search: catalogs, gazetteers, and maps, College and Research Libraries (2007)
[4] Buckland, M., Chen, A., Gey, F.C., Larson, R.R., Mostern, R., Petras, V.: Geographic Search: Catalogs, Gazetteers, and Maps. College & Research Libraries 68(5), 376–387 (2007)
[5] Larson, R.R.: Geographic IR and Visualization in Time and Space. In: ACM SIGIR 2008 Thirty-first annual international ACM SIGIR Conference on Research and Development in Information Retrieval, Singapore, July 20-24, 2008, p. 886 (2008)

Information Environment Metadata Schema Registry

Emma Tonkin and Alexey Strelnikov

UKOLN, University of Bath, UK, BA27AY
e.tonkin@ukoln.ac.uk,
a.strelnikov@ukoln.ac.uk

Abstract. Several metadata schemas focusing on Dublin Core metadata are now in operation around the world. In this demonstration, we support participants in exploring various components that make up the IEMSR metadata schema registry: a desktop tool enabling creation of simple DC vocabularies and application profiles and their addition to the registry; a Web client that enables the registry to be browsed and searched; a series of prototype tools that demonstrate the use of the machine-to-machine SPARQL endpoint for practical scenarios such as internationalization and complex application profile creation using an explicitly stated entity-relationship model.

Keywords: Metadata schema registry, application profile, Dublin Core.

1 Introduction to the IEMSR

The metadata schema registry is a service or application that collects together information from a variety of sources on the subject of metadata schemas, structures, terms/elements, and potentially terminologies and value encodings. Schema registries are designed and used for a variety of purposes, such as promotion of the reuse of existing vocabulary, standardisation, and discovery of appropriate terms. Human-readable descriptions of vocabulary terms and of their uses in given contexts may be provided alongside machine-readable mappings of equivalency, term hierarchy, etc. Metadata schemas and vocabularies represent work that should be retained, published and curated responsibly; low availability or high barrier to access of existing metadata structures and vocabularies discourages reuse. Johnston[1] identifies a number of functions common to metadata schema registries: disclosure/discovery of information about metadata terms and relationships between metadata terms; mapping/inferencing services connecting terms; verification of the provenance or status of metadata terms; discovery of related resources (aggregations, usage guidelines, bindings etc). Several metadata schema registries focusing on Dublin Core metadata exist, including the NSDL metadata registry[4] and the IEMSR metadata schema registry, a prototype service currently under evaluation at UKOLN[3].

This session is intended to focus primarily on the IEMSR, which consists of several tools: a SPARQL endpoint for data access; a web client; a desktop client; a set of design/prototyping tools. This demonstration session will examine these in detail, ideally via example scenarios provided by participants (standard examples will also

M. Agosti et al. (Eds.): ECDL 2009, LNCS 5714, pp. 487–488, 2009.

be available); for example, development of a sample AP, or exploration of a structured AP. Participants may wish to bring along some documentation, such as an informal set of requirements, describing a scenario they wish to explore. The target audience of this demonstration includes metadata schema, application profile and vocabulary developers; those with an interest in Semantic Web technologies; those making practical use of metadata, particularly DC, such as registry managers and developers; developers interested in making use of information about metadata standards as part of software development in general terms; for example, those with an interest in web engineering; those with a more general interest in metadata standards, including those who do not closely follow developments in the DCMI world. Previous demonstrations have taken place in a light-hearted 'bar camp' atmosphere; participants are encouraged to engage with the concept on their own terms. Because the IEMSR is a network-based application, network access will be required throughout the demonstration. It is possible for the IEMSR to be installed locally or booted from a live CD or USB drive, but access to the primary IEMSR endpoint and websites may be preferred, as this enables access to online documentation and tutorial information.

2 Conclusion

The IEMSR, a pilot service currently in a review phase, continues to develop and change with the changing needs of the environment within which it operates. Potential applications of a metadata registry have themselves evolved greatly in the past few years, and as the amount of available data has increased and expectations on metadata as a technologyhave shifted. Ongoing work in development of application profiles for deployment within national and international repository ecosystems has highlighted the question of appropriate development/evaluation methodologies; the schema registry is viewed as a resource able to support rapid prototyping and toolsets developed for this purpose. We aim to provide participants with an introduction to the IEMSR registry, and to provide a general introduction to characteristics of and potential use cases for schema registries in the more general sense of the term.

References

1. Johnston, P.: What are your terms? Ariadne 53 (2005)
2. Greening, O.: IEMSR Marketing Plan (2006),
   http://www.ukoln.ac.uk/projects/iemsr/phase3/
   IEMSR-marketing-plan.pdf
3. The Information Environment Metadata Schema Registry project,
   http://www.ukoln.ac.uk/projects/iemsr/
4. The NSDL metadata registry, http://metadataregistry.org/

Digital Mechanism and Gear Library – Multimedia Collection of Text, Pictures and Physical Models

Rike Brecht, Torsten Brix, Ulf Döring, Veit Henkel, Heidi Krömker,
and Michael Reeßing

University of Technology Ilmenau, Germany
{r.brecht,torsten.brix,ulf.doering,veit.henkel,heidi.kroemker,
michael.reessing}@tu-ilmenau.de

Abstract. We are presenting a digital engineering library – the Digital Mechanism and Gear Library (DMG-Lib). The existing worldwide knowledge in form of books, drawings, physical models etc. is mostly scattered, difficult to access and does not comply with today's requirements concerning a rapid information retrieval. Therefore the development of a digital, internet-based library for mechanisms and gears is necessary, which makes the worldwide knowledge about mechanisms and gears available: http://www.dmg-lib.org. The Digital Mechanism and Gear Library is of particular importance not only for engineers, product designers and researchers, but also for teachers, students and historians.

Keywords: Multimedia, interactive collection, digital engineering library.

1 Introduction

The Digital Mechanism and Gear Library is a heterogeneous digital library with regard to the resources and media types. More than 3800 text documents, 1200 descriptions of mechanisms and machines, 540 videos and animations and 180 biographies of people in the domain of mechanism and machine science are available in the DMG-Lib in January 2009 and the collection is still growing. The multimedia collection consists of e-books, pictures, videos and animations.

2 Physical Models

The DMG-Lib concept and workflow takes into account that technical knowledge exists in different forms e.g. texts, pictures and physical models (see Figure1) and requires analytical, graphical and physical forms of representation.

Fig. 1. Physical models digitized for DMG-Lib

M. Agosti et al. (Eds.): ECDL 2009, LNCS 5714, pp. 489–490, 2009.

Therefore the focus is not only on textual documents and pictures. Thousands of unique physical models made of wood, Plexiglas or metal still exist illustrating and visualizing kinematic basics and methods. These functional models are digitized and available as videos and interactive animations.

3 Interactive e-Books with Animation

Typically text documents in the field of mechanisms and machine science are containing many figures. To better understand the motion of figured mechanisms and machines DMG-Lib can animate selected figures within e-books. These animations are augmenting original figures and can be displayed in the e-book pages with an integrated Java Applet.

http://www.dmg-lib.org/dmglib/main/portal.jsp?mainNaviState=browsen.docum

4 DMG-Lib History Map

The collection of DMG-Lib contains several items described with time and location based metadata. Persons' datasets contain metadata like place and date of birth as well as the other events of persons' CVs e.g. their appointment to university. Mechanism descriptions contain information on date of manufacturing or the actual abode. Metadata belonging to text documents are publication date and place. These metadata are presented in several maps using Google maps application interface. For example, DMG-Lib History Map illustrates the CVs of persons in the 1950s or shows the locations of all mechanism collections in Germany. The map can also be used to locate search results within the map in addition to a result list.

Acknowledgments

The DMG-Lib project (funded by DFG, German Research Foundation) started in 2003 with partners from TU Ilmenau, RWTH Aachen and TU Dresden. Further partners could be enlisted, e.g. the German committee of IFToMM.

Demonstration of User Interfaces for Querying in 3D Architectural Content in PROBADO3D

René Bernd[1], Ina Blümel[2], Harald Krottmaier[1], Raoul Wessel[3],
and Tobias Schreck[4]

[1] Graz University of Technology
[2] German National Library of Science and Technology TIB
[3] University of Bonn
[4] Technische Universität Darmstadt

Abstract. The PROBADO project is a research effort to develop Digital Library support for non-textual documents. The main goal is to contribute to all parts of the Digital Library workflow from content acquisition over semi-automatic indexing to search and presentation. PROBADO3D is a part of the PROBADO framework designed to support 3D documents, with a focus on the Architectural domain. This demonstration will present a set of specialized user interfaces that were developed for content-based querying in this document domain.

1 Interfaces for Querying in 3D Architectural Data

PROBADO3D supports search in metadata space, as well as in content-based space in 3D architectural data comprising models of buildings, and interior and exterior elements. Content-based search relies on domain specific indexing services generating descriptors of the building models during an offline indexing stage. The descriptors include global shape properties as well as connectivity information which describes the layout of rooms within the buildings [1] (cf. Figure 1(a) for an example). For querying in these indexes, specific interfaces have been developed to graphically specify queries for similar content. The following interfaces are currently considered.

Querying for Similar Global Shape. Queries for the overall shape of building models are supported by sketching of 3D volumes, or by 2D floor plans. Two editors allow specification of volumes or floor plans. One interface provides a 2D/3D editor based on the Generative Modeling Language (GML [2]) (cf. Figure 1(b)). The other uses the modeling capabilities of Google Sketchup [3] (cf. Figure 1(c)).

Querying for Room Configurations. The configuration of rooms inside a building is also an important property which architects like to search for (cf. Figure 1(d)). A graph editor interface allows to input an abstract specification of room connectivity structure (cf. Figure 1(e)). A plan-based interface allows to edit a room sketch (cf. Figure 1(f)).

M. Agosti et al. (Eds.): ECDL 2009, LNCS 5714, pp. 491–492, 2009.
© Springer-Verlag Berlin Heidelberg 2009

The input provided by the user via these interfaces is then used to formulate content-based queries. A result list containing the models which best match the query are presented to the user for further inspection and browsing. The user can then refine the query by using any of the answer models as a query-by-example key.

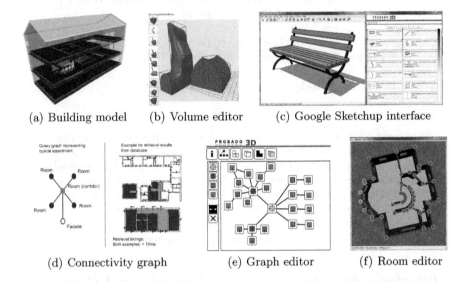

(a) Building model   (b) Volume editor   (c) Google Sketchup interface

(d) Connectivity graph   (e) Graph editor   (f) Room editor

Fig. 1. Specialized graphical query interfaces support retrieval in the 3D Architectural model domain within PROBADO3D

Acknowledgments

We gratefully acknowledge support from the German Research Foundation DFG. The PROBADO project started in February 2006 with a tentative duration of five years. Partners are the University of Bonn, Technische Universitaet Darmstadt, the German National Library of Science and Technology in Hannover, and the Bavarian State Library in Munich. Cooperation partners are Graz University of Technology and Braunschweig Technical University. For further information, please refer to the PROBADO website at http://www.probado.de/. The work presented in this paper was partially supported under grants INST 9055/1-1, 1647/14-1, and 3299/1-1.

References

1. Wessel, R., Blümel, I., Klein, R.: The room connectivity graph: Shape retrieval in the architectural domain. In: Proc. Int. Conf. in Central Europe on Computer Graphics, Visualization and Computer Vision (2008)
2. GML: Generative modeling language (2009),
   http://www.generative-modeling.org/
3. Google: Google sketchup (2009), http://sketchup.google.com/

Hoppla - Digital Preservation Support for Small Institutions

Stephan Strodl, Florian Motlik, and Andreas Rauber

Vienna University of Technology, Vienna, Austria
{strodl,motlik,rauber}@ifs.tuwien.ac.at

Abstract. Digital information is of crucial value to a range of institutions, from memory institutions of all sizes, via industry and SME down to private home computers containing office documents, valuable memories, and family photographs. While professional memory institutions make dedicated expertise and resources available to care for their digital assets, SMEs and private users lack both the expertise as well as the means to perform digital preservation activities to keep their assets available and usable for the future.

This demo presents the Hoppla archiving system[1] providing a digital preservation solution specifically for small institutions and small home/office settings. The system combines bit-stream preservation with logical preservation. It hides the technical complexity and outsource required knowledge and expertise in digital preservation.

Hoppla Concepts

Hoppla[2] - (Home and Office Painless Persistent Long-term Archiving) presents a new approach to automated digital preservation systems that are suited to the needs of small institutions. It considers the abovementioned issues and tackles the emerging challenges to ensure the accessibility and availability of digital objects in the future. The conceptual design of the archiving system was presented in [1].

The underlying principle of the system is providing a best effort solution with respect to the available technology and skills of the users. With Hoppla we are currently developing a solution that combines back-up and fully automated migration services for data collections in small institution settings. The logical preservation obtains practical preservation solutions based on expert advise via an automated web-service-based update of preservation plans. This allows outsourcing of required digital preservation expertise for the institutions.

The modular architecture of the Hoppla system is influenced by the OAIS reference model(ISO 14721:2003). As shown in Figure 1, the six core modules of the system are: acquisition, ingest, data management, preservation, access and storage.

---

[1] Part of this work was supported by the European Union in the 6th framework Program, IST, through the PLANETS project, contract 033789.

[2] http://www.ifs.tuwien.ac.at/dp/hoppla

M. Agosti et al. (Eds.): ECDL 2009, LNCS 5714, pp. 493–494, 2009.

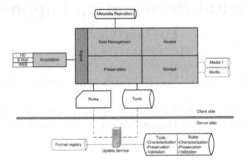

Fig. 1. Architecture of the Hoppla System

The application flow of Hoppla starts with the ingest of new objects from source media, the object's formats are identified and a collection profile is created. Hoppla supports the acquisition of data from different sources via an API for acquisition plugins. The use of plugins allows to support all kinds of storage media and current as well as future data sources.

Based on the collection profile suitable preservation rules are recommended by the web update service. For privacy reasons, the user can define the level of detail of the collection profile that is provided to the web update service. This way the user has strict control, which and how much metadata is sent to the server. New preservation rules and services are downloaded from the web service and preservation actions are performed on the client side. Depending on the configuration all migration steps can run fully automated and transparent for the user, or allow individual interaction and fine-tuning.

The new objects including the resulting objects from preservation actions are ingested into the collection and stored on storage systems. A plugin infrastructure is used to support different storage media, e.g. online and off-line media in both write-once as well as rewritable forms.

By not just using bit storage, but also migration with corresponding web update services HOPPLA can help SOHO institutions and private users to have safe backups of all their data. Not just the physical layer of a file is preserved, but also its logical layer. The fully modular development of the Hoppla software provides facilities to further extend specific modules of the system or integrate other existing system (e.g. storage systems) into the Hoppla system.

References

1. Strodl, S., Motlik, F., Stadler, K., Rauber, A.: Personal & SOHO archiving. In: Proc. of the 8th JCDL 2008, Pittsburgh PA, USA. ACM, New York (2008)

Author Index